Controlling Chaos
and **Bifurcations**
in
Engineering Systems

Controlling Chaos and **Bifurcations** in Engineering Systems

edited by
Guanrong Chen

CRC Press
Boca Raton   London   New York   Washington, D.C.

Library of Congress Cataloging-in-Publication Data

Controlling chaos and bifurcations in engineering systems / edited by
  Guanrong Chen.
      p.   cm.
    Includes bibliographical references and index.
    ISBN 0-8493-0579-9 (alk. paper)
    1. Systems engineering. 2. Chaotic behavior in systems.
  3. Bifurcation theory. I. Chen, G. (Guanrong)
  TA168.C64 1999
  620'.001'1857—dc21   99-16242
                                                                                       CIP

This book contains information obtained from authentic and highly regarded sources. Reprinted material is quoted with permission, and sources are indicated. A wide variety of references are listed. Reasonable efforts have been made to publish reliable data and information, but the author and the publisher cannot assume responsibility for the validity of all materials or for the consequences of their use.

Neither this book nor any part may be reproduced or transmitted in any form or by any means, electronic or mechanical, including photocopying, microfilming, and recording, or by any information storage or retrieval system, without prior permission in writing from the publisher.

The consent of CRC Press LLC does not extend to copying for general distribution, for promotion, for creating new works, or for resale. Specific permission must be obtained in writing from CRC Press LLC for such copying.

Direct all inquiries to CRC Press LLC, 2000 N.W. Corporate Blvd., Boca Raton, Florida 33431.

Trademark Notice: Product or corporate names may be trademarks or registered trademarks, and are only used for identification and explanation, without intent to infringe.

© 2000 by CRC Press LLC

No claim to original U.S. Government works
International Standard Book Number 0-8493-0579-9
Library of Congress Card Number 99-16242
Printed in the United States of America 1 2 3 4 5 6 7 8 9 0
Printed on acid-free paper

Foreword

Although the phenomenon of *chaos* can be traced back to Poincaré more than a century ago, the recognition of its ubiquity and practical significance within the engineering community is of a relatively recent origin. After all, it is not the engineer's business to fool around with chaos, let alone create it. In fact, the advent of chaos in *engineering systems* was ushered in only in the early eighties by the laboratory observation of chaos from "real" electronic circuits, in contrast to computer simulations which had heretofore been greeted by engineers as pathological if not numerical artifacts.

If we identify the eighties as the "age of discovery" of chaos in engineering systems, then the nineties can be labeled as the "age of enlightenment" where chaos had been extensively investigated experimentally and rigorously proven mathematically to be a robust phenomenon that can emerge in many man-made systems, especially over time, when component aging and parameter drift give rise to *bifurcations* and *chaos*. Enlightened and armed with tools from the theory of *nonlinear dynamics*, engineers and scientists have since advanced from their earlier passive role of *analysis* to the active role of *control over chaos*. In particular, intense research activities are presently being directed not only at suppressing chaos, but also at exploiting its immense potential. For example, chaos theory is being applied to arrest arrhythmia in diseased hearts, and to suppress convulsions from epileptic patients. Chaos can be exploited for many mundane applications, such as the mixing of paints and chemicals, and even in the more exotic setting of future interplanetary travels where fuel consumption can be greatly reduced by coasting along an optimum chaotic trajectory derived straight out of the theory of nonlinear dynamics. Even more exciting is the recent discovery of the *local activity* dogma which asserts that the phenomena of *complexity* and *emergence* must evolve on the *edge of chaos*.

Now that the flood gate of chaos has opened, we are poised, at the dawn of the 21st century, for major advances in the controlling and synchronization of chaos for novel applications in all fields of science and technology. It is this vision of exciting exploitation of chaos that makes this book extremely timely and relevant.

Leon O. Chua

Preface

Chaos control refers to any form of manipulation of the chaotic dynamical behavior exhibited by complex nonlinear systems. This new technology promises to have a major impact on many novel, time- and energy-critical engineering applications, such as high-performance circuits and devices (e.g., delta-sigma modulators and power converters), liquid mixing, chemical reactions, biological systems (e.g., the brain and heart), crisis management (e.g., in power electronics), secure information processing, and critical decision-making in political, economic, as well as military systems. This new and challenging research and development area has become a scientific interdiscipline involving systems and control engineers, theoretical and experimental physicists, applied mathematicians, and physiologists.

There are many practical reasons for controlling or ordering chaos. In a system where chaotic response is undesired or harmful, it should be reduced as much as possible, or totally suppressed. Examples of this include avoiding fatal voltage collapse in power networks, eliminating deadly cardiac arrhythmias, guiding disordered circuit arrays (e.g., multi-coupled oscillators and cellular neural networks) to reach a certain level of desirable pattern formation, regulating dynamical responses of mechanical and electronic devices (e.g., diodes, laser machines, and machine tools), and organizing a multiagency corporation to achieve optimal performance.

Ironically, recent research has shown that chaos can actually be useful under certain circumstances, and there is growing interest in utilizing the very nature of it, particularly in some novel time- and/or energy-critical applications. A salient observation about this is that chaos enables a system to explore its every dynamical possibility due to its ergodicity. When chaos is controllable, it can provide the system designer with an exciting variety of properties, richness of flexibility, and a cornucopia of opportunities. Traditional engineering design attempts to completely eliminate such "irregular" dynamical behavior of a system. However, such overdesign is usually accomplished at the high price of loss of flexibility in achieving optimal performance near the stability boundaries, or at the expense of radically modifying the original system dynamics which in many cases is unnecessary.

It has been shown that the sensitivity of chaotic systems to small perturbations can be used to rapidly direct system trajectories to a desired target using minimal control energy. This may be crucial for example in interplanetary space navigation. A suitable manipulation of chaotic dynamics, such as stability conversion or bifurcation delay, not only can significantly extend the operational range of machine tools and aircraft engines, but also may enhance the artificial intelligence of neural networks, as well as increase coding/decoding efficiency in signal and image communications.

Fluid mixing is another example in which chaos is not only useful but actually very desirable. The objective here is to thoroughly mix together two or more fluids while minimizing the energy required. For this purpose, fluid mixing turns out to be much simpler to achieve if the particle motion dynamics are strongly chaotic. Otherwise, it is difficult to obtain rigorous mixing properties due to the possibility of invariant two-tori in the flow. This has been one of the main subjects in fluid mixing, known as *chaotic advection*. Chaotic mixing is also important in applications involving heat transfer, such as in plasma heating within a nuclear fusion reactor. In this process, heat waves are injected into the reactor, and the best result is obtained when the convection inside the reactor is chaotic.

Within the context of biological systems, chaos control may be a crucial mechanism employed by the human brain in carrying out many of it tasks. Additionally, some recent laboratory studies reveal that the complex dynamical variability in a variety of normal-functioning physiological systems demonstrates features reminiscent of chaos. Some medical evidence lends support that control of certain chaotic cardiac arrhythmias may soon lead to the design of a safe and highly effective intelligent pacemaker.

Motivated by many potential real-world applications, current research on control and anti-control (chaotification) of chaos has proliferated in recent years. With respect to theoretical considerations, chaos control poses a substantial challenge to both system analysts and control engineers. This is due to the extreme complexity and sensitivity of chaotic dynamics, which in turn is associated with reductions in long-term predictability and short-term controllability of chaotic systems. A controlled chaotic system is inherently nonautonomous, and in most cases cannot be converted to an autonomous system since the required time-dependent controller has yet to be effectively designed. Possible time-delay, noise, and coupling influences often make a controlled chaotic system Lyapunov-irregular and extremely complex topologically. As a result, many existing theories and methodologies for autonomous systems are no longer applicable. On the other hand, chaos control poses new challenges to circuit designers and instrumentation engineers. A successful circuit implementation in a chaotic environment is generally difficult to achieve due to the extreme sensitivity of chaos to parameter variations and noise perturbations, and to the nonrobustness of chaos with respect to the structural stability of the physical devices involved.

Notwithstanding many technical obstacles, both theoretical and practical developments in this area have experienced remarkable progress in the last decade. This edited volume presents current achievements in this challenging field at the forefront of research, with emphasis on the engineering perspectives, methodologies, and potential applications of chaos and bifurcation controls. It is intended as a combination of overview, tutorial and technical reports, reflecting state-of-the-art research of significant problems in this field. The anticipated readership includes university professors, graduate students, laboratory researchers and industrial practitioners, as well as applied mathematicians and physicists in the areas of electrical, mechanical, physical, chemical and biomedical engineering and science.

I received enthusiastic assistance from several individuals in the preparation of this book. In particular, I am very grateful to Michael E. Brandt, Leon O. Chua and, in particular, Tetsushi Ueta. I would also like to thank Robert Stern and Susan Zeitz, editors of the CRC Press, for their continued support and kind cooperation. Finally, I wish to express my sincere thanks to all of the authors whose significant scientific contributions have directly led to the publication of this valuable treatise.

Guanrong Chen
Houston, Texas
Spring, 1999

Contents

Foreword   v

Preface   vii

1 Reconstructing Input-Output Dynamics from Time Series   1
(A. I. Mees)
- 1.1 Introduction 2
- 1.2 Embeddings for Systems with Inputs 3
- 1.3 Selecting Modeling Variables 6
- 1.4 Model Types 9
- 1.5 An Approach to Reconstructing Dynamics of Input-Output Systems 14
- 1.6 Examples 15

2 Black and Grey-Box Modeling of Nonlinear Systems: Identification and Analysis From Time Series   23
(L. A. Aguirre)
- 2.1 Introduction 24
- 2.2 Modeling of Nonlinear Dynamics: Representations and Techniques 24
- 2.3 NARX Polynomial Models 28
- 2.4 Static Nonlinearities in NARX Models 31
- 2.5 Identification of a Buck Converter 32
- 2.6 Model-Based Analysis of Breathing Patterns 36
- 2.7 Final Remarks 40

3 Design and Implementation of Chaos Control Systems   45
(M. J. Ogorzałek)
- 3.1 Why Chaos Control 46
- 3.2 Conditions for Implementation of Chaos Controllers 51
- 3.3 Short Description of the OGY Technique 52
- 3.4 Implementation Problems for the OGY Method 54

3.5	Brief Review of the OPF Controller	59
3.6	Improved Chaos Controller for Autonomous Circuits	62
3.7	Working Laboratory Chaos Control Systems	63
3.8	Conclusions	65

4 Chaos in Mechanical Systems and Its Control 71
(T. Kapitaniak, J. Brindley, and K. Czolczynski)

4.1	Introduction	72
4.2	Methods of Chaos Control	72
4.3	Nonlinearities in Mechanical Systems	75
4.4	Control Through Operating Conditions	78
4.5	Control by System Design	83
4.6	Discussion	85

5 Utilizing Chaos in Control System Design 89
(T. L. Vincent)

5.1	Introduction	90
5.2	Chaotic Control Algorithm	92
5.3	Inverted Pendulum	95
5.4	Discussion	103

6 Control and Synchronization of Spatiotemporal Chaos 107
(J.-Q. Fang and M. K. Ali)

6.1	Introduction	108
6.2	SESs Modeled by PDEs	108
6.3	SESs Modeled by Coupled ODEs	115
6.4	SESs Modeled by Coupled Map Lattices	119
6.5	Some Characteristic Quantitates and Mechanisms	123
6.6	New Research Outlook	125
6.7	Summary	126

7 Chaotic Vibration of the Wave Equation by Nonlinear Feedback Boundary Control 131
(G. Chen, S.-B. Hsu, and J. Zhou)

7.1	Introduction	132
7.2	Chaos Induced by Boundary Feedback	134
7.3	Feedback of Polynomial Type at the Right Endpoint	141
7.4	Concluding Remarks	150

8 Sensitivity to Initial Conditions of Chaos in Electronics 155
(G. Q. Zhong, Z. F. Liu, K. S. Tang, K. F. Man, and S. Kwong)

8.1	Introduction	156
8.2	Phase Model of PLL	157
8.3	PLL with Nonlinear Control	159

	8.4	Chaotic Characteristics of the Controlled PLL	164
	8.5	Synchronization with PLLs	167
	8.6	Conclusion	176

9 Frequency Domain Methods for Chaos Control 179
(M. Basso, R. Genesio, L. Giovanardi, and A. Tesi)
- 9.1 Introduction . . . 180
- 9.2 System Setup . . . 182
- 9.3 Existence of Periodic Solutions . . . 185
- 9.4 Stability of Periodic Solutions . . . 187
- 9.5 Application to Chaos Control . . . 195
- 9.6 Conclusions . . . 200

10 Controlling Limit Cycles and Bifurcations 205
(G. Calandrini, E. Paolini, J. L. Moiola, and G. Chen)
- 10.1 Introduction . . . 206
- 10.2 Harmonic Balance and Curvature Coefficients . . . 208
- 10.3 Normal Forms and Limit Cycles . . . 214
- 10.4 Controlling the Multiplicity of Limit Cycles . . . 217
- 10.5 Two Illustrative Examples . . . 219
- 10.6 Conclusions . . . 227

11 Theory and Experiments on Nonlinear Time-Delayed Feedback Systems with Application to Chaos Control 233
(P. Celka)
- 11.1 Introduction . . . 234
- 11.2 Delay-Differential Equations as Models . . . 234
- 11.3 From Continuous to Discrete-Time Models . . . 238
- 11.4 Control of DDE Using Delayed Self-Feedback . . . 242
- 11.5 Conclusions . . . 249

12 Time Delayed Feedback Control of Chaos 255
(X. Yu, Y. Tian, and G. Chen)
- 12.1 Introduction . . . 256
- 12.2 Chaos Control with Linear TDFC . . . 257
- 12.3 Chaos Control by Sliding Mode Based TDFC . . . 260
- 12.4 TDFC Design Based on an Optimal Principle . . . 263
- 12.5 Estimation of Delay Time τ . . . 267
- 12.6 Simulation Studies . . . 268
- 12.7 Conclusions . . . 272

13 Impulsive Control and Synchronization of Chaos 275
(J. A. K. Suykens, T. Yang, J. Vandewalle, and L. O. Chua)
- 13.1 Introduction 276
- 13.2 Basic Theory of Impulsive Differential Equations 277
- 13.3 Impulsive Synchronization of Lur'e Systems 281
- 13.4 Impulsive Control to Periodic Motions 287
- 13.5 Experimental Confirmation, Secure Communications, and $(CD)^2MA$ 292
- 13.6 Conclusions 292

14 Control and Anticontrol of Bifurcations with Application to Active Control of Rayleigh-Bénard Convection 299
(H. O. Wang and D. S. Chen)
- 14.1 Introduction 300
- 14.2 Anticontrol of Bifurcations 302
- 14.3 Amplitude Control of Bifurcations 310
- 14.4 Bifurcation Control of Rayleigh-Bénard Convection 312
- 14.5 Conclusions 320

15 Delay Feedback Control of Cardiac Activity Models 325
(M. E. Brandt and G. Chen)
- 15.1 Introduction and Background 326
- 15.2 TDF Control of a Quadratic Map Model of Cardiac Chaos . 328
- 15.3 TDF Control of a Circle-Map Cardiac Model 332
- 15.4 Linear TDF Control of a Cardiac Conduction Model 339
- 15.5 Discussion 343

16 Bifurcation Stabilization with Applications in Jet Engine Control 347
(G. Gu and A. Sparks)
- 16.1 Introduction 348
- 16.2 Local Stability and Stabilization for Hopf Bifurcations . . . 349
- 16.3 Multi-Mode Moore-Greitzer Model 353
- 16.4 Rotating Stall Control 359
- 16.5 Simulation Results and Discussions 363

17 Bifurcations of Control Systems in Normal Form 369
(W. Kang)
- 17.1 Introduction 370
- 17.2 Problem Formulation 371
- 17.3 Normal Forms and Invariants 374
- 17.4 Bifurcations of System with Quadratic Degeneracy 378
- 17.5 Bifurcations of System with Cubic Degeneracy 382
- 17.6 Application Example of Bifurcation Control 384

17.7 Conclusions 386

18 Controlling Bifurcations in Nonsmooth Dynamical Systems   **391**
(M. di Bernardo and G. Chen)
18.1 Introduction 392
18.2 Some Typical Dynamical Phenomena in PWS Systems . . . 393
18.3 Two Examples: Chua's Circuit and the Buck Converter . . 397
18.4 Control of Border-Collision Bifurcations 401
18.5 Feedback Control of PWS Chaotic Systems 404
18.6 Other Control Techniques 410
18.7 Conclusions 412

19 Adaptive Observer–Based Synchronization   **417**
(A. L. Fradkov, H. Nijmeijer, and A. Y. Pogromsky)
19.1 Introduction 418
19.2 General Definition of Synchronization 419
19.3 Adaptive Observers 421
19.4 Adaptive Synchronization of Lur'e Systems 425
19.5 Signal Transmission and Reconstruction 429
19.6 Conclusions 432

20 Discrete-Time Observers and Synchronization   **439**
(H. J. C. Huijberts, H. Nijmeijer, and A. Y. Pogromsky)
20.1 Introduction 440
20.2 Preliminaries and Problem Statement 441
20.3 Systems in Lur'e Form 442
20.4 Transformation into Lur'e Form 445
20.5 Transformation into Extended Lur'e Form 448
20.6 Observers for Perturbed Linear Systems 451
20.7 Conclusions 453

21 Separating a Chaotic Signal from Noise and Applications **457**
(H. Dedieu, T. Schimming, and M. Hasler)
21.1 Introduction 458
21.2 Definition of the Problem 459
21.3 Optimal Solution without Dynamic Constraint on the Estimator 460
21.4 Optimal Solution with Dynamic Constraint on the Estimator 463
21.5 Practical Algorithms for Noise Cleaning: Deterministic Approach 465
21.6 Practical Algorithms for Noise Cleaning: Probabilistic Approach 467
21.7 Communication Applications 471

21.8 Conclusions 474

22 Digital Communications Using Chaos 477
(M. P. Kennedy and G. Kolumbán)
22.1 Motivation 478
22.2 What is Chaos? 479
22.3 Potential Benefits of Chaotic Basis Functions in Digital Communications 479
22.4 Digital Communications Using Chaos 480
22.5 Survey of Noncoherent Chaotic Communication Schemes . . 484
22.6 Summary 496
22.7 Open Problems and Expected Developments 497

23 Synchronization in Arrays of Coupled Chaotic Circuits and Systems: Theory and Applications 501
(C. W. Wu)
23.1 Introduction 502
23.2 Notation and Terminology 502
23.3 Synchronization of Chaotic Circuits and Systems 503
23.4 Static Coupling 506
23.5 Dynamic Coupling 511
23.6 Discrete-Time Systems 516
23.7 Lyapunov Exponents Approach 519
23.8 Dynamics at Synchronization 522
23.9 Synchronization of Clusters 523
23.10 Applications to Graph Coloring 524

24 Chaos in Phase Systems: Generation and Synchronization 529
(V. D. Shalfeev, V. V. Matrosov, and M. V. Korzinova)
24.1 Introduction 530
24.2 A PLL System as a Generator of Chaotic Oscillations . . . 531
24.3 Chaotic Regimes in an Ensemble of Two Coupled PLLs . . 537
24.4 Synchronization of Chaotic Oscillations 543
24.5 Application of Chaotic PLLs to Transmission of Information 553
24.6 Conclusion 554

25 Chaos and Bifurcations in Feedback Control Systems 559
(J. Alvarez and F. Verduzco)
25.1 Introduction 560
25.2 Prediction of Chaos 560
25.3 Linear Plants with Classical Controllers 563
25.4 PD-Controlled Robot Manipulators 569
25.5 Conclusions 578

26 Chaos and Bifurcations in Coupled Networks and Their Control — 581
(T. Ueta and G. Chen)
- 26.1 Introduction 582
- 26.2 Some Simple Equilibria in the System 584
- 26.3 Periodic Solutions and Chaos 586
- 26.4 Controlling to Unstable Periodic Orbits 596
- 26.5 Concluding Remarks 599

27 Return Map Modulation in Nonautonomous Relaxation Oscillator — 603
(T. Saito and H. Torikai)
- 27.1 Introduction 604
- 27.2 Unit Shape Function 605
- 27.3 Integrate-and-Fire Model with Three Inputs 607
- 27.4 Basic Return Map by the First Prime Input 609
- 27.5 Role of the Second Base Input 613
- 27.6 Role of the Threshold Input 616
- 27.7 Conclusions 620

28 Controlling Chaos in Discrete-Time Computational Ecosystems — 625
(T. Ushio, T. Imamori, and T. Yamasaki)
- 28.1 Introduction 626
- 28.2 The Discrete Time Hogg-Huberman Model 627
- 28.3 Analysis of Fixed Points 629
- 28.4 Net Bias and Transient Behavior 630
- 28.5 Application to a Routing Problem 636
- 28.6 Conclusions 641

Index — 645

1
Reconstructing Input-Output Dynamics from Time Series

Alistair I. Mees

Isaac Newton Institute, Cambridge University
and
Centre for Applied Dynamics and Optimization,
The University of Western Australia
Nedlands, Perth, WA 6907, Australia
alistair.mees@uwa.edu.au

Abstract

The art of reconstructing nonlinear dynamics from models of autonomous systems is now relatively well-developed. Less work has been done on generalizing reconstruction theory to handle systems with inputs, which are central to the problem of controlling nonlinear dynamics. If an input-output system is poorly modeled by conventional methods, some generalization of reconstruction is needed. This contribution describes recent work on finding good embeddings and reconstructing dynamics for input-output systems.

1.1 Introduction

Over the past decade, a body of theory and practice has developed around the *reconstruction* of dynamical systems from their outputs [16, 13, 28]. That is, nonlinear dynamical systems have been constructed to match observed data [3] rather than from prior theory. A number of criteria have been used to fit models, but the intention is generally to capture significant features of the dynamics, with goals in mind such as prediction or control, or obtaining a deeper understanding of the nature of the system dynamics by discovering features of the reconstruction such as homoclinicities [19, 22].

Reconstruction grew out of embedding theory [39, 36], which was developed for autonomous nonlinear dynamical systems, so naturally the reconstruction techniques have mostly been concerned with autonomous systems also. Indeed, so great was the influence of autonomous systems with a scalar (univariate) measured quantity that sometimes the real modeling problem was made unnecessarily difficult by neglecting other variables, whether inputs or not. A case in point is the infamous blow-fly population data. This has generally been reconstructed as if all that were available were the population estimates at various times, when in fact the larval populations were also available and make it possible to build better models [44].

The interest of the present volume is largely in systems with inputs. (Of course, parameters can be treated as inputs and this is especially interesting when control is to be exerted by varying the parameters.) Generalization of the theory for autonomous systems to the input-output case is not trivial: the application of the embedding theorem, the central issue in nonlinear modeling, is far from clear in the input-output case. Even when the input can be taken as produced by a deterministic system, so the embedding theorem might be thought to apply to the Cartesian product of the original system and the system that generates the input. It is not in fact apparent that the theorem does apply, because it contains a genericity requirement which is not satisfied in this case.

Interesting explorations of this problem were made several years ago [20, 14] but it is only recently that a satisfactory theory has been developed [38] and that some of the problems of selecting embeddings have begun to be overcome [10, 12]. In this chapter, we summarize some of the theory and practice and give some examples. We concentrate on problems unique to the modeling of input-output systems; this means that we do not do full justice to the many different techniques for nonlinear fitting that are used with greater or lesser success in reconstructing autonomous systems.

The problems we face when trying to reconstruct the dynamics of an input-output system include the selection of a useful subset of variables; the choice of (possibly different) embedding dimensions and lags for different variables; and the fitting criteria to be used. For example, if we have several inputs and outputs, should we use all of them to model some particular output? Given the still-rudimentary nature of nonlinear modeling methods, it may be advantageous to use the smallest subset of inputs that is capable of representing the output. How should we identify this subset? Likewise, once we have chosen the variables, how should they be embedded? At present, we are aware of no entirely satisfactory generalization to input-output systems of the method of false nearest neighbors [4] which is widely used to identify embedding dimensions for autonomous systems, although a natural generalization seems to give good results in some cases [11]. And finally, is minimizing the sum of squares error between the model and measured outputs the most appropriate fitting criterion? The answer is no: it is well-known that even in linear systems this introduces biases in most interesting cases.

Here we will consider some pragmatic methods for solving the first two problems, namely variable selection and choice of embeddings. We will concentrate on geometrically-based methods inspired from dynamical systems theory. There is an existing theory of nonlinear reconstruction for input-output systems using extensions of the standard control-theoretic identification methods [45] and it may be more appropriate in some cases, especially where the bias caused by simple-minded least squares fitting is important. However, the recent development of a satisfactory theory for nonlinear input-output systems means that there is likely to be rapid improvement in dynamics-based approaches. This chapter should be regarded as a spur to new work and as a tentative identification of some of the problems more than as a description of definitive solutions.

1.2 Embeddings for Systems with Inputs

We assume we are given a discrete-time input-output system with system equations

$$x_t = f(x_{t-1}, u_t), \qquad (1.2.1)$$

$$y_t = g(x_t, u_t) \qquad (1.2.2)$$

where $x_t \in M^d$, some unknown smooth manifold of dimension d, and the dynamics f and the readout map g are unknown but assumed smooth. All real systems are affected by noise, but we neglect it here. Much of the work on modeling noisy autonomous systems [31, 29, 35] still has relevance for noisy input-output systems, but would be too lengthy to describe in sufficient detail here.

The input vector u and the output vector y are assumed accessible to observation. The system itself is a black box: we do not know the states

x_t, the dynamics, or even the state-space dimension. In practice, at least some components of u are likely to be controls and so available to improve modeling, but let us take them as given; that is, we can measure one or more time series $\{u_t\}_{t=1}^T$, $\{y_t\}_{t=1}^T$ and the object is to find a nonlinear system that reproduces the given outputs from the given inputs, and has other desirable properties. Often, though not always, the intention is to use the model system as part of a controller for the real-world system, or at least to use it to guide the design of such a controller.

Since we do not have access to x, we cannot directly estimate f and g from the given data. Our job is nevertheless to find a dynamical system that models this system in the sense that the future of y can be predicted from its present and past values, and from present and past values of u.

1.2.1 Autonomous Systems

For an autonomous system, u is not present and the equations become

$$x_t = f(x_{t-1}),$$
$$y_t = g(x_t).$$

The embedding theorem [39] shows (after a trivial extension) that under relatively mild, but essentially uncheckable, conditions there is a smooth map ρ such that

$$y_t = \rho(z_t) \tag{1.2.3}$$

where

$$z_t = (y_{t-\tau}, y_{t-2\tau}, \ldots, y_{t-k\tau}) \tag{1.2.4}$$

is the k-dimensional *embedded state* and τ is the *lag*. The vectors z_t, $t = k\tau + 1, \ldots, T$ are all known and any of a number of approximation methods can be used to find an approximation to the function ρ that defines the output and hence defines the dynamics in the embedding space.

It is necessary for k to be large enough for a proper embedding; the most important feature is that there should be a one-to-one correspondence between the unknown original states x_t and the known embedded states z_t, which turns out to be the case, generically at least, if $k > 2d$ where d is the unknown dimension of the state space. It should be realized that in many cases a much smaller embedding dimension will work, and smaller dimensions are highly desirable for good models. The method of false nearest neighbors [24] is one way of attempting to find the lowest embedding dimension k for which one-to-one requirement appears to hold.

The above discussion makes no reference to the approximation of ρ and it can be argued [23] that approximating ρ and choosing the embedding should be done together; this fits naturally with another extension, *non-uniform embeddings,* in which

$$z_t = (y_{t-\tau_1}, y_{t-\tau_2}, \ldots, y_{t-\tau_k})$$

and it is not necessarily the case that $\tau_j = j\tau$ for some fixed τ. It leads further to ideas of *variable embeddings* in which the embedding dimension k can also vary over a larger containing space. Variable embeddings are closely linked to so-called *contexts* in symbol strings [32, 43] which were developed for data compression but have been shown to be a new and powerful method of obtaining embeddings and other information in dynamical systems [25, 27].

1.2.2 Input-Output Systems

Until recently, there has been no fully satisfactory embedding theory for input-output systems. Oversimplifying somewhat, we can say that Stark et al [38] have now shown that generically, if $k > 2d$ and $\ell > 2d$ then (1.2.1) can be embedded in the following way. There is a map $\rho : \mathbf{R}^k \times \mathbf{R}^\ell \to \mathbf{R}$ such that

$$y_t = \rho(z_t) \qquad (1.2.5)$$

where

$$z_t = (y_{t-\tau}, \ldots, y_{t-k\tau}, u_t, u_{t-\tau}, \ldots, u_{t-(\ell-1)\tau}) \qquad (1.2.6)$$

Notice that both the input and the output have to be embedded, and that both may require up to $2d$ past values. This is true even for random inputs, or inputs coming from deterministic systems which might have dimension much greater than $2d$. Although the author is not aware of any formal work to this effect, we would expect it to be straightforward to extend this to handle non-uniform and variable embeddings, as discussed in Section 1.2.1.

This was essentially the conclusion reached informally by several previous workers in this area [20, 14, 12], but the importance of a satisfactory underlying theory should not be underestimated: for example, it is not intuitively clear whether the above informally-stated result applies when some or all of the inputs are not deterministic.

The inequalities $k > 2d$ and $\ell > 2d$ are (generically) sufficient but not necessary, so even if d were known, which is not usually the case, it would still be necessary to find some method of estimating good values for these embedding dimensions. The problem now therefore becomes how to choose τ, k and ℓ. It is clear that all of the familiar issues from the autonomous embedding theory, about choice of lag and dimension, and more generally about non-uniform embeddings, still arise in the input-output case. There are additional issues: for example, it may be that some inputs have negligible effect on the outputs of interest, and given the difficulty of nonlinear model-building it would be better not to include these variables in any models we make.

1.3 Selecting Modeling Variables

We can think of both different measurement or input channels and the various candidates for entering an embedding vector as different variables that are candidates to be used for modeling; in this way, we can attempt to answer simultaneously the questions of selection of a subset of given time series and selection of embedding.

A natural method of identifying important relationships between variables is to calculate the mutual information [15] between them. If the mutual information between two variables is large, then they may not both be needed for modeling; conversely, if it is small, both may be needed. Assuming the object is to model some particular output $y^{(i)}$, we should try to find a subset of other variables, each of which has high mutual information with $y^{(i)}$ but low mutual information with all the others. This idea was first introduced by Fraser [17], who showed that in principle it could be used to address most of the questions about embedding. In practice [26], it appears that one needs very large amounts of noise-free data to calculate anything more than the lag in a scalar output autonomous system, and its present main use is precisely in calculating that lag [1]. Some of the difficulties may be related to the difficulty of calculating mutual information by crude histogram methods, and recent work [8] attempts to improve the mutual information estimates in the hope that they become more reliable for smaller amounts of data. It may be that in the future, Fraser's program can be extended to answer the many difficult and important questions of input-output system embedding, but given the present state of the art, we prefer a more direct approach.

If one is building a system model such as (1.2.5–1.2.6), the embedding is part of the model and it may be better to address the modeling and embedding questions simultaneously [23]. A straightforward, but surprisingly effective, method turns out to be to try a large number of extremely simple nonlinear models, which can be constructed very quickly, and choose the embedding corresponding to the best of these. Then we fit a more sophisticated nonlinear model to the selected embedding. This kind of fast model building and testing is described in detail elsewhere [10] but a brief outline here may be useful. The method we describe was in fact derived as a simplification of an averaged version of the false-nearest-neighbors approach; readers may also wish to consider the use of the unsimplified version, as previously published [12].

1.3.1 Locally Constant Models

The simplest type of nonlinear model, and probably the fastest to build, is a *locally constant* model. We are given data $\{y_t\}$ and $\{z_t\}$, $t = \max\{k, \ell\}\tau + 1, \ldots, T$ where $y_t \in \mathbf{R}$ and $z_t \in \mathbf{R}^{k+\ell}$ and we assume there is a map ρ such that for each t, $y_t = \rho(z_t)$. We define an approximation $\hat{\rho}$ to ρ by

$$\hat{\rho}(z) = y_{s(z)} \qquad (1.3.7)$$

where $s(z)$ is the time index s corresponding to the closest point z_s to z. That is,

$$s(z) = \operatorname{argmin}_{s \in \{1, \ldots, T\}} |z - z_s|$$

where $|\cdot|$ is the Euclidean or other norm in $\mathbf{R}^{k+\ell}$.

Finding $\hat{\rho}(z)$ means finding the nearest data point to z, something that can be done very quickly by a number of methods, most notably using k-D trees. We call models constructed in this way *piecewise constant* or *locally constant*. The one-dimensional version of $\hat{\rho}$ has a graph that is a stair-step graph with each data point z_t at the center of its step (Figure 1.1). Locally constant models can be thought of as a simplification of the more familiar locally linear models [16]. They perform surprisingly well with low-noise data, especially when the data is plentiful. If there is significant noise, a piecewise linear model will be better and can be used in the same way except that it is computationally more expensive.

Surprisingly, locally constant models can be used to estimate derivatives with similar accuracy to locally linear models: just join the dots in Figure 1.1! This has been used to design feedback controllers for chaotic systems [5].

1.3.2 Embeddings for Single Input-Output Systems

If we now select some model fitting error ε such as root mean square error,[1] we can compare different locally constant models. The models can only be different if they have different definitions for z_t; that is, if they correspond to different embeddings.

First suppose that the input and output are scalar time series. Let

$$z_t^{(k,\ell)} = (y_{t-\tau}, \ldots, y_{t-k\tau}, u_t, \ldots u_{t-(\ell-1)\tau}). \qquad (1.3.8)$$

Define $\hat{\rho}^{(k,\ell)}$ to be the locally constant approximation to ρ corresponding to $z^{(k,\ell)}$ and let the corresponding fitting error be $\varepsilon^{(k,\ell)}$. Then we compute

---

[1] As well as the usual caveat that this particular choice may introduce biases, we should mention that a maximum error may be more appropriate. This is because an embedding dimension which is too small can give good predictions except in a few places where the improperly embedded state manifold intersects itself. Using a max-error criterion will tend to avoid such self-intersections.

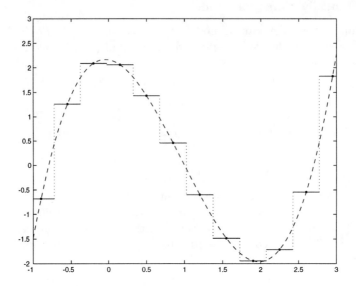

FIGURE 1.1
A locally constant approximation to a nonlinear function.

$\varepsilon^{(k,\ell)}$ for many different values of k and ℓ and select those giving the smallest value of $\varepsilon^{(k,\ell)}$. For example, we might decide to test all pairs of k and ℓ values in the ranges $k = 1, \ldots, k_{\max}$, $\ell = 1, \ldots, \ell_{\max}$, where $k_{\max}$ and $\ell_{\max}$ have to be determined by experimentation. In practice, this will often involve too much computation (especially if we are trying to use non-uniform embeddings) and it may be necessary to choose some subset of possible embedding values.

To summarise, there may be some art or experience required in choosing candidate embeddings to test, but once the candidates are chosen, it is straightforward to build a locally constant approximation for each, then select the (k, ℓ) pair that gives the smallest value of $\varepsilon^{(k,\ell)}$.

1.3.3 Embeddings for Multiple Input-Output Systems

All of the above can be carried out for the multivariate case, too. If either or both of input and output is vector-valued, we have different embedding dimensions k^i for each output $i = 1, \ldots, I$ and ℓ_j for each input u^j, $j = 1, \ldots, J$ and we have a (much) larger search space. It may be necessary to use something more sophisticated than brute-force search, though such sophistication is not always required [10].

1.3.4 Using the Selected Embedding

Locally constant models were chosen because they can be constructed very rapidly. As we shall see in Section 1.4, there are many other modeling methods which will usually be more accurate. A pragmatic approach is to fix on the embedding selected by the locally constant models, but then to construct a more accurate model. This is a conservative approach: we have chosen an embedding that makes the error small using a less accurate method, so other methods should also give small errors with that embedding. We now turn to the modeling problem.

1.4 Model Types

We have already described the locally constant modeling method, which provides a simple and rapid method of model construction. An analogy to an everyday prediction problem may help in understanding the operation of this method. Think of a weather map as a state, analogous to our state z_t. A reasonable way to predict tomorrow's weather is to find a weather map for some day s in the past which is very similar to today's, then assert that tomorrow's weather will likely be similar to that of day $s + 1$. If we embed by also considering weather maps on several days before today, and finding a succession of days in the past with matching maps, we expect to do even better.

Likewise, if we now allow interpolation between several similar sets of weather maps we might do better still: this is the locally linear modeling method. Instead of simple interpolation between the most similar maps, one could envisage a set of standard maps (basis functions) which could be compared to today's and put together in a weighted sum, or in some nonlinear combination, to produce an estimate of tomorrow's weather map: this is the foundation of methods such as the radial basis function approach.

1.4.1 Ad-hoc Local Linearization

The locally linear modeling technique is probably the most popular method for nonlinear dynamical reconstruction at present. It was first introduced for time series reconstruction by Farmer and Sidorowich [16] but a vast amount of work has been done on it since. In essence, one asserts that since the dynamics are assumed to be smooth, they will appear linear in a small enough neighborhood about any point in embedding space[2]. So to

---

[2] It is important to realize that the linearization is local in state-space, not in time. This is a fundamental difference between the nonlinear dynamics approach and earlier time series methods.

find an approximation to $\rho(z)$ we find several points near z and fit a linear model to them:
$$\hat{\rho}(z) = \alpha(z)^T z + \beta(z)$$
where $\alpha(z)$ and $\beta(z)$ are obtained by fitting
$$y_{t_i} = \alpha(z)^T z_{t_i} + \beta(z)$$
to points (z_{t_i}, y_{t_i}), $t_i \in I(z)$ in the selected neighborhood. It is common to use weighted least squares, with points further from z being given smaller weights. There are many additions and alterations to the method to be found in the literature: a recent highly tuned version is described by Yu et al [46].

The choice of neighborhood size can be important. If $|I(z)| = 1$, then we have the locally constant method, and if $|I(z)| = T$, we are fitting an autoregressive linear model to the whole of the data. Somewhere in between, presumably, there is an optimal neighborhood size, where the improved accuracy in estimating α and β over a large number of points is balanced by the decreased accuracy in assuming linearity over too large a region[3]. Most workers seem to recommend a fixed number of points in the region, though this is at best a rule of thumb. More usefully, Abarbanel [1] recommends that a locally quadratic (or higher order) model be fitted. If the quadratic terms are small, the neighborhood is small enough; if they are large, either a smaller neighborhood should be used, or the locally quadratic model is actually required.

It is obvious that locally linear models and their generalizations are as easy to apply to input-output system models as they are to apply to autonomous systems. The only difference is that z_t is defined via (1.2.6) rather than by (1.2.4).

1.4.2 Continuous Local Models

The major limitation of locally constant and heuristic neighborhood-based locally linear models is that they are discontinuous. Imagine moving z through the space. Points enter and leave the neighborhood, and as they do so, the value of $\rho(z)$ changes discontinuously. The effect is that "free-run" simulations (where the model is iterated forward from time T without reference to any additional data) do not work properly; that is, the dynamics of the model is seldom anything like the dynamics of the original system. Since the aim is to model dynamics, this is a severe disadvantage for most applications.

---

[3] Casdagli is usually credited with the observation that building locally linear models with a range of neighborhood sizes and looking at the fitting error as a function of size is a good test for nonlinearity: if smaller neighborhood sizes result in better fits, the dynamics are better fit by nonlinear than by linear models.

It is possible to produce approximations that are locally linear but are continuous. This results in models that have very good behavior in the sense that their dynamics tends to mimic the original system rather well, as well as in the more conventional sense of fitting the data one timestep ahead [28]. Consider first a noise-free system with a two dimensional embedding, resulting in some embedded data points $\{z_t\}$ in the plane, as shown in Figure 1.2. There is an essentially unique triangulation, the Delaunay triangulation [41], which tends to make the triangles as nearly equilateral as possible. That is, it minimizes the diameters of the triangles on average [37]. There are efficient algorithms for carrying out the triangulation [41].

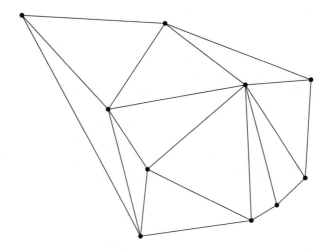

FIGURE 1.2
A Delaunay triangulation of a few points in the plane. In dimension $d > 2$ the triangles are replaced by simplices with $d + 1$ vertices. Any point z within the convex hull of the data can be expressed as a convex combination of the $d + 1$ vertices of the triangle containing z and the same convex combination can be used to define a *continuous* piecewise-linear approximation to a map sampled at the vertices.

Consider a point z within the convex hull of the embedded data. It lies in a triangle with its vertices in an index set $I(z)$ and can be expressed in the form

$$z = \sum_{i \in I(z)} \lambda_i(z) z_i \qquad (1.4.9)$$

where $\sum \lambda_i = 1$ and $\lambda_i \geq 0$. The variables $\lambda_i(z)$ are called *local coordinates* for z and they are continuous functions of z. (For example, as z leaves a triangle, $\lambda_i(z) = 0$ where i indexes the vertex that is being left behind, and as it enters the next triangle, it picks up a $\lambda_j(z)$ which is initially zero.)

At each of the vertices z_t, there is a known image point y_t where we are assuming $y = \rho(z)$. The approximation

$$\hat{\rho}(z) = \sum_{i \in I(z)} \lambda_i(z) y_i$$

is easily seen to have error $O(\epsilon^2)$ where ϵ is the diameter of the triangle defining $I(z)$. Thus, the Delaunay triangulation will tend to give good approximations in some average sense. More importantly for nonlinear system reconstruction, it will be accurate in regions where the data is dense, which will correspond for chaotic systems to regions where the invariant measure of the system has most of its weight. In other words, the approximation will be best where it matters most[4].

The generalization to higher dimensions is straightforward, the index set $I(z)$ now containing $d+1$ points rather than 3, though for computational reasons triangulation reconstruction is probably best confined to dimension $d = 5$ or less.

Triangulations are particularly powerful in understanding nonlinear dynamics. For example, they can be used to estimate invariant measures [18] and to compute all of the periodic points up to a given order [6]. The corollary that they have good dynamical behavior is, however, of greatest interest to us here.

Triangulation methods also work for noisy data [7]. In this case, one selects the vertices using an information-theoretic criterion rather than using all the data points. The criterion is essentially the same as that used in radial basis subset selection as described below.

1.4.3 Global Modeling Methods

An alternative nonlinear approximation scheme is to use a set of basis functions, whether multinomials, wavelets, radial basis functions or other functions that can be used to form a complete basis. Some approximation methods have better behavior than others as the dimension increases; these include neural nets, wavelets, and radial basis functions. All of these have been used for nonlinear reconstruction. Neural nets are powerful and widely used, but the models they produce tend to be hard to interpret in any geometrical sense; they have nevertheless been used with some success [42]. Wavelets can be very powerful in some applications but do not at

---

[4]This is also true for ad-hoc local approximations, and indeed for most of the geometrically-inspired nonlinear reconstruction methods.

present appear particularly good for dynamical reconstruction; they tend to oscillate wildly away from the data points [9].

Radial basis functions were introduced into dynamical reconstruction by Casdagli [13]. A radial basis approximation to a function can be thought of as a neural net with only one hidden layer, and with a different philosophy for choosing the basis functions. Given a set of *centers* $c_i \in \mathbf{R}^{k+\ell}$, $i = 1, \ldots, m$, define

$$\hat{\rho}(z) = \alpha^T z + \beta + \sum_{i=1}^{m} \lambda_i \phi(|z - c_i|) \qquad (1.4.10)$$

where $\phi : \mathbf{R}_+ \to \mathbf{R}$ is called the basis function. Notice that $\rho(z)$ only depends on the distances of z from the various centers: hence the name "radial basis" approximation.

Popular choices for ϕ include the identity $\phi(r) = r$, the spline $\phi(r) = r^2 \log r$, the cubic $\phi(r) = r^3$, and the Gaussian $\phi(r) = \exp(-r^2/\sigma^2)$. The first three can be thought of as approximating a function ρ by fitting an elastic sheet to data points, with the stiffness of the sheet (more correctly, the number of continuous derivatives at each data point) increasing with the power of r. The Gaussian approximates ρ as a weighted sum of bumps.

In each case, we select the weights α, β, λ_i, the centers c_i and, where relevant, the scales σ_i, to fit the data points (z_i, y_i). If the fitting criterion is least squares, then finding λ is an easy linear problem. We are left with the problem of finding the centers and scales, together with the often overlooked problem of finding m. A review of work on this is given by Mees [30]. More detail is contained in Judd and Mees [21]. In brief, an effective approach is to generate a large number of candidate centers and scales from the data, then select a subset of them (using an efficient heuristic scheme based on a pivoting-like operation applied to the QR algorithm) of some size m. The value of m is chosen to minimize the model description length [33], which has the effect of simultaneously estimating the noise level and avoiding over-fitting.

1.4.4 Models Derived from Information Theory

Information theory plays an important role in reconstruction because of the need to avoid over-fitting. Minimizing description length, which is an estimate of effective model size, is the most effective criterion for building good models [29, 31, 21].

A direct application of information theory to model building is in the construction of context trees, which are designed for data compression from finite source alphabets [34, 43]. It has recently been shown that constructing context trees provides a powerful method of determining variable embeddings while simultaneously constructing high-quality models [27, 25]. The fact that finite alphabets are assumed does not seem to be a significant limitation for many practical problems. However, at present the theory and practice only extend to autonomous systems, so more work is required before they become applicable to most of the problems of interest in this volume.

1.5 An Approach to Reconstructing Dynamics of Input-Output Systems

Here we summarize a possible approach to reconstructing input-output dynamics. Again, we emphasize that this is by no means the last word on this problem, and that advances in calculation of mutual information [8] and successful generalization to input-output systems of variable embeddings [23] may result in considerable improvements or simplifications.

1. Inspect the data and identify time-scales for all inputs and outputs.
2. Using the identified time-scales, or other methods such as mutual information, specify lags $\tau^{(j)}$ and $\tau^{(i)}$ for each of the inputs $u^{(j)}$ and outputs $y^{(i)}$. For example, a lag of about a quarter of a period is a good place to start.
3. Again using the data's natural time-scales, possibly in conjunction with Lyapunov exponent estimates [1], specify maximum delay times $D_{\max}^{(i)}$ and $D_{\max}^{(j)}$, and hence maximum embedding dimensions

$$k_{\max}^{(j)} = \lceil D_{\max}^{(j)} / \tau_{\max}^{(j)} \rceil$$

and

$$\ell_{\max}^{(i)} = \lceil D_{\max}^{(i)} / \tau_{\max}^{(i)} \rceil,$$

where the ceiling function $\lceil \theta \rceil$ is defined to be the smallest integer m such that $m \geq \theta$.
4. Build locally constant models as in Section 1.3.1 for a large number of different embeddings up to the maximum dimensions above. (If the noise level of the data is not small, it may be better to use locally linear models instead.)
5. Select a modeling error criterion such as least squares and choose the embedding giving the smallest error.
6. Build a model of the system using the chosen embedding, with a nonlinear reconstruction method appropriate to the system. (Ad-hoc locally linear modeling is often adequate.)

The choice of reconstruction method in the last step above will depend on factors such as the expected noise level, the types of functions believed to be most appropriate for fitting, and the kind of questions one wishes to ask. For example, triangulation methods are well-adapted to answering geometrical questions, but if inputs are random or are not produced by a deterministic system (including a deterministic feedback controller) there is no adequate geometrical theory, and therefore, ad-hoc local linear fitting may be all that is required.

1.6 Examples

Applications of approaches similar to those of Section 1.5 are described in detail by Cao et al [10, 12]. Both artificial and real data are considered: for example, a physiological application to leg movement is described in detail. A similar approach was taken by Abarbanel et al [5, 2], where locally constant models were used throughout, even for controller design; real-time control of an electronic circuit was demonstrated.

An interesting earlier application of input-output reconstruction was to the Milankovic theory of ice ages: the inputs were the Earth's orbital parameters such as its changing orbital eccentricity and the precession of its axis, and the output was total insolation. It was claimed [20] that a relatively high percentage of the climatic variation could be explained in this way.

Given the existence of these convincing applications, let us restrict attention here to an artificial example. This is more suitable as a tutorial because it is easier for the reader to reproduce and to experiment with.

We examine the Ikeda map with a random scalar input u that appears in a highly nonlinear way. The equations are

$$p_t = 1 + u_t(p_{t-1}\cos(r_{t-1}) - q_{t-1}\sin(r_{t-1})),$$
$$q_t = u_t(p_{t-1}\sin(r_{t-1}) + q_{t-1}\cos(r_{t-1})),$$

where

$$r = 0.4 - 6/(1 + p^2 + q^2).$$

We assume variables u_t and p_t are measured but q_t is unknown. The object is to reconstruct the dynamics of q given its past, and the past and present values of u. The model is constructed from 5000 data points. The input $u_t = 0.2 + 0.8\gamma_t$ where γ is random, consisting of independent realizations of a uniform $[0, 1]$ random variable. Short segments of the graphs u and p are shown in Figure 1.3.

We select lags τ of 1 for both variables, either by observing that there is no obvious larger time-scale or by examining mutual information or autocorrelation. We select maximum embedding dimensions of 6 for each of u and q, and fit a locally constant model to the 5000 data points. Using either a Euclidean or a maximum error norm, the best fit to the time series of q values is obtained with the model

$$q_t = \hat{\rho}(q_{t-1}, q_{t-2}, q_{t-3}, u_{t-1})$$

which is plausible from what we actually know about the system.

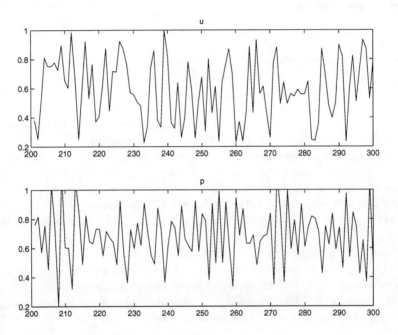

FIGURE 1.3
A segment of the random input to the Ikeda system and the corresponding output.

Examples

To make this chapter self-contained, we use a rather crude radial basis function model which does not require describing a detailed algorithm. Using the known data points, we construct the embedded points $z_t = (q_{t-1}, q_{t-2}, q_{t-3}, u_{t-1})$ and select a random subset of 100 of them to use as centers. We choose a cubic radial basis function and fit the weights λ_i, α and β in (1.4.10) by least squares. This defines our approximation $\hat{\rho}$.

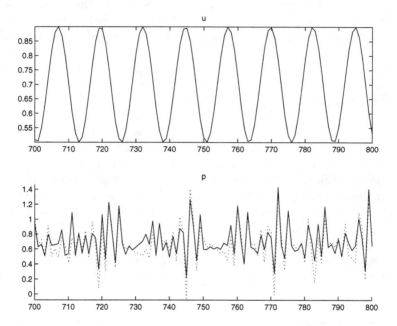

FIGURE 1.4
One-step prediction using local linear model with embedding discovered via zero-order model. Data from Ikeda map with random input was used to discover the embedding and to build the model but the actual input was the sinusoid shown. Dotted line shows output from true system with the same input.

Now let us test the reconstruction by applying it to a case where the input is very different. Figure 1.4 shows a segment of a sinusoidal input and the corresponding output, for both the real Ikeda system and the system we have just constructed from a random input. The reconstructed system is only predicting one step ahead: that is, z_t for the model is constructed from the sinusoidal input and from the embedded outputs from the true system with that input. It can be seen that the one-step prediction is good, even though the input is entirely different from that used to build the model. The fact that the reconstructed system undershoots on peaks and overshoots on troughs is due to the biases introduced by overly simple-minded least-squares fitting.

Because of the relatively large Lyapunov exponent of the Ikeda map, and the crudeness of this estimate, longer term predictions are poor. The error becomes large after about 3 iterates, although the model output and the true system output tend to re-synchronize when the input is near 0. If we require better models, we can use triangulations or less crude radial basis function models, as described in Sections 1.4.2 and 1.4.3. A better fitting criterion is also required; we should be using a better noise model [40].

Although this example dealt with a system with significant noise on the input, the modeling task did not require us to be overly concerned with noise since there is neither internal (and unmeasurable) dynamical noise, nor measurement errors. All of the noise was directly observed, so we were able to use low-noise methods, such as locally constant models. It is only when the observations are obscured by observational noise, or there is internal dynamical noise, that we require a more robust modeling process.

Acknowledgments

This work was done while I was a participant in the program "Nonlinear and Non-stationary Signal Processing" at the Isaac Newton Institute, Cambridge University, whom I thank for financial support and for an outstanding working environment. I also thank The University of Western Australia for leave and financial support. This work was partially supported by a grant from the Australian Research Council.

References

[1] H. D. I. Abarbanel, *Analysis of Observed Chaotic Data*, Springer, New York, 1996.

[2] H. D. I. Abarbanel, L. Korzinov, A. I. Mees, and I. M. Starobinets, "Optimal control of nonlinear systems to given orbits," *Systems and Control Letters*, vol. 31, pp. 263–276, 1997.

[3] H. D. I. Abarbanel, R. Brown, J. J. S. Sidorowich, and L. S. Tsimring, "The analysis of observed chaotic data in physical systems," *Reviews of Modern Physics*, vol. 65, pp. 1331–1392, 1993.

[4] H. D. I. Abarbanel and M. B. Kennel, "Local false nearest neighbors and dynamical dimensions from observed chaotic data," Technical report, Department of Physics, University of California, San Diego, 1992.

[5] H. D. I. Abarbanel, L. Korzinov, A. I. Mees, and N. F. Rulkov, "Small force control of nonlinear systems to given orbits: Experimental and theoretical results," *IEEE Trans. Circ. Sys. I*, vol. 44, pp. 1018–1023, 1997.

[6] S. Allie and A. I. Mees, "Finding periodic points from short time series," *Physical Review E*, vol. 56, pp. 346–350, 1997.

[7] S. Allie and A. I. Mees, "Reconstructing noisy dynamical systems by triangulation," *Physical Review E*, vol. 55, pp. 87–93, 1997.

[8] D. Allingham, A. I. Mees, and K. Judd, "Reliability of mutual information estimates," Technical report, CADO, The University of Western Australia, 1999.

[9] D. Allingham, M. West, and A. I. Mees. "Wavelet reconstruction of nonlinear dynamics," *Int. J. of Bifurcation and Chaos*, vol. 9, pp. 2191-2201, 1998.

[10] L. Cao, A. I. Mees, and K. Judd, "Dynamics from multivariate time series," *Physica D*, 2062, 1998.

[11] L. Cao, A. I. Mees, K. Judd, and G. Froyland, "Identification of variable relations from multivariate time series data," *Proc. of NOLTA '97*, 1997.

[12] L. Cao, A. I. Mees, K. Judd, and G. Froyland, "Determining the minimum embedding dimensions of input-output time series data," *Int. J. of Bifurcation and Chaos*, vol. 8, pp. 1491–1504, 1998.

[13] M. Casdagli, "Nonlinear prediction of chaotic time series," *Physica D*, vol. 35, pp. 335–356, 1989.

[14] M. Casdagli, "A dynamical systems approach to modeling input-output systems," in M. Casdagli and S. Eubank (Eds.), *Nonlinear Modeling and Forecasting*, pp. 265–281. Addison-Wesley, New York, 1992.

[15] T. Cover, *Elements of Information Theory*. Wiley Interscience, New York, 1991.

[16] J. D. Farmer and J. J. Sidorowich, "Predicting chaotic time series," *Phys. Rev. Letters*, vol. 59, pp. 845–848, 1987.

[17] A. M. Fraser and H. L. Swinney, "Independent coordinates for strange attractors from mutual information," *Physical Review A*, vol. 33, pp. 1134–1140, 1986.

[18] G. Froyland, K. Judd, A. I. Mees, K. Murao, and D. Watson, "Constructing invariant measures from data," *Int. J. of Bifurcation and Chaos*, vol. 5, pp. 1181–1192, 1995.

[19] J. Glover and A. I. Mees, "Reconstructing the dynamics of Chua's circuit," *J. of Circuits, Systems and Computers*, vol. 3, pp. 201–214, 1992.

[20] N. Hunter, "Application of nonlinear time-series models to driven systems," in M. Casdagli and S. Eubank (Eds.), *Nonlinear Modeling and Forecasting*, pp. 467–491. Addison-Wesley, New York, 1992.

[21] K. Judd and A. I. Mees, "On selecting models for nonlinear time series," *Physica D*, vol. 82, pp. 426–444, 1995.

[22] K. Judd and A. I. Mees, "Modeling chaotic motions of a string from experimental data," *Physica D*, vol. 92, pp. 221–236, 1996.

[23] K. Judd and A. I. Mees, "Embedding as a modeling problem," *Physica D*, vol. 120, pp. 273–286, 1998.

[24] M. B. Kennel, R. Brown, and H. D. I. Abarbanel, "Determining embedding dimension for phase-space reconstruction using a geometrical construction," *Physical Review A*, vol. 45, pp. 3403–3411, 1992.

[25] M. B. Kennel and A. I. Mees, "Stationarity of dynamics from time series," Technical report, INLS, UC San Diego, November 1998.

[26] J. M. Martinerie, A. M. Albano, A. I. Mees, and P. E. Rapp, "Mutual information, strange attractors and optimal estimation of dimension," *Phys. Rev. A*, vol. 45, pp. 7058–7064, 1992.

[27] L. Mason, A. I. Mees, and K. Judd, "Context trees and embeddings," Technical report, Centre for Applied Dynamics and Optimization, The University of Western Australia, 1997.

[28] A. I. Mees, "Dynamical systems and tesselations: Detecting determinism in data," *Int. J. of Bifurcation and Chaos*, vol. 1, pp. 777–794, 1991.

[29] A. I. Mees, "Parsimonious dynamical reconstruction," *Int. J. of Bifurcation and Chaos*, vol. 3, pp. 669–675, 1993.

[30] A. I. Mees, "Nonlinear dynamical systems from data," in F. P. Kelly (Ed.), *Probability, Statistics and Optimisation*, pp. 225–237, Wiley, Chichester, England, 1994.

[31] A. I. Mees and K. Judd, "Parsimony in dynamical modeling," in Y. Kravtsov and J. Kadtke (Eds.), *Predictability of Complex Dynamical Systems*, pp. 123–142. Springer, Berlin, 1996.

[32] J. Rissanen, "A universal data compression scheme," *IEEE Trans. Info. Theory*, vol. 29, pp. 656–664, 1983.

[33] J. Rissanen, *Stochastic Complexity in Statistical Inquiry*, World Scientific, Singapore, 1989.

[34] J. Rissanen, "A universal regression model," Technical report, IBM Almaden Research Center, 1996.

[35] T. Sauer, "A noise reduction method for signals from nonlinear systems," *Physica D*, vol. 58, pp. 193–201, 1992.

[36] T. Sauer, J. A. Yorke, and M. Casdagli, "Embedology," *J. of Stat. Phys.*, vol. 65, pp. 579–616, 1992.

[37] R. Sibson, "Locally equiangular triangulations," *Computer Journal*, vol. 21, pp. 243–245, 1978.

[38] J. Stark, D. S. Broomhead, M. E. Davies, and J. Huke, "Takens embedding theorems for forced and stochastic systems," *Nonlinear Analysis*, vol. 30, pp. 5503–5314, 1997.

[39] F. Takens, "Detecting strange attractors in turbulence," in D. A. Rand and L. S. Young (Eds.), *Dynamical Systems and Turbulence*, pp. 365–381, Springer, Berlin, 1981.

[40] D. M. Walker and A. I. Mees, "Reconstructing nonlinear dynamics by extended Kalman filtering," *Int. J. of Bifurcation and Chaos*, vol. 8, pp. 557–569, 1997.

[41] D. F. Watson, "Computing the n-dimensional Delaunay tesselation with application to Voronoi polytopes," *Computer Journal*, vol. 24, pp. 167–172, 1981.

[42] A. S. Weigend, B. A. Huberman, and D. E. Rumelhart, "Predicting the future: A connectionist approach," *Int. J. of Neural Systems*, vol 1, pp. 193–209, 1990.

[43] F. M. J. Willems, Y. M. Shtarkov, and T. J. Tjalkens, "The context tree weighting method: Basic properties," *IEEE Trans. Info. Theory*, vol. 41, pp. 653–664, 1995.

[44] P. Young, private communication.

[45] P. Young, S. Parkinson, and M. Lees, "Simplicity out of complexity in environmental modelling: Occam's razor revisited," *J. of Applied Statistics*, vol. 23, pp. 165–210, 1996.

[46] D. Yu, W. Lu, and R. G. Harrison, "Phase-space prediction of chaotic time series," *Dynamics and Stability of Systems*, vol. 13, pp. 219–236, 1998.

2

Black and Grey-Box Modeling of Nonlinear Systems: Identification and Analysis From Time Series

Luis Antonio Aguirre

Laboratório de Modelagem
Análise e Controle de Sistemas Não Lineares
Departamento de Engenharia Eletrônica
Universidade Federal de Minas Gerais
Av. Antônio Carlos 6627
31270-901 Belo Horizonte, M.G., Brazil
aguirre@cpdee.ufmg.br

Abstract

This chapter describes the modeling of nonlinear dynamical systems starting from time series. After a brief overview of some mathematical representations for nonlinear systems, the NARMAX polynomial model is described in some detail. An important view that permeates the chapter is that if the model structure is not too complex and is dynamically compatible with the data, the estimation of a model-to-fit-the-data need not be the ultimate goal. On the contrary, such a model should be useful to reveal dynamical properties of the system. Two examples using real data illustrate the use of global models in the modeling and analysis of nonlinear systems. The first example uses data from a Buck voltage converter and points out how to use prior knowledge about the system to help determine an adequate structure for the model. The second example uses a set of biomedical time

series taken from a patient who suffers from sleep apnea. It is
shown how models obtained from data can be used to quantify
fixed point stability. The results suggest that the appearance of
apnea is associated with the loss of stability of the fixed point
estimated from the blood oxygen time series.

2.1 Introduction

One of the great challenges of mathematical modeling and analysis is to
develop tools for building models and analyzing systems simply from measured data.

One of the first decisions to be made is to choose the mathematical representation. Linear representations have received much attention but in
many applications the use of such models is limited. This has prompted
many researchers to develop nonlinear representations and tools for obtaining global nonlinear models.

This chapter provides a very brief introduction to the problem of modeling nonlinear systems. Irrespective of which representation is being used, a
crucial problem is that of how to choose the model structure. An algorithm
for automatically choosing the structure of a class of models is described
in detail.

A number of questions remain to be answered in the realm of nonlinear
modeling. For instance, having estimated a nonlinear model from a set of
data, how can information (such as static nonlinearities, stability properties, etc.) about the system be obtained from the model? Also, if some
information about the system is known *a priori*, how can such information
be used to obtain better models? This chapter discusses some of these ideas
and points to future directions in the exciting field of nonlinear modeling
and analysis.

2.2 Modeling of Nonlinear Dynamics: Representations and Techniques

2.2.1 Reconstruction of Nonlinear Dynamics: Preliminaries

Given a measured time series $y(1)$, $y(2)$, ..., $y(N)$ which lies on a D-dimensional attractor of an n th-order deterministic dynamical system, an
important question in modeling is how to learn the dynamics of the D-dimensional attractor from such a time series. The starting point for most
current methods is to obtain an *embedding*, that is, a reconstructed space

which is diffeomorphic to the original state space. A convenient, though not unique, representation is achieved by using *delay coordinates* [28, 31, 30]. Other coordinates include the *singular value* [12, 7] and *derivatives* [8, 19]. A framework for the comparison of several reconstructions has been developed in [15].

A *delay vector* has the following form

$$\mathbf{y}(k) = [y(k) \; y(k-\tau) \; \ldots \; y(k-(d_e-1)\tau)]^{\mathrm{T}} , \qquad (2.2.1)$$

where d_e is the *embedding dimension* and τ is the *delay time*. Clearly, $\mathbf{y}(k)$ can be represented as a point in the d_e-dimensional *embedding space* also referred to as the *state space*, for convenience. Takens has shown that embeddings with $d_e > 2n$ will be faithful generically so that there is a smooth map $f_T : \mathbf{R}^{d_e} \to \mathbf{R}$ such that [31]

$$y(k+T) = f_T(\mathbf{y}(k)) \qquad (2.2.2)$$

for all integers k, and where the *forecasting time* T and the delay τ are also assumed to be integers. A consequence of Taken's theorem is that the attractor reconstructed in $\mathbf{R}^{d_e}$ is diffeomorphic to the original attractor in state space and therefore, the former retains dynamical and topological characteristics of the latter. Recently, it has been shown that it is typically sufficient to take $d_e > 2D$ for equation (2.2.2) to hold [30]. These results hold for infinite and noise-free time series of a generic measuring function. In practice, with short and noise-corrupted data, it is usually the case that some variables are better suited than others for state-space reconstruction [24].

Taken's theorem gives sufficient conditions for equation (2.2.2) to hold; however, no indication is given as to how to estimate the map f_T. A number of papers have been devoted to this goal and such methods can be separated into two major groups, namely *local* and *global* approximation techniques.

The local approaches usually begin by partitioning the embedding space into neighborhoods $\{\mathcal{U}_i\}_{i=1}^{N_n}$ within which the dynamics can be appropriately described by a linear map $g_T : \mathbf{R}^{d_e} \to \mathbf{R}$ such that

$$y(k+T) \approx g_{T\,i}(\mathbf{y}(k)) \quad \text{for } \mathbf{y}(k) \in \mathcal{U}_i, \; i=1,\ldots,N_n . \qquad (2.2.3)$$

Several choices for g_T have been suggested in the literature such as linear polynomials [17] which can be interpolated to obtain an approximation of the map f_T. Simpler choices include *zeroth-order approximations*, also known as *local constant predictors* [21] and a *weighted predictor* [25].

A common difficulty of such approaches is that the data have to be separated into neighborhoods and that the effort required to accomplish this grows exponentially with the embedding dimension. One way of avoiding the need for constructing neighborhoods is to fit global models to the data.

The use of global models, however, has two major difficulties, namely i) the choice of a representation for the model which should be sufficiently complex to approximate the dynamics of f_T, and ii) the selection of the correct model structure or basis within the chosen representation. Some mathematical representations are briefly described in the next section.

2.2.2 Representation of Nonlinear Systems

Functional representations. The output $y(t)$ of a nonlinear system with input $u(t)$ can be represented by the so-called Volterra series

$$y(t) = \sum_{j=1}^{\infty} \int_{-\infty}^{\infty} \cdots \int_{-\infty}^{\infty} h_j(\tau_1, \ldots, \tau_j) \prod_{i=1}^{j} u(t - \tau_i) \, d\tau \; . \qquad (2.2.4)$$

The Volterra series and other related functional representations were among the first models to be used in nonlinear approximation. A well known difficulty with such representations is the enormous amount of parameters required in order to approximate simple nonlinearities. Related techniques seem to suffer from the same problem and, in addition, tend to require very large data sets in order to estimate the parameters [18].

Discrete polynomial models. One of the most attractive representations of dynamical models is the polynomial form. A polynomial model can be represented as

$$y(t) = \sum_{i} c_i \prod_{j=1}^{n_y} y(k - n_j) \prod_{r=1}^{n_u} u(k - n_r) \; . \qquad (2.2.5)$$

Apart from being easy to interpret, simulate, and operate, algorithms for the estimation of the parameters of polynomial models are currently widely available. One of the disadvantages of global polynomials, however, is that even for polynomial models of moderate order, the number of terms can become impractically large. This type of representation will be described in detail in the next section.

Continuous polynomial models. Suppose the continuous-time signal is measured $y(t)$. A continuous polynomial model can be estimated from the data by choosing $\dot{X} = Y = y(t)$, $\dot{Y} = Z$ and $\dot{Z} = F(X, Y, Z)$. The regressors (X, Y, Z) are, as can be seen, $y(t)$ and derivatives $\dot{y}$ and $\ddot{y}$. F can be approximated by a polynomial composed by monomials made up of linear and nonlinear combinations of $y(t)$, $\dot{y}$ and $\ddot{y}$. This type of mathematical representation can be written as [19]

$$X = y(t); \quad Y = \dot{y}(t); \quad Z = \sum_{l=1}^{n_\theta} \theta_l p_l , \qquad (2.2.6)$$

where $p_l = X^i Y^j Z^k$ and $i, j, k \in \mathbf{N}$. There are some problems which are common to this and the discrete polynomial representation such as the choice of model structure. The representation (2.2.6) results in *global* continuous models as opposed to *local* continuous models which can be obtained using tesselations [26].

Discrete rational models. A rational model can be represented as the ratio of two polynomials each with the form of equation (2.2.5) [9]. Compared to polynomial models, for the same number of parameters, a rational model can produce a wider variety of dynamical regimes and, in this sense, they can be considered parsimonious [32]. On the other hand, rational models are far more sensitive to noise in the data, therefore, it is more difficult to estimate stable models. Also, structure selection of rational models is more involved than for polynomial representations.

Radial basis functions. A *radial basis function* (RBF) expansion implements a mapping of the type

$$f(\mathbf{y}) = \omega_0 + \sum_i \omega_i \phi(\| \mathbf{y} - \mathbf{c}_i \|) , \qquad (2.2.7)$$

where $\mathbf{y} \in \mathbf{R}^{d_e}$, $\| \cdot \|$ is the Euclidean norm, $\omega_i \in \mathbf{R}$ are weights, $\mathbf{c}_i \in \mathbf{R}^{d_e}$ are the RBF centres and $\phi(\cdot) : \mathbf{R}^+ \to \mathbf{R}$ is a function which is usually assumed to be given.

The RBF approach is a global interpolation technique with good localization properties, and it is easy to implement as the algorithm is essentially independent of the dimension [14]. However, performance of radial basis functions depends critically upon the centers which should therefore be chosen with care [16]. For a few hundred data points, the choice of the centers is a difficult task. This representation is a special case of a more general type, the artificial neural networks, which is linear in the parameters and this facilitates parameter estimation.

Piecewise linear approximations. Local approximations are concerned with the mapping of a set of neighboring points in a reconstructed state space into their future values. A major problem here is to select the neighborhoods. The size of the neighborhoods depends on the noise level and the complexity of the dynamics. In addition, local predictors are discontinuous and suffer from undesirable behavior when long term intervals are computed. Piecewise linear models have been found to be unreliable indicators of the underlying dynamics in some cases [11]. Thus, local predictors may not always be suitable for predicting invariant measures [13].

2.3 NARX Polynomial Models

This section briefly presents the main ideas behind NARX polynomial models. It should be noticed that if the structure of a polynomial model is not chosen carefully, in most applications, such models would simply diverge when iterated. However, if the model structure is judiciously chosen, polynomial models can successfully represent the dynamics underlying the data in many applications. In fact, all model predictions presented in this chapter were obtained from *free-run simulations* as opposed to the commonly used *one-step-ahead predictions*.

2.3.1 Representation and Estimation

Consider the NARMAX model [23]

$$y(k) = F^\ell[y(k-1), \ldots, y(k-n_y), u(k-d), \ldots$$
$$\ldots, u(k-d-n_u+1), e(k), \ldots, e(k-n_e)] , \qquad (2.3.8)$$

where n_y, n_u and n_e are the maximum lags considered for the output, input and noise terms, respectively, and d is the delay measured in sampling intervals, T_s. Moreover, $u(k)$ and $y(k)$ are respectively the input and output signals. $e(k)$ accounts for uncertainties, possible noise, unmodeled dynamics, etc. and $F^\ell[\cdot]$ is a polynomial-type function with nonlinearity degree $\ell \in \mathbf{Z}^+$. In order to estimate the parameters of this map, equation (2.3.8) should be expressed in prediction error form as

$$y(k) = \psi^{\mathrm{T}}(k-1)\hat{\Theta} + \xi(k) , \qquad (2.3.9)$$

where $\psi(k-1)$ is the regressor vector which contains linear and nonlinear combinations of output, input, and noise terms up to and including time $k-1$. The parameters corresponding to each term in such matrices are the elements of the vector $\hat{\Theta}$. Finally, $\xi(k)$ are the residuals which are defined as the difference between the measured data $y(k)$ and the one-step-ahead prediction $\psi^{\mathrm{T}}(k-1)\hat{\Theta}$. The parameter vector Θ can be estimated by orthogonal least-squares techniques.

One of the many advantages of such algorithms is that the Error Reduction Ratio (ERR) can be easily obtained as a by-product [10, 22]. This criterion provides an indication of which terms to include in the model by ordering all the candidate terms according to a hierarchy which depends on the relative importance of each term. After the terms have been ordered by the ERR, information criteria can be used to help decide a good cut-off point. Based on previous results [2], it appears that while such criteria

2.3.2 Term Clustering

The deterministic part of a polynomial NARMAX model can be expanded as the summation of terms with degrees of nonlinearity in the range $1 \leq m \leq \ell$. Each mth-order term can contain a pth-order factor in $y(k - n_i)$ and a $(m - p)$th-order factor in $u(k - n_i)$ and is multiplied by a coefficient $c_{p,m-p}(n_1, \ldots, n_m)$ as follows

$$y(k) = \sum_{m=0}^{\ell} \sum_{p=0}^{m} \sum_{n_1, n_m}^{n_y, n_u} c_{p,m-p}(n_1, \ldots, n_m) \prod_{i=1}^{p} y(k - n_i) \prod_{i=p+1}^{m} u(k - n_i) ,$$
(2.3.10)

where

$$\sum_{n_1, n_m}^{n_y, n_u} \equiv \sum_{n_1=1}^{n_y} \cdots \sum_{n_m=1}^{n_u} ,$$
(2.3.11)

and the upper limit is n_y if the summation refers to factors in $y(k - n_i)$ or n_u for factors in $u(k - n_i)$.

For the sake of presentation, suppose that T_s is short enough such that $y(k - 1) \approx y(k - 2) \approx \ldots \approx y(k - n_y)$ and $u(k - 1) \approx u(k - 2) \approx \ldots \approx u(k - n_u)$, then equation (2.3.10) can be rewritten as

$$y(k) \approx \sum_{n_1, n_m}^{n_y, n_u} c_{p,m-p}(n_1, \ldots, n_m) \sum_{m=0}^{\ell} \sum_{p=0}^{m} y(k-1)^p u(k-1)^{m-p} .$$
(2.3.12)

It is pointed out that the assumption made above is for the purpose of presentation only. In practice, cluster analysis can be performed without assuming oversampling. Also, considering an asymptotically stable model in steady-state, term clustering occurs *exactly* irrespective of T_s [5, 27].

DEFINITION 2.1 [3]. In equation (2.3.12), $\sum_{n_1, n_m}^{n_y, n_u} c_{p,m-p}(n_1, \ldots, n_m)$ are the coefficients of the *term clusters* $\Omega_{y^p u^{m-p}}$, which contain terms of the form $y(k-i)^p u(k-j)^{m-p}$ for $m = 0, \ldots, \ell$ and $p = 0, \ldots, m$. Such coefficients are called *cluster coefficients* and are represented as $\Sigma_{y^p u^{m-p}}$.

In words, a term cluster is a set of terms of the same type and the respective cluster coefficient is obtained by the summation of the coefficients of all the terms of the respective cluster which are contained in the model. In practice, it will be helpful to notice that terms of the same cluster explain the same type of nonlinearity.

2.3.3 NARMAX and Embedding Techniques: Similarities

There are some similarities between the approach described in section 2.3.1 and the more general setting stated in section 2.2.1. In particular, the NARMAX approach, used throughout this chapter, does not involve finding neighborhoods, hence $N_n = 1$ and all the data belong to a unique neighborhood, that is, $\mathbf{y}(k) \in \mathcal{U}_1$ $k = d_e, \ldots, N$. This reduces the number of data required to estimate the dynamics. Moreover, the delay time is taken to be equal to the sampling period, $\tau = T_s$.

There are, however, a number of important differences. Firstly, a NARMAX model includes input terms. This enables fitting data from non-autonomous systems and therefore estimating input/output maps, see [20] and the chapter by Alistair Mees in this volume for related ideas on the subject. An immediate consequence of this is that for input/output systems, it is not required that the output be on any particular attractor. Once an input/output model has been estimated, a particular input can be used to generate data on a specific attractor.

Another important difference is the presence of noise terms, that is, the moving average part of the model. It should be noted that equation (2.2.2) will only hold in the unlikely case when noise is absent. Any noise in the data or any imperfection in the estimate of the map f_T will result in an extra term in the right hand side of equation (2.2.2). Such a term would be responsible for representing the mismatch introduced by the noise and unmodeled dynamics. It is a well known result in the theory of system identification that if such a term is omitted from the model structure, the estimate of the map f_T will become biased during parameter estimation and nonlinear models are no exception to this rule.

It seems that when the noise is white and enters the system as a purely additive component, the division of the data into neighborhoods and subsequent estimation reduces the bias. This will not be the case however if the model is global or if the noise is correlated. Hence, in order to reduce bias, a model for the noise and uncertainties, $\Psi^T_{yu\xi}(k-1)\hat{\Theta}_{yu\xi} + \Psi^T_{\xi}(k-1)\hat{\Theta}_{\xi}$, is included in the model structure before proceeding to parameter estimation that is now achieved using prediction error methods. Once parameters have been estimated, only the deterministic part of the model is used, namely $\Psi^T_{yu}(k-1)\hat{\Theta}_{yu}$. This procedure can handle moderate amounts of white and correlated noise.

Summarizing, equation (2.3.9) is a hybrid model since it is composed of a deterministic part and a stochastic component. The latter is only used during parameter estimation in order to reduce bias on the former. Therefore, in this chapter the deterministic component of the identified models is an approximation to the dynamics, that is, $f_T \approx \Psi^T_{yu}(k-1)\hat{\Theta}_{yu}$ where $T = T_s$.

2.4 Static Nonlinearities in NARX Models

In this section, formulae will be given which relate the model terms and coefficients to the static nonlinearities; that is, in this case the gain is represented as a function of some other variable in the model.

Assume that the static nonlinearity is of the general form

$$K_{\mathrm{s}}(\bar{x}_{K_1}, \ldots, \bar{x}_{K_{n_K}}) = K_0 + \sum_{i=1}^{n_K} K_i \bar{x}_{K_i}, \qquad (2.4.13)$$

where K_i, $i = 0, \ldots, n_K$ are constants; and $\bar{x}_{K_i}$, $i = 1, \ldots, n_K$ are the steady-state value of signals such as inputs, outputs, or combinations of these. The overbar indicates the steady-state value of the respective signal. For example, suppose the static nonlinearity of a hypothetical system is $K_{\mathrm{s}}(\bar{u}^2, \bar{y}) = 2 + 0.5\bar{u}^2 + 3\bar{y}$. In this case, $n_K = 2$, $K_0 = 2$, $K_1 = 0.5$, $K_2 = 3$, $\bar{x}_{K_1} = \bar{u}^2$ and $\bar{x}_{K_2} = \bar{y}$.

A third-order NARX model with $n_u = n_y = 3$ can be written as

$$y(k) = c_0 + \sum_{j=1}^{3} \tau_0^j y(k-j) + \sum_{j=1}^{3} K_0^j u(k-j)$$

$$+ \sum_{j=1}^{3} \sum_{i=1}^{n_\tau} \tau_i^j y(k-j) x_{\tau_i} + \sum_{j=1}^{3} \sum_{i=1}^{n_K} K_i^j x_{K_i} u(k-j). \qquad (2.4.14)$$

The model above need not have all the terms indicated. If any term is missing, then the corresponding coefficient is zero. Notice that the coefficients in model (2.4.14) are not the same as those in equation (2.4.13) and the difference is indicated by the use of superscripts in (2.4.14). Actually, the coefficient in equation (2.4.13) will not be estimated, but rather the coefficients in the model will be used to form an estimate of $K_{\mathrm{s}}(\cdot)$. In the model (2.4.14), all nonlinear terms are either related to x_{τ_i} or to x_{K_i}.

Using the definition of cluster coefficients, an estimate of the static gain can be obtained for the general case as

$$\hat{K}_{\mathrm{s}}(\bar{x}_{\tau_i}, \bar{x}_{K_i}) = \frac{\bar{y}}{\bar{u}} = \frac{c_0/\bar{u} + \Sigma_u}{1 - \Sigma_y - \sum_{i=1}^{n_\tau} \Sigma_{yx_{\tau_i}} \bar{x}_{\tau_i}} + \frac{\sum_{i=1}^{n_K} \Sigma_{x_{K_i} u} \bar{x}_{K_i}}{1 - \Sigma_y - \sum_{i=1}^{n_\tau} \Sigma_{yx_{\tau_i}} \bar{x}_{\tau_i}}. \qquad (2.4.15)$$

It should be noted that in equation (2.4.15) the numbers $\Sigma_{yx_{\tau_i}}$ and $\Sigma_{x_{K_i} u}$ are the coefficients of the clusters $\Omega_{yx_{\tau_i}}$ and $\Omega_{x_{K_i} u}$ and should not be confused with the summations. As illustrated in the next section, equation (2.4.15) can be used, given a dynamical model, to determine *analytically* an estimate of the system static nonlinearity. Analogous results for the case the original system is a map, have been discussed in [1].

2.5 Identification of a Buck Converter

The Buck converter used in this work is shown in Figure 2.1. The MOSFET IRF840 is switched by actuating on the gate G. The duty cycle, defined as $D = T_{\text{on}}/T$, was varied by the integrated circuit LM3524 using Pulse Width Modulation (PWM) techniques at a rate of $1/T = 33\,\text{kHz}$, which results in a continuous mode operation.

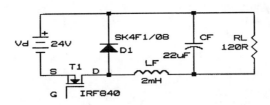

FIGURE 2.1
Buck DC-DC converter.

In order to test the converter, a Pseudo Random Binary Signal (PRBS) was designed, generated by microcomputer and applied to the Buck via a D/A converter. During the tests, the output was always the voltage across the load while the input was the DC signal used to generate the PWM that actually commands the MOSFET switch. Also, the DC voltage supply was kept constant at $V_{\text{d}} = 24\,\text{V}$.

Figure 2.2 shows the outcome of a typical dynamical test as measured and recorded by a digital oscilloscope. These data are clearly oversampled, and were decimated by a factor of 12 to yield a working sampling time of $T_{\text{s}} = 120\mu s$. The data corresponding to the first 1000 samples shown in the figure were used as identification data.

2.5.1 Use of *a priori* Information

The steady-state voltage relation of the implemented Buck converter is

$$\overline{V_{\text{o}}} = (1-D)V_{\text{d}} = \left(1 - \frac{\bar{u}-1}{3}\right)V_{\text{d}} = \frac{4V_{\text{d}}}{3} - \frac{V_{\text{d}}}{3}\bar{u}\,, \qquad (2.5.16)$$

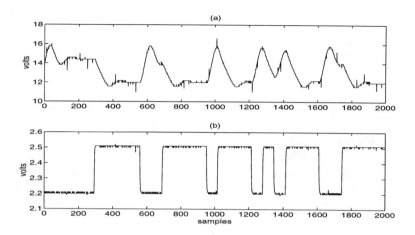

FIGURE 2.2
Test data. (a) Buck output voltage, (b) input signal $u(k)$. These data were decimated by a factor of 12 to yield working data.

where $\overline{V_o}$ is the steady-state voltage across the load, D is the duty cycle, V_d is the constant voltage supply (see Figure 2.1) and $\bar{u}$ is the steady state value of the model input $u(k)$. An important observation is that the steady-state gain $\overline{V_o}/\bar{u}$ is *not* constant, but rather depends on the operating point defined by $\bar{u}$. Therefore, a linear model would be inadequate to correctly represent the converter dynamic and static characteristics over a wide operating range. Hence, it would be a most welcome feature if the estimated nonlinear models incorporated such a steady-state gain. If this were the case, it would mean that the estimated models would yield the correct output steady-state voltage for any constant input within the range of validity. Unfortunately, there is at the moment no clear way in which the relation (2.5.16) can be "imposed" on the model because the corresponding steady-state voltage relation depends on the estimated parameter values [1]. It will be shown that by constraining the model structure, it is possible to "induce" the overall shape of the static nonlinearity. The exact location of such a function is determined by the estimated parameters.

The first one thousand data points shown in Figure 2.2 were decimated by a factor of 12 thus producing 83 samples that were actually used to identify models. The best nonlinear model identified in a 'black box' fashion was [4]

$$y(k) = 1.1101 y(k-1) - 3.5707 y(k-2) + 5.5123 \times 10^{-2} y(k-1)^2 - 2.6830 u(k-1)^2$$
$$-9.8706 \times 10^{-2} y(k-1) y(k-3) + 9.2757 \times 10^{-2} y(k-2) y(k-3)$$

$$-8.1247 \times 10^{-4} y(k-3)u(k-1) + 2.7632 \times 10 + 7.8813 \times 10^{-1} y(k-2)u(k-1)$$
$$+ \sum_{j=1}^{20} \hat{\theta}_j \xi(k-j) + \xi(k) \; , \tag{2.5.17}$$

where $\xi(k)$ is the sequence of residuals and the summation indicates the twenty moving average terms included during parameter estimation to reduce bias and is *not* used in the simulations. $y(k)$ is the converter output voltage $\overline{V_o}$. By 'black box' identification, it is meant that there was no intervention during structure selection or parameter estimation. Hence, after generating a comprehensive set of candidate terms, the ERR algorithm, mentioned in section 2.3.1, was used to *automatically* select the 'best' terms to compose the model.

The validation of model (2.5.17) is shown in Figure 2.3. As can be seen, the performance of model (2.5.17) in explaining two different sets of data is excellent. Unfortunately, this model is valid only over a rather narrow range of values of $u(k)$. In fact, it is not possible to express the static nonlinearity of (2.5.17) as $\overline{V_o} = f(\bar{u})$ and for values of the input outside the range $2.0\,\mathrm{V} < u(k) < 2.6\,\mathrm{V}$, the model performance is poor and for a range slightly wider, the model becomes unstable.

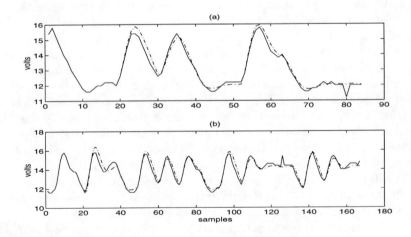

FIGURE 2.3
Validation of model (2.5.17). (—) measured data, (-··-) model free-run prediction. (a) The measured data were obtained by decimating the second half of Figure 2.2, and (b) the measured data here were obtained from a totally different and independent test.

2.5.2 Modeling with Prior Knowledge

The key point to realize in the modeling of this converter is that whatever the model might be, it should somehow contain the steady-state voltage relationship indicated by equation (2.5.16). A careful observation of that equation promptly reveals that a linear model cannot possibly incorporate such a static relationship.

The static relation of model (2.5.17) *cannot* be obtained analytically, suggesting that the model structure is probably inadequate. Moreover, it becomes clear that the respective static relation has a denominator that becomes zero for constant values of $u(k)$ within the operation range. This explains the fact that although model (2.5.17) performs remarkably well over windows of data used in the validation, such a model is valid only over a limited operating range around $u = 2.3\,\mathrm{V}$.

The discussion in the previous paragraph suggests that it is desirable to obtain a nonlinear model from which a steady-state voltage relation could be obtained *analytically*, and such a relation would not vanish for constant values of $u(k)$ within the operating range. As will be seen, the price paid for this benefit is a slight degradation in forecasting ability of the resulting model. It is important to realize that because prior knowledge is being used (the overall shape of equation (2.5.16)), the resulting identification procedure can be classified as a grey-box procedure.

The identified model should have a static nonlinearity that approximates (2.5.16), which has the form $\overline{V}_\mathrm{o} = f(\bar{u}) = a_0 + a_1\bar{u}$. There are, of course, a few alternatives. One such alternative which has extra degrees of freedom is $\overline{V}_\mathrm{o} = a_0 + a_1\bar{u} + a_2\bar{u}^2 + a_3\bar{u}^3$. It should be noted that terms of the form $y(k-i)^p u(k-j)^{m-p}$ for $m = 0,\ldots, \ell - p$ and $p = 2, \ldots, m$ and $i = 1,\ldots,n_y$; $j = 1,\ldots,n_u$ cannot be included in the model. Taking these simple but nonetheless vital pieces of information into account, a candidate set of terms was generated with terms of the following clusters: constant, Ω_y, Ω_u, Ω_{u^2} and Ω_{u^3}. This procedure increases the chances of obtaining models with the desired characteristics. In fact, the following model was obtained using the ERR algorithm to automatically select the terms from the chosen clusters

$$\begin{aligned}
y(k) = {} & 1.2013 y(k-1) - 2.6082 \times 10^{-1} y(k-2) + 6.2479 - 2.6783 u(k-1)^3 \\
& - 2.0807 \times 10^{-1} y(k-3) + 8.8399 u(k-1)^2 u(k-3) + 3.6636 u(k-3)^3 \\
& - 6.1623 \times 10^{-1} u(k-1) u(k-3) - 9.7707 u(k-1) u(k-3)^2 \\
& + \sum_{j=1}^{20} \hat{\theta}_j \xi(k-j) + \xi(k)\ .
\end{aligned} \qquad (2.5.18)$$

For this model, the following can be readily found: $\Sigma_y = 0.7324$; $\Sigma_u = 0$,

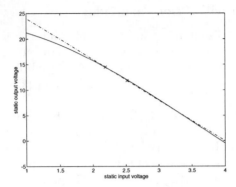

FIGURE 2.4
Static relationship between output and input voltages. (-··-) equation (2.5.16), (—) of model (2.5.18). The crosses indicate the range covered by the data during dynamic testing.

$c_0 = 6.2479$, $\Sigma_{u^2} = -6.1623 \times 10^{-1}$, $\Sigma_{u^3} = 5.4500 \times 10^{-2}$, $n_\tau = 0$, $n_K = 2$, $\bar{x}_{K_1} = \bar{u}$, $\bar{x}_{K_2} = \bar{u}^2$. Using (2.4.15) the estimated static nonlinearity is $\overline{V}_o = 23.35 - 2.30\bar{u}^2 + 0.204\bar{u}^3$ and is compared to the theoretical one in Figure 2.4. Comparison suggests that the static characteristic of model (2.5.18) is quite accurate in the range $2\,\text{V} < \bar{u} < 4\,\text{V}$, and that in the range $1\,\text{V} < \bar{u} < 2\,\text{V}$ the maximum error is around 10%. The predictions made with this model over the two previous validation data sets are shown in Figures 2.5a and 2.5b.

2.6 Model-Based Analysis of Breathing Patterns

The data used in this section were recorded from a 49-year-old male in the Sleep Laboratory of Boston's Beth Israel Hospital and have been described in detail by Rigney and co-workers [29].

The subject apparently suffers from sleep apnea which is a life-threatening disorder. In short, during sleep apnea, the subject stops breathing. The periods during which there is little or no respiration can vary from twenty seconds up to a minute. As a consequence, the oxygen saturation drops and "suddenly the subject takes about four deep breaths and then stops breathing again. The process repeats itself over and over, and for this reason, this abnormal pattern of respiration is called periodic breathing" [29]. On the other hand, during normal breathing, the oxygen saturation is practically constant. Of course, in between these two extremes there are many

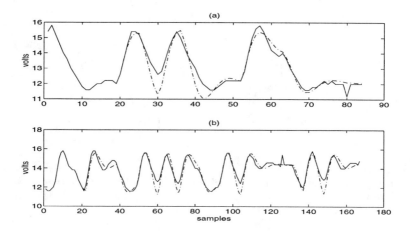

FIGURE 2.5
(—) measured data, (-·-) model (2.5.18) free-run prediction.
(a) second half of Figure 2.2, (b) data from a different test.

not-so-well-defined breathing patterns that will be referred to informally as intermittent apnea.

The data set consists of three time series, namely heart rate (HR), respiration (R), and blood oxygen saturation (BOS). The R and BOS time series were simultaneously recorded during 4 hours and 43 minutes at a rate of 250 Hz and subsequently these data were averaged over windows of 0.08 seconds and decimated at a rate of 2 Hz and provided with this sampling rate [29]. The HR were obtained taking the reciprocals of each R-R interval and assigned to the beginning of each such interval. Subsequently, interpolation techniques are used to obtain "samples" at each 0.5s, for details and discussion on possible artifacts, see [29]. The working sampling time in this section will be $T_s = 0.5\,\mathrm{s}$ resulting in each time series having around 34,000 points.

Several models were identified from windows of data corresponding to normal, periodic, and intermittent breathing. A number of cross-validation tests were performed and described to assess model adequacy. Such models have BOS as output, $y(k)$ and HR and R as inputs, respectively $u_1(k)$ and $u_2(k)$.

Figures 2.6a-c show three rather short windows of the BOS time series represented in phase space using delay coordinates with a lag of eight samples. Such plots respectively correspond to normal breathing, intermittent apnea and, finally, periodic breathing (fully developed apnea).

These plots suggest that the transition from normal respiration to apnea might be associated with the loss of stability of the fixed point seen in

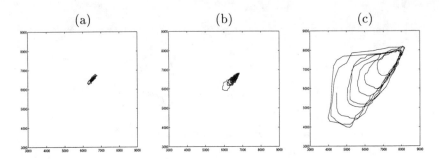

FIGURE 2.6
State phase reconstruction of windows of the BOS time series.
(a) normal breathing, (b) intermittent breathing, and (c) periodic breathing (apnea). x-axes are $y(k)$ and the y-axes are $y(k-8)$.

Figure 2.6a. It was then wondered if the models for normal breathing and apnea had incorporated this feature, namely that the dynamics of the former seems to be related to a stable fixed point while the dynamics of the latter apparently can be very roughly thought of as an unstable fixed point surrounded by a stable limit cycle. In order to verify this, the following procedure was used.

1. An estimated model and a window of data of interest were taken. For each data point, the respective values of $u_1(k)$ and $u_2(k)$ were substituted into the model. The resulting equation is of the form $y(k) = g[c\ y(k-1) \ldots y(k-n_y)]$, where $g[\cdot]$ is a polynomial, usually nonlinear and c is a constant.

2. Determine the Jacobian of the equation obtained in the first step, $\partial g[\cdot]/\partial y$, where $y = [y(k-1)\ \ldots\ y(k-n_y)]^\top$. If the equation is linear, the Jacobian will be a constant matrix.

3. If the equation is nonlinear, the Jacobian depends on lagged values of $y(k)$. In this case, estimate the location of the fixed point $\bar{y}$ from the data [6].

4. Replace the estimated value into the Jacobian, that is, $y(k-1) = \ldots = y(k-n_y) = \bar{y}$. This yields a constant matrix whose n_y eigenvalues are then computed and plotted.

5. Repeat steps 1 to 4 until the end of the data window.

Figures 2.7a–d show the eigenvalues estimated as outlined above. The plots in these figures were obtained by using models of normal breathing on normal breathing data windows and periodic breathing models on periodic breathing data windows. In these figures, the stability boundary is the unit circle, hence the closer the eigenvalues get to such a boundary, the more unstable is the fixed point at which the Jacobian was evaluated.

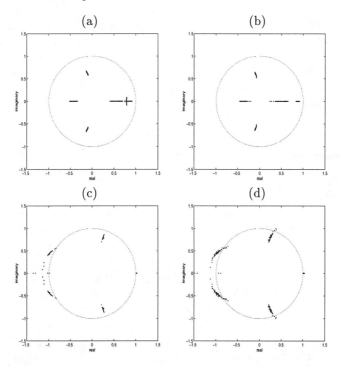

FIGURE 2.7
Eigenvalues in the complex plane for normal breathing model and data (a) normal breathing data window used in the identification and (b) a different normal breathing data window. Eigenvalues for periodic breathing model and data (c) periodic breathing data window used in the identification and (d) a different periodic breathing data window.

Such figures suggest that, in fact, the model estimated from the normal breathing data seems to indicate that the underlying dynamics are far more stable than for the periodic breathing pattern. It seems adequate to conjecture that such dynamics are associated with a stable fixed point whereas the periodic breathing dynamics can be characterized by an unstable fixed point surrounded by a stable limit cycle or alternatively a fixed point which is stable but not asymptotically stable.

The same models used to produce Figure 2.7 were used on different windows of data with qualitatively identical results. Therefore, the results suggest that estimated models can be used to quantify the stability of fixed points in real physiological data. In the present case, the fixed point clearly loses stability as the subject stops breathing normally and begins to breathe periodically (apnea). Although the models do not provide any insight into the physiology of this phenomenon, an issue of interest would be to verify if such a loss of stability can be detected with the use of modeling techniques before the patient actually stops breathing normally. Conceivably, the eigenvalues of the Jacobian matrix could be monitored. In this case, the parameters of the identified model could be updated using standard recursive algorithms. The migration of the updated Jacobian eingevalues towards the stability limit could be an indication that the patient is drifting towards apnea.

2.7 Final Remarks

This chapter has briefly mentioned mathematical representations used in modeling nonlinear dynamical systems. Each representation has different properties and it seems reasonable to make a choice based on the properties which better suit the application at hand.

Two applications have been described. In both cases, NARMAX polynomials were used and different properties of this representation were explored. The first example considered a Buck voltage converter. As many other voltage converters, this device has a steady-state voltage relation with known overall shape. This information has been used to restrict the types of terms admitted in the model. Also, equation (2.4.15) has been used to determine analytically the steady-state voltage relation estimated by each model. This procedure in which some prior knowledge about the system is systematically used in the modeling process is known as *grey-box modeling*. In particular, the best model identified using black-box techniques fits the data much better, but has a very limited domain of validity. On the other hand, the counterpart obtained by grey-box techniques does not fit the data as well but is valid over a very wide range of operating points. These concepts have been discussed and illustrated using the steady-state voltage relations of the real converter and of the estimated models which can be easily obtained from NARMAX polynomials.

The second example used biomedical data. In short, multivariable NARMAX polynomials were used not only to model the underlying dynamics but also, and especially, were used to characterize the system local stability. The results discussed suggest that the appearance of sleep apnea can be quantified by the stability of the fixed point estimated from the blood oxygen time series. If correct, this conclusion will be relevant in predicting the onset of apnea.

Perhaps the main concern of this chapter has been to point out that the estimation of a model-to-fit-the-data is not the ultimate goal in nonlinear identification. On the contrary, if the models obtained are nonlinear, there are a number of open problems and applications concerning system analysis. In particular, it is desirable, given a model obtained from data, to be able to extract from such a model information about the system. This chapter has illustrated how to use estimated models to obtain static nonlinearities and fixed-point stability properties of the original systems.

Acknowledgments

The author is grateful to CNPq and PRPq/UFMG for financial support. The author is indebted to C.R.F. Jácome for assistance in obtaining the results described in section 2.4 and to V.C. Barros and A.V.P. Souza for assistance in section 2.6. Some preprints and data sets related to this chapter are available at http://www.cpdee.ufmg.br/~MACSIN.

References

[1] L. A. Aguirre, "Recovering map static nonlinearities from chaotic data using dynamical models," *Physica D*, vol. 100, pp. 41–57, 1997.

[2] L. A. Aguirre and S. A. Billings, "Dynamical effects of overparametrization in nonlinear models," *Physica D*, vol. 80, pp. 26–40, 1995.

[3] L. A. Aguirre and S. A. Billings, "Improved structure selection for nonlinear models based on term clustering," *Int. J. Control*, vol. 62, pp. 569–587, 1995.

[4] L. A. Aguirre, P. F. Donoso-Garcia, and R. Santos-Filhos, "Use of *a priori* information in the identification of global nonlinear models for a Buck converter," *IEEE Trans. Circuits Syst. I*, 1999, to appear.

[5] L. A. Aguirre and E. M. Mendes, "Nonlinear polynomial models: structure, term clusters and fixed points," *Int. J. Bifurcation and Chaos*, vol. 6, pp. 279–294, 1996.

[6] L. A. Aguirre and A. V. P. Souza, "An algorithm for estimating fixed points of dynamical systems from time series," *Int. J. Bifurcation and Chaos*, vol.8, pp. 2203–2213, 1999.

[7] A. M. Albano, J. Muench, C. Schwartz, A. I. Mees, and P. E. Rapp, "Singular-value decomposition and the Grassberger-Procaccia algorithm," *Phys. Rev. A*, vol. 38, pp. 3017–3026, 1988.

[8] E. Baake, M. Baake, H. G. Bock, and K. M. Briggs, "Fitting ordinary differential equations to chaotic data," *Phys. Rev. A*, vol. 45, pp. 5524–5529, 1992.

[9] S. A. Billings and S. Chen, "Extended model set, global data and threshold model identification of severely nonlinear systems," *Int. J. Control*, vol. 50, pp. 1897–1923, 1989.

[10] S. A. Billings, S. Chen, and M. J. Korenberg, "Identification of MIMO nonlinear systems using a forward-regression orthogonal estimator," *Int. J. Control*, vol. 49, pp. 2157–2189, 1989.

[11] S. A. Billings and W. S. F. Voon, "Piecewise linear identification of nonlinear systems," *Int. J. Control*, vol. 46, pp. 215–235, 1987.

[12] D. S. Broomhead and G. P. King, "Extracting qualitative dynamics from experimental data," *Physica D*, vol. 20, pp. 217–236, 1986.

[13] R. Brown, P. Bryant, and H. D. I. Abarbanel, "Computing the Lyapunov spectrum of a dynamical system from an observed time series," *Phys. Rev. A*, vol. 43, pp. 2787–2806, 1991.

[14] M. Casdagli, "Nonlinear prediction of chaotic time series," *Physica D*, vol. 35, pp. 335–356, 1989.

[15] M. Casdagli, S. Eubank, J. D. Farmer, and J. Gibson, "State space reconstruction in the presence of noise," *Physica D*, vol. 51, pp. 52–98, 1991.

References

[16] S. Chen, S. A. Billings, C. F. N. Cowan, and P. M. Grant, "Practical identification of NARMAX models using radial basis functions," *Int. J. Control*, vol. 52, pp. 1327–1350, 1990.

[17] J. D. Farmer and J. J. Sidorowich, "Predicting chaotic time series," *Phys. Rev. Lett.*, vol. 59, pp. 845–848, 1987.

[18] M. Giona, F. Lentini, and V. Cimagalli, "Functional reconstruction and local prediction of chaotic time series," *Phys. Rev. A*, vol. 44, 3496–3502, 1991.

[19] G. Gouesbet and C. Letellier, "Global vector field reconstruction by using a multivariate polynomial l_2 approximation on nets," *Phys. Rev. E*, vol. 49, pp. 4955–4972, 1994.

[20] N. Hunter, "Application of nonlinear time-series models to driven systems," in M. Casdagli and S. Eubank (Eds.), *Nonlinear Modeling and Forecasting*, pp. 467–491, Addison-Wesley, New York, 1992.

[21] M. B. Kennel and S. Isabelle, "Method to distinguish possible chaos from coloured noise and to determine embedding parameters," *Phys. Rev. A*, vol. 46, pp. 3111–3118, 1992.

[22] M. Kortmann, K. Janiszowski, and H. Unbenhauen, "Application and comparison of different identification schemes under industrial conditions," *Int. J. Control*, vol. 48, pp. 2275–2296, 1988.

[23] I. J. Leontaritis and S. A. Billings, "Input-output parametric models for nonlinear systems part II: Stochastic nonlinear systems," *Int. J. Control*, vol. 41, pp. 329–344, 1985.

[24] C. Letellier, J. Maquet, L. Le Sceller, and L. A. Aguirre, "On the nonequivalence of observables in phase-space reconstructions from recorded time series," *J. of Phys. A*, vol. 31, pp. 7913–7927, 1998.

[25] P. S. Linsay, "An efficient method of forecasting chaotic time series using linear regression," *Phys. Lett. A*, vol. 153, pp. 353–356, 1991.

[26] A. I. Mees, "Dynamical systems and tesselations: detecting determinism in data," *Int. J. Bifurcation and Chaos*, vol. 1, pp. 777–794, 1991.

[27] E. M. A. M. Mendes and S. A. Billings, "On overparametrization of nonlinear discrete systems," *Int. J. Bifurcation and Chaos*, vol. 8, pp. 535–556, 1998.

[28] N. H. Packard, J. P. Crutchfield, J. D. Farmer, and R. S. Shaw, "Geometry from a time series," *Phys. Rev. Lett.*, vol. 45, pp. 712–716, 1980.

[29] D. R. Rigney, A. L. Goldberger, W. C. Ocasio, Y. Ichimaru, G. B. Moody, and R. G. Mark, "Multi-channel physiological data: Description and analysis," in A. A. Weigend and N. A. Gershenfeld (Eds.), *Time Series Prediction*, pp. 105–129, Addison-Wesley, New York, 1994.

[30] T. Sauer, J. A. Yorke, and M. Casdagli, "Embedology," *J. of Statistical Physics*, vol. 65, pp. 579–616, 1991.

[31] F. Takens, "Detecting strange attractors in turbulence," in D. A. Rand and L. S. Young (Eds.), *Dynamical Systems and Turbulence*, Lecture Notes in Mathematics, vol. 898, pp. 366–381, Springer-Verlag, Berlin, 1980.

[32] Q. M. Zhu and S. A. Billings, "Parameter estimation for stochastic nonlinear rational models," *Int. J. Control*, vol. 57, pp. 309–333, 1993.

3

Design and Implementation of Chaos Control Systems

Maciej J. Ogorzałek

Department of Electrical Engineering
University of Mining and Metallurgy
Kraków, Poland
maciej@zet.agh.edu.pl

Abstract

> We analyze general possibilities for electronic implementations of chaos controllers. First, we consider general questions of control of chaotic systems and conditions for implementation of chaos controllers. Next, we consider two particular classes of chaos controllers: i) based on the OGY approach; and ii) OPF controllers. Advantages and disadvantages of both classes of controllers in real implementations are discussed. Some examples of laboratory-scale chaos control systems are also reviewed.

3.1 Why Chaos Control

Deterministic systems considered in classic papers and textbooks on system dynamics and control display three types of behavior of their solutions: they approach constant solutions, they converge toward periodic solutions or they converge toward quasi-periodic solutions.

During the last decade it has been confirmed that almost every physical system can also display behaviors which do not belong to any of the above-mentioned classes — they become aperiodic (chaotic) if their parameters, internal variables, or external signals are chosen in a specific way.

We will assume that the following specific properties qualify dynamical behavior as chaotic:

1. the solutions exhibit sensitive dependence on initial conditions (trajectories are unstable in the Lyapunov sense) but remain bounded in space as time elapses (they are stable in the Lagrange sense);
2. trajectory moves over a strange attractor — a geometric invariant object which can possess fractal dimension. The trajectory passes arbitrarily close to any point of the attractor set — as we say there is a dense trajectory or that the system is ergodic.
3. chaotic behavior appears in the system as via a "route" to chaos which typically is associated with a sequence of bifurcations — qualitative changes of observed behavior when varying one or more of the parameters.

Sensitive dependence on initial conditions means that trajectories of a chaotic system starting from nearly identical initial conditions will eventually separate and become uncorrelated (but they will always remain bounded in space). In Figure 3.1, we give an example of two trajectories starting from initial conditions differing by 0.001 — after some period when they remain close to each-other, they eventually separate.

This property has important practical consequences. In real systems and also in numerical simulations, one can specify the initial conditions only with some finite accuracy ε. If two initial conditions are closer to each-other, then they are not distinguishable in measurements. The trajectories of a chaotic system starting from such initial conditions will after a finite time diverge and become uncorrelated. For any precision we use in measurements (experiments), the behavior of trajectories is not predictable – the solutions look virtually random despite the fact they are produced by a deterministic system. There is also another consequence of this property which may be appealing for the control purposes — one can notice that very small stimulus in the form of tiny change of parameters can have very large effect for the system behavior.

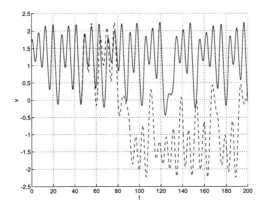

FIGURE 3.1
Two trajectories of Chua's oscillator starting from nearly identical initial conditions eventually separate resulting in different behavior — the system displays sensitive dependence on initial conditions.

The second property can be explained looking at Figure 3.2. The trajectory "fills" out some part of the phase-space. If we choose a point within this region of space and a small ball of radius ε around it, the trajectory will eventually pass through this ball after a finite time (which might be however very long!).

Talking about routes to chaos and bifurcations one should mention that in many cases creation of trajectories that are observable in experiments (stable) via bifurcation is accompanied by creation of unstable orbits which are invisible in experiments. Many of such unstable orbits also persist within the chaotic attractor — many authors consider as fundamental the property of existence of a countable (infinite) number of unstable periodic orbits within an attractor. The observed chaotic trajectory passes arbitrarily close to any of these orbits. Using software tools it is possible to detect some of such orbits in numerical experiments.

These fundamental properties of chaotic solutions (trajectories, systems) are the basis of the chaos control techniques.

3.1.1 What We Mean by Chaos Control?

Chaos, so commonly encountered in physical systems, represents rather a peculiar type of behavior commonly considered as causing malfunctioning, disastrous and thus unwanted in most applications. In real systems, we would like to avoid situations like fibrillation or arrhythmias in medicine or hurricanes and other atmospheric disasters believed to be associated with

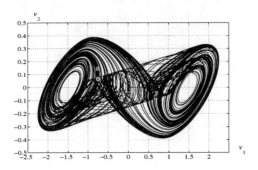

FIGURE 3.2
Two-dimensional projection of the Double scroll attractor observed in Chua's circuit.

large-scale chaotic behavior.

Thus, the most common goal of control for chaotic systems is suppression of oscillations of the erratic, bizzare kind and influencing it in such a way that it will produce a prescribed, desired motion. The goals vary depending on a particular application. The most common goal is to convert chaotic motion into a stable periodic or constant one. It is not at all obvious how such a goal could be achieved as one of the fundamental features of chaotic systems - the sensitive dependence on initial conditions seems to contradict any stable system operation.

Recently several applications have been mentioned in the literature where the desired state of system operation is chaotic — the control problems in such cases are defined as: to convert unwanted chaotic behavior into another kind of chaotic motion with prescribed properties. This is the goal of chaos synchronization and many chaos based signal transmission systems where specific types of chaotic behaviors are required as carriers or spreading codes. We can also imagine situations like mixing of components in a chemical reactor which would be much better in a chaotic state than in any other one. One can imagine a requirement for changing periodic behavior into chaotic motion (which might be the goal in the case of removal of epileptic seizures). The last-mentioned type of control is often referred to as *anti-control* of chaos.

Many chaotic systems display so-called multiple basins of attraction and fractal basin boundaries. This means that, depending on the initial conditions, trajectories can converge to different steady states. (Trajectories in nonlinear systems may possess several different limit sets and thus exhibit a variety of steady-state behaviors depending on the initial condition which could be chaotic or not.)

In many cases, the sets of initial states leading to a particular type of behavior are inter-wound in a complicated way forming fractal structures.

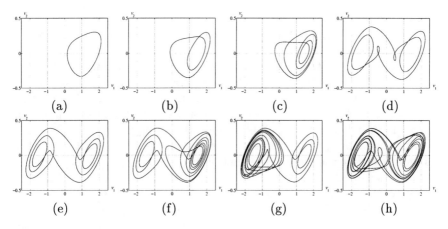

FIGURE 3.3
Unstable periodic orbits uncovered from the double scroll attractor shown in the previous figure. The shape of these orbits resemble the overall structure of the attractor.

Thus, elimination of multiple basins of attraction could be yet another kind of control goal.

Considering the possibilities of influencing the dynamics of a chaotic circuit, one can distinguish four basic approaches:

1. variation of an existing accessible system parameter,
2. change in the system design - modification of its internal structure,
3. injection of an external signal(s),
4. introduction of a controller (classical PI, PID, linear or nonlinear, neural, stochastic etc.).

Due to very rich dynamic phenomena encountered in typical chaotic systems, there exist a variety of approaches to controlling such systems [2].

3.1.2 Properties of Chaotic Systems and Goals of the Control

As already mentioned, systems displaying chaotic behavior possess specific properties. We will exploit these properties when attacking the control problem. In what way a chaotic system differs from any other object of control—what are its specific properties which could be useful?

Route to chaos via a sequence of bifurcations has important implications for chaos control: firstly, it gives an insight into other accessible behaviors that can be obtained by changing parameters (this may be used for redesigning the system); secondly, stable and unstable orbits that are created or annihilated in bifurcations may still exist in the chaotic range and constitute potential goals for control.

Three fundamental properties of chaotic systems are of potential use for control purposes.

For a long time, this fundamental property has been considered as the main obstacle for control — How could one visualize successful control if the dynamics may change drastically with small changes of initial conditions or parameters? How could one produce a prescribed kind of behavior if errors in initial conditions will be exponentially amplified?

This instability property does not however necessarily mean that control is impossible! It has been shown that despite the fact that nearby starting trajectories diverge, they can be convergent to another prescribed kind of trajectory—one simply has to employ a different notion of stability. In fact, we do not need that the nearby trajectories converge—the requirement is quite different—the trajectories should merely converge to some goal trajectory $g(t)$, namely:

$$\lim_{t\to\infty} |x(t) - g(t)| = 0 \qquad (3.1.1)$$

Depending on a particular application, $g(t)$ could be one of the solutions existing in the system or any external waveform we would like to impose. Extreme sensitivity may even be of prime importance as control signals are in such cases very small.

The second important property of chaotic systems which will be exploited is existence of a countable infinity of unstable periodic orbits within the attractor, already considered earlier. These orbits, although invisible during experiments, constitute a skeleton of the attractor. Indeed, the trajectory passes arbitrarily close to every such orbit. This invisible structure of unstable periodic orbits plays a crucial role in many methods of chaos control—using specific methods, the chaotic trajectory can be perturbed in such a way that it will stay in the vicinity of a chosen unstable orbit from the skeleton.

These two fundamental properties of chaotic signals and systems offer some very interesting issues for control not available in other classes of systems. Namely:

1. due to sensitive dependence on initial conditions it is possible to influence the dynamics of the systems using very small perturbations; moreover, the response of the system is very fast, and

2. the existence of a countable infinity of unstable periodic orbits within the attractor offers extreme flexibility and a wide choice of possible goal behaviors for the same set of parameter values.

3.2 Conditions for Implementation of Chaos Controllers

During the last few years, dozens of chaos control techniques have been proposed (see e.g. [2]). Analyzing these techniques one can easily realize that most of them are of purely academic interest - their implementation in real systems would be extremely difficult if not impossible. There is a general lack of easily implementable chaos control approaches. Looking at the possible applications alone, it becomes obvious that chaos control techniques and their possible implementations will greatly depend on the nature of the process under consideration. Looking from the control implementation perspective, real systems exhibiting chaotic behavior show many differences. The main ones are:

1. speed of the phenomenon (frequency spectrum of the signals),
2. amplitudes of the signals,
3. existence of corrupting noises, their spectrum and amplitudes,
4. accessibility of the signals to measurement,
5. accessibility of the control (tuning) parameters,
6. acceptable levels of control signals

Looking for an implementation of a particular chaos controller, we must first look at the above system-induced limitations: How can we measure and process signals from the system? Are there any accessible system variables and parameters which could be used for the control task? How to choose the ones that offer the best performance for achieving control? At what speed do we need to compute and apply the control signals? What is the lowest acceptable precision of computation? Can we achieve control in real-time?

A slow system like a bouncing magneto-elastic ribbon (with eigen-frequencies below 1 Hz) is certainly not as demanding as a telecommunication channel (running possibly at GHz) or a laser when it comes to control!

Considering electronic implementations, one must look at several closely linked areas: sensors (for measurements of signals from a chaotic process), electronic implementation of the controllers, computer algorithms (if computers are involved in the control process) and actuators (introducing control signals into the system). External to the implementation (but directly involved in the control process and usually fixed based on the measured signals) is finding of the goal of the control.

However, many methods have been developed and described in the literature [2, 16, 17], most of them are still only of academic interest because of lack of success in implementation. A control method cannot be accepted as successful if computer simulation experiments are not followed by further laboratory tests and physical implementations. Only very few results of such tests are known - among the exceptions are: control of a green light laser [28], control of a magneto-elastic ribbon [7] and a few other examples.

An obvious question arises at this point - why, despite a wealth of developed methods, have so few successful implementations and real applications appeared?

In this chapter, we try to answer this question, at least partially, by looking at the two most appealing methods - namely the OGY technique and OPF control. Among the approaches and methodologies for chaos control described in the literature [2, 15, 16, 17], these two approaches are of interest because they use specific properties of chaotic systems [20]:

1. First, a chaotic attractor contains an infinite number of unstable periodic orbits embedded within it;

2. Second, there exist dense orbits in the sense that a typical trajectory on the attractor passes arbitrarily close to any point on it (it also passes arbitrarily close to any of the unstable periodic orbits);

3. In addition, these methods require very small signals to achieve control and thus, are more realistic for implementation purposes.

From the implementation point of view, these two methods are very different. OGY works on the basis of measured signals and uses a computer to find the goal of the control and make necessary calculations of the control signals — thus, all the signals used in calculations are discretized both in time and space. OPF is purely analog, all operations are implemented in hardware. To consider the implementation limitations, let us first look at the principles of operation of both methods.

3.3 Short Description of the OGY Technique

The OGY control method developed by Ott, Grebogi and Yorke [21, 22] in 1990 uses the two above-mentioned properties. The goal of control is to stabilize one of the unstable periodic orbits by perturbing a chosen (accessible) system parameter over a small range about some nominal value.

To explain in some detail the action of the OGY method, let us assume for simplicity that we have a three-dimensional continuous-time system of first-order autonomous ordinary differential equations:

$$\frac{d\mathbf{x}}{dt} = \mathbf{F}(\mathbf{x}, p), \qquad (3.3.2)$$

where $\mathbf{x} \in \mathbb{R}^3$ is the state and $p \in \mathbb{R}$ is a system parameter which we can change. We also assume that parameter p can be modified within a small interval around its nominal value p_0 ($p \in [p_0 - \delta p_{\max}, p_0 + \delta p_{\max}]$, where $\delta p_{\max}$ is the maximum admissible change in the parameter p). We choose a two-dimensional Poincaré surface Σ which defines a Poincaré map $\mathbf{P}$ (for

$\xi \in \Sigma$, we denote by $\mathbf{P}(\xi)$ the point at which the trajectory starting from ξ intersects Σ for the first time). Since the vector field $\mathbf{F}$ depends on p, the Poincaré map $\mathbf{P}$ also depends on this parameter p. Thus, we have

$$\mathbf{P}\colon \mathbb{R}^2 \times \mathbb{R} \ni (\xi, p) \longrightarrow \mathbf{P}(\xi, p) \in \mathbb{R}^2 \qquad (3.3.3)$$

Let us assume that $\mathbf{P}$ is differentiable. Say we have selected one of the unstable periodic orbits embedded in the system's attractor as the goal of our control because, for example, it offers an improvement in system performance over the original chaotic behavior. This could be the case, for example, of a chaotic laser intensity which is clearly an unwanted phenomenon and the effective power of the laser beam can be enhanced using control to stabilize or eliminate chaotic behavior [28]. Another example of unwanted chaotic behavior is fibrillation, where the heart pumps blood in an inefficient manner. In this application, controlling the heart-beat into a nearly periodic regime is of paramount importance [7]. For simplicity, we assume that this is a period-1 orbit (a fixed point of the map $\mathbf{P}$).

Let us denote by ξ_F an unstable fixed point of $\mathbf{P}$ for $p = p_0$ ($\mathbf{P}(\xi_F, p_0) = \xi_F$). Let the first-order approximation of $\mathbf{P}$ in the neighborhood of (ξ_F, p_0) be of the form

$$\mathbf{P}(\xi, p) \approx \mathbf{P}(\xi_F, p_0) + \mathbf{A} \cdot (\xi - \xi_F) + \mathbf{w} \cdot (p - p_0), \qquad (3.3.4)$$

where $\mathbf{A}$ is a Jacobian matrix of $\mathbf{P}(\cdot, p_0)$ at ξ_F, and $\mathbf{w} = \frac{\partial \mathbf{P}}{\partial p}(\xi_F, p_0)$ is the derivative of $\mathbf{P}$ with respect to the parameter p.

Stabilization of the fixed point is achieved by realizing feedback of the form

$$p(\xi) = p_0 + \mathbf{c}^T(\xi - \xi_F). \qquad (3.3.5)$$

In the original description of the OGY method [21], the vector $\mathbf{c}$ is computed using the expression

$$\mathbf{c} = -\frac{\lambda_u}{\mathbf{f}_u^T \mathbf{w}} \mathbf{f}_u^T, \qquad (3.3.6)$$

where λ_u is the unstable eigenvalue and $\mathbf{f}_u$ is the unstable contravariant corresponding left eigenvector of $\mathbf{A}$.

Thus, the OGY method relies on a local linearization of the Poincaré map in the neighborhood of the chosen unstable fixed point and local linear stabilizing feedback.

An advantage of the OGY method is that all of the necessary calculations can be performed off-line on the basis of measurements (e.g., finding the unstable periodic orbits, fixing one of them as the goal of the control, computing the variables and parameters necessary for calculation of the control signal).

Once the goal of the control (unstable orbit to be stabilized) has been selected, the control signal is applied only when the observed trajectory passes close to the fixed point (where the linearization is valid). The assumption about the existence of a dense orbit guarantees that eventually the trajectory will enter the control window. However, the time one has to wait before starting and achieving control might be very long.

It should be mentioned here that Dressler and Nitsche [4] have proposed a variant of the OGY method in which only one variable is measured in the system and other variables needed for control are reconstructed using the delayed coordinate method.

3.4 Implementation Problems for the OGY Method

When implementing the OGY method for a real world application, one has to do the following series of elementary operations:

1. data acquisition — measurement of a (usually scalar) signal from the chaotic system under consideration. This operation should be performed in such a way as not to disturb the existing dynamics. For further computerized processing, measured signals must be sampled and digitized (A/D conversion);
2. selection of appropriate control parameter;
3. finding unstable periodic orbits using experimental data (measured time series) and fixing the goal of control;
4. finding parameters and variables necessary for control (as described above);
5. application of the control signal to the system — this step requires continuous measurement of system dynamics in order to determine the moment at which to apply the control signal, i.e., the moment when the actual trajectory passes in a small vicinity of the chosen periodic orbit, and immediate reaction of the controller (application of the control pulse) in such an event.

In computer experiments, it has been confirmed that all the above-mentioned steps of OGY can be carried out successfully in a great variety of systems, achieving stabilization of even long-period orbits.

There are several problems which arise when attempting to build an experimental setup. Despite the fact that the variables and parameters can be calculated off-line, one has to consider that the signals measured from the system are usually corrupted due to noise, several nonlinear operations associated with the A/D conversion (possibly rounding, truncation, finite word-length, overflow correction, etc.). Using corrupted signal values and the introduction of additional errors by the computer algorithms and linearization used for the control calculation may result in a general failure of the method. Additionally, there are time delays in the feedback loop (e.g., waiting for reaction of the computer, interrupts generated when sending and receiving data, etc.)

3.4.1 Effects of Calculation Precision

In a simple example below, we consider the case of calculating control parameters to stabilize a fixed point in the Lozi map and show how the A/D conversion accuracy and the resulting calculations of limited precision affect the possibilities for control. In the tests described below, we consider the quality of computations alone, without looking at other problems like time delays in the control loop.

To be able to compare the results of digital manipulations in [19] we first computed the interesting parameters using analytical formulas and compared the results with computations based on measured time-series and finite precision calculations in the case of Henon map. Let us look at three cases compared below with the analytical results:

Reference case - analytically calculated parameters:
Coordinates of the fixed point: $(0.8879418373, 0.8879418373)$
Control vector g: $[0.4038961828, 0.4038961828]$
Jacobian eigenvalues: $-1.913225419, -0.1553417742$
Stable direction: $[0.1535007507, 09881485105]$
Unstable direction: $[0.8880129457, -0.459818393]$
Possibilities of control: successful

1. fixed point representation, 12bit precision, rounding
Coordinates of the fixed point: $(0.8831, 0.8810)$
Control vector g: $[0.4352, 0.4656]$
Jacobian eigenvalues: $-1.9221, -0.0315$
Stable direction: $[0.1156, 0.9933]$
Unstable direction: $[0.8829, -0.4696]$
Possibilities of control: successful

2. fixed point representation, 10bit precision, rounding
Coordinates of the fixed point: $(0.883, 0.881)$
Control vector g: $[0.350, 0.362]$
Jacobian eigenvalues: $-1.899, -0.021$
Stable direction: $[0.137, 0.991]$
Unstable direction: $[0.891, -0.455]$
Possibilities of control: often fails

3. fixed point representation, 8bit precision, rounding
Coordinates of the fixed point: $(0.89, 0.89)$
Control vector g: $[0.0, 0.0]$
Jacobian eigenvalues: $0, 0$

Stable direction: Impossible to determine
Unstable direction: Impossible to determine
Possibilities of control: impossible

Comparing the results of computations summarized above, we can easily see that if we are able to achieve an accuracy of two to three decimal digits, the calculations are precise enough to ensure proper functioning of the OGY algorithm in the case of the Lozi system. To have some safety margin and robustness in the algorithm, the acceptable A/D accuracy cannot be lower than 12 bit and probably it would be best to apply 16bit conversion. This kind of accuracy is nowadays easily available using general purpose A/D converters even at speeds in the MHz range. Implementing the algorithms, one must consider the cost of implementation — with growing precision and speed requirements, the cost grows exponentially. This issue might be a great limitation when it comes to IC implementations.

3.4.2 Approximate Procedures for Finding Periodic Orbits

Another possible source of problems in the control procedure is the errors introduced by algorithms for finding periodic orbits (goals of the control). Using experimental data, we can only find approximations to unstable periodic orbits [1, 14, 24].

Commonly used is a simple technique proposed by Lathrop and Kostelich [14] for recovering unstable periodic orbits from an experimental time series. This procedure assumes that we have a series of successive points $\{x_i\}$, $i = 0, 1, ...N$ on the system trajectory and taking any of these points x_m, we search forward for the smallest positive integer k, such that $||x_{m+k} - x_m|| < \varepsilon$, where ε is the specified accuracy. It is further claimed that the orbit detected in this manner lies close to the unstable periodic orbit whose period is approximated by that of the detected sequence.

This approach has several drawbacks. First, the results strongly depend on the choice of ε and the length of the measured time series. Further, they depend on the choice of norm and the number of state variables analyzed; Second, the stopping criterion ($||x_{m+k} - x_m|| < \varepsilon$) in the case of discretely sampled continuous-time systems is not precise enough. This means that one can never be sure of how many orbits have been found or whether all orbits of a given period have been recovered.

In many applications, however, it is sufficient to find only some of the unstable periodic orbits embedded in the attractor and choose one of them.

We have developed a set of computer programs [20] for detecting unstable periodic orbits. The Lathrop-Kostelich procedure has been refined by means of a stopping condition based on the distance of the initial point x_m from the evolving trajectory — not from distinct points belonging to it which are sampled discretely in time (D/A conversion of measured signals) or computed via numerical integration. (Thus we avoid the problem of not detecting an orbit when x_m falls between two points on the trajectory.) This slows down the computations slightly but the results are more reliable - the problem of not finding some of the orbits due to distance mismatch can be avoided.

In our experiments [20], we varied ε between 0.000001 and 0.001 and fixed the threshold for distinguishing between orbits at 0.001. Although with greater ε more orbits with given period were detected, most of them were later recognized as identical — there was no significant difference in the number and shape of different unstable periodic orbits found (compare Figure 3.3 which was obtained using these simple procedures).

As this step is typically carried out off-line, it does not badly affect the whole control procedure. It has been found in experiments that when the tolerances for detection of unstable orbits were chosen too large, the actual trajectory stabilized during control showed greater variations and the control signal had to be applied every iteration to compensate for inaccuracies. Clearly, making the tolerance large could cause failure of control.

Some new methods have been proposed recently [25] which could possibly improve localization procedures for unstable periodic orbits. Particularly interesting are interval arithmetics methods [5, 12] based on the interval Newton method which enable precise calculation of periodic orbit's position for systems with known mathematical models.

3.4.3 Effects of Time Delays

Several elements in the control loop may introduce time delays that can be detrimental to the functioning of the OGY method. Although all calculations may be done off-line, two steps are of paramount importance:

1. detection of the moment when the trajectory passes the chosen Poincaré section
2. determination of the moment at which the control signal should be applied (close neighborhood of chosen orbit).

When these two steps are carried out by a computer with a data acquisition card, at least a few interrupts (and therefore a time delay) must be generated in order to detect the Poincaré section, take the decision of being in the right neighborhood, and to send the correct control signal.

Most experiments with OGY control of electronic circuits have been able to achieve control when the systems were running in the 10-100 Hz range. We found out that for higher frequency systems, time delays become a crucial point in the whole procedure. The failure of control was mainly due to the late arrival of the control pulse –the system was being controlled at a wrong point in state space where the formulas used for calculations were probably no longer valid; trajectory was already far away from the section plane when the control pulse arrived.

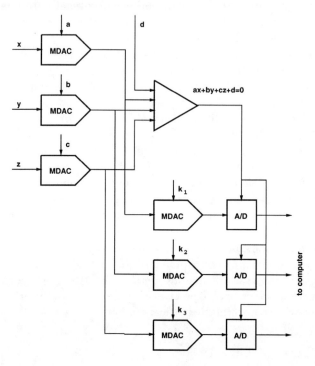

FIGURE 3.4
Fast Poincaré section detector for improved OGY implementation.

To compensate for some of the delays, we have proposed a hardware solution for a detector of the Poincaré section and vicinity detector. Block diagrams of these two pieces of equipment are shown in Figs. 3.4 and 3.5. The Poincaré section system here uses all state variables (three in our application) to simplify detection. To implement this function using just one variable, delay coordinates must be introduced — realization in hardware would become much more complicated in this case if possible (one could think of calculating suitable time delay by a computer algorithm and saving the necessary time-delayed samples in special-purpose registers).

3.5 Brief Review of the OPF Controller

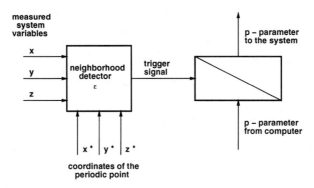

FIGURE 3.5
ε-comparator for detection of the vicinity of a desired periodic orbit.

The occasional proportional feedback (OPF) technique [10, 28] can be considered as a one-dimensional version of the OGY method.

Let us describe the action of the OPF controller for the case of stabilizing a fixed point of the Poincaré map.

In the OPF method, the control signal is computed using only one variable, for example ξ_1:

$$p(\xi) = p_0 + c(\xi_1 - \xi_F). \tag{3.5.7}$$

Adjusting the values of c for which ξ_F is a stable fixed point of the system $\xi \mapsto \mathbf{P}(\xi, p(\xi))$ ensures proper functioning of the algorithm. In [6] we have described some theoretical results concerning the choice of coefficients and possibilities for successful OPF control. The best results can be obtained if the unstable eigenvector is parallel to the coordinate which is used to compute the control signal, and the possibility of control using the OPF technique depends on the form of the linear approximation of the system's behavior in the neighborhood of the periodic orbit.

A schematic of Hunt's implementation of the OPF control method is shown in Figure 3.6. The window comparator, taking the input waveform, gives a logical high when the input waveform is inside the window. This is then compared with the delayed output from the external frequency generator. This logical signal drives the timing block which triggers the sample-and-hold and then the analog gate. The output from the gate, which represents the error signal at the sampling instant, is amplified and applied to the interface circuit which transforms the control pulse into a perturbation of the parameter p. The frequency, delay, control pulse width,

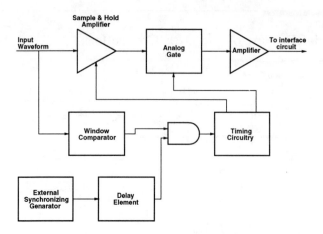

FIGURE 3.6
Hunt's implementation of the OPF method.

window position, width and gain are all adjustable — they fix the position of the section plane, values of p_0 and c. The interface circuit depends explicitly on the chaotic system under control.

One of the major advantages of Hunt's controller over OGY is that the control law depends on just one variable and does not require any complicated calculations (as was necessary in the case of the OGY scheme) in order to generate the required control signal. All the operations can easily be performed by hardware function blocks.

The disadvantage of the OPF method is that there is no systematic method for finding the embedded unstable orbits (unlike OGY). For comparison with the OGY method, let us summarize the main features of the OPF controller:

1. it uses just one system variable as input;
2. uses the peaks of this system variable to generate a one-dimensional map;
3. a window around a fixed level sets the region where control is applied;
4. peaks are located by means of a synchronizing generator, the frequency of which has to be adjusted by either a trial-and-error procedure or by consulting, for example, measured power spectra of the signals.

Brief Review of the OPF Controller

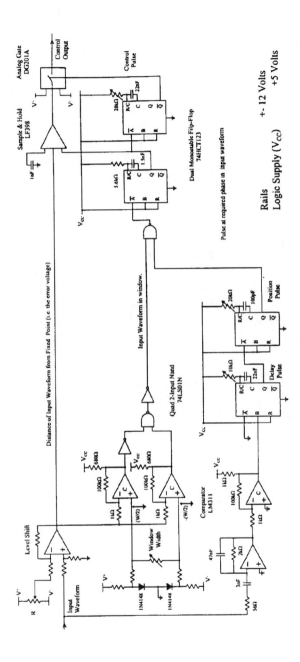

FIGURE 3.7
Improved analog chaos OPF controller without external synchronization.

3.6 Improved Chaos Controller for Autonomous Circuits

Recently, we have developed an improved chaos controller modifying the structure proposed by Hunt [6]. The modified controller uses Hunt's method without the need for an external synchronizing oscillator. Its circuit diagram is shown in Figure 3.7 In the modified controller, the derivative of the input signal generates a pulse when it passes through zero. This pulse replaces the driving pulses from the external oscillator as the "synch" pulse for our Poincaré map. This obviates the need for the external generator and so makes the controller simpler and cheaper to build.

The variable level window comparator is implemented using a window comparator around zero and a variable level shift. Two comparators and three logic gates form the window around zero. The synchronizing generator used in Hunt's controller is replaced by an inverting differentiator and a comparator. A rising edge in the comparator's output corresponds to a peak in the input waveform. We use the rising edge of the comparator's output to trigger a monostable flip-flop. The falling edge of this monostable's pulse triggers another monostable, giving a delay. We use the monostable's output pulse to indicate that the input waveform peaked a fixed time earlier. If this pulse arrives when the output from the window comparator is high, then a monostable is triggered. The output of this monostable triggers a sample-and-hold on its rising edge which samples the error voltage; on its falling edge, it triggers another monostable. This final monostable generates a pulse which opens the analog gate for a specific time (the control pulse width). The control pulse is then applied to the interface circuit, which amplifies the control signal and converts it into a perturbation of one of the system parameters, as required.

The modified controller offers, in our opinion, the simplest and most reliable implementation for chaos control. It has been tested successfully on Chua's circuit and the Colpitts oscillator working in the kHz range [6], enabling stabilization of unstable periodic orbits up to order 8.

Listing its advantages, we can summarize that:

1. it uses just one system variable as input;
2. a window around a fixed level sets the region where control is applied;
3. it overcomes the main drawback of using external synchronization
4. the speed of control is only limited by frequency limitations of the components used. One can easily visualize adapting the scheme for controlling very fast systems
5. the controller implementation is cheap, simple, and easy to build. IC implementation remains one of the possible future issues.

3.7 Working Laboratory Chaos Control Systems

3.7.1 Control of a Magneto-Elastic Ribbon

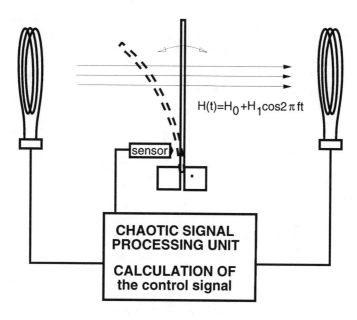

FIGURE 3.8
Chaos control set-up for the magneto-elastic ribbon system.

Already in 1990, soon after publication of the OGY method, scientists from the Applied Chaos Laboratory directed by W. Ditto reported on one of the first laboratory-scale successful real implementation of a chaos-control system [7]. The experimental system (see Figure 3.8) consisted of a gravitationally buckled magneto-elastic ribbon fixed in a vertical position (clamped at the bottom end) and placed in a time-varying magnetic field having $H = H_0 + H_1 cos(2\pi ft)$. A sensor measured the curvature of the ribbon near its base - voltage measured by this sensor was further converted to digital form and transferred to the computer carrying on the control task. H_0 has been chosen as the control parameter. Using OGY technique principles, H_0 perturbations could be calculated to successfully stabilize the low-order oscillations of the beam out of its originally tuned chaotic behavior. Ditto, Rauseo and Spano demonstrated in this experiment that chaos could be controlled in a real physical system using OGY method. It should be mentioned however that the control set-up was highly susceptible to noise and disturbances. To ensure proper functioning, compensation of the Earth's magnetic field had to be placed using three pairs of mutually orthogonal Helmholtz coils. To increase the precision, temperature stabilization was also very important. Successful implementation of chaos control in this particular case was possible because of very slow oscillation frequencies of the system. The eigen-frequency of the ribbon was in the range of single Hz!! In such a case, all effects of time delays were negligible and also computer calculations were fast enough for the continuous robust operation of the system at a low-order stabilized orbit.

3.7.2 Control of a Chaotic Laser

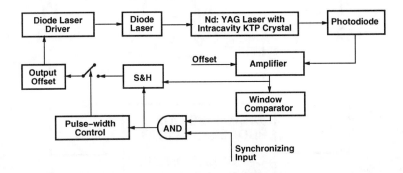

FIGURE 3.9
Schematic diagram of the OPF controller driving the Nd:YAG laser system.

R. Roy and his co-workers [28] reported very interesting results obtained in experimental control of chaotic lasers. The goal of control was here stabilization of the chaotically varying light intensity. Roy adopted the OPF technique and was able to achieve 15-fold increase in power output achieving at the same time stable operation. Control was performed on a fast time-scale of a few microseconds. Thus, OGY method requiring numerical computations could not be applied. As indicated on the schematic diagram (Figure 3.9), the choice of a suitable goal of control can be achieved by adjusting three parameters: the synchronizing frequency, the wave-form offset and the width of the control signal. Several periodic orbits, e.g., padiod-1, period-4 or period-9 could be stabilized out of the chaotic regime. It should be stressed here that the laser itself is a higher-dimensional system and that successful control did not require any knowledge of its model - the control task is performed on the basis of measured signal (from the photo-diode) and all "computations" are done in a fast analog way. It seems that OPF controller is the only implementation alternative here.

Chaotic lasers are one of the deeply investigated chaos control objects

- many research groups world-wide work on stabilization of chaotic laser using various techniques. Interested readers should consult [28].

3.7.3 Chaos-Based Arrhythmia Suppression and Defibrillation

One of the most spectacular applications of chaos control techniques is arrhythmia removal or defibrillation of the heart-beat. Starting from many numerical experiments concerning control of arrhythmias on heart models, research and exploitation of real measured signals [7, 8] in this domain has reached a stage where successful approaches have been patented [26, 27] and clinical tests are under way [13]. In the clinical setup consisted of a quadrupolar electrode catheter which was inserted via the femoral vain (FV) of the patient under test and positioned in the lateral right atrium (RA). During atrial fibrillation, a pair of electrodes measured electrical activity signals which were digitized and processed in the computer. Measured time-series was further used to calculate the unstable fixed point and its stable and unstable manifolds and calculate the desired control pulses. These pulses via the second pair of electrodes were applied to the atrium to control the pumping motion towards stable periodic state. The schematic of this setup is shown in Figure 3.10. The authors [13] report on results of clinical tests performed on 25 patients. 36% consistent results of control were obtained, 40% presented partial chaos control (10% -50% of escapes from the controlled orbit) and in 24% of cases control was unsuccessful. Chaos control offers a promising alternative for quenching arrhythmia and especially fibrillation.

3.8 Conclusions

In classical linear and nonlinear control problems, one did not exploit existence of unstable periodic orbits, sensitive dependence on initial conditions or ergodicity of trajectories on the attractor. Chaos control techniques clearly rely on these properties. In chaos control, a target for tracking is not limited to constant vectors in the state space: it often is an unstable periodic orbit of the given system. This generally requires only tiny control to arrive at, but technically can be quite difficult due to the instability of the target. Moreover, in chaos control, the terminal time is infinite to be meaningful and practical, because most nonlinear dynamical behaviors such as equilibrium states, limit cycles, attractors, and chaos are asymptotic properties. These are theoretical limitations for chaos control. There exist also several limitations imposed by electronic implementation of chaos controllers. These limitations are imposed by the speed of the

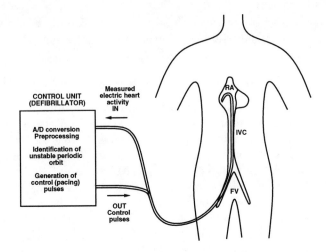

FIGURE 3.10
Schematic diagram of the arrhythmia removal system.

considered chaotic system (highest frequency), accuracy of the measurements, errors introduced by quantization and signal processing algorithms and calculations.

Taking into account these limitations when looking for an implementation of a controller, we must make a trade-off between high speed (analog implementation without possibilities of pre-specification of the goal, and a trial-and-error procedure for achieving the desired behavior) and precise knowledge of the orbit which is interesting as a goal of the control.

The modified OPF controller works well in many high-frequency systems, eliminating chaos but without prior knowledge of attainable orbits.

OGY method is very attractive when precise knowledge of the goal is needed (stabilized orbits offer some optimal type of performance [9]) but it is possible to implement it in very low frequency (slow) systems only.

In this study, we did not consider the actuator design problem which depending on real application might pose specific problems — e.g., application of defibrillating signal to the heart might be of paramount importance far above any of the controller and algorithm design problems.

There are two interesting areas of further research and developments:

1. Hardware (possibly IC) implementation of Poincaré section detectors and possibly higher iterates of return maps;
2. Hardware implementation of goal (unstable orbit) detection for use with the modifications of analog OPF method;
3. development of specific hardware for use with OGY method;

4. implementation of both methods in high-order (possibly hyper-chaotic) systems;
5. IC implementations of complete controllers.

Finally, we should mention that there exist other control schemes, e.g., delayed feedback method introduced by Pyragas [15] where certainly the controller is the cheapest possible (delay line which in some cases might be even a piece of cable!) — when it comes to real implementations problems arise due to limitations imposed by available tunable delay elements. In our opinion, such controllers might also find successful applications (see the chapters by Brandt and Chen, by Celka, and by Yu *et al.* in this book).

Acknowledgment

This research has been supported in part by The University of Mining and Metallurgy grant 11.120.460.

References

[1] D. Auerbach, P. Cvitanović, J.-P. Eckmann, G. Gunaratne and I. Procaccia, "Exploring chaotic motion through periodic orbits," *Phys. Rev. Letters*, vol. 58, pp. 2387-2389, 1987.

[2] G. Chen and X. Dong, *From Chaos to Order — Methodologies, Perspectives and Applications*, World Scientific Pub. Co., Singapore, 1998.

[3] W. L. Ditto, S. N. Rauseo and M. L. Spano, "Experimental control of chaos," *Phys. Rev. Letters*, vol. 65, pp. 3211-3214, 1990.

[4] U. Dressler and G. Nitsche, "Controlling chaos using time delay coordinates," *Phys. Rev. Letters*, vol. 68, pp. 1-4, 1992.

[5] Z. Galias, "Rigorous numerical studies of the existence of periodic orbits for the Henon map," *J. Univ. Comp. Sci.*, vol. 4, pp. 114-124, 1998.

[6] Z. Galias, C.A. Murphy, M.P. Kennedy and M.J. Ogorzałek, "Electronic chaos controller," *Chaos, Solitons and Fractals*, vol. 8, pp. 1471-1484, 1997.

[7] A. Garfinkel, M. L. Spano, W. L. Ditto and J. N. Weiss, "Controlling cardiac chaos," *Science*, vol. 257, pp. 1230-1235, 1992.

[8] A. Garfinkel, J. N. Weiss, W. L. Ditto and M. L. Spano, "Chaos control of cardiac arrhythmias," *Trends in Cardiovasc. Med.*, vol. 5, p. 76, 1995.

[9] B. R. Hunt and E. Ott, "Optimal periodic orbits of chaotic systems," *Phys. Rev. Letters*, vol. 76, pp. 2254-2257, 1996.

[10] E. R. Hunt, "Stabilizing high-period orbits in a chaotic system: The diode resonator," *Phys. Rev. Lett.*, vol. 67, pp. 1953–1955, 1991.

[11] E. R. Hunt, "Keeping chaos at bay," *IEEE Spectrum*, vol. 30, pp. 32-36, 1993.

[12] R. B. Kearfott and M. Novoa, "Algorithm 681: INTBIS, A portable interval Newton/bisection package," *ACM Trans. Math. Software*, vol. 16, pp. 152-157, 1990.

[13] J. J. Langberg, A. Bolmann, K. McTeague, M.L. Spano, V. In, J. Neff, B. Meadows and W.L. Ditto, "Chaos control of human atrial fibrillation," preprint, 1997.

[14] D. P. Lathrop and E. J. Kostelich, "Characterization of an experimental strange attractor by periodic orbits," *Phys. Rev. A*, vol. 40, pp. 4028-4032, 1989.

[15] M. J. Ogorzałek, "Taming chaos: Part 2 — control," *IEEE Trans. Circ. Sys. I*, vol. 40, pp. 700–706, 1993.

[16] M. J. Ogorzałek, "Chaos control: How to avoid chaos or take advantage of it,", *J. Franklin Inst.*, vol. 331B, pp. 681-704, 1994.

[17] M. J. Ogorzałek, "Controlling chaos in electronic circuits," *Phil. Trans. R. Soc. Lond. A*, vol. 353, pp. 127-136, 1995.

[18] M. J. Ogorzałek, *Chaos and Complexity in Nonlinear Electronic Circuits*, World Scientific Pub. Co., Singapore, 1997.

[19] M. J. Ogorzałek, "Design considerations for electronic chaos controllers," *Chaos, Solitons and Fractals*, vol. 9, pp. 295-306, 1998.

[20] M. J. Ogorzałek and Z. Galias, "Characterization of Chaos in Chua's oscillator in terms of unstable periodic orbits," *J. Circ. Syst. Comp.*, vol. 3, pp. 411-429, 1993.

[21] E. Ott, C. Grebogi, and J.A. Yorke, "Controlling chaos," *Phys. Rev. Lett.*, vol. 64, pp.1196–1199, 1990.

[22] E. Ott, C. Grebogi, and J. A. Yorke, "Controlling chaotic dynamical systems, " in D.K.Campbell (Ed.), *Chaos: Soviet-American Perspectives on Nonlinear Science*, Amer. Inst. of Physics, New York, 1990.

[23] R. Roy, T. W. Murphy, T. D. Maier, Z. Gillis, and E. R. Hunt, "Dynamical control of a chaotic laser: Experimental stabilization of a globally coupled system," *Phys. Rev. Lett.*, vol. 68, pp. 1259–1262, 1993.

[24] I. B. Schwartz, "Estimating regions of existence of unstable periodic orbits using computer-based techniques", *SIAM J. Numer. Anal.*, vol. 20, pp. 106-120, 1983.

[25] P. So, E. Ott, S. J. Schiff, D. T. Kaplan, T. Sauer and C. Grebogi, "Detecting unstable periodic orbits in chaotic experimental data," *Phys. Rev. Letters*, p. 4705, 1996.

[26] M. L. Spano, W. L. Ditto, A. Garfinkel and J. N. Weiss, "Real time cardiac arrhythmia stabilizing system," US Patent no. 5, 342, 401 (8/30/94).

[27] M. L. Spano, W. L. Ditto, A. Garfinkel and J.N. Weiss, "Real time cardiac arrhythmia stabilizing system," US Patent no. 5, 447, 520 (9/05/95).

[28] Theme Issue: Control of Chaos – New Perspectives in Experimental and Theoretical Nonlinear Science. Int. J. Bif. Chaos, vol. 8, no. 8/9, 1998.

4

Chaos in Mechanical Systems and Its Control

T. Kapitaniak[1], J. Brindley[2], and K. Czolczynski[1]

[1]Division of Dynamics
Technical University of Lodz
Stefanowskiego 1/15
90-924 Lodz, Poland
tomaszka@ck-sg.p.lodz.pl

[2]Department of Applied Mathematical Studies
University of Leeds, Leeds LS2 9JT, UK

Abstract

Though in a few cases it is possible to model real mechanical engineering systems by linear equations, much more generally nonlinearities have a fundamental role in characterizing the qualitative, as well as quantitative, behavior of the system. The nonlinearities may arise from material properties or from system design. This nonlinearity means that chaos is possible for some parameter regions. We suggest that, in smooth systems (i.e., systems with a differentiable vector field), chaos occurs only for vibration amplitudes too large to be permitted in practical cases. For non smooth systems (e.g., systems having impacts or dry friction, we may expect chaos at much smaller amplitudes. The two main control methods are by discrete or continuous feedback, or by modifications of system design or operating conditions. In mechanical systems, feedback controllers are likely to be large and costly, whereas, since it is often the case that the dynamics is known, modifications to system design or operating conditions may be relatively straightforward and cheap. We review the effectiveness of these methods through two specific examples.

4.1 Introduction

Chaos occurs widely in applied mechanical systems; historically, it has usually been regarded as a nuisance and designed out if possible. It has been noted only as irregular or unpredictable behavior, and often attributed to random external influences. More recently, the idea of controlling chaos [20, 32, 12, 4] has suggested the potential usefulness of chaotic behavior, and we describe some in mechanical engineering.

First of all, we discuss the nature of chaotic behavior in applied mechanical systems and review a number of methods by which undesirable chaotic behavior may be controlled or eliminated. More speculatively, we indicate ways in which the existence of chaotic behavior may be directly beneficial or exploitable.

4.2 Methods of Chaos Control

We can divide chaos controlling approaches into two broad categories; first, those in which the actual trajectory in the phase space of the system is monitored, and some feedback process is employed to maintain the trajectory in the desired mode, and second, nonfeedback methods in which some other property or knowledge of the system is used to modify or exploit the chaotic behavior.

Feedback methods do not change the attractors of the controlled systems and commonly operate to stabilize an unstable periodic orbit on a strange chaotic attractor. In contrast, nonfeedback methods slightly change the attractors of the controlled system, mainly by a small permanent shift of a control parameter, changing the system behavior from chaotic to periodic orbit which is located in the neighborhood of the initial attractor. Main ideas of both methods are illustrated in Figure 4.2 (grey arrows indicate the goals of the control).

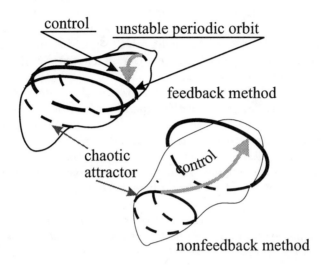

FIGURE 4.1
General idea of feedback and nonfeedback methods.

Ott, Grebogi and Yorke [20, 22, 25] have, in an important series of papers, proposed and developed a feedback method by which chaos can always be suppressed by shadowing one of the infinitely many unstable periodic orbits (or perhaps steady states) embedded in the chaotic attractor. The idea is to start with any initial condition, wait until the trajectory falls into a target region around the desired periodic orbit, and then apply feedback control.

More specifically, the procedure is to first identify some suitable orbit, then to apply some (small) control so as to stabilize this orbit. The effectiveness of the OGY method is increased if it is combined with an appropriate targeting procedure [26, 27] in which the actual trajectory is directed into the neighborhood of an appropriate unstable periodic orbit.

Later developments of the OGY approach have established connections between control magnitude and length of chaotic transient preceding steady periodicity, and have indicated the important (and possibly damaging) effects of low levels of noise which may occasionally provoke large excursions from the desired orbit. For example, they present results for the control of the double rotor map, a four-dimensional system that describes a particular impulsively periodically forced mechanical system [8].

The OGY approach has stimulated a good deal of research activity, both theoretical and experimental. The efficiency of the technique has been demonstrated [6] in a periodically forced system, converting its chaotic behavior into period-one and period-two orbits, and the application of the method to stabilize higher periodic orbits in a chaotic diode resonator have been demonstrated by [10]. Another interesting application of the method is the generation of a desired aperiodic orbit [16], and again [30] has been able to demonstrate controlled transient chaos. Related work [7] uses time delay techniques to control chaos.

Though the OGY theory has been proposed in the context of low order dynamical systems, and most of the experimental or observation investigations have been concerned with clearly low order mechanical or electrical contexts, the interesting experiments [29] demonstrate its potential for fluid (and perhaps fluid-solid) mechanical phenomena. The experiments succeeded in achieving regular laminar flow in previously unstable thermal convection loops by use of a thermostat-type feedback.

Generally, experimental application of the OGY method requires a permanent computer analysis of the state of the system. The changes of the parameters, however, are discrete in time since the method deals with the Poincaré map. This leads to some serious limitations. The method can stabilize only those periodic orbits whose maximal Lyapunov exponent is small compared to the reciprocal of the time interval between parameter changes. Since the corrections of the parameter are rare and small, the fluctuation noise leads to occasional bursts of the system into the region far from the desired periodic orbit, especially in the presence of physical noise.

Recent research has extended the OGY approach to the control of impact systems [3, 9, 1], but since its applications need knowledge of position and velocity at each impact, its usefulness in real mechanical systems is likely to be limited.

A different approach to feedback control which avoids the above mentioned problems, the method of a time-continuous control, was proposed by Pyragas [23, 24]. This method is based on the construction of a special form of a time-continuous perturbation, which does not change the form of the desired unstable periodic orbit, but under certain conditions can stabilize it. Two feedback controlling loops have been proposed. A combination of feedback and periodic external force is used in the first method while the second method does not require any external source of energy and is based on a self controlling delayed feedback.

Chaos can also be controlled by a variety of feedback methods known from the classical control theory [4].

Feedback methods have been extensively researched and reviewed [28, 18, 19, 20, 15, 12, 4], and we will not repeat such a review here. Rather we remark that feedback controllers for mechanical systems are often large (sometimes larger than the controlled system!) and expensive, and it is therefore more practicable to use knowledge of the dynamics of the system to design nonfeedback methods.

The rest of this chapter is devoted to a review of some of these methods.

4.3 Nonlinearities in Mechanical Systems

In almost every branch of machine or mechanical system design, we encounter problems connected with vibrations. Their solution, sometimes very difficult to arrive at, is an indispensable requirement in order to achieve effective and safe operation of the structure being designed.

Many practical problems can be modeled with satisfactory accuracy by means of a linear model in which the resisting (damping) force and the elasticity force depend linearly on the velocity or the position of the vibrating mass (linear characteristics). Such an assumption is possible at small actual deformations of structural elements or when the displacements occurring during vibrations in actual structures are small with respect to all the dimensions of structural elements.

Although it is not possible to discount the role of linear models in basic analysis, their usefulness is very limited for the description of many phenomena. In a more detailed description of the dynamics of machines, nonlinear models have to be taken into account. For instance, the analysis of vibration at high amplitudes, in the case where the systems are treated as nonlinear, can lead to completely new phenomena, impossible to reproduce in any linear system.

Nonlinear vibrations occur when the vibrating system is characterized by nonlinear characteristics of the elasticity force, damping force or both at the same time. In almost every mechanical system, numerous such nonlinearities occur. They have various characters, and generally they can be divided into two groups, geometrical (where the configuration of the system or trajectory of its movement implies nonlinear equations of motion), and material (where the materials from which the system is build have nonlinear characteristics).

All the materials to which the Hooke's law does not apply produce material nonlinearities. They can be divided into two groups again:

(i) materials for which, along with an increase in the deformation, the elasticity force grows faster than this deformation,

(ii) materials for which, along with an increase in the deformation, the elasticity force grows slower than this deformation.

Diagrams showing the relation between the elasticity force and deformation for the first and second group of materials are presented in Figures 4.2(a) and 2(b), respectively. In these figures, S denotes force and x deformation. The broken lines mark the respective linear characteristics. It is customary to call the characteristic curve of the first type a stiff (hard) characteristic, whereas the characteristic curve of the second type is called a soft one. The first group includes such materials as rubber, leather, etc., the second one – cast iron, concrete, etc. For systems with stiff characteristics, the frequency of self-vibrations for nonlinear vibrations is greater than the frequency of self-vibrations of these systems in the linear frame. For systems with soft characteristics, an opposite phenomenon takes place. An inclusion of such nonlinear elements into a vibrating system can decrease the effects of resonance to a significant degree. An increase in the amplitude will be followed by a change in the frequency, which will cause the system to escape the resonance.

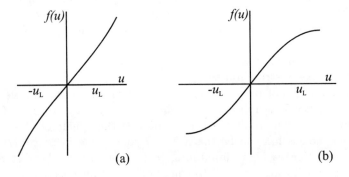

FIGURE 4.2
Typical characteristics of the stiffness of the spring.

Nonlinearities in mechanical systems occur widely, and the nature of the problem varies from very big systems (whose analysis is based mainly on the finite element method, where the number of degrees of freedom can be equal to even several hundred thousands) to very small systems (with one or two degrees of freedom). Despite the fact that many issues connected with nonlinearities are currently being investigated and many studies of the quality of the motion of such systems are being carried out, major problems concerning the analysis of systems with nonlinearities, especially when different kinds of nonlinearities occur in one system, leading to a chaotic behavior, still remain unsolved.

Chaotic phenomena generally occur in systems with strong nonlinearities; they have been observed in mechanical systems with

(i) nonlinear elastic or spring elements,
(ii) nonlinear friction damping,

(iii) impacts due to the clearance,
(iv) fluid related forces,
(v) nonlinear boundary conditions,
(vi) nonlinear feedback control forces in servo-systems.

For examples, see [31, 17, 11].

However, most of the examples of chaotic behavior motivated by smooth mechanical systems are unrealistic from the point of view of real engineering application. Chaotic behavior can be observed only when the amplitudes of motion are too big to be allowed in real operating machines with often huge oscillating mass. One such example is the case of oscillation of the centre of the mass of a rotor supported in a bearing. During chaotic vibrations of the nonlinear elastic rotor, the vibration amplitude can reach a very high value, exceeding the maximum value determined by the clearance between the rotor and its casing. This often implies catastrophic failure of the machine, and the possibility of its occurrence must be "designed out" at an early stage. Similarly in the case of a (stiff) rotor supported in oil or gas bearings, there is a possibility of occurrence of chaotic motion in instability regimes of the rotor static equilibrium position. Investigation of rotor behavior in this region, at rotational velocity exceeding the critical value for stability of the static equilibrium position, has a limited meaning, however. The loss of stability of the static equilibrium position is accomplished by a Hopf bifurcation and the rotor trajectories tend to stable limit cycles. Since the radius of the limit cycle exceeds the bearing clearance even at the threshold of instability, any attempt to observe the chaotic behavior is foiled by the destruction of the bearing as a result of collision of the rotor or journals with the bearing [5].

This restriction of chaotic behavior to what are in some sense large oscillations does not apply when the chaos is associated with discontinuous forcing, as for example, when dry friction or impacts occur in the system, and in our opinion, chaotic oscillations in mechanical systems of industrial importance will be significant only when dry friction and (or) impacts are present. Such a chaotic motion can occur without large amplitudes of oscillations and takes place usually at the micro scale, inside joints, bearings, etc.

For this reason, in what follows, we discuss the problem of controlling chaos in systems with dry friction and impacts.

4.4 Control Through Operating Conditions

Virtually all engineering systems are subjected during operation to external forcing. This forcing will contain (and hopefully be dominated by) planned and intentional components; it will also almost invariably contain unintentional "noise." Judicious design and control of this forcing is often able to annihilate or shift to a harmless region of parameter space an unwanted chaotic behavior (in some circumstances, as we remark later, exactly the reverse process may be desirable, so that we may wish to produce chaotic behaviour).

We describe this controlling method using the example of the impact force generator. In numerous industrial machines, the impact of movable parts is either the basic objective of their operation, or an effect which improves their operating efficiency. The classic examples of such machines or devices are: a pneumatic hammer, impact dampers, or heat exchangers [33]. Among the factors that contribute to intensification of the heat exchange process are disturbances in the rotational velocity of the rotor of the heat exchanger [34]. One type of disturbance which can result in intensification of the heat exchange is step disturbance of the rotational velocity. The simplest way to generate such disturbances is to employ the phenomenon of impact which causes, according to the Newton's hypothesis, step variations of the velocity of the bodies impacting on each other.

The object of consideration is a system composed of two main parts (Figure 4.3): a rotor equipped with a fender, driven by an electric motor, and a hammer in the form of a cylinder with a semicircular end, mounted on the end of a cantilever beam of length l_s. During the operation of the generator, the rotor fender impacts on the hammer, which causes vibrations of the hammer on one hand and the desired step variations of the rotational velocity of the rotor on the other hand. ϕ_r is the angular position of the fender at the moment of impact. The geometry of the generator means that two kinds of impact occur during its operation: I - the fender collides with the cylindrical part of the hammer, II - the fender collides with the spherical part of the hammer. Equations of motion of the generator, and equations of impact have been presented in detail in [2].

In order to evaluate the generator's performance, the most important information is the average value of its angular velocity ω_a, together with the value of the velocity before the impact ω_{bi} and after the impact ω_{ai}. Figure 4.4 presents a map of impacts for many generators which differ in the length of the cantilever beam l_s. The values of the angular velocity of the rotor before ω_{bi} and after ω_{ai} impacts and, additionally, the curves showing the values of the fractions of the basic frequency of free vibrations of the hammer: $1/2\alpha_1$, $1/3\alpha_1$, $1/4\alpha_1$ and $1/5\alpha_1$ have been shown on this map. Here, a close relation between the variations of both the velocities

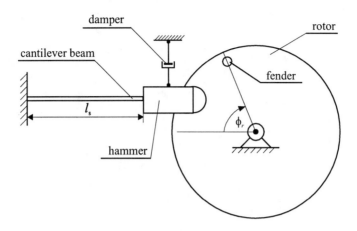

FIGURE 4.3
Schematic diagram of the generator.

ω_{ai} and ω_{bi} and the function of α_1 can be easily observed. For instance, when $l_s=0.07$ m, the rotor velocity ω_{bi} (close to its average angular velocity ω_a), is equal to approximately 1/3 of the value of the basic frequency of free vibrations of the hammer α_1. This means that the impact forcing of these vibrations has a subharmonic frequency.

An increase in the cantilever beam length causes a decrease in the value of $1/3\alpha_1$. Despite this, however, the hammer "wants" the impact forcing of its free vibrations to have a subharmonic character: the impacts are stronger and stronger, which, as a consequence, causes the decrease of both ω_{bi} and ω_{ai}. This way of affecting the angular velocity by the hammer has been called the self synchronization of the system. When l_s exceeds the value 0.0875 m, the conditions for easy generation of the hammer vibrations with the next subharmonic frequency $1/2\alpha_1$, arise and the situation repeats.

In the majority of cases presented in Figure 4.4, distinct, sharp contours of the markers can be seen. This means the system motion is fully regular in these cases. In the vicinity of the first subharmonic resonance ($l_s=0.0925$ m), blurred columns of the markers can be seen, which indicates that the system motion has become irregular.

Figure 4.5 depicts a global map of impacts, from which it can be easily observed that the values of ω_{bi} and ω_{ai} during individual impacts are different, and, moreover, that impacts do not occur at all during some rotations of the rotor (e.g., rotation No. 526, 528, 530). In order to investigate the character of the system motion with $l_s=0.0925$ m in detail, maps of impacts showing the impacts occurring in the subsequent time intervals of the gen-

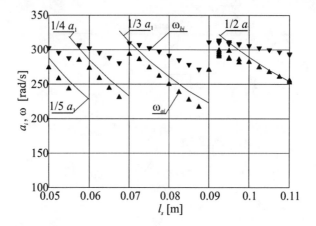

FIGURE 4.4
Impacts map for various l_s; (in numerical calculations, the restitution coefficient $k = 0.9$ was considered and inertial damping of cantilever beam was described by logarithmic decrement $\Delta = 1.5$).

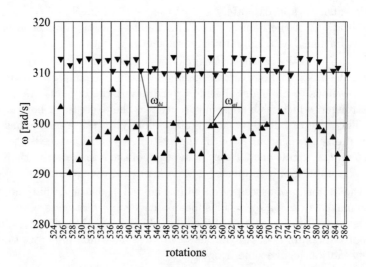

FIGURE 4.5
Global map of impacts; (in mumerical calculations the restitution, coefficient $k = 0.9$ was considered and inertial damping of cantilever beam was described by logarithmic decrement $\Delta = 1.5$).

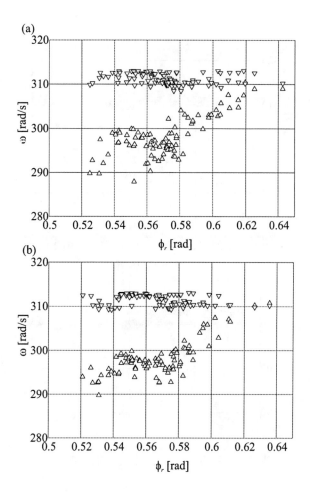

FIGURE 4.6
Local maps of impacts; (in numerical calculations, the restitution coefficient $k = 0.9$ was considered and inertial damping of cantilever beam was described by logarithmic decrement $\Delta = 1.5$): (a) t=0–3 s; (b) t=3–6 s; (c) t=6–9 s; (d) t=9–12 s.

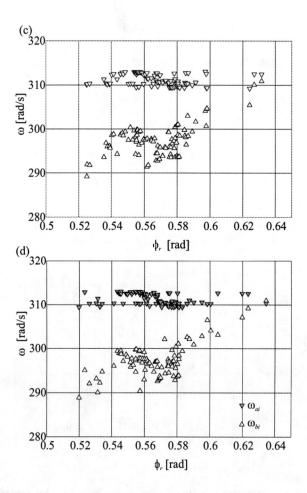

FIGURE 4.6
Continued.

erator motion have also been drawn (Figure 4.6). As can be seen, as time passes, the motion does not become stable but rather assumes a chaotic character. Such a character of the motion is caused here by the occurrence of impacts of each of kinds I and II by the fact that the hammer and the fender often miss each other without impact.

In this example forcing, determined by the frequency of free vibration of the hammer, may be modified to obtain desired changes both in the average rotational velocity of the generator rotor and in the intensity of impacts. Moreover, when the forcing is almost synchronous, it is possible to generate chaotic motion. The example illustrates the fact that knowledge of the dynamics of the system (as is common in mechanical systems) allows easy control of behavior, including chaos, by small modifications to system design (length l_s) or forcing (magnitude of ω_a). In practical terms, this is likely to be much preferred to costly and difficult feedback procedures.

4.5 Control by System Design

In this section we explore further the idea of modifying or removing chaotic behavior by appropriate system design. It is clear that, to a certain extent, chaos may be "designed out" of a system by appropriate modification of parameters, perhaps corresponding to modification of mass or inertia of moving parts. Equally clear, there exist strict limits beyond which such modifications cannot go without seriously affecting the efficiency of the system itself.

More usefully, it may be possible to effect a control by joining the chaotic system with some other (small) system [14, 13]. The idea of this method is similar to that of the so-called dynamical vibration absorber. A dynamical vibration absorber is a one-degree of freedom system, usually a mass on a spring (sometimes viscous damping is also added), which is connected to the main system as shown in Figure 4.7. The additional degree of freedom introduced shifts resonance zones, and in some cases can eliminate oscillations of the main mass. Although such a dynamical absorber can change the overall dynamics substantially, it need usually only be physically small in comparison with the main system, and does not require an increase of excitation force. It can be easily added to the existing system without major changes of design or construction. This contrasts with devices based on feedback control, which can be large and costly.

To explain the role of dynamical absorbers in controlling chaotic behavior, let us consider the nonlinear oscillator with dry friction damping, coupled with an additional linear system:

$$\ddot{x} + a\dot{x} + bx + cx^3 + d(x - y) = B_0 + B_1 \cos \Omega t \qquad (4.5.1)$$

$$\ddot{y} + \mu \, \text{sgn} \dot{y} + e(y - x) = 0, \quad (4.5.2)$$

where a, b, c, d, e, B, μ and Ω are constant. Here d and e are the characteristic parameters for the absorber, and we take e as control parameter. The parameters of equations (4.5.1-4.5.2) are related to those of Figure 4.7 in the following way: $a = c/m\Omega$, $b = k/m\Omega^2$, $c = k_c/m\Omega^2$, $d = k_a/m\Omega^2$, $e = k_a/m_a\Omega^2$, $\mu = F_f/m_a\Omega^2$, $B_0 = F_0/m\Omega^2$ and $B_1 = F_1/m\Omega^2$. It should be noted here that parameters d and e are related to each other through the absorber stiffness k_a. For simplicity in the rest of this section, we assume that d is constant and consider e as control parameter, i.e., we take constant stiffness k_a and allow the absorber mass, m_a, to vary.

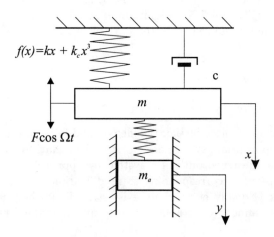

FIGURE 4.7
Schematic diagram of the main system and the dynamical absorber.

The parameters of the Eq. (4.5.1) (i.e., system without the dynamical absorber) have been chosen so as to show chaotic behavior; the controlling method provides an easy way of switching between chaotic and periodic behavior.

From Figure 4.8, it is clear that, for fixed Ω, we can obtain different types of periodic behavior by making slight changes in e. As an example, consider a system with $\Omega = 0.98$. For $e < 0.04$, the system shows chaotic behavior, but by changing e from 0.04 to 0.14, it is possible to easily obtain T, 2T, 4T, 8T periodic orbits. Theoretically orbits of higher periods are also possible, but their narrow range of existence makes them difficult to find either experimentally or numerically. What is of vital significance is that values of the parameter $e \in [0.04, 0.14]$ can be obtained with an absorber mass m_a approximately 100 times smaller than the main mass

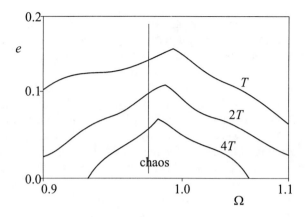

FIGURE 4.8
Behavior of the nonlinear oscillator for different values of e and Ω: $a = 0.077$, $b = 0$, $c = 1.0$, $B_0 = 0.045$, $\mu = 0.01$ and $B_1 = 0.16$.

(Figure 4.7).

A similar controlling effect can be obtained by varying the absorber stiffness, i.e., by simultaneous changes of parameters d and e.

4.6 Discussion

We have described several methods by which chaotic behavior in a mechanical system may be modified, displaced in parameter space, or removed. The OGY method is extremely general, relying only on the universal property of chaotic attractors that they have embedded within them infinitely many unstable periodic orbits (or even static equilibria). On the other hand, the method requires us to follow the trajectory and employ a feedback control system which must be highly flexible and responsive; such a system in the mechanical configuration may be large and expensive as it requires a number of sensors and data acquisition system for permanent monitoring of the system behavior. It has the additional disadvantage that small amounts of noise may cause occasional large departures from the desired operating trajectory.

The nonfeedback approaches are inevitable, much less flexible, and require more prior knowledge of equations of motion. On the other hand, to apply such methods, we do not have to follow the trajectory. The control procedures can be applied at any time and we can switch from one periodic orbit to another without returning to the chaotic behavior, although after each switch, transient chaos may be observed. Lifetime of this transient chaos strongly depends on initial conditions. Moreover in a nonfeedback method, we do not have to wait until the trajectory is close to an appropriate unstable orbit; in some cases, this time can be quite long. The dynamic approach can be very useful in mechanical systems, where feedback controllers are often very large (sometimes larger than the controlled system). In contrast, a dynamical absorber having a mass of order 1% of that of the control system is able to convert chaotic to periodic behavior over a substantial region of parameter space. Indeed, the simplicity by which chaotic behavior may be changed in this way, and possibility of easy access to different periodic orbits, may actually motivate the search for, and exploitation of, chaotic behavior in practical mechanical systems. This prompts us to pose a final question - how can we exploit chaos in mechanical systems? The OGY method, at least in theory, gives access to the wide range of possible behavior encompassed by the unstable periodic (and other) orbits embedded in a chaotic attractor. Moreover, the sensitivity of the chaotic regime to both initial conditions and parameter values means that desired effects may be produced by fine tuning. Thus, we might actually wish to design chaos into a system in order to exploit this adaptability. Nonfeedback methods can, in principle, give us advice on the design, whether we wish to design chaos out or in. Additionally, they enable us to choose regions of design parameter space or operating parameter space within which chaos will occur and be acceptable. An example of practical use might be the minimization of metal fatigue by switching from a necessary strictly periodic operation of the fully loaded conditions, which repeated stresses applied at the same places, to a noisy periodicity (rather like a healthy heartbeat) under idling conditions.

References

[1] E. Barreto, F. Casas, C. Grebogi, and E. J. Kostelich, "Control of chaos: Impact oscillators and targetting," in D. H. van Campen (Ed.), *Interaction between Dynamics and Control in Advanced Mechanical Systems*, Kluver Academic Pub., Dordrecht, 1997, pp. 17-26.

[2] B. Blazejczyk-Okolewska, K. Czolczynski, and A. Jach, "Dynamics of the impact force generator," *Mechanics and Mechanical Engineering*, vol. 1, pp. 103-109, 1997.

[3] F. Casas and C. Grebogi, "Control of chaotic impacts," *Int. J. Bifur. Chaos*, vol. 7, pp. 951-955, 1997.

[4] G. Chen and X. Dong, *From Chaos to Order: Methodologies, Perspectives and Applications*, World Scientific Pub. Co., Singapore, 1998.

[5] K. Czolczynski, T. Kapitaniak, and J. Brindley, "Controlling Hopf bifurcation in mechanical systems," in D. H. van Campen (Ed.), *Interaction between Dynamics and Control in Advanced Mechanical Systems*, Kluver Academic Pub., Dordrecht, 1997, pp. 83-90.

[6] W. L. Ditto, S. W. Rauseo, and M. L. Spano, "Experimental control of chaos", *Phys. Rev. Lett.*, vol. 65, pp. 3211-3216, 1991.

[7] U. Dressler and G. Nitsche, "Controlling chaos using time delay coordinates," *Phys. Rev. Lett.*, vol. 68, pp. 1-5, 1992.

[8] C. Grebogi, and Y.-C. Lai, "Controlling chaos in high dimension," *IEEE Trans. on Circ. Sys.*, vol. 44, pp. 971-975, 1997.

[9] E. Gutierrez, J. M. Zalvidar, D. K. Arrowsmith, and F. Vivaldi, "Analysis, modelling and control of impact oscillators," preprint.

[10] E. H. Hunt, "Stabilizing high-periodic orbits in a chaotic system: the diode resonator," *Phys.Rev. Lett.*, vol. 67, pp. 1953-1959, 1991.

[11] T. Kapitaniak, *Chaotic Oscillations in Mechanical Systems*, Manchester University Press, Manchester, 1991.

[12] T. Kapitaniak, *Controlling Chaos*, Academic Press, New York, 1996.

[13] T. Kapitaniak and J. Brindley, "The control of chaos by dynamical absorber," in J. M. T. Thompson and S.R. Bishop (Eds.), *Nonlinearity and Chaos in Engineering Dynamics*, Wiley, Chichester, 1994, pp. 163-169.

[14] T. Kapitaniak, L. Kocarev, and L. O. Chua, "Controlling chaos without feedback and control signals," *Int'l J. of Bifur. Chaos*, vol. 3, pp. 459-468, 1993.

[15] M. Lakshmanan and K. Murali, *Chaos in Nonlinear Oscillators: Controlling and Synchronization*, World Scientific Pub. Co., Singapore, 1996.

[16] N. Mehta and R. Henderson, "Controlling chaos to generate aperiodic orbit," *Phys. Rev. A*, vol. 44, pp. 4861-4869, 1991.

[17] F. C. Moon, *Chaotic Vibrations*, Wiley, Chichester, 1987.

[18] M. J. Ogorzalek, "Taming chaos, II: Control," *IEEE Trans. on Circ. Sys.*, vol. 40, pp. 700-706, 1993.

[19] M. J. Ogorzalek, "Chaos control techniques: A study using Chua's circuit," in A. C. Davies and W. Schwarz (Eds.), *Nonlinear Dynamics of Electronic Systems: Proc. of Workshop NDES'93*, World Scientific Pub. Co., Singapore, 1993, pp. 89-101.

[20] M. J. Ogorzalek, "Chaos control: How to avoid chaos or take advantage of it," *J. of Franklin Institute*, vol. 331B, pp. 681-704, 1994.

[21] E. Ott, C. Grebogi, and J. A. Yorke, "Controlling chaos," *Phys. Rev. Lett.*, vol. 64, pp. 1196-1199, 1990.

[22] E. Ott, C. Grebogi, and J. A. Yorke, "Controlling chaotic dynamical systems," in D. K. Campbell (Ed.), *Chaos: Soviet-American Perspectives on Nonlinear Science*, Amer. Inst. of Phys., New York, 1990, pp. 153-172.

[23] K. Pyragas, "Continuous control of chaos by self-controlling feedback," *Phys. Lett. A*, vol. 170, pp. 421-427, 1992.

[24] Z. Qu, G. Hu, and B. Ma, "Note on continuous chaos control," *Phys., Lett. A*, vol. 178, pp. 265-272, 1993.

[25] F. Romeiras, E. Ott, C. Grebogi, and W. P. Dayawansa, "Controlling chaotic dynamical system", *Physica D*, vol. 58, pp. 165-180, 1992.

[26] T. Shinbrot, E. Ott, C. Grebogi, and J. A. Yorke, "Using chaos to direct orbits to targets," *Phys. Rev. Lett.*, vol. 65, pp. 3215-3218, 1990.

[27] T. Shinbrot, E. Ott, C. Grebogi, and J. A. Yorke, "Using chaos to direct orbits to targets in systems describable by a one-dimensional map," *Phys. Rev. A*, vol. 45, pp. 4165-4168, 1992.

[28] T. Shinbrot, C. Grebogi, E. Ott, and J. A. Yorke, "Using small perturbations to control chaos," *Nature*, vol. 363, 411-417, 1993.

[29] J. Singer, Y.-Z. Wang, and H. H. Bau, "Controlling a chaotic system," *Phys. Rev. Lett.*, vol. 66, pp. 1123-1126, 1991.

[30] T. Tel, "Controlling transient chaos," *J. Phys. A.*, vol. 24, pp. l1359-l1365, 1991.

[31] J. M. T. Thompson and B. Stewart, *Nonlinear Dynamics and Chaos*, Wiley, Chichester, 1986.

[32] T. L. Vincent and J. Yu, "Control of a chaotic system," *Dynamics and Control*, vol. 1, pp. 35-52, 1991.

[33] W. Wawszczak, "Wysokoobrotowy wymiennik ciepła," (in Polish) *Zeszyty Naukowe PL*, vol. 705, pp. 1-207, 1995.

[34] W. Wawszczak and B. Jagiello, "Generation and analysis of torsional vibration," *Mechanics and Mechanical Engineering*, vol. 1, pp. 43-60, 1997.

5
Utilizing Chaos in Control System Design

Thomas L. Vincent

Aerospace and Mechanical Engineering
University of Arizona
Tucson, Arizona, 85721 USA
vincent@u.arizona.edu

Abstract

The full potential of control system design is often overlooked since there exists a strong prejudice, in classical control text books, to focus on linear systems. Considering the way control theory has been developed and the way it has been applied in practice, this is perhaps a natural result. Since a major focus of control system design is to achieve equilibrium point asymptotic stability and since nonlinear systems are generally "linearized" before applying linear control design methods, chaotic or other motion which can only occur in nonlinear systems is not generally considered to be a part of the design process. However, for years, controls engineers have been using traditional linear control methods to control systems which have the potential for chaos. This is not thought of as "controlling chaos" as chaos or any motion other than asymptotic stability or tracking is suppressed. This chapter will consider turning this approach on its head. Controllers will be designed to produce chaotic motion in order to take advantage of the random like "free ride" the chaotic attractor provides. The idea is stay with this free ride until the system moves into a neighborhood of some desirable equilibrium point. Once in this neighborhood, state variable feedback control is used to provide asymptotic stability for the equilibrium point. A basic requirement with this approach is in the determination of a neighborhood which lies on the domain of attraction to the equilibrium point under state variable feedback control. In addition, this neighborhood must be large enough so that the time it takes for the system to reach it, under chaotic control, is not unreasonably large. After addressing the question as to why this might be a desirable approach for nonlinear control system design, the focus of this paper is on the presentation of a general method for applying chaotic control and on the demonstration of its use.

5.1 Introduction

Poincaré [20] noted in 1892 that certain mechanical systems could display chaotic motion. However, the notion that deterministic models of discrete or continuous nonlinear dynamical systems could behave chaotically did not attract wide attention until Lorenz [13] in 1963, May [14] in 1976, and others reported chaotic behavior in very simple dynamical models. Lorenz was seeking to describe meteorological phenomena with just three nonlinear differential equations and May was dealing with biological modeling using simple nonlinear difference equations. Even the one dimensional logistic map is capable of producing chaos. Since these equations are deterministic, noise can not account for the chaotic behavior. Rather, chaos is due to the sensitivity of solutions to initial conditions. Trajectories which start close together separate exponentially fast in a bounded region of state space. This makes it impossible to predict long term behavior since initial conditions can never be known exactly. However, the bounds on the motion can be determined by solving the equations numerically. The region of state space occupied by the long term motion is called the chaotic attractor.

In 1990, Ott, Grebogi, and York [16] published the first paper to point out that chaos could be advantageous in achieving control objectives. Their method, now called the OGY method, involves stabilizing one of the unstable periodic orbits embedded in the chaotic attractor using small time dependent perturbations of a system parameter. Chaotic motion allows this method to work since all of the unstable periodic orbits will eventually be visited. One simply waits until the chaotic motion brings the system near a neighborhood of the proper unstable periodic orbit, at which time the small control is applied. Many variants of this method have appeared in the literature [4]. A more traditional control approach to regulating a chaotic system appeared in 1991 [29]. A control input was added to the Lorenz equations, the system was linearized about an equilibrium point, and two state variable feedback controllers were designed based on the linear system. When control was applied to the original nonlinear system, the equilibrium point of interest could easily be made asymptotically stable. However when the control effort was made small, the approach to the equilibrium point became interesting with chaotic-like long term transients. This lead to applying a closed loop control only in a small ball about the equilibrium point, allowing chaotic motion to bring the system into the ball where it could be captured.

Since that time, there have been a number of other papers dealing with the control of chaotic systems [3, 9, 18, 19, 10, 24] including feedback control of the Lorenz equations [1, 23]. These references tend to focus on the problem of designing a stabilizing controller for systems which, without control, would be chaotic. However, since chaos can be useful in moving a system to various points in state space, the systems of interest here are not necessarily chaotic but as in [28, 25, 26] chaotic motion can be created as a part of the total control design. We make use of the fact that for many nonlinear systems, chaos is easy to create using open-loop control. In particular, see [22] for a discussion of producing chaos in the driven pendulum system.

In the *chaotic control algorithm* given below, two essential ingredients are needed: a chaotic attractor and a controllable target. It is assumed that chaos can be created using open loop control. A *controllable target* is any subset of the domain of attraction to a equilibrium point, under a corresponding feedback control law, that has a non-empty intersection with the chaotic attractor. Thus, the equilibrium point itself need not be on the chaotic attractor. If we start the system at any point within the basin of attraction of the chaotic attractor, the resulting chaotic motion will ultimately arrive in the controllable target. The chaotic control algorithm simply has to keep track of when the system enters the controllable target. When it does, the open loop control used to create chaos is turned off and at the same time the closed-loop feedback control is turned on. The closed-loop controller may be designed using the linearized version of the nonlinear system about a specified equilibrium point. In applying this algorithm, it is assumed that all points on the chaotic attractor are out of harms way. This need not be the case in general.

The chaotic control algorithm has some distinct advantages in designing controllers for nonlinear systems. Its main advantage is simplicity. Consider for a moment one of the alternatives. Optimal control [12, 2, 27] is well suited for nonlinear problems and numerical methods are available [6] for solving complex problems. However, optimal control solutions for nonlinear problems, obtained by applying Pontryagin's maximum principle are generally open loop. One could use this open loop control to drive the system to a controllable target in a direct fashion. However, such a control program would not be robust in the sense that if a perturbation were to drive the system outside of the controllable target, one could not simply turn the open loop control back on. The only alternative would be to start over. Optimal closed loop solutions for non-linear problems are very difficult to obtain, including approximate ones generated by means of neural networks [7].

5.2 Chaotic Control Algorithm

The method presented here [26] can be applied to either discrete systems modeled by difference equations or continuous systems modeled by differential equations. In this chapter, we will consider only dynamical systems, subject to control, which can be described by nonlinear differential equations of the form

$$\dot{\mathbf{X}} = \mathbf{F}(\mathbf{X}, \mathbf{U}) \qquad (5.2.1)$$

where $\mathbf{F} = [F_1 \cdots F_{N_X}]$ is an N_X dimensional vector function of the state vector $\mathbf{X} = [X_1 \cdots X_{N_X}]$, and control vector $\mathbf{U} = [U_1 \cdots U_{N_U}]$. Current time is indicated by t and the dot ($\dot{\;}$) denotes differentiation with respect to time. The functions F_i are assumed to be continuous and continuously differentiable in their arguments. The control will, in general, be bounded and it is assumed that at every time t, the control $\mathbf{U}$ must lie in a subset of the control space $\mathcal{U}$ defined by the inequalities

$$U_{i_{\min}} \leq U_i \leq U_{i_{\max}}$$

for $i = 1 \cdots N_U$.

The control input $\mathbf{U}$ is either a specified function of time, $\mathbf{U}(t)$ (open loop) or a specified function of the state, $\mathbf{U}(\mathbf{x})$ (closed loop). Assume that for all t there exists a specified open-loop control input $\widehat{\mathbf{U}}(t)$ such that the system has a chaotic attractor. Furthermore, assume that for a specified constant control, $\mathbf{U}(t) \equiv \bar{\mathbf{U}}$, there is a corresponding equilibrium point of interest which is near the chaotic attractor. The equilibrium point satisfies

$$\mathbf{F}\left(\bar{\mathbf{X}}, \bar{\mathbf{U}}\right) = 0.$$

Given the above assumptions, the controllable target is obtained using a method of linearization and Lyapunov function estimates [25, 27, 26]. The nonlinear system is first linearized about the equilibrium point. The resulting linear system is assumed to be controllable and the LQR method [15, 16] is used to design a full state variable feedback controller that will guarantee the origin for this system is asymptotically stable. This, in turn, implies that for the nonlinear system, the equilibrium point will be asymptotically stable in some neighborhood containing the equilibrium point. An under estimate for this neighborhood is determined using a Lyapunov function obtained from the linear system. This under estimate is used as the controllable target.

Linearizing (5.2.1) about the equilibrium point results in the perturbation equations

$$\dot{\mathbf{x}} = \mathbf{A}\mathbf{x} + \mathbf{B}\mathbf{u} \qquad (5.2.2)$$

where

$$\begin{aligned} \mathbf{x} &= \mathbf{X} - \bar{\mathbf{X}} \\ \mathbf{u} &= \mathbf{U} - \bar{\mathbf{U}} \\ \mathbf{A} &= \left.\frac{\partial \mathbf{F}}{\partial \mathbf{X}}\right|_{\bar{X},\bar{U}} \\ \mathbf{B} &= \left.\frac{\partial \mathbf{F}}{\partial \mathbf{U}}\right|_{\bar{X},\bar{U}}. \end{aligned}$$

The origin is now the equilibrium point for equations (5.2.2) under the control $\mathbf{u}(t) \equiv \mathbf{0}$. The stability properties of the origin is determined by the eigenvalues of the $\mathbf{A}$ matrix. If the origin is unstable or is only weakly stable, the control input is used to provide desired stability properties. The LQR method determines gains $\mathbf{K}$ such that under full state feedback of the form

$$\mathbf{u}(\mathbf{x}) = -\mathbf{K}\mathbf{x} \qquad (5.2.3)$$

a quadratic performance index is minimized. The performance index is taken to be an infinite integral of the quadratic form $\mathbf{x}^T\mathbf{Q}\mathbf{x} + \mathbf{u}^T\mathbf{R}\mathbf{u}$ where $\mathbf{Q}$ and $\mathbf{R}$ are symmetric positive definite matrices to be chosen as part of the control design process. For the continuous system, the gain matrix $\mathbf{K}$ is given by

$$\mathbf{K} = \mathbf{R}^{-1}\mathbf{B}^T\mathbf{S}$$

and the matrix $\mathbf{S}$ is determined by solving the Riccati equation

$$\mathbf{S}\mathbf{A} + \mathbf{A}^T\mathbf{S} - \mathbf{S}\mathbf{B}\mathbf{R}^{-1}\mathbf{B}^T\mathbf{S} = -\mathbf{Q}.$$

The design process is greatly facilitated through the use of numerical routines available for solving these equations. The linearized system under full state feedback control is then given by

$$\dot{\mathbf{x}} = \widehat{\mathbf{A}}\mathbf{x} \qquad (5.2.4)$$

where

$$\widehat{\mathbf{A}} = \mathbf{A} - \mathbf{B}\mathbf{K}.$$

A Lyapunov function of the form

$$V(\mathbf{x}) = \mathbf{x}^T\mathbf{P}\mathbf{x} \qquad (5.2.5)$$

may now be determined for the linear stable controlled system (5.2.4) using the Lyapunov equation

$$\mathbf{P\hat{A} + \hat{A}^T P = -\hat{Q}}$$

where $\mathbf{\hat{Q}}$ is a positive definite matrix. For the stable linear system, starting from any point in state space, the solution obtained for $\mathbf{P}$ will result in the property that $\dot{V} < 0$ for every point of the linear system (5.2.4) except at the origin where $V = 0$.

The Lyapunov function (5.2.5) may now be used to determine a controllable target for the nonlinear system (5.2.1) under the LQR control given by (5.2.3). However, since the control must satisfy the control constraints

$$u_{i_{\min}} \leq u_i \leq u_{i_{\max}}$$

where

$$u_{i_{\min}} = U_{i_{\min}} - \bar{U}_i \quad \text{and} \quad u_{i_{\max}} = U_{i_{\max}} - \bar{U}_i$$

we use instead saturating LQR control defined by

$$\text{sat}\,[u_i(\mathbf{x})] := \begin{cases} u_{i_{\max}} & \text{if } u_i(\mathbf{x}) > u_{i_{\max}} \\ u_i(\mathbf{x}) & \text{if } u_{i_{\min}} \leq u_i(\mathbf{x}) \leq u_{i_{\max}} \\ u_{i_{\min}} & \text{if } u_i(\mathbf{x}) < u_{i_{\min}} \end{cases} \quad (5.2.6)$$

Using this control law, we seek the largest level curve $V(\mathbf{x}) = V_{\max}$ such that

$$\begin{aligned} \dot{V} &= 2\mathbf{x}^T \mathbf{P}\dot{\mathbf{x}} \\ &= 2\left(\mathbf{X} - \bar{\mathbf{X}}\right)^T \mathbf{P} \mathbf{F}(\mathbf{X}, \mathbf{U}) < 0 \end{aligned} \quad (5.2.7)$$

for all points on the level curve. Note that in calculating $\dot{V}$ the original nonlinear equations, under saturating LQR control must be used to evaluate $\mathbf{F}$.

The problem of finding the largest $V(x) < V_{\max}$ estimate for the domain of attraction for the feedback control law can be determined by solving a related optimization problem. The region $V(x) < V_{\max}$ that we seek is the largest region for which $\dot{V}$ is negative definite. As a consequence, the $V(x) = V_{\max}$ and $\dot{V} = 0$ surfaces must be tangent to each other at some point $\mathbf{x}^*$. The corresponding optimization problem is given by [27, p. 245]

$$\min V(\mathbf{x}) \quad \text{subject to } \dot{V} \geq 0$$

If this problem has a solution $\mathbf{x}^* \neq \mathbf{0}$, and if the region $V(x) < V(\mathbf{x}^*) = V_{\max}$ is bounded with $\dot{V}(\mathbf{x}) < 0$ in the region and not identically zero except at the origin, then the region $V(x) < V_{\max}$ is the largest estimate for the domain of attraction, based on the Lyapunov function $V(\mathbf{x})$. By definition, we are seeking a global minimum. However, this optimization problem will generally have multiple local minima. Any numerical method which finds

Inverted Pendulum

only local minima must be set up to find (hopefully!) all of them before the global minimum can be identified. In other words, solving this problem is not a trivial task. If the problem is two dimensional, the process can be greatly simplified by choosing a value for V_{max}, set up differential equations for integrating around the level curve $V(x) = V_{max}$ [27, p. 27] and then calculating $\dot{V}$ at every integration step. If the appropriate quantity is negative everywhere, a larger level curve is used and the process repeated. By this procedure one can obtain an x close to the minimal solution. This value can then be used as a starting point in a numerical routine to further refine the result.

It follows from the definition of **u** that the actual control used inside the region $V(x) < V_{max}$ is given by

$$U_i(\mathbf{x}) = \text{sat}\,[u_i(\mathbf{x})] + \bar{U}_i$$

The chaotic control algorithm given by the following code segment:

$$
\begin{array}{|l|}
\hline
\text{if } V > V_{max} \\
\quad \mathbf{U} = \widehat{\mathbf{U}}(t) \\
\text{else} \\
\quad U_i = \text{sat}\,[u_i(\mathbf{x})] + \bar{U}_i \\
\text{end} \\
\hline
\end{array}
\qquad (5.2.8)
$$

and is applied at every time t. Under (5.2.8), if the system is started at a point where $V > V_{max}$ then the system will run under the chaotic control $\widehat{\mathbf{U}}(t)$ until it reaches on a point where $V(\mathbf{x}) \leq V_{max}$ at this time the control is switched to saturating LQR control.

5.3 Inverted Pendulum

Consider the inverted pendulum, attached to a DC motor as shown in Figure 5.1. The pendulum is free to rotate through all angles so that it has stable downward equilibrium positions at $\theta = \pi(1 \pm 2n)$, $(n = 0, 1, 2...)$ and unstable upright equilibrium position at $\theta = 2n\pi$, $(n = 0, 1, 2...)$. Let $x_1 = \pi - \theta$ be the angle of the pendulum as measured from the downward position, $\dot{x}_1$ be the rate of change of this angle, and bu be the torque applied by the motor. Positive values are in the counter clockwise direction. In terms of these variables, the equations of motion are given by

$$
\begin{aligned}
\dot{x}_1 &= x_2 \\
\dot{x}_2 &= a_1 \sin x_1 + a_2 x_2 + bu,
\end{aligned}
\qquad (5.3.9)
$$

where a_1, a_2, and b are constants associated with the system. The term $a_1 \sin x_1$ is the torque provided by gravity, $a_2 x_2$ is damping provided by friction and back EMF of the motor, bu is the control torque provided by a DC motor, and u is the voltage applied to the motor, with $|u| \leq u_{\max}$. The particular system that we will examine here has [8]

$$\begin{aligned} a_1 &= -17.627 \text{rad/sec}^2 \\ a_2 &= -0.187 \text{sec}^{-2} \\ b &= 0.6455 \text{rad/volt} - \text{sec}^2 \\ u_{\max} &= 18 \text{volts}. \end{aligned}$$

Our objective is to stabilize the inverted pendulum in the vertical upright position.

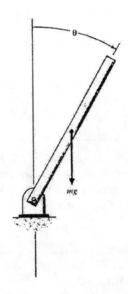

FIGURE 5.1
Inverted pendulum attached to a DC motor.

If we linearize this system about $\bar{\mathbf{X}} = \begin{bmatrix} \pi & 0 \end{bmatrix}^T$, then we can use LQR design with $\mathbf{Q}$ and $\mathbf{R}$ identity matrices to obtain the feedback gains

$$\mathbf{K} = \begin{bmatrix} 54.6333 & 12.7624 \end{bmatrix},$$

with the eigenvalues for the linearized controlled system given by $\lambda_1 = -4.5397$ and $\lambda_2 = -3.8855$. Under LQR design, the feedback control law is given by

$$u(\mathbf{x}) = -\mathbf{K}\left(\mathbf{x} - \bar{\mathbf{X}}\right). \tag{5.3.10}$$

Inverted Pendulum

However, since there are bounds on the control, the actual feedback control law used is that of saturating feedback control sat$[u(\mathbf{x})]$ as defined by (5.2.6).

We can find the domain of attraction under saturating LQR control as applied to the nonlinear system (5.3.9) by integrating backward from points very close to the equilibrium points obtained when the control is set equal to $\pm u_{\max}$. The results are illustrated in Figure 5.2. One way to "see" the domain of attraction in this figure, is to first pinpoint the star located inside the ellipse. This represents $\bar{\mathbf{X}}$. The domain of attraction, to this point is the set of all points obtained by "flooding" with a color, from the star, or by drawing all continuous curves from the star which do not cross any of the curves contained in the figure other than the ellipse. Observe that, in places, the domain of attraction narrows down and becomes tubular.

Note that the x_1 and x_2 axis are in multiples of π. The equilibrium point $\bar{\mathbf{X}}$ and the first 2π multiple to the right and left are marked with a star, the squares locate stable equilibrium points, and the circles locate the equilibrium solutions obtained if saturating LQR control is used outside the domain of attraction.

Consider any one of the lob-like objects which contain a square and a small circle. If the system, under saturating LQR control is started anywhere inside the lob, including its tubular extension, it would remain in them, ultimately arriving at the equilibrium point contained in the lobe. Clearly, these regions are not in the domain of attraction to $\bar{\mathbf{X}}$. In fact, the domain of attraction to $\bar{\mathbf{X}}$ is simply all other points. Note that these other points include the stars located at 2π multiples of $\theta = 0$. Saturating LQR controller will not stabilize the system to these points since it does not recognize any vertical equilibrium position other than the one corresponding to $\theta = \dot{\theta} = 0$. In other words, if the system were started at the upright position $x_1 = 3\pi$, $x_2 = 0$, the LQR controller would "unwind" the pendulum to bring it to $x_1 = x_2 = 0$.

5.3.1 Controllable Targets

Under LQR control

$$\hat{\mathbf{A}} = \begin{bmatrix} 0 & 1 \\ -17.6388 & -8.4251 \end{bmatrix}.$$

Solving the Lyapunov equation with $\mathbf{Q} = \mathbf{I}$, results in

$$\mathbf{P} = \begin{bmatrix} 1.3450 & 0.0283 \\ 0.0283 & 0.0627 \end{bmatrix}.$$

Since the problem is two dimensional, a value for $V_{\max}$ may be obtained by choosing a sufficiently small value for $V_{\max}$ so that integrating around

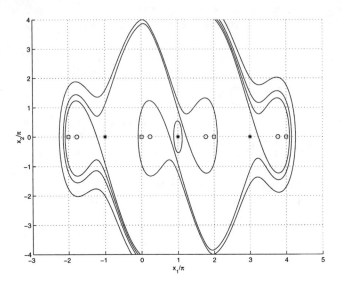

FIGURE 5.2
Domain of attraction for the single pendulum.

the ellipse defined by

$$\mathbf{x}^T \mathbf{P} \mathbf{x} = V_{\max} \tag{5.3.11}$$

results in $\dot{V} < 0$ everywhere on this curve. A larger value for $V_{\max}$ may then be chosen and the process repeated until the inequality is satisfied with $\dot{V}$ very close to zero. Using this procedure results in $V_{\max} = 0.18$ for this problem. One advantage of this method for finding $V_{\max}$ is that it provides a numerical verification that the region inside the ellipse defined by (5.3.11) is a region of guaranteed asymptotic stability for the equilibrium point $x_1 = \pi$, $x_2 = 0$. The ellipse of Figure 5.2 is the one given by (5.3.11).

As we will see shortly, the chaotic attractor is going to be very large in comparison with the controllable target just obtained. This implies that the waiting time between chaotic control and feedback control may be large. There are several possibilities that can be used to improve the odds. One possibility would be to choose feedback gains so that when the Lyapunov equation is solved, the largest possible ellipse is obtained that will just fit into the domain of attraction [5]. However, this procedure requires difficult computations and we will pursue an alternate approach of simply increasing the number of targets. One obvious way of doing this is to introduce the ellipse just obtained at each of the points indicated by the stars. At each of these points, the pendulum is in the upright position

Inverted Pendulum

at zero velocity. Hence if one were to linearize the system about each of these points, determine a saturating LQR feedback control as above, and find the resulting domain of attraction, it would look just like Figure 5.2 with the curves shifted to the right or left. Assuming that we do this for only the three star points indicated in the figure, the resulting feedback control would be of the form of (5.2.6) with $u(\mathbf{x})$ given by (5.3.10) where

$$\bar{\mathbf{X}} = \begin{cases} \begin{bmatrix} 3\pi & 0 \end{bmatrix}^T & \text{if } x_1 > 2\pi \\ \begin{bmatrix} \pi & 0 \end{bmatrix}^T & \text{if } 0 \leq x_1 \leq 2\pi \\ \begin{bmatrix} -\pi & 0 \end{bmatrix}^T & \text{if } x_1 < 0. \end{cases}$$

The domain of attraction under this feedback control law for each of the three equilibrium points indicated by the stars is now much more complicated and it is left as an exercise for the interested reader.

5.3.2 Chaotic Attractor

One might argue: how is it possible to produce chaotic motion for the inverted pendulum system (5.3.9) since it is only two dimensional? Indeed, under state feedback control, the system would remain as an autonomous two dimensional system and chaotic motion would not be possible. However under an appropriate open loop control, chaotic motion is possible since time plays the role of a third state variable. In seeking an open loop control which will provide chaotic motion, it must be able to swing the pendulum to the upright vertical position from any given starting condition. It is at this point that the system displays sensitivity with respect to initial conditions. An easy way to do this is to apply a sinusoidal voltage to the motor. For example, applying the control

$$u = 11 \cos 3t \tag{5.3.12}$$

to the system (5.2.1), starting from $\mathbf{x}(0) = \begin{bmatrix} 0.1 & 0 \end{bmatrix}^T$, integrating for 250 seconds and plotting the state every 0.1 seconds yields the results shown in Figure 5.3. Several observations need to be made. First, driving the system "over the top" is a necessary but not sufficient condition for chaos. While there are many amplitude-frequency combinations which will produce chaos, a small change in one of these values may result in motion which is not chaotic [22]. It is not proven here that the control given by (5.3.12) does actually produce chaos. What is depicted may be a long chaotic transient [29] with the possibility that after a sufficiently long time period, the trajectory could settle down to a limit cycle. Second, the attractor extends beyond the figure in both directions. Third, while the targets appear to have received many "hits" (especially the left one), there are only 21 points which lie inside the ellipses out of the 2,500 points (which lie inside the limits of the figure and outside the limits of the figure). Finally, the results shown in Figure 5.3 are only representative. It is possible that changing the initial conditions only slightly, using the same integration time, could result in more hits, no hits, or with most of the points lying outside the figure.

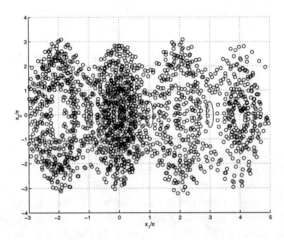

FIGURE 5.3
Chaotic attractor for the inverted pendulum.

Let us now deal with these observations. As long as (5.3.12) produces a long term chaotic transient (if not true chaos) over a range of starting conditions of interest, it remains a viable chaotic controller. Given a sufficiently long running time, a chaotic controller will wind up the pendulum for many revolutions in both directions, moving past the origin many times in a random way. Hence, in actual practice, it would be wise to include additional elliptical targets at additional 2π multiples of $\bar{\mathbf{X}}$ in exactly the same manner that the first two were added. In order to illustrate the effect of different starting conditions, consider the results shown in Figure 5.4 which is exactly the same run as in Figure 5.3 except that the run time was reduce to 150 sec. In this case, there are no points inside the ellipses; in fact, the system seems to be avoiding them. It is clear that with the starting conditions given, one would have to wait longer that 150 second before the pendulum would be captured in the upright position.

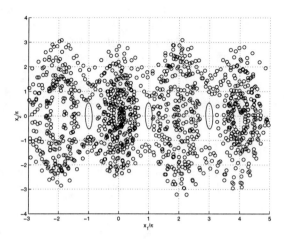

FIGURE 5.4
Chaotic attractor for the inverted pendulum.

5.3.3 Intermediate Targets

From the results obtained so far, it is evident that it would be desirable to add additional targets into regions where the points are more closely clustered. It has been previously shown that the waiting period required with the OGY method can be substantially reduced by using a "targeting method" [11]. This procedure uses intermediate targets in moving the system to a final target. We will use a similar procedure here. One way to arrive at additional intermediate targets is to simply integrate backward, under saturating LQR control from the neighborhood of the original equilibrium point $\bar{\mathbf{X}} = \begin{bmatrix} \pi & 0 \end{bmatrix}^T$ using the initial conditions $\mathbf{x}(0) = \begin{bmatrix} \pi & \varepsilon \end{bmatrix}^T$ and $\mathbf{x}(0) = \begin{bmatrix} \pi & -\varepsilon \end{bmatrix}^T$ for a period of time, t_s. This results in two stopping points $\mathbf{x}_s^+$ and $\mathbf{x}_s^-$. At each of these stopping points, we know that there exists some neighborhood about $\mathbf{x}_s^\pm$ such that if we integrate the system forward in time under saturating LQR control, the system will be returned to $\bar{\mathbf{X}}$. Here we will choose t_s to be relatively small (e.g., so that the approximation $\sin x_1 = x_1$ is valid) and use the controllable target ellipse centered at $\mathbf{x}_s^\pm$ to be an estimate of this additional neighborhood. If the system were indeed linear, the flow of all trajectories though the new intermediate target would converge to the line described by the retro trajectory. Unfortunately, for nonlinear systems there is no guarantee that all points in these intermediate targets will be controllable to the equilibrium point. This possibility does not defeat the method provided that we allow for it in the control algorithm. One way to do this is to make sure that chaotic control will be used in a small neighborhood of all equilibrium points other than the stars. In this way, if the system enters a intermediate target, saturating LQR control is turned on, and the resultant trajectory does not arrive at a star, the chaotic control sequence will begin again. Clearly, the majority of points in the intermediate target, under saturating LQR control must drive the system to a star in order for an intermediate target to be of any use.

5.3.4 Chaotic-LQR control

A sample run using three control targets at the stars and two intermediate targets is illustrated in Figure 5.5. The system starts at the origin under the chaotic control

$$u = 11\sin(3t)$$

and remains under this control until it intersects the upper intermediate target. At this point, saturating LQR control is applied which brings the system to the star equilibrium point. In this case, since the chaotic trajectory is relatively short, it is shown as a solid line.

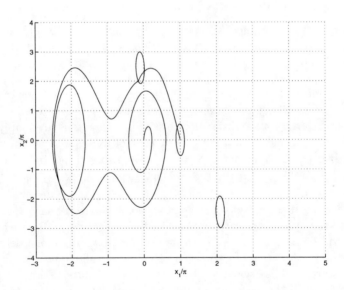

FIGURE 5.5
Utilizing chaos in the control of the inverted pendulum.

The actual control algorithm used is given by the following Matlab code fragment.

```
Xbar = [-pi, 0; pi, 0; 3*pi, 0; Xsup1; Xsup2]';
for i = 1:5
  QF(i) = (x-Xbar(i,:))'*P*(x-Xbar(i,:))
end
Test = QF<Vmax;
if sum(Test) == 1 | ControlOn
  j = find(Test); ControlOn = 1;
  if j = 1|2|3, u = -K*(x-Xbar(j)); end
  if j = 4|5, u = K*(x-Xbar(2)); end
  if abs(u) > Umax, u = Umax*sign(u)
else
  u = 11*sin(3*t)
end
```

The location of the two intermediate targets has been previously determined and are contained in the row vectors Xsup1 and Xsup1. The variable *ControlOn* is a flag which has previously been set equal to zero. The first line defines all the star equilibrium points and intermediate points into a matrix. The next three lines calculate the value of the quadratic functions centered at the points in **Xbar** using the previously determined **P** matrix at the current position **x**. The next line checks to see if **x** is inside any of the ellipses defined by the previously determined $V_{\max}$. The variable *Test* will contain 5 logic zeros if the system is outside all the ellipses and will contain a single logic one if the system is inside one of the ellipses. The next line checks to see if *Test* contains a logic one or if the control has been previously turned on (this is necessary since the saturating LQR control must remain on after leaving a intermediate target). The next line determines which ellipse has been entered and the remaining lines apply the appropriate saturating LQR control. Note that if the **if** statement is not satisfied, then chaotic control is used. This algorithm can be extended in a more efficient and obvious way to include additional star targets and intermediate targets.

5.4 Discussion

The inverted pendulum example was chosen to illustrate the method. There are other ways of approaching the control of this problem which are more efficient than the method presented here. The usefulness of the method becomes more apparent with a more complicated problem. The double link pendulum is one such example [21, 26]. In this case the system is defined by four first order, highly nonlinear differential equations. Under no control, there are 3 unstable equilibrium positions and one stable equilibrium position. Designing an LQR controller to any one of the unstable equilibrium positions is relatively easy. However displaying its domain of attraction is not and indeed, not necessary. The method used here to control the inverted pendulum is applicable for higher dimensional problems. This approach has been successfully applied to an actual double link pendulum system.

Clearly, neither the nonlinear models for real system nor the measurements taken will ever be exact, yet the laboratory system performed well under a chaotic control algorithm. Applying the algorithm requires full state information to evaluate the Lyapunov function and to apply the state variable feedback control. In order to avoid reliance on a model, when applying the results to the double link laboratory system, all state variables associated with position were measured directly and all state variables associated with velocity were obtained using a digital filter. Thus, the model was used only to determine a Lyapunov function. During the chaotic waiting mode under open loop control, errors are of little consequence. During the closed loop control mode, robustness is maintained by using only output information. Thus, there are only two major sources of error: (a) the overlap between the controllable target as determined from the model and the actual domain of attraction under the closed loop control, and (b) measurement error. If these errors are large, then the closed loop control may be turned on many times without capturing the target. However, unless the errors are gross, the algorithm still works, albeit with longer waiting periods.

References

[1] J. Alvarez, "Nonlinear regulation of a Lorenz system by feedback linearization techniques," *Dynamics and Control*, vol. 4, pp. 277-298, 1994.

[2] A. E. Bryson and Y. C. Ho, *Applied Optimal Control*, Blaisdell, Waltham, Massachusetts, 1969.

[3] G. Chen and X. Dong, "From chaos to order - perspectives and methodologies in controlling chaotic nonlinear dynamical systems," *Int. J. of Bifur. Chaos*, vol. 3, pp. 1363-1409, 1993.

[4] G. Chen and X. Dong, "Control of chaos - A survey," *Proc. of the 32nd IEEE Control and Decision Conference*, San Antonio, Texas, Dec. 15-17, 1993.

[5] M. E. Fisher, T. L. Vincent, and D. Summers, "Under-estimates of reachable sets for linear control systems," *Dynamics and Control*, vol. 5, pp. 241-251, 1995.

[6] C. J. Goh and K. L. Teo, *MISER: An Optimal Control Software, Theory and User Manual*, Industrial and Systems Engineering, National University of Singapore, Singapore, 1987.

[7] C. J. Goh, N. J. Edwards, and A. Y. Zomaya, "Feedback control of minimum-time optimal control problems using neural networks," *Optimal Control Appl. & Methods*, vol. 14, pp. 1-16, 1993.

[8] W. J. Grantham and T. L. Vincent, *Modern Control Systems Analysis and Design*, Wiley, New York, 1993.

[9] T. T. Hartley and F. Mossayebi, "A classical approach to controlling the Lorenz equations," *Int. J. of Bifur. and Chaos*, vol. 2, pp. 881-887, 1992.

[10] K. Judd, A. I. Mees, K. L. Teo, and T. L. Vincent, *Control and Chaos*, Birkhäuser, Boston, 1997.

[11] E. J. Kostelich and E. Barreto, "Targeting and control of chaos," in K. Judd, A. I. Mees, K. L. Teo, and T. L. Vincent (Eds.), *Control and Chaos*, Birkhäuser, Boston, 1997.

[12] G. Leitmann, *An Introduction to Optimal Control*, McGraw-Hill, New York, 1966.

[13] E. N. Lorenz, "Deterministic non-periodic flow," *J. of Atmos. Sci.*, vol. 20, pp. 130-141, 1963.

[14] R. M. May, "Simple mathematical models with very complicated dynamics," *Nature*, vol. 261, pp. 459-467, 1976.

[15] K. Ogata, *Discrete-Time Control Systems*, Prentice-Hall, New Jersey, 1987.

[16] K. Ogata, *Modern Control Engineering*, Prentice-Hall, New Jersey, 1997.

[17] E. Ott, C. Grebogi, and J. A. York, "Controlling chaos," *Phys. Rev. Letts.*, vol. 64, pp. 1196-1199, 1990.

[18] M. Paskota, A. I. Mees, and K. L. Teo, "Stabilizing higher periodic orbits," *Int. J. of Bifur. Chaos*, vol. 4, pp. 457-460, 1994.

[19] M. Paskota, A. I. Mees, and K. L. Teo "On local control of chaos, the neighborhood size," *Int. J. of Bifur. Chaos*, vol. 6, pp. 169-178, 1996.

[20] H. Poincaré, *Les Méthodes Nouvelles de la Méchanique Celeste*, Gauthier-Villars, Paris, 1892, in English, NASA Translation TTF-450/452, U.S. Federal Clearinghouse, Springfield, VA, 1967.

[21] P. H. Richter and H. J. Scholz, "Chaos in classical mechanics: The double pendulum," in *Stochastic Phenomena and Chaotic Behavior in Complex Systems*, Springer, Berlin, 1984.

[22] H. G. Schuster, *Deterministic Chaos – An Introduction*, VCH, Weinheim, 1988.

[23] J. Singer, Y. Z. Wang, and H. H. Bau, "Controlling a chaotic system," *Phys. Rev. Letts.*, vol. 66, pp. 1123-1125, 1991.

[24] T. Tél, "Controlling transient chaos," *J. Phys. A: Gen.*, vol. 24, pp. L1359-L1368, 1991.

[25] T. L. Vincent, "Controllable targets near a chaotic attractor," in K. Judd, A. I. Mees, K. L. Teo, and T. L. Vincent (Eds.), *Control and Chaos*, Birkhäuser, Boston, 1997.

[26] T. L. Vincent, "Control using chaos," *IEEE Control Systems*, vol. 17, pp. 65-76, 1997.

[27] T. L. Vincent and W. J, Grantham, *Nonlinear and Optimal Control Systems*, Wiley, New York, 1997.

[28] T. L. Vincent, T. J. Schmitt, and T. L. Vincent, "A chaotic controller for the double pendulum," in R. S. Guttalu (Ed.), *Mechanics and Control*, Plenum Press, New York, pp.257-273, 1994.

[29] T. L. Vincent and J. Yu, "Control of a chaotic system," *Dynamics and Control*, vol. 1, pp. 35-52, 1991.

6

Control and Synchronization of Spatiotemporal Chaos

Jin-Qing Fang[1] and **M. Keramat Ali**[2]

[1]China Institute of Atomic Energy
P.O.Box 275, Beijing, 102413, P. R. China
fjq96@mipsa.ciae.ac.cn

[2]Department of Physics
The University of Lethbridge
Lethbridge, Alberta, T1k 3M4, Canada

Abstract

Control and synchronization of spatiotemporal chaos in spatially extended systems are among of the most challenging topics. Spatially extended systems are modeled by partial differential equations, coupled ordinary differential equations, or coupled map lattices. Simple control and synchronization methods for lower-dimensional systems may be generalized to some spatially extended systems that have only one unstable state. These methods usually fail to apply to those systems with multiple unstable states. Some new strategies and methodologies for controlling and synchronizing higher-dimensional spatiotemporal chaos have recently been developed. These techniques are reviewed, compared, and commented on in this chapter, along with some discussions on their possible applications and further research developments.

6.1 Introduction

Extraordinarily diverse spatiotemporal structures exist in spatially extended systems (SESs), including a variety of spatiotemporal chaos (STC), periodic structures, intermittence, turbulence, pattern formation, etc. [4]. Describing and understanding STC raises many basic and important questions for both theoretical and experimental studies, particularly for SES. The pioneering works of Ott, Grebogi and York [16] on chaos control and of Pecora and Carroll [18] on chaos synchronization have been naturally extended to SESs lately (see, for example, [3, 26, 16]).

The problem of controlling turbulence and STC is much more complicated since the chaos in such systems are typically within higher, even infinite dimensional states, involving numerous stable and unstable spatial modes. Some of STC may be controlled and synchronized by relatively simple methods, particularly in lower-dimensional systems, such as parameter perturbations and time-delay feedbacks to propagating fronts with highly correlated spatial modes, periodic behavior in myocardium tissue, hippocampal baghbrain tissue, etc. [19]. However, simple control approaches typically fail when the unstable state has more than one unstable direction. Some new strategies and methodologies for controlling and synchronizing higher-dimensional spatiotemporal chaos have recently been developed. These techniques are reviewed, compared, and commented on in this chapter, along with some discussions on their possible applications and further research developments.

This presentation is not intended to be exhaustive due to the page limit. Therefore, our main focus is on spatiotemporal chaos in systems described by partial differential equations (PDEs), coupled ordinary differential equations (CODEs) and coupled map lattices (CMLs). There are many other spatiotemporal systems, such as those described by integral and integral-differential equations, variable functional equations, abstract operator equation, point processes, and cellular automata, but these are outside of the scope of this short chapter.

6.2 SESs Modeled by PDEs

Typical spatiotemporal systems are governed by PDEs, such as Navier-Stokes equations, chemical reaction-diffusion equations, lasers equations, and complex Ginzburg-Landau equations (CGLEs), etc. Under certain conditions, such PDEs can exhibit chaos both in space and in time.

STC often occurs in continuous nonlinear systems when different types of motions, excited in local regions of SES, interact. These interactions can destroy the spatial coherence of the system concurrent with the onset of the temporal chaos. Since nonlinear PDEs are usually not analytically solvable [13], controlling and synchronizing STC in PDEs turn out to be quite difficult. Nevertheless, as pointed out in [3], "the infinite-dimensional nature of distributed systems gives rise to many new theoretical and practical challenges that need to be addressed by more advanced mathematical theories and methods." The current studies of this subject have combined theoretical analysis with numerical simulation as well as experiments, which are first reviewed in this section.

Nonlinear Feedback Control of STC Based on Time Series

Although linear or linearized control methods are still the dominant approaches in chaos control, there are many limitations for them to handle STC and complex systems. Most nonlinear systems cannot be exactly linearized via diffeomorphism and smooth feedback. In practice, when the target state is far from the attractor of the unperturbed system, linear control and synchronization methods fail because they do not correctly describe large feedback perturbations that are necessary for control [19]. For example, when a steady state loses stability through a Hopf bifurcation, it becomes separated from the originally stable limit cycle by the nonlinear vector flow. Large perturbations have to be applied in order to move the system orbit from the limit cycle to the unstable steady state through a large region of the nonlinear vector field. Nonlinear targeting can accomplish the task in some cases with the help of linear methods, if the linearized equations can provide a good approximation of the system dynamics. In regimes where nonlinearity of the vector flow is strong, however, nonlinear control methods (NCM) must be used [19].

In general, nonlinear controls have many advantages, such as improving transition time, extending the range of of attraction basin, and more robust to noise, etc. Recently, some effective NCMs have been proposed for STC control. Sliding mode control and switching manifold control are two typical examples [3, 16]. In the second method, for instance, a switching manifold containing the control targets is first found. Then a controller is designed to force the system states to move toward the manifold. Finally, the controller drives the system states that are nearby the target in the manifold to slide onto it.

A new type of NCM is recently proposed by Petrov et al. in [19], which has been applied to PDE models and model-independent experiments. This NCM constructs the control law as a multi-dimensional manifold in the time-delayed space obtained directly from a time series of the system response to input perturbations. The Gray-Scott model for cubic autocatalysis in a flow reaction was used for demonstration:

$$\begin{aligned}\frac{\partial \alpha}{\partial t} &= (1-\alpha)/T_{res} - \alpha\beta^2, \\ \frac{\partial \beta}{\partial t} &= (\beta_0 - \beta)/T_{res} + \alpha\beta^2 - k_2\beta.\end{aligned} \qquad (6.2.1)$$

With $\beta_0 = 1/15$ and $k_2 = 0$, the model has one unstable and two stable stationary states over a range of reciprocal residence time, $\frac{1}{T_{res}} = 0.23-0.354$. Transitions from one stable state to the other are induced by applying appropriate perturbations to $1/T_{res}$. The key idea here is how to control the transitions between the stable and the unstable states using this NCM, with a manifold constructed from time series.

For a 1-D system, the control manifold is constructed by observing the transitions from an initial state x_I to a desired final state $x_F(T+\tau)$ that result from the application of a perturbation p on the sampling interval τ. The collected triplets of values (x_I, x_F, p) lie on a manifold in a 3-D space. This nonlinear manifold,

$$p_{I \to F} = C(x_I, x_F),$$

defines the perturbation that moves the system from the initial state to the desired final state.

Feedback control of multi-dimensional, nonlinear, single-input single-output systems is formulated in terms of an invariant hypersurface in the delayed state space of a system (its observable and control parameter). The manifold is increased directly from the response of the system to random perturbations, leading to a model-independent NCM. The NCM can be used to stabilize unstable state or to drive the system to any particular target state in a minimum number of steps [19].

Because this NCM is simple and robust, it has been successfully applied in experiments of stabilizing periodic states in a liquid bridge. However, stabilization of the steady state was not possible with the controller in a scalar form. Thus, an extended vector form for this purpose was also proposed [19]. The improved NCM works quite well for Maragoni numbers near the onset of an oscillatory flow. Moreover, the ability to control a SES with four unstable degrees of freedom has verified the robustness of the proposed NCM. This NCM can also improve the quality of the zone refining processes in industry if high-power feedback elements are used.

Nonlinear Diffusion Control

CGLEs are quite instructive since they can exhibit several types of STC. These models can account for slow modulations in space and time of the oscillatory state of a physical system that undergoes a Hopf bifurcation. Because of its generality, the CGLEs have been extensively studied in both of regular and turbulent parameter domains. Predictions based on these models are widely used for qualitative interpretation of experimental data, such as in catalytic surface chemical reactions and the Belousov-Zhabotinskki reaction.

A typical CGLE is a prototypical equation for a complex field A that exhibits STC and shows several types of STC. Montagne et al. [15] proposed a nonlinear diffusion control method for the STC in a CGLE of the form

$$\frac{\partial A}{\partial t} = A + (1 + ic_1)\frac{\partial^2 A}{\partial x^2} A - (1 + ic_2)|A|^2 A, \qquad (6.2.2)$$

along with periodic boundary conditions. This equation admits plane-wave solutions. The proposed method is not based on feedback but on nonlinear diffusion control effects. The controlled system is

$$\frac{\partial A}{\partial t} = A + [1 + ic_1 + i\gamma|A|^2/|A_{PW}|^2 - 1]\frac{\partial^2 A}{\partial x^2} A - (1 + ic_2)|A|^2 A, \quad (6.2.3)$$

in which $i\gamma|A|^2/|A_{PW}|^2 - 1$ is the nonlinear diffusion controller. This method has proven to be an effective method of stabilizing unstable plane-wave solutions in different parameter regions of the CGLE where STC, phase turbulence, spatiotemporal intermittence, bichaos and defect turbulence co-exist.

Nonlinear Bidirectional Coupling

Another method of nonlinear control is to use nonlinear coupling. For example, Kocarev et al. [11] used unidirectional coupling for the Garry-Scott equation, and Amengual et al. [1] performed chaos synchronization for a CGLE using a nonlinear bidirectional control (NBC):

$$\begin{array}{l}\frac{\partial A_1}{\partial t} = \mu A_1 + (1 + i\alpha)\frac{\partial^2 A_1}{\partial x^2} - (1 - i\beta)(|A_1|^2 + \gamma|A_2|^2)A_1 \\ \frac{\partial A_2}{\partial t} = \mu A_2 + (1 + i\alpha)\frac{\partial^2 A_2}{\partial x^2} - (1 - i\beta)(|A_2|^2 + \gamma|A_1|^2)A_2,\end{array} \qquad (6.2.4)$$

where μ, α, β are parameters and γ is a coupling constant.

Nonlinear Control by Oscillatory Time-Delay

Schuster and Stemmler [21] controlled the chaotic states of the Kuramoto-Sivashinsky equation to periodic behavior by using oscillatory time-delay in the following form:

$$u_t = -uu_x - u_{xx} - u_{xxxx} - \epsilon^t u_t^{t-\tau}, \quad x \in [0, L], \qquad (6.2.5)$$

where ϵ^t is time-dependent and τ is a constant time delay. The method could be applied to experiments where a time series is monitored in a phase space spanned by Fourier coefficients, if the parameter is accessible for small variations.

Controlling STC by Delayed Global Feedback

Feedback is the most popular approach for STC in SESs [3]. A controlled CGLE for local complex amplitude $A(\mathbf{r}, \mathbf{t})$ can be rewritten in the following dimensionless form:

$$\frac{\partial A}{\partial t} = (1 - i\omega)A - (1 + i\beta)|A|^2 + (1 + i\varepsilon)\nabla^2 A + G(t), \qquad (6.2.6)$$

where β characterizes the nonlinear frequency shift of individual oscillators, ϵ is the dispersion of traveling waves, ω is the linear oscillation frequency of individual oscillators in the system, $\omega \gg 1$ is assumed, amd $G(t)$ is the delayed global feedback (GDF) controller

$$G(t) = \mu e^{i\chi_0} \bar{A}(t - \tau), \bar{A}(t) = \frac{1}{S}\int_{(S)} A(\mathbf{r}, t)d\mathbf{r}. \qquad (6.2.7)$$

Here, $\bar{A}(t)$ is the global average of local A at a delay time, $t - \tau$, when the DGF is added, S is the total area of the system, μ specifies the intensity of the DGF, χ_0 characterizes the phase shift between the delay average oscillation amplitude and the control single, and τ represents the delay time.

STC or turbulence in a 2-D CGLE can be controlled by this DGE method and by adjusting the feedback intensity as well as the delay time. Time delays here modify the phase shifts between the control signal and the oscillating pattern. By adjusting the delay, the synchronization window can always be reached at sufficiently high feedback intensities.

Controlling STC in Laser Systems via Tracking

In nonlinear laser systems, STC commonly refers to an optical turbulence when the chaos is fully developed in both space and time. The STC is attributed to the interaction, through diffraction of optical fields excited by local dipoles. A general 2-D mode description of nonlinear optics is [13]

$$\frac{\partial \mathbf{q}}{\partial t} = \mathbf{N}(\mathbf{q}, \mu) + iD\nabla_\perp^2 \mathbf{q} \qquad (6.2.8)$$

where $\mathbf{q}$ is a set of vector variables, $\mathbf{N}$ is a nonlinear function, t is time, $\nabla_\perp^2$ is the transverse Laplacian, μ is the control parameter of the system, and D is the matrix of diffractive coefficients.

Lu et al. [13] have shown that when weakly perturbed, unstable periodic solutions of a 2-D SES can be stabilized by local feedback without spatial coupling. Stable solutions can be driven into unstable and STC regimes by feedback controls, and the range of stable traveling waves in the system can be greatly extended through one such algorithm. Martin et al. [14] developed a Fourier space technique for stabilizing and tracking unstable patterns in a mean-field model for a two-level medium in an optical cavity. Montagne et al. [15] used an extension of a weak spatial periodic perturbation to mimic the target pattern. Petrov et al. [19] proposed a tracking method for a chemical system exhibiting STC in the Kuramoto-Sivashinsky equation with three stable and six unstable modes arising from the interaction between Turing and Hopf bifurcations.

In case of weak perturbations to periodic solutions, the above evolution equation is reduced to [13]

$$\frac{\partial \delta q_p}{\partial t} = i\omega \delta q_p - iD(\vec{k} + \vec{p})^2 \delta q_p + N'\delta q_p + g_p(t), \tag{6.2.9}$$

where g_p is the feedback depending on p. In the $\vec{r}$ space, the feedback is given by

$$G(\vec{r}, t) = e^{i(\vec{k} \cdot \vec{r} - \omega t)} \int_{-\infty}^{+\infty} f_p(t) e^{\vec{p} \cdot \vec{r}} d\vec{p}. \tag{6.2.10}$$

The feedback required for this case is local in the transverse space. When a state q is close enough to a targeted traveling wave solution of a distributed system, this state can be stabilized to the solution by the extended OGY method [2], occasional proportional feedback, and continuous delayed feedback (CDF) [16].

Controlling STC by Weak Spatial Perturbations

This method is a nonfeedback technique and similar to the approaches in controlling temporal chaos by modulating a system parameter or adding a weak periodic force to the system. This method has been applied to system (6.2.9), where the spatial perturbation is exerted by [15]

$$\mu = \mu_0[1 + \alpha f(x, y)], \tag{6.2.11}$$

in which μ_0 is the unperturbed control parameter, $\alpha < 1$ is the amplitude of the perturbation, and $f(x, y)$ is the spatial perturbation function. Here, $f(x, y)$ should be designed to reflect the signature of the target pattern. For a demonstration, they considered the mean-field model for a two-level atomic medium in an optical cavity, where the pump field E_I and the optical absorptivity C were chosen as control parameters. The perturbed pump field is

$$E_I = E_{I0}[1 + \alpha f(x, y)], \tag{6.2.12}$$

where E_{I0} is the unperturbed pump field. Using different forms of $f(x, y)$, chosen as that of the basic harmonics of the target pattern, they realized stabilization, selection and tracking unstable pattern, such as unstable rolls, squares, hexagons, and honeycombs. Instead of this control, the tasks can also be realized by applying a spatial perturbation to the phase of the pump field: $E_I = E_{I0}e^{i\alpha f(x,y)}$.

Synchronizing STC by a Single Scalar Driving Signal

As mentioned above, STC synchronization between a pair of one-way coupled PDEs was realized by Kocares et al. [11], where Gray-Scott equations (GSE-A) driving a similar GSE-B was discussed as a typical example. The GSE-B is described by

$$\begin{aligned} \frac{\partial v_1}{\partial t} &= -v_1 v_2^2 + a(1 - v_1) + D_1 \nabla_\perp^2 v_1, \\ \frac{\partial v_2}{\partial t} &= v_1 v_2^2 - (a+b)v_2 + D_2 \nabla_\perp^2 v_2 + G(x,t). \end{aligned} \quad (6.2.13)$$

The GSE-A has the same form without $G(x,t)$, and with v_i being replaced by u_i, $i = 1,2$. Let $T > 0$ and $X > 0$ be a real number, and $v_2(t-0)$ be the value of the signal v_2 immediately prior to the time t. The driving function $G(x,t)$ influences the response system in the following way: at each instant $t = kT$, $M = [L/X]$ space points $x = 0, X, 2X, \ldots, (M-1)X$ are simultaneously driven and their corresponding v_2 variables are set to new values $v_2(kT) = v_2(kT - 0) + \epsilon(v_2(kT) - v_2(kT - 0))$. Thus, T denotes the time distance between the occurrence of the driving impulses, and X is the space distance between the driven space points. When $X + T = 0$ and $\epsilon = 1$, this driving method becomes the Pecora-Carrol approach for synchronization of PDEs. Moreover, they found that there exists a *critical value* X_{cr} such that for all $X < X_{cr}$ both systems are synchronized. This remarkable result shows that synchronization of two infinite-dimensional STC systems can be achieved by a single scalar signal alone, at a finite number of space points. This approach can be useful for practical applications whenever one needs to synchronize SES. This method has been extended to coupled ODEs (CODEs) and CMLs. In addition, STC synchronization can be achieved using *random signals* [17].

Some Comments and Remarks

An examination of the control and synchronization of PDEs leads to the following comments: (a) the strategies and methods of control and synchronization for spatially extended systems are not unique: some are linear and some are nonlinear; some are local but spatially distributed but some are global or space averaged; some are model-independent but some are model specific. (b) In the regimes where nonlinearity of the vector flow is strong, nonlinear control methods must be used.

6.3 SESs Modeled by Coupled ODEs

While it is true that a complete description of STC systems and turbulence involve continuous-time and continuous-space models (i.e., PDEs), another amazing discovery is that some PED models are actually equivalent to a set of CODEs that correspond to a higher but finite dimensional system. Indeed, many nonlinear phenomena as well as STC can be modeled by an array of diffusively oscillators [1-4,12]. A real SES accurately described by a finite set of CODEs can provide insights about the nature of STC, which motivates the current study of CODEs.

Synchronization of STC

Synchronizing STC in CODEs has been extensively studied. To demonstrate STC synchronization of arrays, Kocarev et al. [11] have taken as a typical example an array of N diffusively coupled Lorenz systems (CLS-A):

$$\begin{aligned}
\dot{x}_{1i} &= \sigma(y_{1i} - x_{1i}) + D(x_{1i} - 2x_{1i} + x_{1i-1}), \\
\dot{y}_{1i} &= \rho x_{1i} - y_{1i} - x_i z_{1i}, \\
\dot{z}_{1i} &= x_{1i} y_{1i} - b z_{1i},
\end{aligned} \quad (6.3.14)$$

where $i = 1, \cdots, N$. Periodic boundary conditions, $x_0 = x_N$ and $x_{N+1} = x_1$, are used. The Lorenz system is in chaotic state when the value of the parameters are fixed to be ($\sigma = 10, \rho = 23, b = 1$).

The CLS exhibits STC with Lyapunov dimension $D_L = 69.3$ if $D = 6$ and $N = 100$. The array (6.3.14) drives a similar array, named CLS-B, of the same form as CLS-A, with x_{1i}, y_{1i} and z_{1i} being replaced by x_{2i}, y_{2i} and z_{2i}, respectively. The coupling between the two arrays are synchronized by a discrete time coupling of individual cells of the CLS [11]. In fact, either sporadic coupling in time or sporadic coupling in space works very well.

Many control methods are based on the properties of chaotic dynamics. One of the properties of oscillations generated by SESs is their ability to be synchronized [16, 11]. CLA-A and CAL-B can be synchronized even when they are coupled via another cell. This is due to the use of two basic properties [11]: (i) the dissipative character of the interaction between the cells in the arrays and (ii) the synchronization properties of the sporadically driven cells. It is very important that the knowledge about the local dynamics of each cell is used to synchronize systematically the behavior of the global dynamics of the two CODEs.

Synchronization in ALGC by Local Feedback Pinning

Yang et al. [26] presented a local feedback pinning method for synchronizing STC in CODEs. Very high controlling efficiency was found for asymmetric large gradient coupling (ALGC) CODEs with many positive Lyapunov exponents. They considered a class of discrete biased reaction-diffusion equations with N species, that is, N identical CODEs with feedback pinnings G_j, as follows:

$$\dot{u}_j = f(u_j) + (\epsilon+r)\Gamma(u_{j+1} - u_j) + (\epsilon-r)\Gamma(u_{j-1} - u_j) + G_j, \quad (6.3.15)$$

where $j = 1, \cdots, N$. The pinning has the form

$$G_j = c\Gamma[s(t) - u_j] \sum_{k=1}^{N/m} \delta_{j,mk},$$

where $u_j \in R^n$, f is a nonlinear function capable of exhibiting STC solutions, ϵ and r are scalar diffusive and gradient coupling parameters, respectively, Γ is an $N \times N$ constant matrix linking coupled variables, m is the distance between two neighbors pinnings, $s(t)$ is the synchronous state, c is the pinning strength, and $\delta_{j,mk} = 1$ for $j - mk = 0$ and $\delta_{j,mk} = 0$ otherwise. A large m corresponds to a low pinning density, and so represents a high control efficiency.

To study this approach, they took coupled Duffing oscillators as an example. It was shown that high control efficiency is achieved at strong gradient coupling. For SESs with strong gradient couplings the local injection approach may be very effective. This point is very useful in practical control problems. The same ideas apply to general CODEs.

Controlling Spiral Waves in 2D ACC by Feedback Pinning

Chua's circuit is a well-known chaotic system [4]. An array of Chua's circuits (ACC) can be used for studying control and synchronization of STC. A 2D model of ACC is described by [26]

$$\begin{aligned}
\dot{x}_{i,j} &= \alpha[y_{i,j} - x_{i,j} - g(x_{i,j})] \\
&\quad + D[x_{i+1,j} + x_{i-1,j} + x_{i,j+1} + x_{i,j-1} - x_{i,j}], \\
\dot{y}_{i,j} &= x_{i,j} - y_{i,j} - z_{i,j}, \\
\dot{z}_{i,j} &= -\beta y_{i,j}, \quad (i,j = 1, \cdots, N),
\end{aligned} \quad (6.3.16)$$

where

$$g(x) = (1/2)[(s_1+s_2)x + (s_0-s_1)(|x-B_1|-|B_1|) + (s_2-s_0)(|x-B_2|-|B_2|)].$$

Free boundary conditions are used. In a certain parameter region, the ACC has both synchronized oscillation and spiral wave attractor. Hu et al. [26] used feedback pinnings to migrate the system from the spiral wave state to the coherent oscillation. It was shown that some properly designed control schemes may reach very high control efficiency by injecting very few cells.

Controlling STC Using Optimal Disorders

It has recently demonstrated how disorder can organize STC. Lindner et al. [12] studied optimal disorders for controlling STC numerically. They considered a coupled, torqued, damped, nonlinear pendulum, which is governed by

$$ml_n^2 \ddot{\theta}_n + \gamma \dot{\theta}_n = -mg\sin\theta_n + \tau' + \tau \sin\omega t$$
$$+k(\theta_{n+1} - \theta_n) - k(\theta_{n-1} - \theta_n), \qquad (6.3.17)$$

where $n = 1, \cdots, N$, the boundaries are free, $\theta_0 = \theta_1$ and $\theta_n = \theta_{n+1}$. Here, γ represents the viscous damping, k is the coupling, τ, τ', ω parameterize the global external torque, and l_n are the (possibly different) pendulum lengths. The coupling is both local (nearest-neighbor) and linear.

This equation can also be used to model a parallel array of inductively coupled Josephson junctions. The evolution of the array in its $(2N+1)$-dimensional phase space can be characterized by $2N+1$ Lyapunov exponents. It was shown that the pendulum lengths can be in three different ways: random disorder, alternate binary disorder, and linear disorder. Numerical results showed that the largest Lyapunov exponent is negative for "optimal" disorder and the probability of periodic behavior is maximized where the average leading lyapunov exponent is minimized.

Nonlinear Feedback Control of Halo-Chaos in the Linacs

High current ion linacs have recently become a focus of investigation [6], due to possibly important applications (in ion beam drivers, production of tritium, heavy ion fusion, etc.).

In many cases involving high peak current, the distribution spins off a cluster of particles in the form called "halo" surrounding a dense core. Nonlinear resonances and chaotic behavior in the envelope oscillations of an intense beam propagating through a periodic focusing field are found [6]. The halo-chaos is essentially turbulent/STC motion in actual linacs. This is because a matched beam which enters a region of mismatch will undergo very complicated turbulent motion that apparently ejects particles from the core into some sort of halo. Many efforts to remove such halo-chaos by collimation have been unsuccessful since the halo is almost always regenerated.

Recently, Fang et al. [16] have considered to remove some sort of halo-chaos by applying nonlinear feedback control (NFC). The idea is that the linac can be regarded as an array of many periodic focusing channels (PFCs), similar to the CODEs, the halo-chaos would be suppressed by some effective NFC.

Consider a 2D breathing mode round beams in high current ion PFC [6], in which a dimensionless envelope equation of beam-core through the periodic solenoidal focusing field $B(r,s)$ is described for the Kapchinsky-Vladimirsky (KV) distribution by [6]

$$\frac{d^2 r_b}{ds^2} + k_z(s)r_b - \frac{K}{r_b} - \frac{1}{r_b^3} = 0, \qquad (6.3.18)$$

where r_b is the beam radius, $s = z = \beta_b ct = z/S$ is the axial coordinate, β_b is the average axial velocity of the beam particles, and c is the speed of light. The periodic function $k_z(s) = k_z(s+S) = q^2 B_z^2(s)/4\gamma_b^2 \beta_b^2 m^2 c^4$ characterizes the strength of the focusing field, where $B_z(s) = B_z(0)$ is the magnetic field on the z-axis, S is the periodicity length for a periodic solenoidal focusing channel, q and m are the particle charge and rest mass, respectively. The vacuum phase advance over one axial period of such a focusing field is given approximately by $\sigma_0 = [S \int_0^S k_z(s) ds]^{1/2} = [\eta^2 k_z(0)]^{1/2}$. The normalized beam perveance $K = 2q^2 N_b / \gamma_b^3 \beta_b^2 mc^2$ is a measure of the beam self-field intensity, where N_b is the number of particles per unit axial length of the beam, where $\gamma_b = (1 - \beta_b^2)^{-1/2}$ is the relativistic mass factor of the beam particles. The particle beam self-field is calculated from Poission's equation of potential $\phi(\mathbf{r}, \mathbf{z})$. The self-field force acted on a particle then is $F_r = -q \nabla \varphi(\mathbf{r}, \mathbf{z})$. The radial space-charge field of an axisymmetric beam can be calculated from Gauss law by counting the number of particles in cells of a finite radial grid which extends up to 5 times of the beam matched radius in multiparticle simulation using the 2D Particle in Cells (PIC) method.

In order to prevent activation of the beam pipe walls and components of a high-power accelerator, beam loss must be minimized. To do so, we have applied a feedback controller, G, to the self-field force, that is,

$$F_{\mathbf{r}} = -q \nabla \varphi(\mathbf{r}, z) + G,$$

where G is the NFC in the form of

$$G = -k \sin(r_{max} - a_m),$$
$$G = -k (r_{max} - a_m)^2, \quad etc.,$$

where a_m is the match radius of the beam envelope. The gain is $k = -(0.05 - 0.5))$ and the r_{max} is the maximum radius over all particles.

As a control measure of halo-chaos, a halo intensity factor, h, is defined as the number of particles outside the boundary $r_b = 1.75 r_b(0)$ divided by all particles participated in the simulation. The smaller h, the better control performance. The primary results show that the above NFC works well for suppressing (or at least reducing) halo-STC since h is much smaller under control ($h = 0.05 - 0.15$ at 1200 PFC); otherwise $h > 0.6$. For example, we consider a mismatched beam with Gaussian distribution and mismatch parameter μ is 1.5, the vacuum phase advance with space charge $\sigma_0 = 105^0$ and the tune depression $\eta = 0.8$. A good control of halo-STC with $h = 0.1078$ and $r_{max}/a_m = 2.912$ is achieved if $G = -0.15(r_{max}-a_m)^2$ is used for the above beam as sporadically feedback in space, each 5-pulse PFC.

Recently, Barnard et al. [6] have performed 3D PIC simulations on beams with longitudinal mismatches. The suppression of the envelope-particle resonance is achieved by the sinusoidal rf field under certain conditions. For example, mismatch amplitudes below 0.5 have been observed through the rf bucket. But this suppression seems to be impractical since it takes place only at very low or very high currents. Moreover, even when the parametric resonance is suppressed, there are still other growing resonance and emittance.

Coupled Delay-Differential Equations

CODE also includes a class of SESs modeled by coupled delay-differential equations (CDDEs). The general form of a CDDE is described by

$$\dot{x}_i(t) = \alpha\big[f_i(x_1(t-\tau),\ldots,x_N(t-\tau))\big] + G_i,$$

where G_i is a feedback function, α is a positive parameter, f_i's are nonlinear functions, τ is the time delay, and $T = S\tau$ is the period. It has been shown that this network of CDDEs can store and retrieve a total number of $M = Q \times S$ spatial patterns, where Q is the number of STC patterns. This network provides a considerable gain in the storage capacity over traditional neural networks. Time-delay control of Mackey-Glass CDDEs and synchronization of two Mackey-Glass equations have both been achieved [20].

6.4 SESs Modeled by Coupled Map Lattices

In general, PDEs and CODEs give accurate descriptions of SESs. However, nonlinear PDEs and CODEs are difficult to solve and hence are not tractable. Yet many gross macroscopic behaviors of SESs can be generated by the discrete-time (n) and discrete-space (i) dynamics of properly designed CMLs, which are much simpler to analyze. Therefore, and because CMLs provide course-grained approximations to the continuous fields of SESs, they have become valuable tools in the study of control and synchronization of SESs [2, 10].

Typical Forms of CMLs

A variety of special forms of CMLs have been studied, such as:

1. One-way coupled map (OCML) [10]

$$\mathbf{X}_{n+1}(i) = (1 - \epsilon)\,\mathbf{F}(\mathbf{X}_n(i)) + \epsilon\,\mathbf{F}(\mathbf{X}_n(i-1)), \quad i = 1, \cdots, L$$

2. Diffusively coupled map (DCML) [10]

$$\mathbf{X}_{n+1}(i) = (1 - \epsilon)\,\mathbf{F}(\mathbf{X}_n(i)) + \frac{\epsilon}{2}\,[\mathbf{F}(\mathbf{X}_n(i+1)) + \mathbf{F}(\mathbf{X}_n(i-1))]$$

3. Globally coupled map (GCML) [10]

$$\mathbf{X}_{n+1}(i) = (1 - \epsilon)\,\mathbf{F}(\mathbf{X}_n(i)) + \frac{\epsilon}{L}\sum_{j=1}^{L}\mathbf{F}(\mathbf{X}_n(j))$$

4. Symmetrically coupled map (SCML) [26]

$$\mathbf{X}_{n+1}(i) = (1 - \epsilon)\,\mathbf{F}(\mathbf{X}_n(i)) + \frac{\epsilon}{2L}\sum_{j=1}^{L}[\mathbf{F}(\mathbf{X}_n(i-j)) + \mathbf{F}(\mathbf{X}_n(i+j))]$$

5. Asymmetrically coupled map (ACML) [17]

$$\mathbf{X}_{n+1}(i) = (1 - \epsilon_1 - \epsilon_2)\mathbf{F}(\mathbf{X}_n(i)) + \epsilon_1 \mathbf{F}(\mathbf{X}_n(i-1)) + \epsilon_2 \mathbf{F}(X_n(i+1))$$

6. Randomly coupled map with random neighbors and constant coupling coefficients (CMLR) [16]

$$\mathbf{X_{n+1}(i)} = \gamma\,(1 - \epsilon)\mathbf{F}(\mathbf{X}_n(i)) + \frac{\epsilon}{\nu}\sum_{j=1, j\neq i}^{\nu}\mathbf{F}(\mathbf{X}(m_{ij}))$$

7. Randomly coupled maps with random neighbors and random coupling coefficients (CMLRR) [16]

$$\mathbf{X_{n+1}(i)} = \gamma\,(1 - \epsilon)\mathbf{F}(\mathbf{X}_n(i)) + \frac{1}{\nu}\sum_{j=1, j\neq i}^{\nu}\alpha_{ij}\mathbf{F}(\mathbf{X}(m_{ij}))$$

The current understanding of STC has come, to a great extent, from the studies of CMLs. The GCMLs have been used in diverse areas, such as neurodynamics, fluid dynamics, charge density waves, Josephson junction arrays, optics, and economics [22, 10]. Studies of GCMLs with the logistic map for the local dynamics have revealed that they can manifest features including coherent, clustering, partially ordered and turbulent phases. Novel spatiotemporal features have been observed in a hierarchy of GCMLs.

CMLs can be controlled and synchronized by a large number of ways including the common methods applied to PDEs and CODEs, in which linear and nonlinear feedback controls are often used.

Linear Feedback Control

Most control methods are based on linearized models and linear feedback since linear control is easy to realize in engineering and experiment. A shortcoming of employing tracking error in linear feedback is the use of chaotic or even unstable reference signals, which are difficult to be physically used as input. This, however, may be overcame by using a time-delay or self-turning feedback approach [3].

Local feedback control

Local feedback control usually takes the form of

$$G_n(i) = \gamma(X_n(i) - \frac{1}{N}\sum_{i=1}^{N} X_{n-1}(i)),$$

or

$$G_n(i) = \gamma(X_n(i) - X_F),$$

where X_F is the target fixed point and γ is the gain to be designed. This control is able to stabilize the periodic clustered state for ACMLs, such as stabilized the system to a period-four target [17].

Global feedback control with uniform feedback

Based on the condition for linearized stability in the CMLs with the logistic map, a relation between the minimum coupling and the distance between controllers and parameters of the local chaotic maps are obtained in [7, 21]. The maximal distance between the pinnings depends on the strength of noise in the system, and can be estimated analytically. Moreover, the introduction of nonlinearity into the control scheme can decrease the control time by enlarging the capture region. Typical forms of global feedback control include:

(i) $G_n(i) = -\frac{\gamma}{N}[\sum_{i=1}^{N} X_{n-1}(i) - \sum_{i=1,N} X_n(i)]$.

Using this feedback control, different clustered periodic state can be stabilized but with a nonvanished control signal for different values of γ.

(ii) $G_n(i) = -\gamma[\frac{1}{N}\sum_{i=1}^{N} X_{n-1}(i) - X_F]$.

where X_F is the target fixed point. This control can stabilize homogeneous states of the CMLs. Moreover, suppression of STC was achieved via stabilization of clustered states with a nonvanishing control signal.

(iii) $G_n(i) = -\gamma \frac{1}{N} \sum_{i=1}^{N} X_{n-1}(i)$.

This is a global delayed feedback control which can control turbulence in CGLEs and CMLs.

(iv) $G_n(i) = -\gamma [\frac{1}{N} \sum_{i=1} N X_{n-1}(i) - \frac{1}{N} \sum_{i=1} N X_n(i)]$.

The main advantage of global feedback control is the ease of control values acquired from experiments.

With nonuniform feedback pinning

$$G_n(i) = \delta(i - i_p)p, \quad i = 1, \cdots, N,$$

where i_p is the inverse of the pinning density, p_d: if $\delta(i - i_p) = 1$, the ith site is to be pinned and takes a finite value p, or else $p_n(i) = 0$. Either regularly spaced pinned sites or random distribution of pinned sites can be used [7].

It is possible to suppress STC in CMLs models both locally and globally by applying linear feedback even using constant pinning signals. It is noted that spatially localized superposition of STC without disturbing the rest of the lattice can be effectively achieved to the desired dynamics by choosing appropriate pinning strengths. This feature can have important applications to a variety of physical and biological systems.

Nonlinear Feedback Control

Linear controls are local and may not be suitable for controlling states with strong nonlinearity and being far away from the target. These situations lead to the search for nonlinear controls. There are also cases in which a combination of linear and nonlinear feedbacks is desirable.

CMLs can be controlled and synchronized by a large variety of NFC functions [16]. For examples, the CMLR and CMLRR models were used with the logistic map describing the local dynamics. The NFC function was applied at every site as well as sporadically in space. The robustness of the control against noise was observed.

Some smooth NFC functions for CMLR and CMLRR are:

$$G_n(i) = -(X_{dn}(i) - X_n(i))(2X_{dn}(i) - 1),$$

$$G_n(i) = -\frac{1}{2}(X_{dn}(i) - X_n(i))(kX_n(i)^2 - 1), \quad k = 3, 4,$$
$$G_n(i) = -(X_{dn}(i) - X_n(i))tanh[(2X_n(i) - 1)],$$
$$G_n(i) = -(2X_{dn}(i) - 1)sin(X_{dn}(i) - X_n(i)), \quad etc.,$$

where $X_{dn}(i)$ is a desired target, which can be an unstable periodic orbit (UPO) or an unstable periodic pattern (UPP) in the CML. These NFCs are suitable for controlling and synchronizing UPO/UPP of the CMLs [16].

Control and Synchronization of STC in Asymmetrical and One-way CMLs

Kaneko et al. [10] provided details of various phases that can exist in the dynamics of an OCML. It has been shown [26] that chaos synchronization can be achieved by feeding only the most upstream sites of the two OCMLs using a common chaotic signal K_n ($x_n(0) = y_n(0) = K_n$). The chaotic signal can be taken from a third identical OCML with periodic boundary condition. Xiao et al. [26] suggested that this may serve as a tool for one-key-many-channels secure communication. Shuai [22] showed that an OCML ring can be synchronized even when the conditional Lyapunov exponents are positive. For example, they reported a case in which two OCML synchronized in the presence of 75 positive Lyapunov exponents. Recently, Jiang and Parmananda [17] studied the synchronization of STC in ACMLs. An ACML can reduce to an OCML if the second coupling $\epsilon_2 = 0$. They showed that ACMLs, like the OCMLs, can be synchronized by linking just one site. It is also reported [17] that the difference between ACMLs and OCMLs is that different synchronization schemes may give rise to different dynamical behaviors of the ACMLs. Like the OCMLs, ACMLs can have temporally chaotic and spatially periodic phases. These results in the CMLs may have great potential in applications. Generalized and phase synchronization, and some other types of synchronization in the CMLs, are also being investigated.

6.5 Some Characteristic Quantitates and Mechanisms

Control and synchronization have a common basis but different reference states (goals), which can be described under a unified framework [4, 3, 16]. Some characteristic quantitates and mechanisms of control and synchronization of STC are discussed in this section.

Lyapunov Exponent and KS Metric Entropy

For temporal chaos, the largest Lyapunov exponent, λ, is a very important characteristic quantity for distinguishing different kinds of behaviours in a nonlinear dynamical system: if $\lambda > 0$, it is chaotic; if $\lambda < 0$, it is in a periodic state. What about STC? The answer is that they are very similar to the temporal case but still need to be further investigated.

One mechanism of controlling STC is to force λ to change from being positive to negative, or to ensure all conditional Lyapunov exponent (CLE) become negative. STC is sensitive not only to initial perturbation in time but also to initial perturbation in space. For example, using linearized stability, it was shown [11] that it is possible to have a stable synchronized STC solution if the coupling constant is greater than the largest Lyapunov exponent of the system. Kocarev et al. [11] agree with a suggestion of Gauthier et al. that a proper criterion for high-quality synchronized motion is the negativity of the CLEs for all the UPOs embedded in the STC attractor. Therefore, one mechanism of control and synchronization of STC is to force the largest Lyapunov exponent of the system to change from positive to negative, or the CLE all become negative.

The other characteristic quantity is the Komogrov-Sinai (KS) metric entropy, since there is a tight relation between the Lyapunov exponent and the KS metric entropy. In addition, power spectra both in time and space, (mutual) information entropy density can also characterize STC. Analysis and calculation of these quantities will be useful for STC control and synchronization. Since STC is more complex than temporal chaos, these quantities should be extended to better characterize STC.

Correlation Functions in Time and in Space

Another important characteristic quantity is the correlation function (CF), which is useful for understanding the mechanism of control and synchronization of STC. The CF bears the "memory" along a orbit in both space and time, which decays exponentially to zero in the chaotic regime. STC can be characterized by the exponential decays of CF in space and time.

A typical example is the control of UPO/UPP in the following PDE:

$$\partial E/\partial t = \sigma E + gP_2 + ia \nabla_\perp^2 E + G(x,y,z) \qquad (6.5.19)$$

where E is a varying electric field of laser emission and $G(x,y,z)$ is a feedback controller. The goal is to roll the pattern embedded in the turbulent state of the laser system. Using the real part of the laser field as the control variable and applying a delay feedback, one has

$$G(x,y,t) = c_1[E(x,y,t-t_0) - E(x,y,t)] + c_2[E(x+x_0,y,t) - E(x,y,t)]$$
$$+ c_3[E(x,y+y_0,t) - E(x,y,t)],$$

where c_1 and c_2 are constant coefficients, t_0 is the period in time of the desired roll state in local regions, and $x_0 = 2\pi/k_x$ and $y_0 = 2\pi/k_y$ are the characteristic lengths of the rolls in x and y directions, with k_x and k_y being the wave vector components.

The feedback signal is imposed on the system continuously in time, controlling the time evolution of the UPP in local regions through a time-delayed coupling and organizing the spatial distribution in the same time through a spatial network coupling. The orientation of the roll pattern is controlled by choosing k_x and k_y (or x_0 and y_0). The feedback $G(x, y, t)$ tends to vanish when the STC is synchronized to this UPP. The mechanism of UPP control can be explained by the CF which is defined as

$$\zeta(\rho, t) = <E_r(\mathbf{r} + \rho X, t)>/(<E_r(\mathbf{r}, t)><E_r(\mathbf{r}, t)>), \quad (6.5.20)$$

where $<\cdots>$ is taken over all possible spatial positions. Initially, $t = 0$, and $\zeta(\rho, t)$ shows a typical STC behavior. A sharply exponential decrease of the CF implies an increase of the distance ρ. The time scale of this temporal transition to the targeted UPO is found to be the same as that of the spatial transition to the transverse periodic pattern, indicative of a simulation organization of the system in space and time.

It is interesting that Tziperman et al.(1997) [9] have recently used the careful identification of the spatial CF and have successfully applied the STC control method to a complex PDE *El Niño* model for predicting *El Niño* events in the equatorial Pacific.

The generality of time-delay feedback control method and its robustness to the control parameters has been tested for different turbulent patterns of the system.

6.6 New Research Outlook

Control and synchronization of STC in SESs remains a great challenge for further research.

First, what kinds of mechanisms of control and synchronization of STC in SESs are needed to develop? What are the most important properties of STC that should be handled in control and synchronization? How can STC be characterized completely for these tasks? The current methods for characterizing STC are quite inefficient. Both transitions, from STC to order and from order to STC, are equally important. It has been demonstrated that the concepts of scaling, renormalization, universality, symmetry-breaking bifurcation, etc., in nonlinear dynamics are significant.

Second, how to classify difference kinds of synchronization for STC? What kinds of temporal chaos synchronization can be extended to STC?

For temporal chaos, identical synchronization, generalized synchronization, intermittent synchronization, and lag synchronization of chaos oscillators have been investigated. Can these synchronizations of STC be found in SESs? Are there any new phenomena as yet to be discovered in the control and synchronization of STC? Can these phenomena be confirmed experimentally? Many questions like these can be posed for typical models of PDEs, CODEs, CMLs, etc. [11].

Third, what kinds of optimal STC control and synchronization methods can be developed and applied to a variety of experiments and engineering problems? How to find more efficient strategies and methods, such as nonlinear feedback controls, is one of the key issues. Combination of STC with classical control theory should be a promising direction for further research.

Fourth, one tendency towards large SESs is controlling system complexity and various types of bifurcations [16, 3]. They arise in diverse fields such as biology, chemistry, physics (e.g., fluidized beds), nonlinear optical system, neural networks, and so on. Controlling halo-chaos may be a new type of STC in actual high current ion linacs.

Fifth, anticontrol of chaos has recently received considerable attention both in experiment and in theory [19, 9, 3]. Can these methods be extended to STC?

Sixth, what are prospective applications of STC control and synchronization in SESs? What are the advantages of using STC over temporal chaos? This has been an important topic, specially in secure communication and information storage. Current schemes are worthwhile to be tested in engineering design, yet new techniques should be developed in the future. There are conjectures that STC plays an important role in neural networks and some progress has been made in support of this anticipation.

Seventh, current characterizations of chaos based on system trajectories are unsuitable for semiclassical or quantum systems. It will be remarkable if control and synchronization of temporal chaos and STC can be put on a basis common to classical, semiclassical and quantum systems. One attempt is proposed by Harel and Akulin [9]; they studied a control switching method for Hamiltonian systems and suggested an experimental realization of this method for controlling the translational motion of cold atoms.

6.7 Summary

We have presented an overview of current research activities and directions for further research in the control and synchronization of STC in SESs modeled by PDEs, CODEs and CMLs. The diversity in nonlinear phenomena and in the techniques for controlling and synchronizing STC has been highlighted. Illustrations have been given to demonstrate the basic ideas involved in most of the current strategies and methods. We have also emphasized that nonlinear feedback control from engineering is particularly attractive for tackling STC in SESs.

Beyond the methodology issue, it is most interesting to see how the current findings can be put to use in engineering applications. One area to which such findings are directly relevant to is multichannel secure communication and secure spatial information storage.

Acknowledgments

J.Q.F. is supported by CNNSF and CNNICF, and M.K.A by NSERCC. The authors are grateful to the suggestions and information provided by the Editor G. Chen, and by L. Kocarev, U. Parlitz, V. Petrov, J. F. Lindner, W. Lu, S. Sinha, R. L. Gluckstern, and J. J. Barnard.

References

[1] A. Amenggual, E. Hernandez-Garcia and M. S. Miguel, "Synchronization of spatiotemporal chaos: the regime of coupled spatiotemporal intermittency," *Phys. Rev. Lett.*, vol. 78, pp. 4379-4382, 1997.

[2] D. Auerbach, "Controlling extended systems of chaotic elements," *Phys. Rev. Lett.*, vol. 72, pp. 1184-1187, 1994.

[3] G. Chen and X. Dong, *From Chaos to Order: Methodologies, Perspectives and Applications*, World Scientific Pub. Co., Singapore, 1998.

[4] L. O. Chua (Ed.), Special Issue on Nonlinear Waves, Patterns, and Spatiotemporal Chaos in Dynamic Arrays, *IEEE Trans. on Circ. Sys.*, I, vol. 42, No. 10, pp. 557-820, 1995.

[5] J.-Q. Fang, "Control and synchronization of chaos in nonlinear systems – Methods, principles and applications," *Progress of Phys. in China*, vol. 16, pp. 1-74, pp. 137-202, 1996; J.-Q. Fang and M. K. Ali, "Nonlinear feedback control of spatiotemporal chaos in coupled map lattices," *Disc. Dyn. in Natu. and Soci.*, vol. 1, pp. 283-305, 1998; J.-Q. Fang and M. K. Ali, "Controlling spatiotemoral chaos using nonlinear feedback functional method," *Chi. Phys. Lett.*, vol. 14, pp. 823-826, 1997; J.-Q. Fang, Y. G. Hong, H. S. Qin and G. Chen, "A switching manifold approach to chaos synchronization," *Phys. Rev. E.*, 1999, in press.

[6] R. L. Gluckstrn, "Analytic model for halo formation in high current ion linacs," *Phys. Rev. Lett.*, vol. 73, pp. 1247-1250, 1994; R. L. Gluckstrn, A. V. Fedotov, S. S. Kurennoy and R. D. Ryne, "Halo formation in three-dimensional bunches," *Phys. Rev. E.*, vol. 58, pp. 4977-4990, 1998; J. J. Barnard, S. M. Lund, and R. D. Ryne, "Self-consistent 3D simulations of longitudinal halo in rf-LINACS," *LINAC'98 Conference*, Chicago, IL, 1998.

[7] R. O. Grigoriev, M. C. Cross and H. G. Schuster, "Pinning control of spatiotemporal chaos," *Phys. Rev. Lett.*, vol. 79, pp. 2795-2798, 1997; N. Parekh, S. Parthasarathy and S. Sinha, "Global and local control of spatiotemporal chaos in coupled map lattices," *Phys. Rev. Lett.*, vol. 81, pp. 1401-1404, 1998.

[8] G. Hu, J. H. Xiao, L.O.Chua and L. Pivka, "Controlling spiral waves in a model of two-dimensional arrays of Chua's circuits," *Phys. Rev. Lett.*, vol.80, pp. 1884-1887, 1998; J. H. Xiao, G.Hu and Z. L. Qu, "Synchronization of spatiotemporal chaos and its application to multichannel spread-spectrum communication," *Phys. Rev. Lett.*, vol. 77, pp. 4162-4165, 1996; J. Z. Yang, G. Hu and J. G. Xiao, "Chaos synchronization in coupled chaotic oscillators with multiple positive Lyapunov exponents," *Phys. Rev. Lett.*, vol. 80, pp. 496-499, 1998; G. Hu, J. H. Xiao, J. Z. Yang, F. Xie and Z. Qu, "Synchronization of spatiotemporal chaos and its applications," *Phys. Rev. E*, vol. 56, pp. 2738-2746, 1997.

[9] V. In, M. L. Spano and M. Ding, "Maintaining chaos in high dimensions," *Phys. Rev. Lett.*, vol. 80, pp. 700-703, 1998; E. Tziperman and H. Scher, "Controlling spatiotemporal chaos in a realistic EI Nino prediction model," *Phys. Rev. Lett.*, vol. 79, pp. 1034-1037, 1997; G. Harel and V. M. Akulin, "Complete control of Hamiltonian quantum systems: Engineering of Floquet evolution," *Phys. Rev. Lett.*, vol. 82, pp. 1-4, 1999.

[10] K. Kaneko (Ed.), *Coupled Map Lattices*, World Scientific Pub. Co., Singapore, 1992; F. S. de San Roman, S. Boccaletti, D. Maza, and H. Mancini, "Weak synchronization of chaotic coupled map lattices," *Phys. Rev. Lett.*, vol. 81, pp. 3639-3642, 1998.

[11] L. Kocarev, Z. Tasev and U. Parlitz, "Synchronizing spatiotemporal chaos of partial differential equations," *Phys. Rev. Lett.*, vol. 79, pp. 51-54, 1997; "Synchronizing spatiotemporal chaos in coupled nonlinear oscillators," *Phys. Rev. Lett.*, vol. 77, pp. 2206-2209, 1996; *Phys. Rev. Lett.*, vol. 76, pp. 1816-1819, 1996; L. Kocarev, S. V. Kiril and U. Metodij, "Controlling spatiotemporal chaos in coupled nonlinear oscillators," *Phys. Rev. E*, vol. 56, pp. 1238-1241, 1997; L. Kocarev, Z. Tasev, T. Stojanovski and U. Parlitz, "Synchronizing spatiotemporal chaos," *Chaos*, vol. 7, pp. 635-643, 1997.

[12] J. F. Lindner, B. S. Prusha and K. E. Clay, "Optimal disorders for taming spatiotemporal chaos," *Phys. Lett. A*, vol. 231, pp. 164-172, 1997.

[13] W. Lu, D. Yu and R. G. Harrison, "Control of patterns in spatiotemporal chaos in optics," *Phys. Rev. Lett.*, vol. 76, pp. 3316-3319, 1996; Abid., "Tracking periodic patterns into spatiotemporal chaotic regimes," *Phys. Rev. Lett.*, vol. 78, pp. 4375-4378, 1997.

[14] R. Martin, A. J. Scroggie, G.-L. Oppo, and W. J. Firth, "Stabilization, selection, and tracking of unstable patterns by Fourier space techniques," *Phys. Rev. Lett.*, vol. 77, pp. 4007-4010, 1996.

[15] R. Montagne and P. Colet, "Nonlinear diffusion control of spatiotemporal chaos in the complex Ginzburg-Landar equation," *Phys. Rev. E*, vol. 56, pp. 4017-4023, 1997; P. Y. Wang, P. Xie, J. H. Dai and H. J. Zhang, "Stabilization, selection, and tracking of unstable patterns by weak spatial perturbations," *Phys. Rev. Lett.*, vol. 80, pp. 4669-4672, 1998.

[16] E. Ott, C. Grebogi and J. Yorke, "Controlling chaos," *Phys. Rev. Lett.*, vol. 64, pp. 1196-1199, 1990.

[17] P. Parmananda, M. Hildebrand and M. Eiswirth, "Controlling turbulence in coupled map lattice systems using feedback techniques," *Phys. Rev. E*, vol. 56, pp. 239-244, 1997; P. Parmananda, "Synchronization of spatiotemporal chemical chaos," *Phys. Rev. E*, vol. 56, pp. 1595-1598,1997; Y. Jiang and P. Parmananda, "Synchronization of spatiotemporal chaos in asymmetrically coupled map lattices," *Phys. Rev. E*, vol. 57, pp. 4135-4138, 1998; P. Parmananda and Y. Jiang, "Synchronization of spatiotemporal chemical chaos using random signals," *Phys. Lett. A*, vol. 241, pp. 173-178, 1998.

[18] L. M. Pecora and T. L. Carroll, "Synchronizing chaotic systems," *Phys. Rev. Lett.*, vol. 64, pp. 821-824, 1990.

[19] V. Petrov and K. Showalter, "Nonlinear control of dynamical systems from time serties," *Phys. Rev. Lett.*, vol. 76, pp. 3312-3315, 1996; V. Petrov, et al., "Nonlinear control of remote unstable states in a liquid bridge convection experiment," *Phys. Rev. Lett.*, vol. 77, pp. 3779-3782, 1996; V. Petrov, et al., "Mode-independent nonlinear control algorithm with application to a liquid bridge experiment," *Phys. Rev. E*, vol. 58, pp. 427–433, 1998.

[20] K. Pyragas, "Synchronization of coupled time-delay systems: Analytical estimation," *Phys. Rev. E*, vol. 58, pp. 3067-3071, 1998; V. Ahlers, U. Parlize, and W. Lauterborn, "Hyperchaotic dynamics and synchronization of external-cavity semiconductor," *Phys. Rev. E*, vol. 58, pp. 7208-7213, 1998.

[21] H. G. Schuster and M. B. Stemmler, "Control of chaos by osillating feedback," *Phys. Rev. E*, vol. 56, pp. 6410-6417, 1997; D. Battogtokh, A. Preusser, A. Mikhailov, "Controlling turbulence in the complex Ginzburg-Landau equation II. Two-dimensional systems," *Physica D*, vol. 106, pp. 327-362, 1997.

[22] J. W. Shuai, K. W. Wong and L. M. Cheng, "Synchronization of spatiotemporal chaos with positive conditional Lyapunov exponents," *Phys. Rev. E*, vol. 56, pp. 2272-2275, 1997.

7

Chaotic Vibration of the Wave Equation by Nonlinear Feedback Boundary Control

Goong Chen[1], Sze-Bi Hsu[2] and Jianxin Zhou[1]

[1]Department of Mathematics
Texas A&M University
College Station, TX, 77843-3368 USA

[2]Department of Mathematics
National Tsing Hua University
Hsinchu, Taiwan, Republic of China

Abstract

Spatiotemporal chaos or turbulence in partial differential equations is a vastly open research field. In this paper, we show that imbalance of boundary energy flow due to certain types of nonlinear feedback boundary control can cause chaotic vibration of the one-dimensional wave equation. We first show that if there is a linear amplifier with feedback gain η at the left end point, and if there is cubic velocity feedback damping at the right end point, then spatiotemporal chaos of the gradient (w_x, w_t) will occur as the gain parameter η is tuned. We then show through numerical simulations of concrete examples what may happen if a more general polynomial feedback is applied at the right end point. Chaotic profiles of wave motion are illustrated by computer graphics.

7.1 Introduction

Advances in dynamical systems, particularly in the theory of chaos, are widely regarded as one of the foremost scientific achievements of the late 20th century. During the past ten years, control theorists have attempted to incorporate brand new techniques in that area in order to understand and utilize nonlinearities in systems. A particularly remarkable new development is *anti-control* – how to create, maintain or enhance chaos when it is healthy and useful [7, p. 4]. Examples of such include fluid mixing, chemical reactions, the biological systems of human brain, heart and perceptual processes, secure communications ([7]), etc. So far, great success has been achieved in the study of *lumped* parameter systems; see [8], for example.

For *distributed* parameter systems, remarkable progress has been made during the past three decades in the *boundary* controllability and stabilizability of linear partial differential equations due to the pioneering work of Russell [17]. Boundary feedback control achieves the control effect on the system by propagating the controller action from the boundary into the interior of the domain. This is quite advantageous from the practical design point of view, because the boundary is much more accessible than the interior of the domain. (Obviously, internally distributed actuators and sensors are also feasible and have already been in use, but their fabrication is more difficult.) A large assortment of highly advanced modern mathematical methods and tools was developed ([1, 14, 15]) to handle the mathematical intricacies. Even though control theorists have also made certain success in treating nonlinear distributed parameter systems, much of the existing work becomes inadequate when the nonlinearity in the system is "severe" with a chaotic regime. Such "genuinely nonlinear" distributed parameter systems display the behavior/phenomena/properties such as multiple unstable solutions, bifurcations, fractals, hysteresis, strange attractors and randomness which can only be approached by the contemporary study in dynamical systems and chaos.

Researchers began to examine chaotic behavior in distributed parameter vibration systems in [13, 18], for example. In [13], Holmes and Marsden derived a Smale horseshoe from the motion of a nonlinear beam. In [18], Sharkovsky considered some chaotic motion of hyperbolic equations. However, no feedback control was mentioned in those works. The efforts made by this group of authors is to study the chaotic behavior of the one-dimensional (1D) linear wave equation when nonlinear feedback is applied at a boundary point. The propagation of acoustic waves in a pipe, and the vibration of a string or a rod, satisfy the linear wave equation

$$w_{xx}(x,t) - w_{tt}(x,t) = 0, \qquad 0 < x < L, t > 0, \tag{7.1.1}$$

where the subscripts xx of w_{xx}, e.g., denote twice partial differentiations with respect to the x variable. Let the initial conditions (ICs) be

$$w(x,0) = w_0(x), \quad w_t(x,0) = w_1(x), \quad 0 < x < L. \tag{7.1.2}$$

Let the boundary conditions (BCs) be, respectively,

$$w_t(0,t) = -\eta w_x(0,t), \quad t > 0, \tag{7.1.3}$$

$$w_x(L,t) = \alpha w_t(L,t) - \beta w_t^3(L,t), \quad t > 0, \tag{7.1.4}$$

at the left end $x = 0$ and the right end $x = L$, where α, β and η are positive constants lying in certain parameter ranges. The BC (7.1.4) is associated with the name of van der Pol in that it has the effect of self-regulation or self-excitation [19] and is therefore extremely useful in the design of servomechanisms in automatic control. The cubic nonlinear relation in (7.1.4) can be realized by using tunnel diodes ([2, Appendix C]), for example. The left BC (7.1.3) is linear signifying the presence of an *amplifier* (such as a microphone in acoustics); it injects energy into the system, where the *energy* of wave motion at time t is defined by

$$E(t) = \frac{1}{2}\int_0^L [w_x^2(x,t) + w_t^2(x,t)]dx. \tag{7.1.5}$$

Note that the parameter η in (7.1.3) may be considered as the *feedback gain*; this gain will be the varying parameter in most of discussions to follow. The above statements will be clear if we look at the rate of change of energy of the system:

$$\frac{d}{dt}E(t) = \int_0^L [w_{xt}w_x + w_{tt}w_t]dx$$

$$= \cdots \text{(integration by parts and utilization of (7.1.3) and (7.1.4))}$$

$$= \eta w_x^2(0,t) + w_t^2(L,t)[\alpha - \beta w_t^2(L,t)], \tag{7.1.6}$$

where

(i) $\eta w_x^2(0,t) \geq 0$, meaning that energy is injected into the system from the left endpoint;

(ii)
$$w_t^2(L,t)[\alpha - \beta w_t^2(L,t)] \begin{cases} \geq 0, & \text{if } |w_t(L,t)| \leq (\alpha/\beta)^{1/2}, \\ < 0, & \text{if } |w_t(L,t)| > (\alpha/\beta)^{1/2}. \end{cases}$$

signifying a self-regulation effect.

The *imbalance of boundary energy flow* is evident in (7.1.6), as we see that the effect of the left end BC is trying to increase the total energy of the system, while that of the right end BC is to modulate the total energy change. In [3, 4], we have shown that when α, β and η enter a certain regime, chaotic vibration of the gradient (w_x, w_t) occurs. Note that in (7.1.3), *force feedback* (i.e., the $-\eta w_x(0,t)$ term) is used, while in (7.1.4), *nonlinear velocity feedback* (i.e., $\alpha w_t(L,t) - \beta w_t^3(L,t)$ terms) is used. Also, we mention that in [3, (93)] (which is a differentiated form of (7.1.4)) a combination of *nonlinear displacement and velocity feedback* is used.

In this paper, we consider chaotic vibrations generated by *polynomial velocity feedback* at the right end boundary point. The organization of the paper is as follows:

(1) In Section 7.2, we prove that when the van der Pol BC (7.1.4) is replaced by one with *cubic damping*, chaotic vibration still happens when the parameters enter a certain range. This shows that imbalance of boundary energy flow is a major cause of chaos, and the nonlinearity does not have to be van der Pol in order for chaos to occur.

(2) In Section III, we show through numerical simulations and simple arguments certain new features of nonlinear phenomena when higher order polynomial velocity feedback is used at the right endpoint.

The study in (1) and (2) above leads to many pertinent questions whose detailed treatment must be deferred to a sequel where more space is available.

7.2 Chaos Induced by Interaction of Energy Injection at the Left End and Cubic Damping at the Right End

Let us clearly state the problem under consideration:

$$w_{xx}(x,t) - w_{tt}(x,t) = 0, \quad 0 < x < 1, t > 0, \quad (7.2.1)$$

$$w_t(0,t) = -\eta w_x(0,t), \quad t > 0; \eta > 0, \eta \neq 1, \quad (7.2.2)$$

$$w_x(1,t) = -[\alpha w_t(1,t) + \beta w_t^3(1,t)], \quad t > 0; \alpha \geq 0, \beta > 0, \quad (7.2.3)$$

$$w(x,0) = w_0(x), \quad w_t(x,0) = w_1(x), \quad 0 < x < 1. \quad (7.2.4)$$

Comparing (7.2.3) with (7.1.4), we note that the α here has been given a different sign so that the RHS of (7.2.3) now represents a cubic damping velocity feedback (without the van der Pol self-regulation effect). We also mention that we have set L in (7.1.2) to be 1, because the velocity of wave propagation can be easily incorporated and adjusted in the ensuing analysis. The standard change of variables ([2]–[4])

$$w_x = u + v, \quad w_t = u - v \quad (7.2.5)$$

converts (7.2.1)-(7.2.4) to a diagonalized first order hyperbolic system

$$\frac{\partial}{\partial t}\begin{bmatrix}u\\v\end{bmatrix} = \begin{bmatrix}1 & 0\\0 & -1\end{bmatrix}\frac{\partial}{\partial x}\begin{bmatrix}u\\v\end{bmatrix}, \qquad (7.2.6)$$

$$v(0,t) = G_\eta(u(0,t)) \equiv \frac{1+\eta}{1-\eta}u(0,t), \qquad (7.2.7)$$

$$u(1,t) = F_{\alpha,\beta}(v(1,t)) \equiv v(1,t) + g_{\alpha,\beta}(v(1,t)), \qquad (7.2.8)$$

$$u(x,0) \equiv u_0(x) = \frac{1}{2}[w_0'(x) + w_1(x)], \quad 0 < x < 1,$$

$$v(x,0) \equiv v_0(x) = \frac{1}{2}[w_0'(x) - w_1'(x)], \ 0 < x < 1, \qquad (7.2.9)$$

where in (7.2.8), the nonlinear function $g_{\alpha,\beta}(v)$ is defined implicitly through the cubic equation

$$\beta g_{\alpha,\beta}^3(v) + (1+\alpha)g_{\alpha,\beta}(v) + 2v = 0, \qquad (7.2.10)$$

and $F_{\alpha,\beta}(v) \equiv v + g_{\alpha,\beta}(v)$. By an application of the Implicit Function Theorem and a little extra effort, it is easy to show that $g_{\alpha,\beta}(v)$ is a globally well-defined (single-valued) function. By the tracing of characteristics ([2, 10]), the unique solution to (7.2.6)-(7.2.9) can be given explicitly below: For $t = 2k + \tau$, $k = 0, 1, 2, \ldots, 0 \leq \tau < 2$ and $0 \leq x \leq 1$,

$$u(x,t) = \begin{cases} (F_{\alpha,\beta} \circ G_\eta)^k(u_0(x+\tau)), & \tau \leq 1-x, \\ G_\eta^{-1} \circ (G_\eta \circ F_{\alpha,\beta})^{k+1}(v_0(2-x-\tau)), & 1-x < \tau \leq 2-x, \\ (F_{\alpha,\beta} \circ G_\eta)^{k+1}(u_0(\tau+x-2)), & 2-x < \tau \leq 2; \end{cases}$$

(7.2.11)

$$v(x,t) = \begin{cases} (G_\eta \circ F_{\alpha,\beta})^k(v_0(x-\tau)), & \tau \leq x, \\ G_\eta \circ (F_{\alpha,\beta} \circ G_\eta)^k(u_0(\tau-x)), & x < \tau \leq 1+x, \\ (G_\eta \circ F_{\alpha,\beta})^{k+1}(v_0(2+x-\tau)), & 1+x < \tau \leq 2. \end{cases} \qquad (7.2.12)$$

Here, as a rather standard practice in dynamical systems, we have abused the notation $(F_{\alpha,\beta} \circ G_\eta)^k$, e.g., to denote the kth iterate composition of the function $F_{\alpha,\beta} \circ G_\eta$ with itself. *Spatiotemporal chaotic vibration* of u and v occurs when the map $F_{\alpha,\beta} \circ G_\eta$ and/or $G_\eta \circ F_{\alpha,\beta}$ is/are chaotic. Representations (7.2.11) and (7.2.12) show that u and v are completely determined by the composite reflection relations $G_\eta \circ F_{\alpha,\beta}$, $F_{\alpha,\beta} \circ G_\eta$, and their iterates. But $F_{\alpha,\beta} \circ G_\eta$ and $G_\eta \circ F_{\alpha,\beta}$ are topologically conjugate to each other [3, p. 433], so it is sufficient to consider whether $G_\eta \circ F_{\alpha,\beta}$ alone is chaotic. The map $G_\eta \circ F_{\alpha,\beta}$ is thus a natural *Poincaré section* of the PDE system (7.2.1)-(7.2.4) or (7.2.5)-(7.2.6).

For easy visualization, we display the graphs of $G_\eta \circ F_{\alpha,\beta}$ for certain sample values of α, β and η in Figure 7.1.

(a)

(b)

(c)

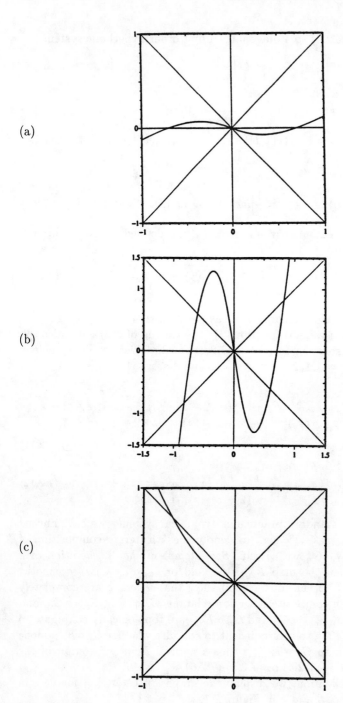

FIGURE 7.1
Graphs of the map $G_\eta \circ F_{\alpha,\beta}$: (a) $\eta = 0$, $\alpha = 0.5$, $\beta = 1$; (b) $\eta = 0.9$, $\alpha = 0.5$, $\beta = 1$; (c) $\eta = 1.5$, $\alpha = 1.2$, $\beta = 1$.

LEMMA 7.1 (Stability of the Origin)
Let $\alpha \geq 0$ and $\beta > 0$. Then

(i) If $\alpha = 0$, then the origin is a neutrally stable fixed point and is (weakly) globally attracting.
(ii) If $\alpha > 0$, then the origin is a globally attracting fixed point of $F_{\alpha,\beta}$.
(iii) If $0 < \alpha < 1$ and either $1 > \eta > \alpha$ or $\eta > 1$, then the origin is a repelling fixed point of $G_\eta \circ F_{\alpha,\beta}$.
(iv) If $\alpha > 1$ and either $0 < \eta < 1$, $\alpha\eta > 1$ or $\alpha > \eta > 1$ then the origin is a repelling fixed point of $G_\eta \circ F_{\alpha\beta}$.

PROOF By differentiating the function $g_{\alpha,\beta}$ in (7.2.10) implicitly, it is easy to obtain

$$F'_{\alpha,\beta}(v) = 1 + g'_{\alpha,\beta}(v) = 1 - \frac{2}{3\beta g^2_{\alpha,\beta}(v) + (1+\alpha)}. \tag{7.2.13}$$

At $v = 0, g_{\alpha,\beta}(0) = 0$, so we obtain

$$F'_{\alpha,\beta}(0) = 1 - \frac{2}{1-\alpha} = \frac{\alpha-1}{\alpha+1}; \quad |F'_{\alpha,\beta}(0)| \leq 1 \text{ for } \alpha \geq 0. \tag{7.2.14}$$

Conclusions in (i) and (ii) above become immediately clear from (7.2.13) and (7.2.14).
Since

$$(G_\eta \circ F_{\alpha,\beta})'(0) = \frac{1+\eta}{1-\eta} \cdot \frac{\alpha-1}{\alpha+1}, \tag{7.2.15}$$

we easily verify the claims in (iii) and (iv). □

LEMMA 7.2 (v-axis Intercepts)
Let $\alpha \geq 0$, $\beta > 0$, $\eta > 0$ and $\eta \neq 1$ be given. Then for $0 \leq \alpha < 1$, the map $G_\eta \circ F_{\alpha,\beta}$ has three distinct v-axis intercepts

$$v = -\sqrt{\frac{1-\alpha}{\beta}}, 0, \sqrt{\frac{1-\alpha}{\beta}}. \tag{7.2.16}$$

For $\alpha \geq 1$, $G_\eta \circ F_{\alpha,\beta}$ has exactly one intercept at $v = 0$.

PROOF Obvious from [3, Lemma 2.3]. □

THEOREM 7.3
(Period-Doubling Bifurcation Theorem for $-G_\eta \circ F_{\alpha,\beta}, 0 < \eta < 1$) Let $\alpha \geq 0$ and $\beta > 0$ be fixed, and let η: $\eta > \alpha$, be a varying parameter such that $1 - \alpha\eta > 0$. Then

(i) $v_0(\eta) \equiv \frac{1+\eta}{2}\sqrt{\frac{\eta-\alpha}{\beta}}$ is a curve of fixed points of $-G_\eta \circ F_{\alpha,\beta}$:

$$-G_\eta \circ F_{\alpha,\beta}(v_0(\eta)) = v_0(\eta). \tag{7.2.17}$$

(ii) *The algebraic equation*

$$\frac{1}{2}\left(\frac{1-\alpha\eta}{3\beta\eta}\right)^{1/2}\left[\frac{1+(3+2\alpha)\eta}{3\eta}\right] = \frac{1+\eta}{2}\sqrt{\frac{\eta-\alpha}{\beta}} \tag{7.2.18}$$

has a unique solution $\eta = \eta_0$ *for any given* α: $0 \leq \alpha \leq 1$ *and* $\beta > 0$. *We have*

$$\frac{\partial}{\partial v}[-G_\eta \circ F_{\alpha,\beta}(v)]\Big|_{\substack{v=v_0(\eta_0) \\ \eta=\eta_0}} = -1. \tag{7.2.19}$$

(iii) *For* $\eta = \eta_0$ *satisfying* (7.2.18), *we have*

$$A \equiv -\left[\frac{\partial^2(G_\eta \circ F_{\alpha,\beta})}{\partial\eta\partial v} - \frac{1}{2}\left(\frac{\partial}{\partial\eta}G_\eta \circ F_{\alpha,\beta}\right)\frac{\partial^2}{\partial v^2}(G_\eta \circ F_{\alpha,\beta})\right]\Big|_{\substack{v=v_0(\eta_0) \\ \eta=\eta_0}}$$

$$= -\frac{[4\alpha(2\alpha-3)+6]\eta_0^3 + (6-4\alpha)\eta_0^2 - 10\eta_0 + 6}{3(1-\eta_0)^3(1+\eta_0)^2}$$

$$\neq 0. \tag{7.2.20}$$

(iv) *For* η_0 *given in* (ii), *we have*

$$B \equiv -\left[\frac{1}{6}\frac{\partial^3(G_\eta \circ F_{\alpha,\beta})}{\partial v^3} - \frac{1}{4}\left(\frac{\partial^2(G_\eta \circ F_{\alpha,\beta}))^2}{\partial v^2}\right)^2\right]\Big|_{\substack{v=v_0(\eta_0) \\ \eta=\eta_0}}$$

$$= \frac{8\beta\eta_0^4\{(1-\eta_0)[5-(1+6\alpha)] + 6\eta_0(1+\alpha\eta_0)\}}{(1-\eta_0)^2(1+\eta_0)^4} > 0. \tag{7.2.21}$$

Consequently, there is period-doubling bifurcation at $(v_0(\eta_0), \eta_0)$. *The stability type of the bifurcated period-2 orbit is attracting.*

PROOF The assertions in (i)-(iv) above are adaptations of [3, Theorem 3.1] by changing α therein to $-\alpha$. The only restrictions we must observe are that $\eta > \alpha \geq 0$ and $1 - \alpha\eta > 0$. It is straightforward to verify (i). We can also confirm (ii)-(iv) with computer-aided proofs. Therefore the Period-Doubling Bifurcation Theorem [16, pp. 220-223] applies. □

Example 7.1
By setting $\alpha = 0.5$ and $\beta = 1$ in (7.2.19) and solve for $\eta = \eta_0$ on the computer, we obtain

$$\eta_0 \approx 0.7676. \tag{7.2.22}$$

This value at which the first period-doubling occurs can be observed in Figure 7.2. □

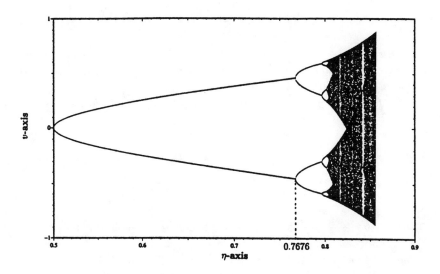

FIGURE 7.2
The orbit diagram of the map $G_\eta \circ F_{\alpha,\beta}$, where we have chosen fixed $\alpha = 0.5$, $\beta = 1$, and let η vary in $[0,1)$. Note that the first period-doubling bifurcation happens at $\eta_0 \approx 0.7676$, agreeing with (7.2.22).

A period-doubling bifurcation may or may not lead to a *full period-doubling cascade* and the consequent chaos, depending on whether the map $-G_\eta \circ F_{\alpha,\beta}$ has the renormalizable property [9, 12]. Here we note that every prime period-2^n orbit of the map $-G_\eta \circ F_{\alpha,\beta}$ corresponds to a unique prime period-2^{n+1} orbit of the map $G_\eta \circ F_{\alpha,\beta}$ [3, Lemma 3.1]. The map $-G_\eta \circ F_{\alpha,\beta}$ is *unimodal* on $[0,\infty)$ and $(-\infty, 0]$, respectively. Therefore, renormalizability is assured. However, the restrictions $\eta > \alpha \geq 0$ and $1 - \alpha\eta > 0$ must also be satisfied all the time. Numerical simulations have indicated that under these restrictions, a full period-doubling cascade will always follow. This is supported by Theorem 2 below, where the existence of homoclinic orbits is shown, which is a surefire way to establish chaos [11, Theorem 1.16.5].

THEOREM 7.4
(Homoclinic Orbits and Homoclinic Bifurcations of the Map $G_\eta \circ F_{\alpha,\beta}$ for $0 \leq \alpha < 1$ with respect to the Varying Parameter η: $0 < \eta < 1$)
Let $0 \leq \alpha < 1, \beta > 0$, and define

$$\underline{\eta}_H = \underline{\eta}_H(\alpha) = \left(1 - \frac{1-\alpha}{3\sqrt{3}}\right) \bigg/ \left(1 + \frac{1-\alpha}{3\sqrt{3}}\right).$$

If η satisfies $1 > \eta \geq \underline{\eta}_H$, then 0 is a repelling fixed point of the map $G_\eta \circ F_{\alpha,\beta}$ having homoclinic orbits. Furthermore, if $\eta = \underline{\eta}_H$, then there are degenerate homoclinic orbits.

Consequently, if $\eta \in [\underline{\eta}_H, 1)$, then the map $G_\eta \circ F_{\alpha,\beta}$ is chaotic on some invariant sets of $G_\eta \circ F_{\alpha,\beta}$.

PROOF   It is easy to check that if $1 > \eta \geq \underline{\eta}_H$, then

$$1 > \eta \geq \left(1 - \frac{1-\alpha}{3\sqrt{3}}\right) \bigg/ \left(1 + \frac{1-\alpha}{3\sqrt{3}}\right) > \alpha,$$

and therefore $\frac{1+\eta}{1-\eta} \cdot \frac{\alpha-1}{\alpha+1} > 1$ the origin becomes a repelling fixed point by (7.2.15) and Lemma 7.1 (iii).

The rest follows easily from an adaptation of [3, Theorem 4.1] by setting α to $-\alpha$ therein. Since the arguments are identical, we refer the reader to [3]. □

Contrary to [3, Theorem 4.2], the origin of the map $G_\eta \circ F_{\alpha,\beta}$ does not have homoclinic orbits when $\eta > 1$. This is also partly evident from Figure 7.1(c).

The chaotic property of the family of maps $G_\eta \circ F_{\alpha,\beta}$ with α and β fixed and varying η is evident from the orbit diagram in Figure 7.2, where $\alpha = 0.5$ and $\beta = 1$ are chosen.

If we break up the overall reflection map $G_\eta \circ F_{\alpha,\beta}$, we have

(i) G_η is a linear expansion map;

(ii) $F_{\alpha,\beta}$ is a nonlinear contraction map.

The interactions of linear expansion and nonlinear contraction can lead to chaos. This is the major conclusion of this section. What can be said about a contrasting case, a composite map $\widetilde{G}_\eta \circ \widetilde{F}_{\alpha,\beta}$, where

(i)' $\widetilde{G}_\eta$ is a linear contraction map;

(ii)" $\widetilde{F}_{\alpha,\beta}$ is a nonlinear expansion map?

We can achieve (i)' by changing BC (7.2.2) to

$$w_t(0,t) = \eta w_x(0,t), \quad \eta > 0,$$

leading to
$$\widetilde{G}_\eta(u) \equiv \frac{1-\eta}{1+\eta}u,$$
and achieve (ii)' by changing BC (7.2.3) to
$$w_x(0,t) = \alpha w_t(0,t) + \beta w_t^3(0,t), \qquad \alpha, \beta > 0, \alpha \geq 1,$$
yielding
$$\widetilde{F}_{\alpha,\beta}(v) = v + g_{\alpha,\beta}(v),$$
where $g_{\alpha,\beta}(v)$ is (uniquely defined as) the solution of
$$\beta g_{\alpha,\beta}^3(v) + (\alpha - 1)g_{\alpha,\beta}(v) - 2v = 0, \qquad v \in \mathbb{R}.$$
Our numerical simulations so far have indicated that the map $\widetilde{G}_\eta \circ \widetilde{F}_{\alpha,\beta}$ is never chaotic throughout the entire parameter range $\eta > 0$, $\alpha \geq 1$ and $\beta > 0$. This seems to be in sharp contrast to the chaos-causing potential of the map $G_\eta \circ F_{\alpha,\beta}$ in the theorems and graphics in this section.

7.3 Feedback of Polynomial Type at the Right Endpoint

The preceding section has shown that the nonlinear feedback does not have to be of the van der Pol type in order for chaotic vibration to occur. This has set the tone in this section for exploring a larger class of nonlinear feedback boundary controls at the right endpoint, which, after interaction with energy injection at the left endpoint, lead to chaos.

The broadest class of nonlinear boundary conditions at the right endpoint can be described by an implicit nonlinear equation
$$\mathcal{F}(w(1,t), w_x(1,t), w_t(1,t)) = 0, \qquad t > 0,$$
involving all the boundary displacement $(w(1,t))$, force $(w_x(1,t))$ and velocity $(w_t(1,t))$ variables. But this class is too general to be useful for control purposes. At this point, it is not clear to the authors what would be "the most natual" choice for $\mathcal{F}$ based upon the physical consideration. Two choices, much narrower but mathematically quite reasonable, would be

(1) a polynomial feedback of velocity to the force, in the form
$$w_x(1,t) = P_n(w_t(1,t)), \tag{7.3.1}$$
where $P_n(\cdot)$ is a polynomial of degree n given by
$$P_n(z) = a_n z^n + a_{n-1} z^{n-1} + \cdots + a_2 z^2 + a_1 z,$$

or

(2) a polynomial feedback of force to the velocity, in the form

$$w_t(1,t) = P_n(w_x(1,t)). \tag{7.3.2}$$

Note that we have not permitted the presence of the displacement variable $w(1,t)$ in either (7.3.1) or (7.3.2) since the corresponding mathematical treatment would be substantially more involved and, so far, few results are available even for very simple $P_n's$.

From the purely mathematical point of view, equations (7.3.1) and (7.3.2) can be handled in the same fashion. For definiteness, let us just study type (7.3.1) throughout the rest of the paper. As before, using (7.2.5) in (7.3.1) and written out in detail:

$$u + v = \sum_{k=1}^{n} a_k(u-v)^k, \qquad (u = u(1,t), v = v(1,t))$$

we obtain

$$u = F_a(v) \equiv v + g_a(v), \tag{7.3.3}$$

where $g_a(v)$ is defined implicitly through the nonlinear equation

$$\sum_{k=2}^{n} a_k g_a^k(v) + (a_1 - 1)g_a(v) - 2v = 0, \qquad v \in \mathbb{R}. \tag{7.3.4}$$

For each given $v \in \mathbb{R}$, (7.3.4) may have as many as n real solutions $g_a(v)$ and, thus, in general, $g_a(v)$ is not a well-defined function of v. (In reality, multivalued $g_a(v)$ is *physically admissible*. What one has is a *hysteresis* situation. Certain special cases have been treated in [4], for example. For general multivalued functions $g_a(v)$, however, the hysteresis behavior will be quite complicated. No systematic study has been done so far.) To avoid this technical difficulty, it is sufficient to assume that

$$P_n'(x) - 1 = \sum_{k=2}^{n} k a_k x^{k-1} + (a_1 - 1) \neq 0, \qquad \forall\, x \in \mathbb{R}. \tag{7.3.5}$$

Under (7.3.5), we have either $P_n'(x) - 1 > 0$ or $P_n'(x) - 1 < 0$ for all $x \in \mathbb{R}$, the Implicit Function Theorem then applies and (with a few extra arguments) we have the single-valuedness of the function $g_a(v)$.

At this point of time, there is no theory available about the dynamic behavior of the composite reflection map $G_\eta \circ F_a$, where G_η and F_a are defined, respectively, in (7.2.7) and (7.3.3). We envision that the development of such a theory will be a major task for us in the next few years. Nevertheless, in what follows we do wish to present two concrete examples to illustrate the novel dynamic features and, perhaps, the advantages of using polynomial feedback boundary control in generating chaotic vibration of the wave equation.

Example 7.2
Assume (7.2.1)-(7.2.4), except that the right end BC (7.2.3) is replaced by

$$w_x(1,t) = -w_t^5(1,t) + 4w_t^4(1,t) - 6w_t^3(1,t) + 0.5w_t(1,t), \quad t > 0. \quad (7.3.6)$$

It is easy to check that here

$$\begin{aligned} P_5'(x) - 1 &= -5x^4 + 16x^3 - 18x^2 - 0.5 \\ &= -x^4 - 4x^2(x-2)^2 - 2x^2 - 0.5 < 0, \quad \forall\, x \in \mathbb{R}, \end{aligned}$$

and, therefore, we have a well-defined function $F_a(\cdot)$; its graph is displayed in Figure 7.3. A distinctive feature observed from Figure 7.3 is that F_a is not unimodal, nor is it an odd function (in contrast to those counterparts studied in [2]–[6]).

We now study the asymptotic dynamic behavior of the composite reflection map $G_\eta \circ F_a$ by varying $\eta \in \mathbb{R}, \eta \neq 1$. The orbit diagram is plotted in Figure 7.4.

From this orbit diagram, we observe the following features:

(a) For larger values of η (i.e., η away from 1), the orbits of $G_\eta \circ F_a$ consist essentially of two branches, with each branch functioning at its own will, for positive v and negative v, separately. For example, the upper branch has already completed the period-doubling cascade (when η decreases leftward toward 1) and is well into chaos, while the lower branch is still undergoing its first period-doubling bifurcation.

(ii) For η in the approximate range [1.751, 1.877], the chaos in the upper branch is suddenly sucked out of existence by attracting periodic orbits of the lower branch, causing "disconnectedness" or "gap" between the two branches.

(iii) The two branches re-attach for $\eta \in (1.488, 1.751)$.

(iv) Another chaotic regime exists for $\eta < 0.633$.

To show some of the chaotic effects of the above spatiotemporally, let us choose $\eta = 1.9417$, with the following ICs:

$$\begin{cases} v_0(x) = 10(x - \tfrac{1}{2}) \cdot \phi_0(x), & x \in [0,1], \\ u_0(x) \equiv 0, & x \in [0,1], \end{cases} \quad (7.3.7)$$

where $\phi_0(x)$ is a C^2-continuous piecewise cubic spline defined by

$$\phi_0(x) = \frac{1}{24} \begin{cases} (x - x_1)^3/h^3, & x_1 \le x \le x_2, \\ 1 + \dfrac{3(x-x_2)}{h} + \dfrac{3(x-x_2)^2}{h^2} - \dfrac{3(x-x_2)^3}{h^3}, & x_2 \le x \le x_3, \\ 1 - \dfrac{3(x-x_4)}{h} + \dfrac{3(x-x_4)^3}{h^2} + \dfrac{3(x-x_4)^3}{h^3}, & x_3 \le x \le x_4, \\ (x_5 - x)^3/h^3, & x_4 \le x \le x_5, \\ 0, & \text{elsewhere,} \end{cases}$$

$$h = 1/6, x_j = j/6, \qquad j = 1, 2, 3, 4, 5.$$

The choice in (7.3.7) implies the corresponding choice of the ICs for w, the original state variable in (7.2.1), to be

$$\begin{cases} w_0(x) = w(x,0) = \int_0^x w_x(\xi, 0) d\xi + C \\ \qquad\qquad\qquad = \int_0^x v_0(\xi) d\xi + C, C = \text{ arbitrary constant,} \\ w_1(x) = w_t(x,0) = u_0(x) - v_0(x) = -v_0(x). \end{cases}$$

According to [3, Theorem 6.1], we have the regularities

$$(u,v) \in [C^2([0,1] \times [0,T])]^2,$$

$$(w, w_t) \in C^3([0,1] \times [0,T]) \times C^2([0,1] \times [0,T]), \text{ for any } T > 0.$$

The initial spatiotemporal profiles of (u,v) and (w_x, w_t) are plotted, respectively, in Figures 7.5 and 7.6, for $(x,t) \in [0,1] \times [0,2]$.

In this example, each *cycle of vibration*, defined to be the time duration required for a wave to travel from $x = 0$ to $x = 1$, reflect at $x = 1$ and return to $x = 0$, is two time units. So let us look at the snapshots of (u,v) during the 50th time cycle, $t = 101.5 \in [50 \cdot 2, 51 \cdot 2] = [100, 102]$, displayed in Figures 7.7(a) and (b). We observe that u and v manifest both chaotic and periodic behavior. This can be easily interpreted with the visual aid of Figure 7.8.

It is known that (w_x, w_t) and (u,v) are topologically conjugate in the sense of [6, §IV]. The profile of w is not displayed here. It has a *fractal* look but does *not* display chaotic behavior; cf. [6, Figure 6.11], for example. In order to make w itself chaotic, the nonlinear BC must contain w; see [3, Theorem 6.2]. □

Example 7.3
Let us return to (7.2.1)-(7.2.4) again, but with (7.2.3) replaced by

$$w_x = -\frac{1}{\mu}(\frac{1}{5}w_t^5 - w_t^4 - \frac{1}{3}w_t^3 + 8w_t^2) - \left(1 - \frac{12}{\mu}\right) w_t; \qquad (\mu = 12.6618)$$
(7.3.8)

at $x = 1$, for all $t > 0$. Then we have

$$P_5'(x) - 1 = -\frac{1}{\mu}(x^4 - 4x^3 - x^2 + 16x - 12) - 2$$

$$= -\frac{1}{\mu}[(x+1)(x+2)(x+3)(x-2)] - 2$$

$$= -\frac{|m|}{\mu}\left\{\frac{1}{|m|}[(x+1)(x+2)(x+3)(x-2)]\right\} - 2$$

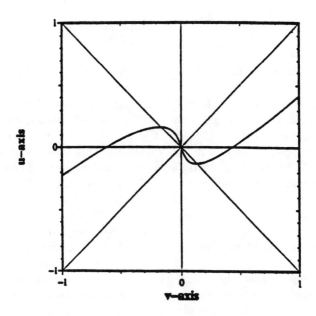

FIGURE 7.3
Graph of the map F_a in Example 7.2.

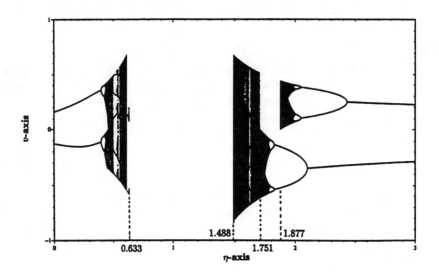

FIGURE 7.4
Orbit diagram of the map $G_\eta \circ F_a$ with η varying in interval $(0, 3)$.

$$< \frac{|m|}{\mu} - 2 < 0, \quad \forall x \in \mathbb{R}, \tag{7.3.9}$$

where

$$m = \inf_{x \in \mathbb{R}} [(x+1)(x+2)(x+3)(x-2)] \approx -24.0572$$

and by the choice of μ we have $\frac{|m|}{\mu} < 2$ and, therefore (7.3.5) is satisfied. The reflection map F_a at the right endpoint is well-defined, which is computed and then shown in Figure 7.9. Comparing it with the counterpart F_a for Example 7.2 in Figure 7.3, we see that Figure 7.9 embodies more features – it has two local maxima and two local minima. (These four local extremal points are somehow "designed" to be related to the four roots of the polynomial $(x+1)(x+2)(x+3)(x-2)$ in (7.3.9).)

The orbit diagram of $G_\eta \circ F_a$ is shown in Figure 7.10 for the varying parameter $\eta < 1$. Here again we see chaos. For η appearing in some range, it is straightforward to *rigorously establish* that the origin is a repelling fixed point *having homoclinic orbits*, and hence chaos ensues. See Proposition 7.5 below. But how about other causes/routes to chaos? The initial period doubling route to chaos seems to have disappeared. Can we still characterize the onset of chaos without period-doubling?

The "irregular" pattern of the orbit diagram in Figure 7.10 leaves many questions waiting to be answered. □

The proof of the following proposition, although stated in a form applicable only to Example 7.3 above, contains ideas which can be used to treat general problems.

PROPOSITION 7.5
Let F_a be the map (7.3.3) defined through the boundary feedback relation (7.3.8) in Example 7.3. Then for η lying in a certain range of the interval $(0,1)$, the origin is an unstable fixed point having homoclinic orbits for the map $G_\eta \circ F_a$.

PROOF Since homoclinic orbits are geometric concept and their existence can be confirmed "visually," an "intuitive proof" suffices provided that all the geometric conditions are met.

First, we observe from Figure 7.9 that the origin (as a fixed point of $G_0 \circ F_a = F_a$) does not have homoclinic orbits. However, as η increases from 0, $\frac{1+\eta}{1-\eta} \cdot m$ begins to decrease past I_ℓ (cf. the caption of Figure 7.9 for the values of m and I_ℓ), and at the same time, the origin becomes a repelling fixed point.

A homoclinic orbit is marked by dotted lines in Figure 7.11, where we have chosen $\eta = 0.45$. □

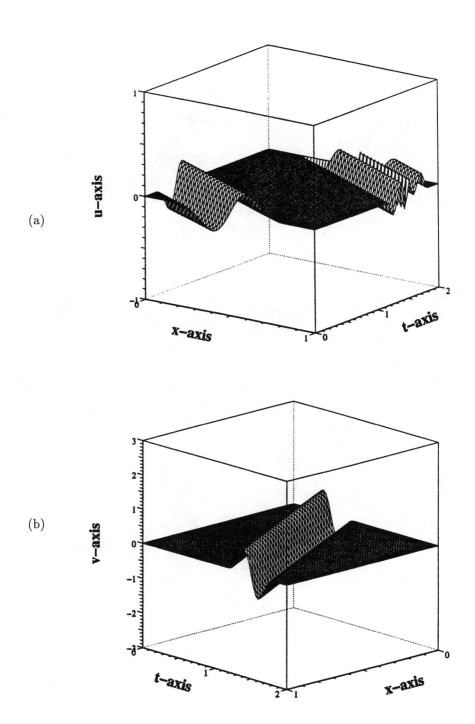

FIGURE 7.5
The initial spatiotemporal profiles of: (a) u and (b) v, for $x \in [0,1]$ and $t \in [0,2]$ in Examples 7.2.

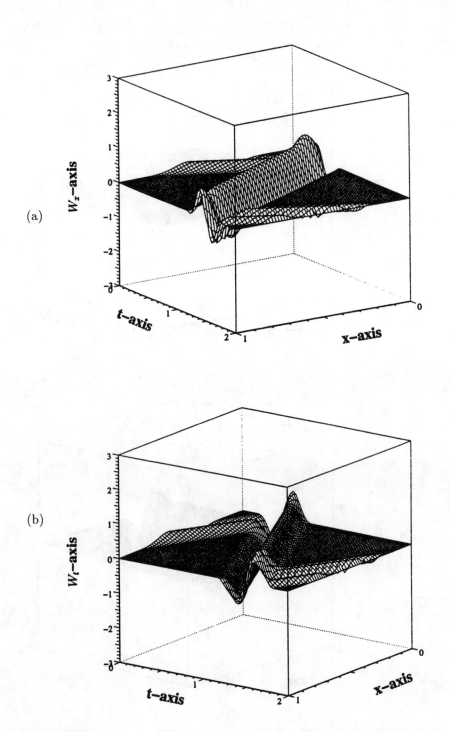

FIGURE 7.6
The initial spatiotemporal profiles of: (a) w_x and (b) w_t, for $x \in [0,1]$ and $t \in [0,2]$ in Example 7.2.

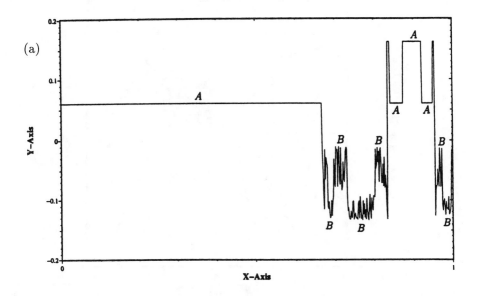

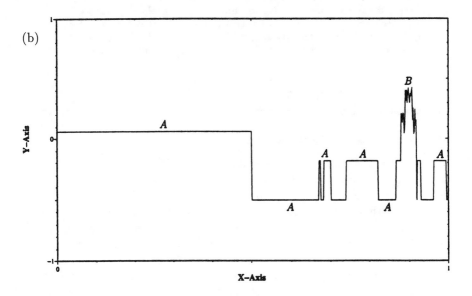

FIGURE 7.7
The snapshots of: (a) u and (b) v, for $x \in [0,1]$ and $t = 101.5 \in$ [100,102] in Example 7.2. Note that the "flat parts" indicated by A correspond to the periodic portion of the solution, while the "highly oscillatory parts" indicated by B correspond to the chaotic portion of the solution. Thus u and v have mixed behavior of being partly periodic and partly chaotic.

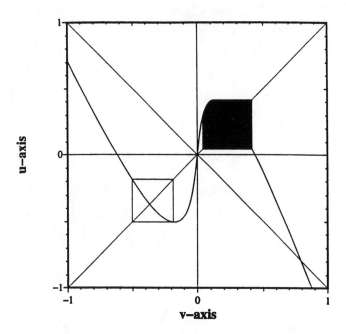

FIGURE 7.8
This graphical analysis shows coexistence of chaos (for $v > 0$) and attracting period-2 orbits (for $v < 0$) of the map $G_\eta \circ F_{\alpha,\beta}$, with $\eta = 1.9417$, for Example 7.2. Note that for $v > 0$, the iterates of $G_\eta \circ F_{\alpha,\beta}$ are chaotic as shown by the denseness of orbits in the middle upper right of the figure, while for $v < 0$ the iterates converge to a period-2 orbit in the middle lower left of of the figure.

7.4 Concluding Remarks

Only the one-dimensional wave equation is treatable so far by the types of nonlinear feedback control (or anti-control) in this chapter. For "similar" equations arising in structural vibration such as the Euler-Bernoulli beam equation, although some reflection relations on the boundary may be defined, the work here cannot be extended to that case when the boundary condition is nonlinear because the wave propagation on an Euler-Bernoulli beam is *dispersive*.

In higher dimensional settings, the study of partial differential equations with nonlinear boundary conditions is a rather difficult subject in itself, let alone that of chaotic effects caused by nonlinear boundary feedback. At present, the best hope seems to be offered by problems on domains with special geometry such as the rectangular, spherical and annular cases.

Concluding Remarks

In any case, the advent of modern dynamical systems and chaos has provided the control theorist with many useful ideas and powerful tools to explore as well as to exploit nonlinearities in distributed parameter control systems, with bright, aplenty future opportunities in this field.

Acknowledgments

G. Chen and J. Zhou are supported in part by NSF Grant 96-10076 and NATO Grant CRG 940369. S.-B.Hsu is supported in part by a grant from the National Council of Science of Republic of China.

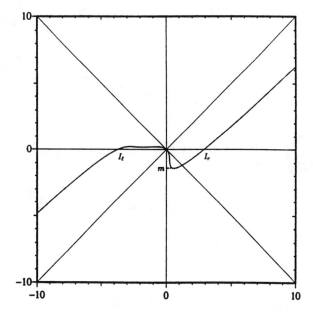

FIGURE 7.9
The graph of the reflection map F_a at $x = 1$ in Example 7.3. In addition to the origin, there are two intercepts as indicated: $I_r \approx 2.8846$ and $I_\ell \approx -3.7260$. A local minimum value $m \approx -1.4321$ is also marked.

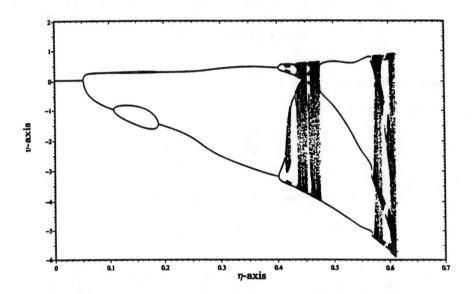

FIGURE 7.10
The orbit diagram of $G_\eta \circ F_a$ with the varying parameter $0 < \eta < 1$ for Example 7.3.

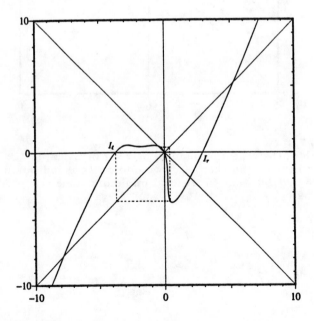

FIGURE 7.11
The dotted lines indicate a homoclinic orbit for the map $G_\eta \circ F_a$ in Example 7.3 and Proposition 7.5, where $\eta = 0.45$ is chosen.

References

[1] G. Chen and J. Zhou, *Vibration and Damping in Distributed Systems*, vols. I & II, CRC Press, Boca Raton, FL, 1993.

[2] G. Chen, S. B. Hsu and J. Zhou, "Chaotic vibrations of the one-dimensional wave equation due to a self-excitation boundary condition, Part I, controlled hysteresis," *Trans. Amer. Math. Soc.*, vol. 350 pp. 4265-4311, 1998.

[3] G. Chen, S. B. Hsu and J. Zhou, "Ibid, Part II, energy injection, period doubling and homoclinic orbits," *Int. J. Bifurcation & Chaos*, vol. 8 pp. 423-445, 1998.

[4] G. Chen, S. B. Hsu and J. Zhou, "Ibid, Part III, natural hysteresis memory effects," *Int. J. Bifurcation & Chaos*, vol. 8 pp. 447-470, 1998.

[5] G. Chen, S. B. Hsu and J. Zhou, "Snapback repellers as a cause of chaotic vibration of the wave equation with a van der Pol boundary condition and energy injection at the middle of the span," *J. Math. Phys.*, vol. 39 pp. 6459-6489, 1998.

[6] G. Chen, S. B. Hsu and J. Zhou, "Nonisotropic spatiotemporal chaotic vibration of the wave equation due to mixing energy transport and a van der Pol boundary condition," submitted.

[7] G. R. Chen, "Chaos: Control and anti-control," *IEEE Circuits and Systems Society Newsletter*, vol. 9 pp. 1-5, 1998.

[8] G. R. Chen and X. Dong, *From Chaos to Order—Methodologies, Perspectives and Applications*, World Scientific Pub. Co., Singapore, 1998.

[9] C. Collet and C. Tresser, "Itération d'endomorphismes et de groupe de renormalisation," *J. de Physique Colloque*, vol. 39 pp. C5-C25, 1978.

[10] R. Courant and D. Hilbert, *Methods of Mathematical Physics*, vol. II, Wiley-Interscience, New York, 1962.

[11] R. L. Devaney, *An Introduction to Chaotic Dynamical Systems*, Addison-Wesley, New York, 1989.

[12] M. Feigenbaum, "Quantitative universality for a class of non-linear transformations," *J. Stat. Phys.*, vol. 21 pp. 25-52, 1978.

[13] P. Holmes and J. Marsden, "A partial differential equation with infinitely many periodic orbits: chaotic oscillators of a forced beam," *Arch. Rat. Mech. Anal.*, vol. 76, 135–165, 1981.

[14] J. E. Lagnese, *Boundary Stabilizability of Thin Plates*, SIAM Studies in Appl. Math., SIAM, Philadelphia, 1989.

[15] J. E. Lagnese and J. L. Lions, *Modeling Analysis and Control of Thin Plates*, Masson, Paris, 1988.

[16] C. Robinson, *Dynamical Systems, Stability, Symbolic Dynamics and Chaos*, CRC Press, Boca Raton, FL, 1995.

[17] D. L. Russell, "Controllability and stabilizability theory for linear partial differential equations, recent progress and open questions," *SIAM Review*, vol. 20 pp. 383-431, 1978.

[18] A. N. Sharkovsky, "Ideal turbulences in an idealized time-delayed Chua's circuit," *Int. J. Bifurcation & Chaos*, vol. 4 pp. 303-309, 1994.

[19] J. J. Stoker, *Nonlinear Vibrations*, Wiley-Interscience, New York, 1950.

8

Sensitivity to Initial Conditions of Chaos in Electronics

G. Q. Zhong, Z. F. Liu, K. S. Tang
K. F. Man and S. Kwong

City University of Hong Kong
Tat Chee Avenue, Kowloon, Hong Kong
eekman@cityu.edu.hk

Abstract

It has been well known that the sensitivity to the initial conditions can lead to unpredictable trajectory. In this chapter, we discuss the issue on the basis of a Phase Locked Loop circuit with an embedded nonlinear controller. A number of well established techniques for the chaos evaluation have been employed, namely bifurcation, Lyapunov exponent spectrum, fractal basin, etc. The investigations cover not only the basic chaos phenomena, but have also been extended to consider the case of synchronization. As a result, it has been found that this particular controller can broaden the phase lock-in range while its freedom of switching the attractors from chaos to order, or vice versa, can potentially lead to industrial applications.

8.1 Introduction

It is well known in chaos study that chaotic motion is very sensitive to its initial states. Even when starting from nearby initial conditions, the corresponding trajectories can quickly become uncorrelated. Because of the phenomenon, the trajectories may either attract to a fixed point, a stable limit cycle or a chaotic attractor. The boundary set points which divide the two distinct attractors, appear to be extremely complicated. These constitute a boundary surface with a fractal dimension between the domains of attraction in the state space [1, 10, 14].

Despite the active research activity into chaos for the last twenty years or so, a practical industrial system based on chaos is far from being developed. Although the fundamental study in chaos is rich, particularly in the area of synchronization and control concepts, practicing engineers have always seen the chaos dynamics as a stumbling block particularly when a slight change in initial conditions can yield totally unpredictable results.

Having said that, there are systems or electronic circuitries which they apply daily, that are intrinsically chaotic in nature. Therefore, due to a lack of appreciation of chaos knowledge, this part of system design has somehow been ignored, if not totally avoided. As a result, the dynamic range of the system activity is limited, and in some cases, completely incomprehensible.

A typical example is the application of Phase Locked Loop (PLL). This circuit or system has been widely used in industry for many decades. It can be found in the area of communications, instrument devices, and motor speed control, etc. However, the chaos phenomenon due to its inherent nonlinearity has received little attention by practicing engineers although the research literature has been quite substantial [5, 7, 8, 9]. The recent study in control and synchronization based on PLL has already attracted considerable attention [8] and the knowledge gained in confining the intrinsic chaotic behavior so as to stabilize and broaden the lock-in range [2] has practical value.

In this chapter, we present a new nonlinear controller for improving the PLL performance. This controller is not only capable of driving the dynamics from chaos to order, and vice versa, but can also be used to tame the extremely sensitive dependence on initial conditions. As a result, it can broaden the lock-in range of the PLL.

The spin-off of this nonlinear control technique also provides a new direction for the concept of "control and anticontrol of chaos" [4] within the PLL because of its richness in bifurcation and the sensitivity to initial conditions. It is also anticipated that the ultimate application of this control formulation will be ideally suited for secure communications within the PLL system.

8.2 Phase Model of PLL

A general PLL system consists of three main components: (i) a phase detector (PD), (ii) a low pass filter (LPF), and (iii) a voltage controlled oscillator (VCO).

Assume that the input signal v_1 of the system is a periodic signal defined by

$$v_1 = f_{in}(\theta_i(t)), \quad \theta_i(t) \in S^1 \bmod 2\pi \quad (8.2.1)$$

where $S^1 \bmod 2\pi$ denotes the phase angle, measured in modulo 2π, i.e., along a circle S^1, and θ_i is the input phase. The output v_4 of VCO is also a periodic signal:

$$v_4 = f_{out}(\theta_o(t)), \quad \theta_o(t) \in S^1 \bmod 2\pi \quad (8.2.2)$$

where θ_o is the output phase. The PD produces an output v_2 proportional to the phase difference $(\theta_i(t) - \theta_o(t))$ between v_1 and v_4, i.e.,

$$v_2 = K_{pd} h(\phi(t)) \quad (8.2.3)$$

where $\phi(t) \equiv \theta_i(t) - \theta_o(t) \in S^1 \bmod 2\pi$ is called the phase error. The actual characteristic of the PD depends mainly on the choice of the waveforms for v_1 and v_4.

Assuming that v_1 and v_4 are sinusoidal, and the PD is a mixer (multiplier), the output of the mixer is $v_2 = K_{pd}[\sin(\theta_i(t) - \theta_o(t)) + \sin(\theta_i(t) + \theta_o(t))]$. The second term has a higher fundamental frequency than the first, and the succeeding loop filter is designed to cut off this component. In so doing, the function $h(\phi)$ becomes a *sinusoid* and the characteristic of PD can then be simplified as:

$$v_2 = K_{pd} \sin(\phi(t)) \quad (8.2.4)$$

Similarly, if we choose an exclusive *OR* logic for the PD and square waves for v_1 and v_4, then $h(\phi)$ becomes *a triangle characteristic* which is defined as:

$$h(\phi) = \begin{cases} \phi, & |\phi| < \frac{\pi}{2} \\ -\phi + \pi, & \frac{\pi}{2} < \phi < \frac{3\pi}{2} \end{cases}$$

where $h(\phi) = h(\phi + 2\pi)$.

Should a flip-flop logic be selected for the PD with pulse waves for v_1 and v_4, then "h" becomes *sawtooth*.

The loop LPF is a time-invariant linear filter whose sole purpose is to cut off the noise components of the input signal v_1. It is usually a first-order filter with the following transfer function:

$$F(s) = \frac{1}{(1+\tau_1 s)}$$

where "s" is an operator denoting "d/dt". Then the filter takes the form:

$$\frac{dv_3(t)}{dt} + \frac{v_3(t)}{\tau_1} = -\frac{v_2(t)}{\tau_1} \qquad (8.2.5)$$

The VCO is an oscillator whose angular frequency changes with the control voltage v_3 in the following manner:

$$\frac{d\theta_o(t)}{dt} = \omega_o + K_v v_3(t) \qquad (8.2.6)$$

where ω_o is the *free running angular frequency* of the VCO, and K_v is the control gain.

Based on 8.2.4-8.2.6, the phase shifts $\tilde{\theta}_i$ and $\tilde{\theta}_o$ can be expressed as follows:

$$\tilde{\theta}_i(t) = \theta_i(t) - \omega_o t \qquad (8.2.7)$$

$$\frac{d\tilde{\theta}_o(t)}{dt} = K_v v_3(t) \qquad (8.2.8)$$

For simplicity, the gains K_{pd} and K_v can be combined to give the total loop gain as $K_0 = K_{pd} K_v$ and embedded within the transfer function $F(s)$ of the loop low-pass filter (LPF). This can now form an non-autonomous second-order nonlinear differential equation as follows:

$$\frac{d^2\phi}{dt^2} + \frac{1}{\tau_1}\frac{d\phi}{dt} + \frac{K_0}{\tau_1}h(\phi) = \frac{d^2\tilde{\theta}_i}{dt^2} + \frac{1}{\tau_1}\frac{d\tilde{\theta}_i}{dt} \qquad (8.2.9)$$

Assuming that the input signal is modulated by a sinusoidal waveform so that $d\theta_i/dt = \omega_c + M\sin\omega_m t$, then it follows from 8.2.7 that

$$\frac{d\tilde{\theta}_i}{dt} = \Delta\omega + M\sin\omega_m t \qquad (8.2.10)$$

where $\Delta\omega \equiv \omega_c - \omega_o$, M is the maximum angular frequency deviation, and ω_m is the modulation angular frequency.

The natural angular frequency (ω_n) and the damping coefficient (ζ) are defined as:

$$\omega_n \equiv \sqrt{K_0/\tau_1} \qquad (8.2.11)$$

$$\zeta \equiv \frac{1}{2}\sqrt{K_0 \tau_1} \qquad (8.2.12)$$

Furthermore, the following parameters are also defined:

$$\beta \equiv \frac{\omega_n}{K_0} = \frac{1}{\sqrt{K_0 \tau_1}} \quad \text{[normalized natural frequency]} \qquad (8.2.13)$$

$$\sigma \equiv \frac{\Delta\omega}{\omega_n} \quad \text{[normalized frequency detuning]} \qquad (8.2.14)$$

$$\Omega \equiv \frac{\omega_m}{\omega_n} \quad \text{[normalized modulation frequency]} \qquad (8.2.15)$$

$$m \equiv \frac{M}{\omega_n} \quad \text{[normalized maximum frequency deviation]} \qquad (8.2.16)$$

By changing the time 't' into

$$\tau = \omega_n t \qquad (8.2.17)$$

the nonlinear differential equation 8.2.9 can be normalized as follows:

$$\ddot{\phi} + \beta\dot{\phi} + h(\phi) = \beta\sigma + \beta m \sin \Omega\tau + m\Omega \cos \Omega\tau \qquad (8.2.18)$$

where $\cdot \equiv d/d\tau$.

For a *sinusoidal* PD characteristics, i.e., $h(\phi) = \sin\phi$, 8.2.18 takes the form:

$$\ddot{\phi} + \beta\dot{\phi} + \sin\phi = \beta\sigma + \beta m \sin \Omega\tau + m\Omega \cos \Omega\tau \qquad (8.2.19)$$

This can be transformed into state-space form using new variables such as $x_1 = \phi$, $x_2 = \dot{\phi}$, then we have

$$\left.\begin{array}{rl} \dot{x}_1 &= x_2 \\ \dot{x}_2 &= -\beta x_2 - \sin x_1 + \beta\sigma + \beta m \sin \Omega\tau + m\Omega \cos \Omega\tau \end{array}\right\} \qquad (8.2.20)$$

Without further circuitry augmentation, this PLL model forms an intrinsic chaotic system. A horseshoe type of chaos can be observed due to the nonlinear $h(\phi)$ for a wide range of parameter values. This chaotic behavior appears not only in the out-of-lock condition, but also surfaces at the lock-in stage [6]. This finding is considered to be extremely hazardous for practical applications since the complexity of the chaotic dynamics governs the system performance. Hence, to bring these dynamics under control offers an interesting challenge.

8.3 PLL with Nonlinear Control

Under a normal control operation, LPF not only facilitates the necessary action for the high frequency cut-out, but also acts as a controller to ensure the overall stability of the PLL. However, it has its limits. The main problem largely hinges on the system nonlinearity. As a result, the present PLL configuration has a small stability margin.

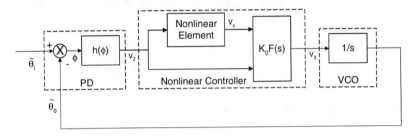

FIGURE 8.1
Phase model of PLL with nonlinear control.

Phase Model of PLL with Nonlinear Control For a classic controller design, the obvious variable for augmentation is ϕ. To maintain this variable at a equilibrium point, the controller has to produce an adequate signal to the VCO in such a way that ϕ tends to zero as time progresses. Since PLL is an intrinsic nonlinear system, the demand placed on a simple LPF to provide the necessary signal to achieve this task can become a burden.

The appropriate action needed to achieve this control objective, is therefore to re-configure the LPF structure in order that the system nonlinearity can be accommodated, while the signal from this controller to VCO is sufficient to maintain the overall system performance.

This design is possible if ϕ is accessible. Then, the LPF can take an extra signal path originating from the basis of ϕ so that its signal to the VCO can be modulated according to the deviation of its equilibrium point. The appropriate function for this extra signal may be in the form of a cubic characteristic, i.e. $v_c = v_2^3 = \phi^3$, which is shown by the cross-line in Figure 8.2.

This function enables the signal to have a substantially large magnitude when ϕ increases. On the other hand, when ϕ decreases, a weak signal is produced. Because of the nature of this cubic function, where both overshoot and transient period can respectively be reduced and shortened, this is, therefore, highly desirable for the control of the chaotic dynamics.

However, the concept cannot be realizable based on ϕ. This variable is not measurable, nor it can be obtained by the inverse of multivalued function of $h(\phi)$ due to its sinusoidal or triangular characteristics. Therefore, a suitable alternative is to model the nonlinear cubic function by taking the form of $v_2 = h(\phi)$, or $v_c = u = Kh^3(\phi)$, where K is a d.c. gain.

Note that the implementation of this circuitry configuration requires two multipliers which can induce a long response time. A simpler electronic version [13], allows the cubic function to be modified with the following characteristic:

$$v_c(t) = K|h(\phi)|h(\phi) \qquad (8.3.21)$$

The characteristics of this nonlinear signal based on the *sinusoidal* and a *triangular* PD output are depicted by the solid lines shown in Figures 8.2(a) and (b), respectively. It can be seen that these characteristics of the proposed nonlinear controller in the domain of $\pm\frac{\pi}{4}$ in particular, are very close to the true cubic function (cross-line).

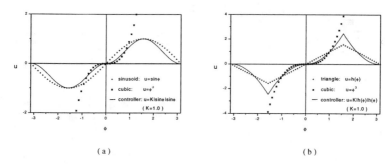

(a)   (b)

FIGURE 8.2
Characteristic curve of nonlinear controller (solid line), curve of a cubic function (cross-line), and PD characteristic (dotted line).
(a) a sinusoidal PD; (b) a triangular PD.

Having established the control action, the nonlinear controller 8.3.21 can now be introduced into the PLL for the control of the phase error of the system as stated in 8.2.18. The behavior of the system shown in Figure 8.1 is governed by the following differential equations:

$$\frac{d^2\phi}{dt^2} + \frac{1}{\tau_1}\frac{d\phi}{dt} + \frac{K_0}{\tau_1}h(\phi) = \frac{d^2\tilde{\theta}_i}{dt^2} + \frac{1}{\tau_1}\frac{d\tilde{\theta}_i}{dt} - \frac{K_0}{\tau_1}v_c(t),$$

where $v_c(t) = K|h(\phi)|h(\phi)$, or its normalized form with 8.2.11-8.2.17:

$$\ddot{\phi} + \beta\dot{\phi} + (K|h(\phi)| + 1)h(\phi) = \beta\sigma + \beta m\sin\Omega\tau + m\Omega\cos\Omega\tau \qquad (8.3.22)$$

The control signal v_c from the nonlinear controller is applied to the loop low pass filter together with the output of PD. It then follows that

$$\frac{v_c(t)}{R} + \frac{v_2(t)}{R} = -C\frac{dv_3(t)}{dt} - \frac{v_3(t)}{R} \qquad (8.3.23)$$

where $R = R_e = R_s$.

The physical circuitry implementation of this nonlinear part is presented in Figure 8.3, where the low pass filter (LPF) is used to cut off the high frequency component of v_2, as mentioned above.

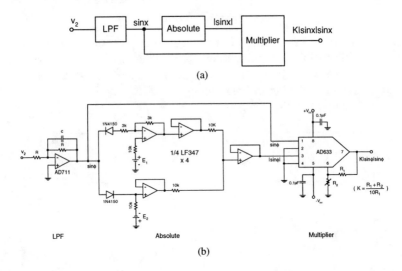

FIGURE 8.3
Block diagram (a) and the circuitry realization (b) of the nonlinear controller (for a sinusoidal PD).

Stability of Controlled PLL To analyze the stability of the controlled PLL, we employ the state-space model of the system by rewriting 8.3.22 for a sinusoidal PD as follows:

$$\dot{x}_1 = f_c(x_1, x_2) = x_2;$$

$$\dot{x}_2 = g_c(x_1, x_2)$$
$$= -\beta x_2 - (K|\sin x_1| + 1)\sin x_1 + \beta\sigma + \beta m \sin \Omega\tau + m\Omega \cos \Omega\tau$$

(8.3.24)

The Jacobian matrix associated with 8.3.24 about an equilibrium point $(\bar{x}_1, \bar{x}_2)$ is derived as follows [3]:

$$J_c = J_c(\bar{x}_1, \bar{x}_2) = \begin{bmatrix} \frac{\partial f_c}{\partial x_1} & \frac{\partial f_c}{\partial x_2} \\ \frac{\partial g_c}{\partial x_1} & \frac{\partial g_c}{\partial x_2} \end{bmatrix}_{(\bar{x}_1, \bar{x}_2)}$$

$$= \begin{bmatrix} 0 & 1 \\ -(2K|\sin \bar{x}_1| + 1)\cos \bar{x}_1 & -\beta \end{bmatrix} \quad (8.3.25)$$

Its characteristic equation is expressed as:

$$\det[sI - J_c] = \det \begin{bmatrix} s & -1 \\ (2K|\sin \bar{x}_1| + 1)\cos \bar{x}_1 & s+\beta \end{bmatrix}$$

$$= s^2 + \beta s + (2K|\sin \bar{x}_1| + 1)\cos \bar{x}_1 = 0$$

(8.3.26)

For the system to be stable, it is necessary to have all the roots of 8.3.26 located in the open left-half of the s-plane. The following conditions are therefore yielded:

$$\left.\begin{array}{r} \beta > 0 \\ (2K|\sin \bar{x}_1| + 1)\cos \bar{x}_1 > 0 \end{array}\right\} \quad (8.3.27)$$

Since $\beta > 0$ and $K > 0$ are true for a physical system, the above two inequalities hold when $-\frac{\pi}{2} < \bar{x}_1 < \frac{\pi}{2}$. Hence, the introduction of the nonlinear controller induces no additional condition to destabilize the PLL.

It should also be noted that the other advantage of this controller is its ability to broaden the phase lock-in range (PLIR). The PLIR is defined as the maximum frequency deviation of the input signal from the free-running frequency of VCO, namely, $(\omega - \omega_0)$, within which the phase-locked state can be maintained, where ω is the input frequency, and ω_0 is the VCO free-running frequency. This nonlinear controller serves as a means to improve this value added property.

This can be done by considering the PLL in the phase-locked state whose maximum frequency deviation m is zero, and the phase error ϕ settles down in a stable equilibrium point ϕ_e, i.e., $\ddot{\phi} = \dot{\phi} = 0$. The state equation 8.3.22 of the PLL system becomes:

$$(K|h(\phi_e)| + 1)h(\phi_e) = \beta\sigma \quad (8.3.28)$$

Since $h(\phi_e) \leq 1$ for a sinusoidal PD, whereas $h(\phi_e) \leq \pi/2$ for a triangular PD, this stable equilibrium point ϕ_e exists for

$$|\sigma| \leq \sigma_{max} \quad (8.3.29)$$

where

$$\sigma_{max} = \frac{1+K}{\beta} \quad \text{for a sinusoidal PD,}$$

and

$$\sigma_{max} = \frac{(1+K\pi/2)\pi}{2\beta} \quad \text{for a triangular PD.}$$

The σ_{max} gives the upper bound for phase-locked state, namely, the lock-in range (σ_L), as defined in literature. Whereas for $K = 0$, the σ_{max} signifies the PLIR for the original uncontrolled PLL. This agrees with the results given by [5]. Obviously, the PLIR is broadened by the controller since $K > 0$.

8.4 Chaotic Characteristics of the Controlled PLL

Having established the fundamental background for the nonlinear controller of PLL, the investigation of this chaotic system can proceed. In order to gain a full understanding of this particular chaos, a number of classical tools for the chaos evaluation, including *bifurcation diagram, Lyapunov exponent* and *basin of attraction* are employed. The knowledge gained from these investigations is useful for subsequent applications.

8.4.1 Bifurcation Map

A bifurcation map often provides an outline for easy understanding of chaotic behavior. In this case, the initial conditions $(x(0), y(0))$ are chosen as $(\pi - \arcsin(\beta\sigma), \beta\sigma)$ for a sinusoidal PD, where $\beta = 0.56$, $\sigma = 1.2749$, $m = 0.8$, $\Omega = 0.7$. For a triangular PD, the initial conditions are chosen as $(\pi - \beta\sigma, \beta\sigma)$ with $\beta = 0.56$, $\sigma = 1.4334$, $m = 0.11$, $\Omega = 3.04$.

A bifurcation process for both sinusoidal and triangular PD with respect to variable K is obtained as shown in Figure 8.4. It should be noted that there are several periodic windows between the chaotic regions for each PD, and the PLL system changes eventually to a period-1 state via a tangent bifurcation as K increases in either case.

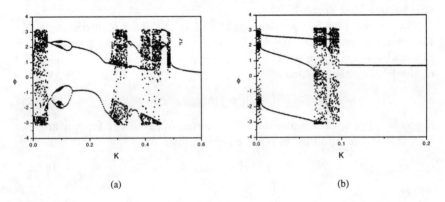

(a) (b)

FIGURE 8.4
One-parameter bifurcation diagram of controlled PLL system for sinusoidal PD and triangular PD.

To further illustrate the power of the nonlinear controller, it can be seen from Figure 8.5 that the chaotic state at $K = 0$ can be quickly switched into periodic state for $K = 0.58$ for the case of a sinusoidal PD, whereas $K = 0.18$ is assigned for a triangular PD. This result matches perfectly to the information depicted in Figure 8.4.

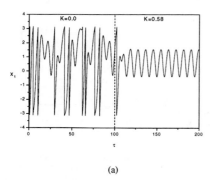 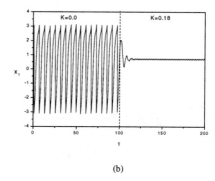

FIGURE 8.5
Normalized time series of phase error ϕ: (a) $K = 0.58$, (b) $K = 0.18$.

8.4.2 Lyapunov Exponent Spectrum

Another useful tool for evaluating a chaotic system is *Lyapunov exponent*. Lyapunov exponents are a generalization of the eigenvalues at an equilibrium point and of characteristic multipliers. They are used to determine the stability of any type of steady state behavior including quasi-periodic and chaotic solutions. The Lyapunov exponent λ_i describes the average rate of contraction ($\lambda_i < 0$) or expansion ($\lambda_i > 0$) in a particular direction near a particular trajectory [11]. For an attractor, the contraction must outweigh the expansion, and hence

$$\sum_{i=0}^{n} \lambda_i < 0.$$

What distinguishes a strange attractor from other types of attractor is the existence of at least one positive Lyapunov exponent. For the purpose of analyzing the controlled PLL, a Maximum Lyapunov exponent (λ_{max}) spectrum diagram with respect to the control gain K is presented in Figure 8.6.

These Lyapunov exponents are calculated on the basis of 1000 points starting from the 501st point to the 1500th points of the trajectory. As the transient, all of the 500 previous points are ignored. It can be seen that the state of the attractors indicated in Figure 8.6 and Figure 8.4 is in agreement with the corresponding values of K for both sinusoidal and triangular PD.

The negative peaks below the $\lambda_{max} = 0$ line, as indicated in Figure 8.6, are the periodic state, otherwise, the PLL is in a chaotic regime. Note that the λ_{max} around the bifurcation points may rapidly switch from positive to negative, or vice versa, as the control gain varies. The threshold for the turning is quite small and the system is sensitive to the initial values of

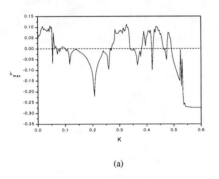 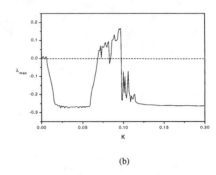

(a) (b)

FIGURE 8.6
Bifurcation diagram of Lyapunov exponent (λ_{max}) with respect to the control gain K for the controlled PLL with (a) sinusoidal PD: ($\beta = 0.56, \sigma = 1.2749, m = 0.8, \Omega = 0.7,$). Initial conditions: ($\pi - \sin^{-1}(\beta\sigma), \beta\sigma$); (b) triangular PD: ($\beta = 0.56, \sigma = 1.4334, m = 0.11, \Omega = 3.04,$). Initial conditions: ($\pi - \beta\sigma, \beta\sigma$).

the parameters. The result shows that the proposed nonlinear controller is effective.

It should be also noted that this spectrum can provide a sure means of determining the appropriate value of K to address the transform of attractors whether it is from chaos to order or vice versa.

8.4.3 Fractal Basin Boundaries for the Controlled PLL

It is reported in [7] that the basin boundaries of the PLL system are a fractal when the parameter values located above the Melnikov's curve are selected. These fractal basin boundaries represent the long chaotic transient well before the system settles down to a periodic attractor. This implies that, over certain regions in the parameter space, the system is virtually unpredictable over a long period of time. Obviously, for a practical system, this kind of behavior is undesirable.

In order to examine the effect of the proposed nonlinear controller, a number of initial condition phase planes are obtained for the controlled PLL. The parameter values $\Omega = 1.6, m = 0.6$ of the PLL were chosen for the task and the results are presented in Figures 8.7(a)–(e) for a variation of K being used.

In the absence of control, i.e. the control gain $K = 0$, the basin boundaries with a fractal structure in $(d\phi(0)/dt, \phi(0))$ plane are shown in Figure 8.7(a), a sinusoidal PD, and Figure 8.8(a), a triangular PD. The black zone represents the perturbed periodic solution of the second type (PS2), which is a solution for winding around the cylindrical coordinate system [7]. This region can give rise to the bifurcation into chaos even for a small external sinusoidal forcing signal. The attractor may or may not be in a phase-locked state, hence the relationship between an initial condition and the associated attractor is unpredictable.

In the case of the other colors, this represents the perturbed periodic solution of the first type (PS1) attractors. PS1 is a solution for trajectory revolving around the focus located at $(\phi_f + 2n\pi, 0)$, where $\phi_f = h^{-1}(\beta\sigma)(0 < \phi_f < \pi)$. The PS1 attractors can be distinguished from each other by their phase difference equal to an integral multiple of 2π [7].

The situation can be improved and the fractal basin boundaries are gradually smoothed out as the control gain K increases. This effect can be observed from Figures 8.7(b)–(e) for the sinusoidal PD, and Figures 8.8(b)–(e) for the triangular PD, respectively. In the case of the smooth basin boundary, the relationship between initial conditions and the associated attractor becomes more predictable.

Note that the required control force to drive the PLL from chaos-order, order-chaos, or order-order is small. From an engineering point of view, the presence of fractal basin boundaries in a PLL means that the pull-in time before synchronization will be extremely long, whereas for a smoothed basin boundary, due to the effect of the nonlinear controller, the PLL performs very well in these regions.

8.5 Synchronization with PLLs

Thus far, the fundamental issues of the chaotic dynamics in PLL have been fully addressed. We have also demonstrated that the newly proposed nonlinear controller is not only capable of taming the chaotic behavior, including the sensitivity of initial conditions, but its ability to broaden the phase lock-in range is also proved to be a value added design.

Given the background knowledge gained from the PLL study, a natural extension of the work is to tackle the synchronization problem using the nonlinear controller.

In this section, we apply the nonlinear controller to control two driven PLLs. The basin boundaries of the initial conditions of the PLLs with and without the controller are studied.

8.5.1 PLLs Model for Synchronization

Following Pecora and Carroll [35], a chaos synchronization system can be constructed as follows:

Consider an n-dimensional dynamical system

$$\dot{u} = f(u), \qquad (8.5.30)$$

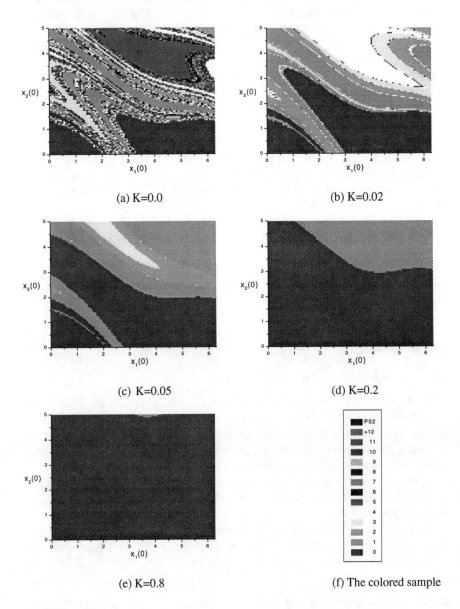

FIGURE 8.7
Basin boundaries of initial conditions of controlled PLL for sinusoidal PD.

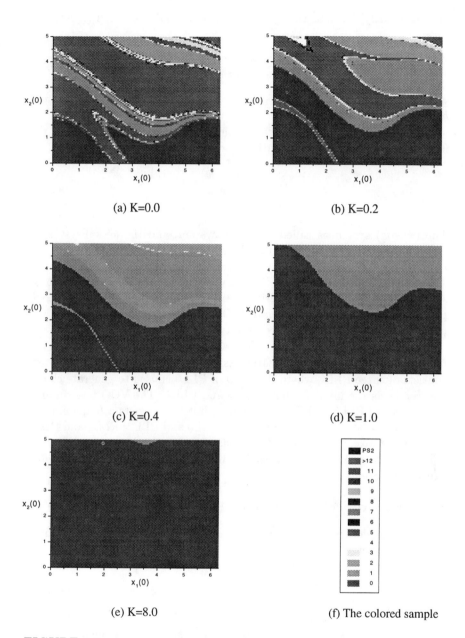

FIGURE 8.8
Basin boundaries of initial conditions of controlled PLL for triangular PD.

and arbitrarily dividing the system u into two subsystems, i.e., let $u = (v, w)$:

$$\begin{aligned} \dot{v} &= g(v,w) \\ \dot{w} &= h(v,w) \end{aligned} \right\} \quad (8.5.31)$$

where $v = (u_1, \cdots, u_m), g = (f_1(u), \cdots, f_m(u)), w = (u_{m+1}, \cdots, u_n)$, and $h = (f_{m+1}(u), \cdots, f_n(u))$.

Create a new subsystem w' which is identical to the subsystem w as the first response system by duplication, such that

$$\begin{aligned} \dot{v} &= g(v,w) \\ \dot{w} &= h(v,w) \\ \dot{w}' &= h(v,w') \end{aligned} \right\} \quad (8.5.32)$$

The $(v - w)$ system is called the drive system, and the w' subsystem is called the response system. Under the right conditions, where $w' - w \to 0$ as $t \to \infty$, i.e., the two subsystems are synchronized with each other. This situation occurs when all of the Lyapunov exponents of the response subsystem w' for a particular driven trajectory (so called conditional Lyapunov exponents, CLE) are negative. Otherwise, even if one of the CLE is positive, synchronization between the two subsystems fails.

Using the above concept, Endo and Chua [8, 9] proposed a method for realizing PLL chaos synchronization as shown in Figure 8.9. The system consists of three PLLs, where the VCO of PLL0 is used to generate chaos, and serves as the input for both PLL1 and PLL2. The VCO free-running frequencies of PLL0 and PLL1 are equal, but there exists a frequency detuning $\Delta\omega_{vco}$ between the VCO of both PLL0 and PLL2. PLL1 and PLL2 are identical except for the VCO free-running frequencies.

Here the loop filters are assumed to be the lag filters with the following transfer functions:

$$F_0(s) = 1/(1 + \tau_0 s), F_1(s) = 1/(1 + \tau_1 s), F_2(s) = 1/(1 + \tau_2 s). \quad (8.5.33)$$

The phase detector characteristics $h(\phi)$ is a 2π-periodic triangular function of the form

$$h(\phi) = \begin{cases} \phi, & |\phi| < \frac{\pi}{2} \\ -\phi + \pi, & \frac{\pi}{2} < \phi < \frac{3\pi}{2} \end{cases}$$

where $h(\phi) = h(\phi + 2\pi)$.

For completeness, the following parameters should be defined:
the phase errors:

$$\phi_0 \equiv \theta_i(t) - \theta_o^0(t), \phi_1 \equiv \theta_{in}(t) - \theta_o^1(t), \phi_2 \equiv \Delta\omega_{vco}t + \theta_{in}(t) - \theta_o^2(t);$$

the detuning between the carrier frequency and the VCO free-running fre-

Synchronization with PLLs

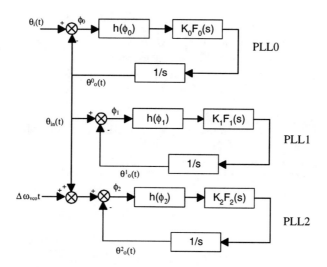

FIGURE 8.9
Configuration of the synchronized PLL system.

quency of the PLL0:

$$\Delta\omega \equiv \omega_c - \omega_{vco0}, \theta_{in}(t) = \theta_o^0(t) = \theta_i(t) - \phi_0(t); \text{ and}$$

the state variables:

$$x_1 = \phi_0, x_2 = d\phi_0/d\tau, x_3 = \phi_1, x_4 = d\phi_1/d\tau, x_5 = \phi_2, x_6 = d\phi_2/d\tau$$

where the time variable $\tau = \omega_n t$.

The normalized differential equations of the phase errors for the system shown in Figure 8.9 can be written as the following *6th-order* system of nonautonomous differential equations:

$$\left.\begin{aligned}
\frac{dx_1}{dt} &= x_2 \\
\frac{dx_2}{dt} &= -\beta x_2 - h(x_1) + \beta\sigma + m\beta\sin\Omega t + m\Omega\cos\Omega t \\
\frac{dx_3}{dt} &= x_4 \\
\frac{dx_4}{dt} &= -r_1\beta x_4 - r_1 h(x_3) + h(x_1) + \\
&\quad \beta(r_1 - 1)(-x_2 + \sigma + m\sin\Omega t) \\
\frac{dx_5}{dt} &= x_6 \\
\frac{dx_6}{dt} &= -r_2\beta x_6 - r_2 h(x_5) + h(x_1) + \\
&\quad \beta(r_2 - 1)(-x_2 + \sigma + m\sin\Omega t) + \beta r_2\delta
\end{aligned}\right\} \quad (8.5.34)$$

where $\omega_n = \sqrt{K_0/\tau_0}$ (natural angular frequency), $\beta = \omega_n/K_0$ (normalized natural angular frequency), $r_1 = \tau_0/\tau_1$ (ratio of the loop filter time constants), $r_2 = \tau_0/\tau_2$ (ratio of the loop filter time constants), $\sigma = \Delta\omega/\omega_n$ (normalized frequency detuning of the input FM carrier frequency), M

(maximum angular frequency deviation), $m = M/\omega_n$ (normalized maximum angular frequency deviation), $\Omega = \omega_m/\omega_n$ (normalized modulation frequency), $\delta = \Delta\omega_{vco}/\omega_n$ (normalized frequency detuning of the PLL2).

This synchronization model works within a 10% difference in initial conditions provided that the VCO free-running frequencies are not large [8, 9].

8.5.2 PLLs Synchronization with Nonlinear Controller

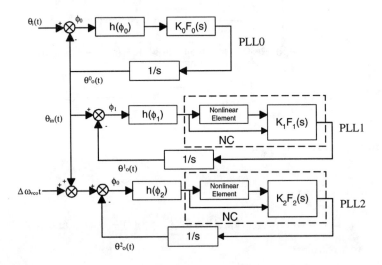

FIGURE 8.10
Block diagram of the synchronized PLL system with control.

To introduce the nonlinear controller, $u = -k|h(\phi)|h(\phi)$, involves no extra effort. All the previous implementations are applied, and the new model is shown in Figure 8.10. The phase error equation 8.5.34 which includes the embedded nonlinear controller (NC) takes the form:

$$\left.\begin{aligned}
\frac{dx_1}{dt} &= x_2 \\
\frac{dx_2}{dt} &= -\beta x_2 - h(x_1) + \beta\sigma + m\beta\sin\Omega t + m\Omega\cos\Omega t \\
\frac{dx_3}{dt} &= x_4 \\
\frac{dx_4}{dt} &= -r_1\beta x_4 - (r_1 + K_1|h(x_3)|)h(x_3) + h(x_1) + \\
&\quad \beta(r_1 - 1)(-x_2 + \sigma + m\sin\Omega t) \\
\frac{dx_5}{dt} &= x_6 \\
\frac{dx_6}{dt} &= -r_2\beta x_6 - (r_2 + K_2|h(x_5)|)h(x_5) + h(x_1) + \\
&\quad \beta(r_2 - 1)(-x_2 + \sigma + m\sin\Omega t) + \beta r_2\delta
\end{aligned}\right\} \quad (8.5.35)$$

where K_1 and K_2 are the gains of NC for PLL1 and PLL2, respectively.

To assess the effect of the nonlinear controller on the synchronization of chaos in the PLL system, a study has been carried out to examine the

TABLE 8.1
RMSE of controlled and uncontrolled PLLs: small-damping case ($\zeta = 0.28$).

parameter variations	uncontrolled system	controlled system
$r_1 = 1.0, r_2 = 1.0, \delta = 0.01$	0.0369	0.0036
$r_1 = 1.0, r_2 = 1.0, \delta = 0.1$	0.2885	0.0364
$r_1 = 1.05, r_2 = 0.95, \delta = 0.0$	0.5599	0.0876
$r_1 = 1.1, r_2 = 0.9, \delta = 0.0$	1.4486	0.1674

TABLE 8.2
RMSE of controlled and uncontrolled PLLs: large-damping case ($\zeta = 0.635$).

parameter variations	uncontrolled system	controlled system
$r_1 = 1.05, r_2 = 0.95, \delta = 0.0$	0.0259	0.0236
$r_1 = 1.0, r_2 = 1.0, \delta = 0.01$	0.0126	0.0059
$r_1 = 1.0, r_2 = 1.0, \delta = 0.1$	1.2615	0.0594
$r_1 = 1.0, r_2 = 1.0, \delta = 0.15$	1.1493	0.0878

performance of synchronization for both control and without control cases.

For small damping ($\zeta = 0.28$), the parameters were $\beta = 2\zeta = 0.56$, $\sigma = 1.43341$, $m = 0.097$, $\Omega = 0.637$, and the initial conditions $x_1(0) = 2.3389$, $x_2(0) = 2.8027$ for the PLL0, $x_3(0) = 0.1$, $x_4(0) = 0.1$ for the PLL1, and $x_5(0) = 0.09$, $x_6(0) = 0.09$ for the PLL2.

For large damping $\zeta = 0.635$, the parameters were $\beta = 1.27$, $\sigma = 1.15$, $m = 0.053$, $\Omega = 0.2$, and the initial conditions $x_1(0) = 1.6811$, $x_2(0) = 3.4605$ for the PLL0, $x_3(0) = 0.1$, $x_4(0) = 0.1$ for the PLL1, and $x_5(0) = 0.09$, $x_6(0) = 0.09$ for the PLL2.

Note that there exists a 10% difference between the initial conditions for the PLL1 and PLL2 in both cases.

Based on the Runge-Kutta-Gill algorithm, we first calculate the Root Mean Square Error (RMSE) of 50 points of trajectories in the Poincaré section starting from the 51st step. The initial 50 points have not been taken into the calculation because of transiency. The results obtained for the systems with and without control are tabulated in Tables 8.1 and 8.2. The controller gains used for the controlled system in both small and large damping cases are: $K_1 = K_2 = 0.9$.

It can be seen that the RMSE in the controlled system is less than that in the uncontrolled system in both the small and large damping cases. Furthermore, for some parameter variations, the uncontrolled system cannot maintain synchronization, while the controlled system maintains its synchronization mode.

The performances observed on a Poincaré section for both the small

($\zeta = 0.28$) and large ($\zeta = 0.635$) damping cases are shown in Figures 8.11 and 8.12, respectively.

The traces of ϕ_1 versus ϕ_2 (X_3 versus X_5) on a Poincaré section in the uncontrolled system are chaotic in form and are shown in Figures 8.11(a) and 8.12(a); while for the controlled system, straight lines are observed and shown in Figures 8.11(b) and 8.12(b), respectively. This vividly indicates that the system is in synchronization mode.

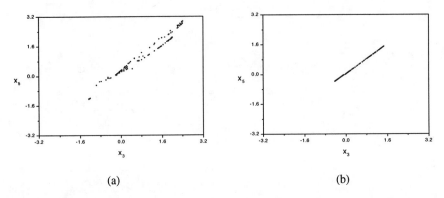

FIGURE 8.11
Traces between ϕ_1 and ϕ_2 observed on a Poincaré section in the small damping case; parameters: ($\zeta = 0.28$), $r_1 = 1.0$, $r_2 = 1.0$, $\delta = 0.1$.
(a) uncontrolled system, $K_1 = K_2 = 0$; (b) controlled system, $K_1 = K_2 = 0.9$.

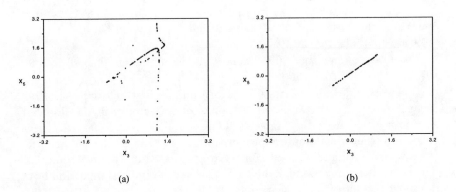

FIGURE 8.12
Traces between ϕ_1 and ϕ_2 observed on a Poincaré section in the large damping case; parameters: ($\zeta = 0.635$), $r_1 = 1.0$, $r_2 = 1.0$, $\delta = 0.15$.
(a) uncontrolled system, $K_1 = K_2 = 0$; (b) controlled system, $K_1 = K_2 = 0.9$.

8.5.3 Sensitive Dependence on Initial Conditions in Synchronization of PLLs with Nonlinear Control

As mentioned above, the sensitive dependence on initial conditions is an important issue to be addressed and form a fundamental property of a chaotic system. This sensitivity in synchronization of PLLs is also studied here. The results are shown in Figure 8.13.

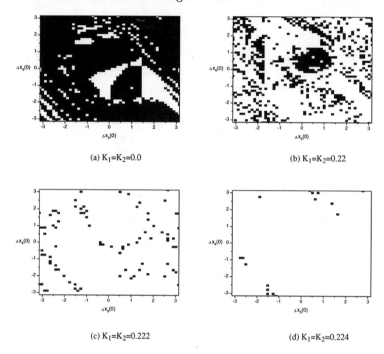

FIGURE 8.13
Basin boundaries of the initial conditions associated with chaos synchronization in PLL system. Parameters: $\zeta = 0.28$, $\beta = 2\zeta = 0.56$, $\sigma = 1.43341$, $m = 0.097$, $\Omega = 0.637$, $r_1 = 1.0$, $r_2 = 1.0$, $\delta = 0.1$; initial conditions: $(X_1(0), X_2(0)) = (2.3389, 2.8027)$, $(X_3(0), X_4(0)) = (0.1, 0.1)$, $(X_5(0), X_6(0)) = (X_3(0) + \Delta X_5(0), X_4(0) + \Delta X_6(0))$.

Similar to the previous case, the state of the system is examined by means of the RMSE calculated starting from the 51st to the 100th points of the trajectory on the Poincaré section. The figures are depicted based on the criterion of $RMSE = 0.28$. The black region represents $RMSE \geq 0.28$ and the white region is for $RMSE < 0.28$. These clearly indicate that the systems are in asynchronization and synchronization modes, respectively.

It is also seen from Figures 8.13(a)–(c) that the basin boundaries in $\Delta x_5(0)$ versus $\Delta x_6(0)$ plane are fractal for an uncontrolled system, but the situation improves with a small increase of controller gains. Hence, we can conclude that the synchronization state in a PLL system is sensitive to initial conditions and the nonlinear controller offers the benefit of broadening the synchronization basin of initial conditions, which eventually enhances the robustness of the system.

8.6 Conclusion

A new nonlinear controller has been proposed in the chapter to contend sensitivity of the initial conditions. To justify this claim, the controlled PLL was repeatedly evaluated by assessing its bifurcation process as well as the confirmation of the Lyapunov exponent spectrum. This was then further verified by its fractal basin of the initial conditions with and without the effect of the controller. It has been aptly illustrated that the sensitivity to initial conditions can influence the phase space of an attractor. But, this can be confined by implementing appropriate controller design.

This knowledge forms a solid foundation for the development of PLL synchronization. The results have indicated that the introduction of this nonlinear controller can also greatly improve the overall performance.

In the light of widely available PLL within the communication industry, this study could lead to the use of chaos for secure communications.

References

[1] F. Ayrom and G-Q. Zhong, "Chaos in Chua's circuit," *IEE Proc.*, Pt. D, vol. 133, pp. 307-312, 1986.

[2] E. Bradley and D. E. Straub "Using chaos to broaden the capture range of a phase-locked loop: Experimental verification," *IEEE Trans. Circ. Sys.*, I, vol. 43, pp. 914-922, 1996.

[3] G. Chen, "Control and anticontrol of chaos," *Proc. 1st International Conference on Control of Oscillations and Chaos*, vol. 2, pp. 181-186, St. Petersburg, Russia, August 1997.

[4] G. Chen and X. Dong, "On feedback control of chaotic continuous-time systems," *IEEE Trans. Circ. Sys.*, I, vol. 40, pp. 591-601, 1993.

[5] T. Endo, "A review of chaos and nonlinear dynamics in phase-locked loops," *J. of the Franklin Institute*, vol. 331B, pp. 859-902, 1994.

[6] T. Endo and L. O. Chua, "Chaos from phase-locked loops," *IEEE Trans. Circ. sys.*, vol. 35, pp. 987-1003, 1988.

[7] T. Endo and L. O. Chua, "Bifurcation diagrams and fractal basin boundaries of phase-locked loop circuits," *IEEE Trans. Circ. Sys.*, vol. 35, pp. 534-540, 1990.

[8] T. Endo and L. O. Chua, "Synchronization of chaos in phase-locked loops," *IEEE Trans. Circ. Sys.*, vol. 38, pp. 1580-1588, 1991.

[9] T. Endo and L. O. Chua, "Synchronizing chaos from electronic phase-locked loops," *Int. J. of Bifur. Chaos*, vol. 1, pp. 701-710, 1991.

[10] B. B. Mandelbrot, *Fractals: Forms, Chance, and Dimension*, W. H. Freeman and Co., 1977.

[11] T. S. Parker and L. O. Chua, "Chaos: A tutorial for engineers," *Proc. IEEE*, vol. 75, pp. 982-1008, 1987.

[12] L. M. Pecora and T. L. Carroll, "Synchronization in chaotic systems," *Phys. Rev. Lett.*, vol. 64, pp. 821-824, 1990.

[13] K. S. Tang and K. F. Man, "An alternative Chua's circuit implementation," *IEEE Int. Symposium on Industrial Electronics Proc.*, pp. 441-444, Pretoria, South Africa, July 1998.

[14] G-Q. Zhong and F. Ayrom, "Experimental confirmation of chaos from Chua's circuit," *Int. J. Circ. Theory Appl.*, vol. 13, pp. 93-98, 1985.

9

Frequency Domain Methods for Chaos Control

Michele Basso, Roberto Genesio, Lorenzo Giovanardi, Alberto Tesi

Dipartimento di Sistemi e Informatica
Università di Firenze
Via S. Marta, 3 - 50139 Firenze, Italy
basso,genesio,giovanardi,atesi@dsi.unifi.it

Abstract

It is well known that the problem of controlling chaos is strictly related to the stabilization of one of the infinite unstable periodic orbits that coexist in the underlying chaotic attractor. This work introduces a general framework based on classical frequency domain tools concerning the stabilization of periodic orbits in an important class of periodically forced nonlinear systems. Some results ensuring both existence and input-output small signal L_2-stability of a family of T-periodic solutions are provided. Such results are exploited for the design of finite dimensional linear time invariant controllers that ensure a larger stability interval of T-periodic solutions without requiring a high control energy. The model of a CO_2 laser is used throughout the chapter as a vehicle to illustrate the features of the approach.

9.1 Introduction

The past two decades have witnessed an increasing interest in nonlinear dynamics across a broad ranges of disciplines. The issues related to bifurcations and chaotic motions have generated a huge number of papers in several old and new journals of engineering, physics, and natural and social sciences. In particular, the problem of controlling chaos has received a lot of attention from both a theoretical and an application point of view [21, 4]. Indeed, the ability of controlling nonlinear dynamics can produce several practical advantages such as pushing the systems to work near their limits of performance by considering oscillatory regimes, reducing the control energy needed to achieve the desired behavior, and so on.

Since the beginning, it became clear that one of the most appealing approaches for controlling chaos was the stabilization of one of the infinite unstable periodic orbits that coexist in the chaotic attractor. This problem, whose main feature is the requirement of a low control energy, was first considered by Ott, Grebogy and Yorke in [25], originating the so-called OGY methods [20]. A distinct approach, known as "delayed feedback" technique, was subsequently proposed by Pyragas in [19], resulting the basic step for other related works [22, 13].

Although based on quite different properties and design procedures, both methods lead to feedback controllers utilizing small perturbation signals and exploiting inherent characteristics of chaos. Due to this last fact, the above methods as well as most of the literature on controlling chaos appear to be somewhat distinct from classical feedback control theory, where the underlying nonlinear dynamics are not fully taken into account in the controller design. Since the considered problem amounts to stabilize a periodic solution, it is however clear that classical feedback control tools can be exploited to some extent, as it has already been done in the case of bifurcation control [1, 24, 6, 2, 5].

The main purpose of the present work is to build a bridge between the problems of controlling complex dynamics and the classical feedback control theory. The basic idea is to introduce a general framework based on classical frequency domain tools for the stabilization of periodic orbits in a general class of systems. Some preliminary attempts in this direction have been reported in [7, 3].

The class of systems under investigation is presented in Section 9.2. Such systems are composed of the feedback interconnection of a given plant, described by a nonlinear time invariant state-space model subject to a periodic forcing input, and a feedback controller, assumed to be finite dimensional linear time invariant, to be designed. A single-mode CO_2 laser is introduced as an application example. When the cavity losses are modulated by a sinusoidal signal of increasing amplitude, the laser exhibits a

Introduction

cascade of period doubling bifurcations leading to chaos [23].

In Section 9.3, some results on the existence of periodic solutions for the considered class of systems are given. Such conditions consider both the uncontrolled case, i.e., the feedback controller is set to zero, and the controlled case.

Several types of stability can be used with reference to periodic solutions. When dealing with forced systems, it is customary to investigate input-output stability in place of asymptotic stability. In Section 9.4, the input-output small signal L_2-stability problem of periodic solutions is considered. Employing a linearization approach, it is first pointed out that this problem can be formulated as a standard input-output L_2-stability problem of the feedback interconnection of a dynamical system, described by a rational transfer function matrix, and a memoryless system, described by a periodic gain. Such a formulation makes it possible to employ frequency criteria in the L_2 setting, such as the multivariable circle criterion [25], the Willems criterion [26], and the most general Integral Quadratic Constraint (IQC) tools [17], to provide sufficient conditions for the sought small signal L_2-stability of the periodic orbits. In particular, we focus on the most well-known circle criterion, though quite similar developments can be pursued for the other criteria. The material of the section is illustrated via the laser application example.

Finally, in Section 9.5, it is shown how existence and stability results of the previous sections can be effectively applied for controlling complex dynamics. In particular, the following scenario is considered: i) the equilibrium point of the unforced system is stable; ii) for an increasing amplitude of the periodic forcing term the system displays a stable periodic solution until a period doubling bifurcation happens for a given amplitude; iii) for larger amplitudes, the system exhibits a full cascade of period doubling bifurcations leading to chaos. The control objective consists in enlarging the stability interval of the T-periodic solutions of the uncontrolled system without requiring a high control energy. Said another way, a finite dimensional linear time invariant controller has to be designed such that the bifurcation diagram simply consists of shifting the period-doubling bifurcation to higher values of the forcing amplitude. A general result ensuring both existence and input-output small signal L_2-stability of family of T-periodic solutions is first given. The design method is based on this result and consists in determining the controller of a given (small control energy) class that makes the family of T-periodic solutions as large as possible. Finally, such design method is applied to the laser example.

Notation

Z: set of integer numbers;
R: set of real numbers;
$\mathbf{R}^n$: space of n-component real (column) vectors;
$\mathbf{R}^{r \times n}$: space of $(r \times n)$ real matrices;
$\mathcal{D}[\cdot] : \mathbf{R}^n \to \mathbf{R}^{n \times n}$: *diagonalization operator* mapping a vector $v \in \mathbf{R}^n$ into a diagonal matrix $M \in \mathbf{R}^{n \times n}$ such that $m_{ii} = v_i$, $i = 1, \ldots, n$;
$\mathcal{S}[\cdot] : \mathbf{R}^{r \times n} \to \mathbf{R}^{r \cdot n}$: *stack operator* mapping a matrix $M \in \mathbf{R}^{r \times n}$ into a vector $v \in \mathbf{R}^{r \cdot n}$ such that $v_{n(i-1)+j} = m_{ij}$, $i = 1, \ldots, r$, $j = 1, \ldots, n$;
$O_n \in \mathbf{R}^{n \times n}$: null matrix of order n;
$I_n \in \mathbf{R}^{n \times n}$: identity matrix of order n;
$e_i \in \mathbf{R}^n$: i-th element of the n-dimensional canonical base;
$x(t) \in \mathbf{R}^n$: n-valued time domain signal;
$\|x\|_\infty \doteq \sup_{t \geq 0} \max_{k \in [1,n]} |x_k(t)|$: ∞-norm of $x(t)$;
$\|x\|_2 \doteq [\int_0^\infty x(t)^\top x(t)\, dt]^{1/2}$: 2-norm of $x(t)$;
L_2: vector space of signals with finite 2-norm;
$X(s)$: Laplace transform of $x(t)$.

9.2 System Setup

The *plant* system Σ_P depicted in Figure 9.1 is the nonlinear system under investigation, possessing a forcing input $w \in \mathbf{R}^{n_w}$, and a corresponding output $z \in \mathbf{R}^{n_z}$ to be controlled. System Σ_P is also provided with a sensed output $y \in \mathbf{R}^{n_y}$ and a control input $u \in \mathbf{R}^{n_u}$.

We restrict the analysis to the class of nonlinear systems Σ_P described

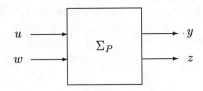

FIGURE 9.1
Plant system Σ_P.

System Setup

by the following state-space representation:

$$\begin{cases} \dot{x} &= f(x) + g(x)w - B\,u \\ y &= C\,x \\ z &= h(x) \end{cases} \quad (9.2.1)$$

where the state vector x takes values in some subset $X \in \mathbf{R}^n$ containing the origin, $B \in \mathbf{R}^{n \times n_u}$ and $C \in \mathbf{R}^{n_y \times n}$ are constant matrices, and

$$f : X \to \mathbf{R}^n \; , \; g : X \to \mathbf{R}^{n_w} \; , \; h : X \to \mathbf{R}^{n_z}$$

are nonlinear functions. This class is actually quite general, comprised of a large number of widely studied chaotic systems such as Duffing, Van der Pol, Chua and Toda oscillators, to name but a few.

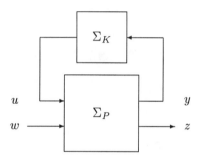

FIGURE 9.2
Controlled system Σ.

Figure 9.2 shows the *controlled* system Σ resulting from the feedback interconnection of Σ_P and the *controller* system Σ_K to be designed. The latter is assumed to be finite dimensional linear time invariant and can therefore be represented by an $(n_y \times n_u)$ rational transfer function matrix $K(s)$. Obviously, Σ_K admits an equivalent state space representation of the form

$$\begin{cases} \dot{x}_K &= A_K x_K + B_K y \\ u &= C_K x_K + D_K y \end{cases} \quad (9.2.2)$$

where $x_K \in \mathbf{R}^{n_K}$ is the state vector and the matrices A_K, B_K, C_K, D_K are such that

$$K(s) = C_K(sI - A_K)^{-1} B_K + D_K \; . \quad (9.2.3)$$

In this work, we are interested in considering the behavior of the system Σ subject to periodic forcing inputs w of varying amplitude. To this purpose, we suppose that w belongs to the following linearly parameterized set of nominal T-periodic signals

$$\mathcal{W} \doteq \{w_\mu(t) = \mu\, \overline{w}(t)\,,\ \mu \geq 0\} \qquad (9.2.4)$$

where the amplitude μ will serve as a bifurcation parameter and $\overline{w}(t) = \overline{w}(t+T)$ is a given non-trivial periodic signal.

Some topological properties of the set X as well as some smoothness conditions on functions f, g and h will be outlined in Sections 9.3 and 9.4. For the moment, we assume w.l.o.g. that $f(0) = 0$, so that $x = 0$ is an equilibrium point of the unforced uncontrolled system ($w \equiv 0$, $u \equiv 0$).

Example 9.1

A *single-mode CO_2 laser* can exhibit a cascade of period doubling bifurcations leading to chaos when the cavity losses are modulated by a sinusoidal signal of increasing amplitude.

We consider the control-relevant model described in [23], rewritten after a shift of the state variables that moves the equilibrium of the unforced system into the origin

$$\begin{cases} \dot{x}_1 &= k_0\,[x_2 + w - u] \\ \dot{x}_2 &= -\Gamma x_2 + \gamma_R x_3 - \eta[(x_2+1)\exp(x_1) - 1] \\ \dot{x}_3 &= -\alpha x_3 + \beta x_2 \\ y &= x_1 \\ z &= x_1\,. \end{cases} \qquad (9.2.5)$$

Here, x_1 is proportional to the logarithm of the laser intensity, x_2 is related to the populations of the two lasing states, x_3 also takes into account the global populations of the manifolds of rotational levels and w is the forcing term belonging to the class of signals

$$\{w_\mu(t) = \mu \cos \tfrac{2\pi}{T} t\,,\ T = 10^{-5}\text{s}\,,\ \mu \geq 0\}.$$

The remaining parameters are set as follows

$$\begin{array}{ll} k_0 = 3.18 \times 10^7 \text{s}^{-1}\,, & \gamma_R = 7.0 \times 10^5 \text{s}^{-1}\,, \\ \Gamma = 7.05 \times 10^6 \text{s}^{-1}\,, & \eta = 9.129 \times 10^4 \text{s}^{-1}\,, \\ \alpha = 6.767 \times 10^5 \text{s}^{-1}\,, & \beta = 6.626 \times 10^6 \text{s}^{-1}\,. \end{array}$$

If we set

$$f(x) = \begin{bmatrix} k_0 x_2 \\ -\Gamma x_2 + \gamma_R x_3 - \eta[(x_2+1)\exp(x_1) - 1] \\ \beta x_2 - \alpha x_3 \end{bmatrix},$$

$$B = [k_0\ 0\ 0]^T\,,\quad C = [1\ 0\ 0]\,,\quad g(x) = B\,,\quad h(x) = x_1\,,$$

the model (9.2.5) has clearly the form (9.2.1) of Σ_P.

9.3 Existence of Periodic Solutions

In this section, we provide some results about the *existence* of periodic solutions of systems described by (9.2.1) with the periodic forcing term (9.2.4). This, together with the related *stability* issue, has been an important research topic for several generations of math scientists, and many, though somewhat disparate, results are available. For an exhaustive review, the reader is referred to [11].

9.3.1 Uncontrolled Case

As a first step, we will consider the uncontrolled system only. This means that in the description (9.2.1) of Σ_P the control input is set to zero, i.e., $u(t) \equiv 0$. Moreover, except for some pathological structures of h, periodicity of $z_\mu(t)$ follows from that of $x_\mu(t)$, so we can concentrate on the state equations only.

Among the methods developed with the aim of establishing existence of periodic solutions, those general enough to have an engineering interest can be classified in two groups, both having their origin in the works of H. Poincaré. The first is the group of *topological methods*, according to which periodic solutions of a generic periodic system

$$\dot{x} = f(x,t) , \quad f(x,t) = f(x, t+T) ,$$

are seen as fixed points of a corresponding *Poincaré-Andronov* operator (see [11] for details). The second group consists of *perturbation methods*, that work under the assumption that the given system is a "perturbation" of a nominal one whose solution is known. Due to the nature of these latter methods, they can be applied to a quite general class of systems, including the one under investigation.

To this purpose, we consider the uncontrolled system subject to a T-periodic forcing input $w_\mu(t) \in \mathcal{W}$. Suppose that the *Jacobian* of f at $x = 0$ exists, i.e., we can define

$$A \doteq \left. \frac{\partial f(x)}{\partial x} \right|_{x=0} .$$

Under the quite mild assumption that the set X is open and connected, the following existence result is obtained [4].

THEOREM 9.1
Suppose that

i) the matrix A has no eigenvalues on the imaginary axis;
ii) $g(x) \neq 0 \ \forall \ x \in X$.

Then, there exist positive constants $\hat{\mu}$ and ρ such that for all $\mu \in (0, \hat{\mu})$ the uncontrolled system subject to the forcing T-periodic input $w_\mu \in \mathcal{W}$ possesses a unique non-trivial T-periodic solution $x_\mu(t)$ with the property that $\|x_\mu(t)\| \leq \rho\mu$.

9.3.2 Controlled Case

Theorem 9.1 can be easily extended to the controlled system Σ, once the controller Σ_K is described by (9.2.2). Indeed, it is enough to replace the matrix A in assumption i) with the matrix

$$\begin{bmatrix} A - BD_K C & -BC_K \\ B_K C & A_K \end{bmatrix}. \quad (9.3.6)$$

It should however be noted that the periodic output solutions $z_\mu(t)$ of the controlled system Σ are in general modified with respect to the uncontrolled system (Σ_P with $u(t) \equiv 0$). Indeed, if the original T-periodic solution must be maintained in Σ, the following constraint on the controller transfer function must be imposed

$$K\left(jk\frac{2\pi}{T}\right) = 0 \quad \forall k \in \mathbf{Z}. \quad (9.3.7)$$

This constraint yields a class of infinite dimensional controllers which can be expressed in the form

$$K(s) = F(s)\left[1 - \exp(-sT)\right], \quad (9.3.8)$$

where $F(s)$ is a transfer function matrix. Such controllers, which are a generalization of the "delayed feedback" controller introduced by Pyragas [19], have been considered in several papers dealing with complex dynamics [22, 13, 7, 3].

Example 9.2
Consider again system (9.2.5). Since

$$A = \begin{bmatrix} 0 & k_0 & 0 \\ -\eta & -\Gamma - \eta & \gamma_R \\ 0 & \beta & -\alpha \end{bmatrix} = \begin{bmatrix} 0 & 31.8 & 0 \\ -0.09129 & -7.141 & 0.7 \\ 0 & 6.626 & -0.6767 \end{bmatrix} \times 10^6$$

has eigenvalues

$$[-0.1905 + 0.4773j, -0.1905 - 0.4773j, -7.4370]^\top \times 10^6$$

and the function $g(x) = B \neq 0$, assumptions i) and ii) of Theorem 9.1 are satisfied, thus guaranteeing existence of periodic solutions for some interval $(0, \hat{\mu})$.

9.4 Stability of Periodic Solutions

When dealing with forced systems, it is a standard approach in control systems theory to investigate input-output stability in place of asymptotic stability.

In this work, we consider the L_2-stability setting. Assume that system Σ is operating in a periodic regime, where (w_μ, z_μ) represents a nominal input-output signal pair of period T, and consider a generic perturbation δw of the input. The following is the standard definition of L_2-stability.

DEFINITION 9.3[1] *The solution $z_\mu(t)$ of system Σ is L_2-stable if there exists a positive constant γ such that $\|z-z_\mu\|_2 \leq \gamma\|\delta w\|_2 \; \forall \; \delta w \in L_2$.*

The above definition concerns global L_2-stability of a given periodic solution. Here we are more interested in local stability properties as captured by next definition.

DEFINITION 9.4 *The solution $z_\mu(t)$ of system Σ is small signal L_2-stable if there exist positive constants γ and ϵ such that $\|z-z_\mu\|_2 \leq \gamma\|\delta w\|_2 \; \forall \; \delta w \in L_2, \; \|\delta w\|_\infty \leq \epsilon$.*

Note that in Definition 9.4 the perturbation δw is required to be uniformly small over the time interval. For more details about such definitions, see [25, 14, 10].

9.4.1 The Linearized System

Hereafter, small signal L_2-stability of the T-periodic solution $z_\mu(t)$ of system Σ will be considered. In the analysis of nonlinear systems, classical linearization techniques often represent a powerful tool for investigating local stability of a given motion. For the problem at hand, assuming that the nonlinear functions f, g and h are sufficiently smooth, linearization of Eqs. (9.2.1) around $x_\mu(t)$ leads to the linear periodic system

$$\begin{cases} \delta\dot{x} &= A\,\delta x - \Delta_\mu(t)\,\delta x + \Phi_\mu(t)\,\delta w - B\,\delta u \\ \delta y &= C\,\delta x \\ \delta z &= \Psi_\mu(t)\,\delta x \end{cases} \quad (9.4.9)$$

---

[1] The definition is also referred to as L_2-stability with finite gain and no bias [25, 14].

where

$$\Delta_\mu(t) = A - \left.\frac{\partial f(x)}{\partial x}\right|_{x_\mu(t)} - \left.\frac{\partial g(x)}{\partial x}\right|_{x_\mu(t)} w_\mu(t) ,$$

$$\Phi_\mu(t) = g[x_\mu(t)] , \qquad (9.4.10)$$

$$\Psi_\mu(t) = \left.\frac{\partial h(x)}{\partial x}\right|_{x_\mu(t)}$$

are T-periodic bounded matrices and δx, δu, δw, δy, δz are appropriate perturbed vectors. With the above choice, notice that $\Delta_0(t) = O_n$.

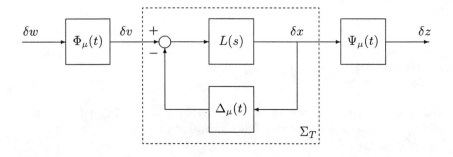

FIGURE 9.3
Block diagram of the linearized system Σ_L.

Since the controller Σ_K is linear, the linearization of the controlled system Σ around $z_\mu(t)$ provides the linear periodic system Σ_L depicted in Figure 9.3, where

$$\delta v = \Phi_\mu(t)\, \delta w ,$$

and

$$L(s) = [sI - A + B\, K(s)\, C]^{-1} . \qquad (9.4.11)$$

Under quite mild assumptions[2] on the nonlinear functions f, g and h, a sufficient condition for the small signal L_2-stability of $z_\mu(t)$ can be obtained by investigating Σ_L [4].

---

[2] It is basically required that the remainder of the linearization is of second order with respect to δx and δw uniformly over the time period T (see, e.g., [14] and [10]).

THEOREM 9.2
If the linear periodic system Σ_L is input-output L_2-stable then the solution $z_\mu(t)$ of system Σ is small signal input-output L_2-stable.

REMARK 9.3 Since $\Phi_\mu(t)$ and $\Psi_\mu(t)$ are bounded and T-periodic, it is clear that L_2-stability of Σ_L is equivalent to L_2-stability of the subsystem Σ_T. Therefore, it is enough to assume L_2-stability of Σ_T for ensuring small signal L_2-stability of $z_\mu(t)$. □

9.4.2 Frequency Domain Criteria

Theorem 9.2 and Remark 9.3 suggest that conditions for small signal L_2-stability of periodic solutions $z_\mu(t)$ can be obtained by investigating L_2-stability of system Σ_T. Since $\Delta_\mu(t)$ is bounded and T-periodic, frequency domain criteria such as the circle criterion [25, 14], the Willems criterion [26], and the more general Integral Quadratic Constraint (IQC) tools [17] can be successfully exploited.

Hereafter, we focus on the multivariable circle criterion, which is widely known as a simple and powerful analysis tool. Nevertheless, quite similar results can be obtained through IQC methodology [12].

To this purpose, we first define the following sector condition for the matrix gain $\Delta_\mu(t)$.

DEFINITION 9.5 The square matrix $\Delta_\mu(t)$ belongs to sector $[\underline{\Delta}, \overline{\Delta}]$ if

i) $\xi^\top [\overline{\Delta} - \Delta_\mu(t)]^\top [\Delta_\mu(t) - \underline{\Delta}] \xi \geq 0 \quad \forall\, t \in [0,T]\,,\ \forall\, \xi \in \mathbf{R}^n$;

ii) $(\overline{\Delta} - \underline{\Delta})$ is symmetric and positive definite.

The next result directly follows from the standard circle criterion, Theorem 9.2 and Remark 9.3.

THEOREM 9.4
Consider the linearized system Σ_L in Figure 9.3. If

i) $L(s)[I + \underline{\Delta} L(s)]^{-1}$ is stable;
ii) $\Delta_\mu(t)$ belongs to sector $[\underline{\Delta}\,,\ \overline{\Delta}]$;
iii) $[I + \overline{\Delta} L(s)]^\top [I + \underline{\Delta} L(s)]^{-1}$ is strict positive real;

then the T-periodic solution $z_\mu(t)$ of Σ is small signal L_2-stable.

The above criterion provides a frequency domain tool for assessing small signal L_2-stability of a T-periodic solution of system Σ, once the matrices $\overline{\Delta}$ and $\underline{\Delta}$ are chosen according to Definition 9.5.

In this respect, it is evident that Theorem 9.4 can result very sensitive to the choice of the matrices $\underline{\Delta}$ and $\overline{\Delta}$. Indeed, such matrices should be chosen in order to identify the tightest sector in Definition 9.5, so as to reduce the inherent conservativeness of the circle criterion. Obviously, tightness of the sector condition strongly depends on the structure of $\Delta_\mu(t)$. From a computational point of view, it is clear that a diagonal structure for $\Delta_\mu(t)$ would be highly desirable. In fact, in this case the sector conditions i) and ii) of Definition 9.5 are satisfied for any couple of diagonal matrices $\underline{\Delta}$ and $\overline{\Delta}$ such that

$$\begin{aligned} \overline{\Delta} - \max_{t\in[0,T]} \Delta_\mu(t) &\geq 0 \\ -\underline{\Delta} + \min_{t\in[0,T]} \Delta_\mu(t) &\geq 0 \end{aligned} \quad (9.4.12)$$

where the max-min operations are componentwise. Note that $\underline{\Delta}$ and $\overline{\Delta}$ can be easily selected according to (9.4.12) even if only some bounds on the T-periodic solutions are known.

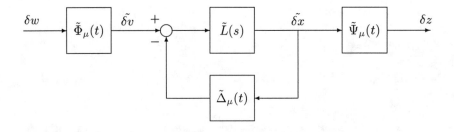

FIGURE 9.4
Block diagram of the diagonalized system $\tilde{\Sigma}_L$.

Motivated by the above discussion, we present a general procedure for arriving at a system, equivalent to Σ_L from the input-output L_2-stability point of view and possessing a diagonal feedback gain. To this purpose, we introduce the matrices

$$Q \doteq \begin{bmatrix} I_n \\ I_n \\ \vdots \\ I_n \end{bmatrix}, \quad P \doteq [\underbrace{e_1 \ldots e_1}_{n \text{ times}} | \underbrace{e_2 \ldots e_2}_{n \text{ times}} | \ldots | \underbrace{e_n \ldots e_n}_{n \text{ times}}]$$

and consider the system $\tilde{\Sigma}_L$ depicted in Figure 9.4, where $\tilde{L}(s)$ is the ($n^2 \times$

Stability of Periodic Solutions

n^2) transfer function matrix defined by

$$\tilde{L}(s) = Q\, L(s)\, P\,, \tag{9.4.13}$$

$\tilde{\Delta}_\mu(t)$ is a diagonal $(n^2 \times n^2)$ matrix gain such that

$$\Delta_\mu(t) = P\, \tilde{\Delta}_\mu(t)\, Q\,, \tag{9.4.14}$$

and $\tilde{\Phi}_\mu(t)$ and $\tilde{\Psi}_\mu(t)$ are given by

$$\tilde{\Phi}_\mu(t) = \tilde{P}\Phi_\mu(t)\,, \quad \tilde{\Psi}_\mu(t) = \Psi_\mu(t)\tilde{Q}\,, \tag{9.4.15}$$

being $\tilde{P}$ and $\tilde{Q}$ matrices of dimension $(n^2 \times n)$ and $(n \times n^2)$, respectively. Note that condition (9.4.14) implies that the diagonal of $\tilde{\Delta}_\mu(t)$ is simply the vector formed by the n ordered rows of $\Delta_\mu(t)$, i.e.,

$$\tilde{\Delta}_\mu(t) = \mathcal{D}[\mathcal{S}[\Delta_\mu(t)]]. \tag{9.4.16}$$

We have the following result.

LEMMA 9.5
For any couple of matrices $\tilde{P}$ and $\tilde{Q}$ such that

$$P\tilde{P} = \tilde{Q}Q = I_n \tag{9.4.17}$$

the following statements are equivalent:

i) Σ_L *is L_2-stable;*
ii) $\tilde{\Sigma}_L$ *is L_2-stable.*

PROOF It is enough to show that Σ_L and $\tilde{\Sigma}_L$ are described by the same equations. Indeed, mixing Laplace transforms and time-domain signals, with some abuse of notation, we can describe $\tilde{\Sigma}_L$ as

$$\begin{aligned}\delta \tilde{x} &= \tilde{L}(s)\tilde{\Phi}_\mu(t)\delta w - \tilde{L}(s)\tilde{\Delta}_\mu(t)\delta\tilde{x} \\ \delta z &= \tilde{\Psi}_\mu(t)\delta\tilde{x}\end{aligned} \tag{9.4.18}$$

Exploiting (9.4.13), (9.4.14), (9.4.15) and (9.4.17), Eqs. (9.4.18) can be rewritten equivalently as

$$\begin{aligned}Q\left[\delta x - L(s)\Phi_\mu(t)\delta w + L(s)\Delta_\mu(t)\delta x\right] &= 0 \\ \delta z &= \Psi_\mu(t)\delta x\end{aligned} \tag{9.4.19}$$

Since Q is a full rank matrix, (9.4.19) are equivalent to

$$\begin{aligned}\delta x &= L(s)\Phi_\mu(t)\delta w - L(s)\Delta_\mu(t)\delta x \\ \delta z &= \Psi_\mu(t)\delta x\end{aligned} \tag{9.4.20}$$

which are the equations describing system Σ_L. □

Since there exist many matrices $\tilde{P}$ and $\tilde{Q}$ satisfying (9.4.17), it is reasonable to exploit such degree of freedom. Indeed, $\Delta_\mu(t)$ is quite commonly a sparse matrix (see the example at the end of the section) and correspondingly $\tilde{\Delta}_\mu(t)$ possesses several null diagonal entries. Therefore, we will look for matrices $\tilde{P}$ and $\tilde{Q}$ allowing to reduce the dimension of the system through elimination of some rows and columns corresponding to null diagonal entries of $\tilde{\Delta}_\mu(t)$.

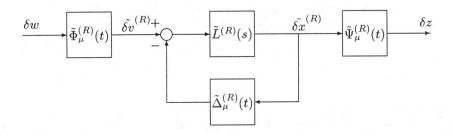

FIGURE 9.5
Block diagram of the reduced system $\tilde{\Sigma}_L^{(R)}$.

To this purpose, consider the sets $\{k_i\}_{i=1}^{r}$, $1 \leq k_1 < \cdots < k_r \leq n^2$ and $\{\overline{k}_i\}_{i=1}^{n^2-r}$, $1 \leq \overline{k}_1 < \cdots < \overline{k}_{n^2-r} \leq n^2$ partitioning the sequence of the first n^2 positive integers, and introduce the matrices

$$E \doteq [e_{k_1} \cdots e_{k_r}] \qquad (9.4.21)$$

and

$$R \doteq [e_{\overline{k}_1} \cdots e_{\overline{k}_{n^2-r}}] \qquad (9.4.22)$$

of dimension $(n^2 \times r)$ and $(n^2 \times (n^2 - r))$ respectively.

Consider now the reduced order system $\tilde{\Sigma}_L^{(R)}$ depicted in Figure 9.5 where $\tilde{L}^{(R)}(s)$, $\tilde{\Delta}_\mu^{(R)}(t)$, $\tilde{\Phi}_\mu^{(R)}(s)$ and $\tilde{\Psi}_\mu^{(R)}(t)$ are obtained via the elimination of rows and/or columns of indexes $k_1, \ldots, k_r$ from $\tilde{L}(s)$, $\tilde{\Delta}_\mu(t)$, $\tilde{\Phi}_\mu(t)$ and $\tilde{\Psi}_\mu(t)$ as follows

$$\begin{aligned}
\tilde{L}^{(R)}(s) &= R^\top \tilde{L}(s) R \\
\tilde{\Delta}_\mu^{(R)}(t) &= R^\top \tilde{\Delta}_\mu(t) R \\
\tilde{\Phi}_\mu^{(R)}(t) &= R^\top \tilde{\Phi}_\mu(t) \\
\tilde{\Psi}_\mu^{(R)}(t) &= \tilde{\Psi}_\mu(t) R \, .
\end{aligned} \qquad (9.4.23)$$

Stability of Periodic Solutions

The following result pertains to $\tilde{\Sigma}_L^{(R)}$.

LEMMA 9.6
Suppose there exist $\tilde{P}$ and $\tilde{Q}$ such that conditions

$$E^\top \mathcal{D}[\mathcal{S}[\Delta_\mu(t)]]E = 0 \qquad (9.4.24)$$

$$E^\top \tilde{\Phi}_\mu(t) = 0 \qquad (9.4.25)$$

$$\tilde{\Psi}_\mu(t)E = 0 \qquad (9.4.26)$$

hold for all $t \in [0,T]$. Then, the following statements are equivalent:

i) $\tilde{\Sigma}_L$ *is L_2-stable;*
ii) $\tilde{\Sigma}_L^{(R)}$ *is L_2-stable.*

PROOF It suffices to observe that if conditions (9.4.24), (9.4.25) and (9.4.26) hold, then the components of $\tilde{\delta x}$ and $\tilde{\delta v}$ of indexes $\{k_i\}_{i=1}^r$ in the equations describing $\tilde{\Sigma}_L$ are identically equal to zero. □

Summing up, we can give the main result of the section.

THEOREM 9.7
Consider the linearized system Σ_L of Figure 9.3, and define P and Q as in (9.4.17). Suppose there exist matrices E, R defined in (9.4.21), (9.4.22) and matrices $\tilde{P}$, $\tilde{Q}$ such that

$$\begin{aligned} P\tilde{P} &= I_n & & \\ \tilde{Q}Q &= I_n & & \\ E^\top \mathcal{D}[\mathcal{S}[\Delta_\mu(t)]]E &= 0 & & \forall\, t \in [0,T] \\ E^\top \tilde{P}\Phi_\mu(t) &= 0 & & \forall\, t \in [0,T] \\ \Psi_\mu(t)\tilde{Q}E &= 0 & & \forall\, t \in [0,T]\,. \end{aligned} \qquad (9.4.27)$$

Furthermore, let $\underline{\Delta}$ and $\overline{\Delta}$ be two diagonal matrices such that

$$\begin{aligned} \overline{\Delta} - \max_{t\in[0,T]} \mathcal{D}[\mathcal{S}[\Delta_\mu(t)]] &\geq 0 \\ -\underline{\Delta} + \min_{t\in[0,T]} \mathcal{D}[\mathcal{S}[\Delta_\mu(t)]] &\geq 0 \end{aligned} \qquad (9.4.28)$$

where the max-min operations are componentwise. If

i) $R^\top QL(s)PR\left[I + R^\top \underline{\Delta} RR^\top QL(s)PR\right]^{-1}$ *is stable;*
ii) $[I + R^\top \overline{\Delta} RR^\top QL(s)PR]^\top [I + R^\top \underline{\Delta} RR^\top QL(s)PR]^{-1}$ *is strict positive real;*

then the T-periodic solution $z_\mu(t)$ of system Σ is small signal L_2-stable.

PROOF Conditions (9.4.27) ensure that Lemmas 9.5 and 9.6 are valid and, therefore, L_2-stability of Σ_L is equivalent to L_2-stability of $\tilde{\Sigma}_L^{(R)}$. Condition (9.4.28) and conditions i) and ii) are simply the equivalent conditions i) – iii) of Theorem 9.4 for system $\tilde{\Sigma}_L^{(R)}$. □

Example 9.6
For the linearized model (9.2.5) we have

$$L(s) = \begin{bmatrix} s+k_0 K(s) & -k_0 & 0 \\ \eta & s+\eta+\Gamma & -\gamma_R \\ 0 & -\beta & s+\alpha \end{bmatrix}^{-1} \doteq \begin{bmatrix} L_{11}(s) & L_{12}(s) & L_{13}(s) \\ L_{21}(s) & L_{22}(s) & L_{23}(s) \\ L_{31}(s) & L_{32}(s) & L_{33}(s) \end{bmatrix},$$

$$\Delta_\mu(t) = \begin{bmatrix} 0 & 0 & 0 \\ k_1(t) & k_2(t) & 0 \\ 0 & 0 & 0 \end{bmatrix}, \quad \Phi_\mu(t) = \begin{bmatrix} k_0 \\ 0 \\ 0 \end{bmatrix}, \quad \Psi_\mu(t) = [1\ 0\ 0],$$

with

$$k_1(t) = \eta\left[(x_{2\mu}(t)+1)\exp(x_{1\mu}(t)) - 1\right], \quad k_2(t) = \eta\left[\exp(x_{1\mu}(t)) - 1\right].$$

Application of Lemma 9.5 results in the system $\tilde{\Sigma}_L$ where

$$\tilde{L}(s) = \begin{bmatrix} L_{11} & L_{11} & L_{11} & L_{12} & L_{12} & L_{12} & L_{13} & L_{13} & L_{13} \\ L_{21} & L_{21} & L_{21} & L_{22} & L_{22} & L_{22} & L_{23} & L_{23} & L_{23} \\ L_{31} & L_{31} & L_{31} & L_{32} & L_{32} & L_{32} & L_{33} & L_{33} & L_{33} \\ L_{11} & L_{11} & L_{11} & L_{12} & L_{12} & L_{12} & L_{13} & L_{13} & L_{13} \\ L_{21} & L_{21} & L_{21} & L_{22} & L_{22} & L_{22} & L_{23} & L_{23} & L_{23} \\ L_{31} & L_{31} & L_{31} & L_{32} & L_{32} & L_{32} & L_{33} & L_{33} & L_{33} \\ L_{11} & L_{11} & L_{11} & L_{12} & L_{12} & L_{12} & L_{13} & L_{13} & L_{13} \\ L_{21} & L_{21} & L_{21} & L_{22} & L_{22} & L_{22} & L_{23} & L_{23} & L_{23} \\ L_{31} & L_{31} & L_{31} & L_{32} & L_{32} & L_{32} & L_{33} & L_{33} & L_{33} \end{bmatrix},$$

$$\tilde{\Delta}_\mu(t) = \begin{bmatrix} 0 & 0 & 0 & 0 & 0 & 0 & 0 & 0 & 0 \\ 0 & 0 & 0 & 0 & 0 & 0 & 0 & 0 & 0 \\ 0 & 0 & 0 & 0 & 0 & 0 & 0 & 0 & 0 \\ 0 & 0 & 0 & k_1(t) & 0 & 0 & 0 & 0 & 0 \\ 0 & 0 & 0 & 0 & k_2(t) & 0 & 0 & 0 & 0 \\ 0 & 0 & 0 & 0 & 0 & 0 & 0 & 0 & 0 \\ 0 & 0 & 0 & 0 & 0 & 0 & 0 & 0 & 0 \\ 0 & 0 & 0 & 0 & 0 & 0 & 0 & 0 & 0 \\ 0 & 0 & 0 & 0 & 0 & 0 & 0 & 0 & 0 \end{bmatrix},$$

$$\tilde{\Phi}_\mu(t) = \tilde{P}\begin{bmatrix} k_0 \\ 0 \\ 0 \end{bmatrix}, \quad \tilde{\Psi}_\mu(t) = [1\ 0\ 0]\,\tilde{Q},$$

being $\tilde{P}$ and $\tilde{Q}$ such that condition (9.4.17) holds.

Note that $\tilde{\Delta}_\mu(t)$ has only two nonzero diagonal elements. Therefore, we apply the order reduction procedure selecting the matrices

$$\tilde{P} = [e_1 \ e_4 \ e_7], \quad \tilde{Q} = [e_1 \ e_2 \ e_3]^\top$$

that yield the fulfillment of conditions (9.4.24), (9.4.25) and (9.4.26) of Lemma 9.6 with

$$E = [e_2 \ e_3 \ e_6 \ e_7 \ e_8 \ e_9], \quad R = [e_1 \ e_4 \ e_5].$$

This leads to the equivalent system $\tilde{\Sigma}_T^{(R)}$ where

$$\tilde{L}^{(R)}(s) = \begin{bmatrix} L_{11}(s) & L_{12}(s) & L_{12}(s) \\ L_{11}(s) & L_{12}(s) & L_{12}(s) \\ L_{21}(s) & L_{22}(s) & L_{22}(s) \end{bmatrix}, \quad \tilde{\Delta}_\mu^{(R)}(t) = \begin{bmatrix} 0 & 0 & 0 \\ 0 & k_1(t) & 0 \\ 0 & 0 & k_2(t) \end{bmatrix},$$

$$\tilde{\Phi}_\mu^{(R)}(t) = \begin{bmatrix} k_0 \\ 0 \\ 0 \end{bmatrix}, \quad \tilde{\Psi}_\mu^{(R)}(t) = [1 \ 0 \ 0].$$

9.5 Application to Chaos Control

This section is devoted to show how the concepts established throughout the previous sections can be effectively applied for the control of complex dynamics. In this respect, a very typical scenario concerning the uncontrolled system (Σ_P with $u(t) \equiv 0$) is as follows:

i) the equilibrium point of the unforced system ($\mu = 0$) is stable;

ii) for small amplitudes $\mu \in (0, \mu^*)$ of the T-periodic forcing input $w_\mu(t)$ the system possesses a corresponding family $z_\mu(t)$ of stable T-periodic solutions;

iii) when the amplitude increases beyond μ^*, these solutions become unstable.

Here, μ^* generically represents a codimension one bifurcation, which — in the most frequent route to chaos — is the first of a full cascade of period-doubling bifurcations leading to chaotic behaviors.

Example 9.7

Figure 9.6 shows the bifurcation diagram for the laser model (9.2.1) with respect to the state variable x_1, from which the value $\mu^* \simeq 0.042$ of the first period doubling bifurcation can be obtained. The corresponding $2T$-periodic solutions remain stable up to $\mu \simeq 0.077$ while chaotic motion occurs for $\mu > 0.1$. This diagram approximates quite satisfactorily the results obtained experimentally (see [23]).

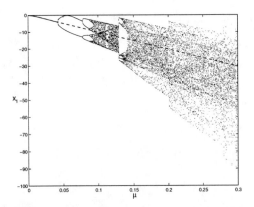

FIGURE 9.6
Bifurcation diagram of the CO_2 laser and unstable T-periodic solutions (dashed).

For many real-world physical systems, it is well known that the extension of the stable T-periodic regimes means enhancing performances and reliability. Furthermore, the control energy effort is very often required to be small, in the sense that the controller action is present only during the transient behavior, while being practically absent in the steady state. These are indeed the motivations of the OGY and "delayed feedback" approaches [25, 20, 19, 22, 13]:

Therefore, a reasonable control objective consists of enlarging the stability interval of the T-periodic solutions of the uncontrolled system, without a hard modification of their shape. Said another way, the modification of the bifurcation diagram simply consists of shifting the period-doubling bifurcation to higher values of the forcing amplitude μ, i.e., the diagram of the controlled system approximately coincides with that of the uncontrolled system if a suitable contraction of the μ-axis is performed.

The above objective has already been investigated in the bifurcation control setting by employing some harmonic balance techniques [1, 24, 6, 2, 5]. We would like to show now that the same objective can be pursued via the results of Sections 9.3 and 9.4. Indeed, Theorem 9.1 and Theorem 9.7 can be exploited to provide a general result ensuring that the controlled system Σ possesses a family of small signal L_2-stable T-periodic solutions.

THEOREM 9.8
Consider the controlled system Σ subject to the family of T-periodic forcing inputs $w_\mu \in \mathcal{W}$ and let $\underline{\Delta}$ and $\overline{\Delta}$ be two diagonal $(n^2 \times n^2)$ matrices such that the null matrix O_{n^2} belongs to the sector $[\underline{\Delta}, \overline{\Delta}]$, i.e.,

$$\underline{\Delta} \leq O_{n^2} \leq \overline{\Delta} . \tag{9.5.29}$$

Suppose there exist matrices $\tilde{P}$, $\tilde{Q}$, E, and R such that the following conditions hold:

$$\begin{aligned} P\tilde{P} &= I_n \\ \tilde{Q}Q &= I_n \\ E^\top \Delta E &= 0 & \forall \, \Delta \in [\underline{\Delta}, \overline{\Delta}] \\ E^\top \tilde{P} g(x) &= 0 & \forall \, x \in X \\ \tfrac{\partial h(x)}{\partial x} \tilde{Q} E &= 0 & \forall \, x \in X . \end{aligned} \tag{9.5.30}$$

If

i) $g(x) \neq 0 \ \forall \, x \in X$;

ii) $L(s)$ *is stable*;

iii) $[I+R^\top \overline{\Delta} RR^\top QL(s)PR]^\top [I+R^\top \underline{\Delta} RR^\top QL(s)PR]^{-1}$ *is strict positive real*;

then there exists a positive constant μ^* such that for all $\mu \in (0, \mu^*)$ system Σ possesses a unique non-trivial small signal L_2-stable T-periodic solution $z_\mu(t)$.

Furthermore, a lower bound of μ^* is given by the smallest positive value of μ such that conditions

$$\begin{aligned} \overline{\Delta} - \max_{t \in [0,T]} \mathcal{D}[\mathcal{S}[\Delta_\mu(t)]] &\geq 0 \\ -\underline{\Delta} + \min_{t \in [0,T]} \mathcal{D}[\mathcal{S}[\Delta_\mu(t)]] &\geq 0 \end{aligned} \tag{9.5.31}$$

no longer hold.

PROOF Conditions i) and ii) ensure that Theorem 9.1 holds for the controlled system Σ. Indeed, stability of $L(s)$ implies that the matrix (9.3.6) has no eigenvalues on the imaginary axis. Therefore, existence of a family of T-periodic solutions $z_\mu(t)$ is proven.

Conditions (9.5.30) ensure that L_2-stability of Σ_L is equivalent to L_2-stability of $\tilde{\Sigma}_L^{(R)}$ for each T-periodic solution $z_\mu(t)$ such that $\mathcal{D}[\mathcal{S}[\Delta_\mu(t)]]$ belongs to the sector $[\underline{\Delta}, \overline{\Delta}]$. Now, since condition (9.5.29) implies that

$\mathcal{D}[\mathcal{S}[\Delta_0(t)]]$ belongs to sector $[\underline{\Delta}, \overline{\Delta}]$, from ii) and iii) it turns out (see Theorem 9.7) that there exists $\mu^* \geq 0$ such that $z_\mu(t)$ is small signal L_2-stable for $\mu \in [0, \mu^*)$.

Finally, the lower bound for μ^* follows directly from the fact that under conditions (9.5.31) Theorem 9.7 can be applied. □

REMARK 9.9 Theorem 9.8 states that if conditions (9.5.29), (9.5.30), and i) – iii) hold for some $\underline{\Delta}$ and $\overline{\Delta}$, then a family of small signal L_2-stable T-periodic solutions $z_\mu(t)$ exists. Moreover, if the T-periodic solutions $z_\mu(t)$ are known, then conditions (9.5.31) provide a lower bound of μ^*. Obviously, the larger is the sector $[\underline{\Delta}, \overline{\Delta}]$ the tighter is the lower bound on μ^*. However, $\underline{\Delta}$ and $\overline{\Delta}$ cannot be chosen too large since in this case condition iii) might be no longer satisfied. □

Now, since μ^* is a function of $K(s)$ via $L(s)$ (see (9.4.11)), it is reasonable to look for the controller that maximizes μ^*, without producing a large modification of the shape of the uncontrolled T-periodic solutions. Unfortunately, the parameter μ^* cannot be computed exactly. Nevertheless, according to Remark 9.9, a larger value of μ^* can be expected by controllers $K(s)$ (and therefore transfer function matrices $L(s)$) such that, for a fixed sector $[\underline{\Delta}, \overline{\Delta}]$, condition ii) of Theorem 9.8 is not invalidated. Indeed, for such controllers Theorem 9.8 holds also for larger sectors and consequently for larger lower bounds of μ^*.

To formulate such control design problem, we first observe that in order to exactly maintain the T-periodic solutions of the uncontrolled system, the control action must vanish on them and, as discussed in Subsection 4.3.2, the controller transfer function must have the "delayed feedback" structure (9.3.8). Such observation suggests the following structure for the finite dimensional controller to be designed

$$K(s) = F(s)\, s \prod_{i=1}^{N} \left(1 + \frac{T^2}{4N^2\pi^2} s^2\right). \qquad (9.5.32)$$

Here, $F(s)$ belongs to some given class $\mathcal{F}$ of rational $(n_y \times n_u)$ transfer function matrices, while N is a given positive integer. Since $K(s)$ satisfies constraints (9.3.7) for $k = 0, \pm 1, \ldots, \pm N$, a very slight modification of the T-periodic solutions of the uncontrolled system is expected. Note that $K(s)$ possesses a washout filter at the working frequencies $\omega = 0, \frac{2\pi}{T}, \ldots, \frac{2\pi N}{T}$ [24, 5, 15, 16].

Another important point is the selection of the diagonal matrices $\underline{\Delta}$ and $\overline{\Delta}$. The basic idea is to exploit the available information on the uncontrolled system, since the above choice of the controller class (9.5.32) guarantees that the behavior of the controlled system Σ is sufficiently close to that of

Application to Chaos Control

the uncontrolled system.

To this purpose, consider the uncontrolled system (Σ_P with $u(t) \equiv 0$ or, equivalently, Σ with $K(s) = 0$) subject to the forcing T-periodic inputs $w_\mu(t)$ for $\mu \in [0, \bar{\mu}]$, where $\bar{\mu}$ is a prespecified maximum amplitude. Simulation of the uncontrolled system provides the corresponding T-periodic solutions $z_\mu(t)$ and, consequently, the T-periodic gain matrices $\Delta_\mu(t)$ (see (9.4.10)), and the bifurcation diagram of the uncontrolled system. Now, if conditions of Theorem 9.8 hold for matrices $\underline{\Delta}$ and $\overline{\Delta}$ chosen as

$$\overline{\Delta} = \overline{\Delta}_{\bar{\mu}} \doteq \max_{\mu \in [0,\bar{\mu}]} \max_{t \in [0,T]} \mathcal{D}[\mathcal{S}[\Delta_\mu(t)]]$$
$$\underline{\Delta} = \underline{\Delta}_{\bar{\mu}} \doteq \min_{\mu \in [0,\bar{\mu}]} \min_{t \in [0,T]} \mathcal{D}[\mathcal{S}[\Delta_\mu(t)]] , \qquad (9.5.33)$$

it follows that $\bar{\mu}$ is a lower bound of μ^*. On the other hand, if the conditions of Theorem 9.8 do not hold for the uncontrolled system, one can look for a controller $K(s)$ such that Theorem 9.8 is valid, so that $\bar{\mu}$ is a lower bound of μ^* for such controlled system.

According to this observation and taking into account that conditions of Eqs. (9.5.29), (9.5.30), and condition i) in Theorem 9.8 are structurally satisfied for any controlled system, the following control design problem can be stated.

Let the class $\mathcal{F}$ and the positive integer N be given. Then, determine the controller $K(s)$ of structure (9.5.32) maximizing the value of $\bar{\mu}$ such that

a) the transfer function

$$[sI - A + B\,K(s)\,C]^{-1} \qquad (9.5.34)$$

is stable;

b) the transfer function

$$\left[I + R^T \overline{\Delta}_{\bar{\mu}} RR^T Q\,[sI - A + B\,K(s)\,C]^{-1} PR\right]^T \times$$
$$\left[I + R^T \underline{\Delta}_{\bar{\mu}} RR^T Q\,[sI - A + B\,K(s)\,C]^{-1} PR\right]^{-1} \qquad (9.5.35)$$

is strict positive real.

Example 9.8

In this example, we solve the above design problem for the laser model (9.2.1). We consider controllers of transfer function (9.5.32) where

$$N = 1, \quad F(s) \in \mathcal{F} = \left\{k_c \cdot \tilde{F}(s)\right\},$$

$$\tilde{F}(s) = \frac{\frac{T}{2\pi}}{\left(\frac{T^2}{2\pi^2}s^2 + \frac{T}{\sqrt{2\pi}}s + 1\right)\left(\frac{5T}{12\pi}s + 1\right)^2}, \quad T = 10^{-5} \text{ s},$$

and the gain k_c is to be designed. A controller structure of this kind has already been applied for laser control in [8].

The matrices $\overline{\Delta_{\bar{\mu}}}$ and $\underline{\Delta_{\bar{\mu}}}$ are obtainable from data used to construct the bifurcation diagram of the uncontrolled system in Figure 9.6. Therefore, we have to select the value of k_c that solves conditions (9.5.34) and (9.5.35) for the largest value of $\bar{\mu}$.

Condition (9.5.34) turns out to be satisfied for $k_c \in [-0.043, +\infty)$, while evaluation of condition (9.5.35) provides the diagram of Figure 9.7 showing the maximum values of $\bar{\mu}$ as a function of k_c. Choosing the optimal $k_c = 0.015$, the bifurcation diagram of Figure 9.8 is obtained. A comparison between Figures 9.6 and 9.8 reveals that the designed controller is effective in shifting the first period doubling bifurcation, and at the same time, it does not modify significantly the shape of the T-periodic solutions, i.e., (as required in the control design problem), the selected closed-loop system needs a small control energy.

9.6 Conclusions

This chapter has focused on stabilization of periodic orbits, an issue that is strictly related to the problem of controlling complex dynamics. Indeed, since a long time it was recognized that one of the most appealing approaches for controlling chaos was the stabilization of one among the infinite unstable periodic orbits embedded in the underlying chaotic attractor.

A general framework based on classical frequency domain tools for the stabilization of periodic orbits in a significant class of periodically forced nonlinear systems has been introduced. Some results ensuring both existence and input-output small signal L_2-stability of a family of T-periodic solution have been given. Such conditions require the verification of stability and strict positive realness of suitable transfer function matrices.

Such existence and stability results have been exploited for the design of finite dimensional linear time invariant controllers guaranteeing stability of periodic solutions for larger amplitudes of the forcing input, without requiring a high control effort. As a typical application, the problem of designing a controller for delaying the first period doubling bifurcation without large modification of the shape of the periodic solutions has been considered.

Finally, the model of a single mode CO_2 laser with modulated cavity losses has been used throughout the chapter to illustrate the features of the approach.

Conclusions

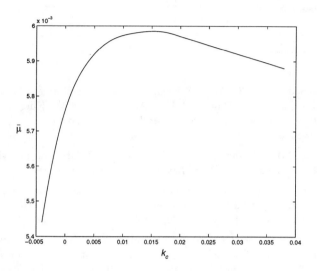

FIGURE 9.7
Performance index $\bar{\mu}$ as a function of the gain k_c.

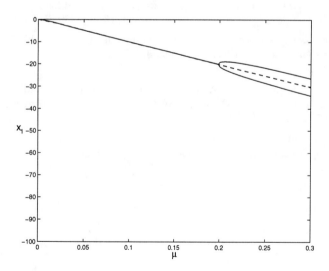

FIGURE 9.8
Bifurcation diagram of the controlled CO_2 laser (solid); T-periodic solutions of the uncontrolled system (dashed).

References

[1] E. H. Abed and H. O. Wang, "Feedback control of bifurcations and chaos in dynamical systems," in N. S. Namachchivaya and W. Kliemann (Eds.), *Recent Developments in Stochastic and Nonlinear Dynamics: Applications to Mechanical Systems*, CRC Press, 1995.

[2] E.H. Abed, H.O. Wang and A. Tesi, "Control of bifurcation and chaos," in *The Control Handbook*, W. S. Levine (Ed.), CRC Press, 1995.

[3] M. Basso, R. Genesio, L. Giovanardi, A. Tesi and G. Torrini, "On optimal stabilization of periodic orbits via time delayed feedback control," *Int. J. of Bifur. Chaos*, vol. 8, pp. 1699-1706, 1997.

[4] M. Basso, R. Genesio, L. Giovanardi and A. Tesi, "L_2 input-output stabilization of periodic orbits in forced nonlinear systems," Technical Report RT 02/99, Dipartimento di Sistemi e Informatica, Università degli Studi di Firenze, Italy, 1999.

[5] M. Basso, R. Genesio, M. Stanghini and A. Tesi, "Subharmonic control of chaos with application to a CO_2 laser," *Chaos, Solitons and Fractals*, vol. 8, pp. 1449-1460, 1997.

[6] M. Basso, R. Genesio and A. Tesi, "A frequency method for predicting limit cycle bifurcations," *Nonlinear Dynamics*, vol. 13, pp. 339-360, 1997.

[7] M. Basso, R. Genesio and A. Tesi, "Stabilizing periodic orbits of forced systems via generalized Pyragas controllers," *IEEE Trans. Circ. and Sys.*, I, vol. 44, pp. 1023-1027, 1997.

[8] M. Basso, R. Genesio and A. Tesi, "Controller design for extending periodic dynamics of a chaotic CO_2 laser," *Systems and Control Letters*, vol. 31, pp. 287-297, 1997.

[9] G. Chen and X. Dong, *From Chaos to Order: Methodologies, Perspectives and Applications*, World Scientific Pub. Co., Singapore, 1998.

[10] C. A. Desoer and M. Vidyasagar, *Feedback Systems: Input-Output Properties*, Academic Press, New York, 1975.

[11] M. Farkas, *Periodic Motions*, Springer-Verlag, New York, 1994.

[12] L. Giovanardi, "Input-output stability of feedback systems. New techniques with application to stabilization of periodic solutions," Laurea Degree Dissertation (in Italian), Electronic Engineering, Università degli Studi di Firenze, Italy, 1997.

[13] W. Just, T. Bernard, M. Ostheirer, E. Reibold and H. Benner, "Mechanism of time-delayed feedback control," *Phys. Rev. Lett.*, vol. 78, pp. 203-206, 1997.

[14] H.K. Khalil, *Nonlinear Systems*, 2nd Ed., Macmillan, New York, 1996.

[15] H. C. Lee, "Robust control of bifurcating nonlinear systems with applications," Ph.D. Dissertation, Electrical Engineering, University of Maryland, College Park, MD, 1991.

[16] H. C. Lee and E. H. Abed, "Washout filters in the bifurcation control of high alpha flight dynamics," *Proc. American Control Conference*, Boston, MA, 206-211, 1991.

[17] A. Megretski and A. Rantzer, "System analysis via integral quadratic constraints," *IEEE Trans. Auto. Control*, vol. 42, pp. 819-830, 1997.

[18] E. Ott, C. Grebogi and J. A. Yorke, "Controlling chaos," *Phys. Rev. Lett.*, vol. 64, pp. 1196-1199, 1990.

[19] K. Pyragas, "Continuous control of chaos by self-controlling feedback," *Phys. Lett. A*, vol. 170, pp. 421-428, 1992.

[20] F. J. Romeiras, C. Grebogi, E. Ott and W. P. Dayawansa, "Controlling chaotic dynamical systems," *Physica D*, vol. 58, pp. 165-192, 1992.

[21] T. Shinbrot, C. Grebogi, E. Ott and J. A. Yorke, "Using small perturbations to control chaos," *Nature*, vol. 363, pp. 411-417, 1993.

[22] J. E. S. Socolar, D. W. Sukow and D. J. Gauthier, "Stabilizing unstable periodic orbits in fast dynamical systems," *Phys. Rev. E*, vol. 50, pp. 3245-3248, 1994.

[23] M. Stanghini, M. Basso, R. Genesio, A. Tesi, M. Ciofini and R. Meucci, "A new three-equation model for the CO_2 laser," *IEEE J. Quantum Electron.*, vol. 32, pp. 1126-1131, 1996.

[24] A. Tesi, E. H. Abed, R. Genesio and H. O. Wang, "Harmonic balance analysis of period doubling bifurcations with implications for control of nonlinear dynamics," *Automatica*, vol. 32, pp. 1255-1271, 1996.

[25] M. Vidyasagar, *Nonlinear System Analysis*, Second Edition, Prentice-Hall, Englewood Cliffs, NJ, 1992.

[26] J. C. Willems, *The Analysis of Feedback Systems*, The Mit Press, Cambridge, MA, 1971.

10

Controlling Limit Cycles and Bifurcations

Guillermo Calandrini[1]**, Eduardo Paolini**[1]
Jorge L. Moiola[1]**, and Guanrong Chen**[2]

[1]Departamento de Ingeniería Eléctrica
Universidad Nacional del Sur
(8000) Bahía Blanca, Argentina
comoiola@criba.edu.ar

[2]Department of Electrical and Computer Engineering
University of Houston, Houston, Texas, 77204 USA
gchen@uh.edu

Abstract

A certain type of degenerate Hopf bifurcations determines the appearance of multiple limit cycles under system parameter perturbations. In the study of these degenerate Hopf bifurcations, computational formulas for the curvature coefficients (i.e., stability indexes) are essential. However, such analytic formulas are very difficult to obtain, and so are usually computed by different approximation methods. Inspired by the feedback control methodology and the harmonic balance approximation technique, higher-order approximate formulas for such curvature coefficients are derived in this chapter in the frequency domain setting. Then, the problem of controlling the multiplicity of periodic solutions of a nonlinear system is addressed. The approach utilizes the system curvature coefficients to build up a suitable configuration of limit cycles. Two examples are presented for illustration: a typical planar cubic system, and a three-dimensional chemical reactor system.

10.1 Introduction

A certain type of degenerate (or singular) Hopf bifurcations determines the appearance of multiple limit cycles under parameter perturbations. In the study of these degenerate Hopf bifurcations, computational formulas for the *curvature coefficients* or *stability indexes* (also called *focal values*) are essential, since they contain important information about the stabilities of periodic solutions emerging from the Hopf bifurcation in a nonlinear dynamical system. On the other hand, bifurcation control involves designing a controller for modification of bifurcative dynamical behaviors of a complex nonlinear system, attempting some beneficial effects caused by the appearance, delay, or change of a certain type of bifurcations. Objectives include delaying the appearance of oscillations or cascaded period-doubling bifurcations, controlling the amplitude of the limit cycle emerging from a Hopf bifurcation, stabilizing bifurcated periodic trajectories, controlling the multiplicity of periodic solutions, and so on. The problems of stabilizing an emerging limit cycle through a Hopf bifurcation and stabilizing a stationary bifurcation were studied in late 1980s [2, 3]. Since then, a considerable number of articles on the subject of bifurcation control and its implication to chaos suppression have been published (see, e.g., [13, 14]).

Perhaps the most focusing topic in the area of bifurcation control is the modification of the stability of the emerging limit cycles, i.e., designing a feedback controller to stabilize an unstable limit cycle (see, e.g., [33, 36, 41]). Other investigations have been centered around the delay of the appearance of period-doubling bifurcations, thereby delaying the appearance of chaos generated via the cascade bifurcation route [39]. In particular, some nontraditional control objectives have been formulated (in, e.g., [15, 1, 4]) by taking into account the consequence of the system dynamics near a degenerate Hopf bifurcation. Here, a degenerate Hopf bifurcation refers to as a certain singular situation where some classical Hopf bifurcation conditions fail [22]. Using a frequency domain approach, we have shown how to control the amplitude and frequency of the limit cycles that emerge near degenerate Hopf bifurcation points [11]. This control objective encompasses those applications in engineering that can tolerate small-amplitude oscillations or permit an extension of the basin of attraction of system stable solutions, within the limit of the feasibility region of the given system parameters.

Closely related to the control of the amplitudes of limit cycles is the issue of controlling the multiplicity of periodic solutions, as pointed out in [7]. In chemical reactor engineering, for example, it is known that by varying the bifurcation-control parameters the system can be shifted to some regions with two, three, or more periodic solutions; some of them can have large amplitudes [37]. These studies greatly motivated the research work presented in this chapter.

To create an appropriate scenario for modifying the system bifurcation parameters, or for designing nonlinear feedback controllers for the intended control task, this work continues our earlier pursuit along the line of bifurcations control. This is an engineering feedback approach formulated in the frequency domain [34], which takes advantage of local predictions obtained by applying different higher-order harmonic balance approximations (HBAs). The proposed control method is based on the computation of the curvature coefficients for degenerate Hopf bifurcations in the system parameter space [35]. The classical normal-form technique is also used, to embed the system dynamics with an appropriate configuration of periodic solutions.

A brief historical introduction to the relevance of the computation of the curvature coefficients can be found in [19]. The first curvature coefficient computed by using the so-called succession functions is given in [8], as well as other topics on creation of multiple cycles, where the approach is limited to two-dimensional systems. Extensions of this method to n-dimensional systems were suggested later in [22, 24, 32], based on quite different mathematical methods.

For the aforementioned control purposes, if the first curvature coefficient vanishes at criticality, the next curvature coefficient (the *second* coefficient) has to be calculated; but if the second coefficient is also zero, then the *third* coefficient needs to be found, and so on. Each stage of these computations involves the knowledge of several partial derivatives of a nonlinear function, evaluated at the equilibrium point of the system, up to the order of $2k + 1$ where k represents the order of the curvature coefficient. Moreover, the complexity in the computation of each curvature coefficient increases considerably from one order to the next higher order.

Computation of the second curvature coefficient was first presented in [24] (see also [23]), which was also derived by using different methods (see, e.g., [22]). The first *three* curvature coefficients were obtained, although only for two-dimensional systems, by using a method employing Lyapunov constants in [21]. Later, the formulation given in [22] was extended in [19], to include the *third* curvature coefficient for n-dimensional systems, which was applied to a classical quadratic planar system for determining the order of a *weak* focus. Here, the order of a weak focus means the order of the least curvature coefficient that has a definite sign at criticality, while all the lower-order coefficients vanish. The order of a weak focus indicates how many limit cycles can bifurcate from this special equilibrium point, under appropriate (small) perturbations of the system parameters.

Related to the topic of the number of limit cycles emerging from a *weak* focus (a local phenomenon), there is one important and interesting problem in mathematics, the Hilbert 16th problem, which refers to finding the maximum number of limit cycles, $H(n)$, arising from an nth-order planar polynomic system in a global sense. This problem is still unsolved; but for *quadratic* planar systems the answer is very close to the end (see, for example, [18]). As a matter of fact, publication on the maximum number of limit cycles in planar polynomic (mainly quadratic and cubic) systems has considerably increased over the last decade. At the beginning, attention was drawn to Russian and Chinese mathematicians (see [9, 38, 28, 40], to cite only a few), yet recently this topic has received many new contributions from Western mathematicians (see, e.g., [12, 31, 42, 43]). For a clearly written survey on this subject, the reader is referred to [29].

The use of symbolic programming allows efficient intensive computations. For instance, with the help of computer symbolic calculations, the curvature coefficients of arbitrary orders can be computed [30, 26]. Also, the birth of multiple Hopf bifurcations arising from an ordinary differential equation can be calculated [25], and degenerate Hopf bifurcations in functional differential equations can be studied (see [20], and also some earlier studies in [5]).

Along the line of studying nonlinear oscillations by employing the engineering feedback system approach, which was first proposed in [6] and then enhanced in [32], the *second, third* and *fourth* curvature coefficients were provided in [35], where the birth of multiple limit cycles was studied from a feedback methodology and harmonic balance approximation technique. In the present chapter, a combination of some analysis for curvature coefficients and a continuation technique for periodic solutions is used to build up a desired configuration of the multiple cycles.

The chapter is organized as follows. In Section 10.2, an iterative formulation for computing the first *four* curvature coefficients for a general n-dimensional system is briefly introduced in the frequency-domain setting. These coefficients are computed in Section 10.3 by using some traditional normal forms for planar vector fields. In Section 10.4, a procedure using these approximation formulas is given for generating a desired set of multiple limit cycles. Then, in Section 10.5, the results obtained are transformed to the context of controlling the multiplicity of limit cycles, where the system dynamics are driven to some more accurate regions in the parameter space. Two examples are discussed: a planar cubic system and a three-dimensional system. Finally, Section 10.6 concludes the study of this chapter.

10.2 Harmonic Balance and Curvature Coefficients

Consider a dynamical system with a linear feedforward path and a general nonlinear feedback path, as shown in Figure 10.1, which is described by

$$\dot{x} = A(\mu, \varepsilon)x + B(\mu, \varepsilon)u,$$
$$y = -C(\mu, \varepsilon)x,$$
$$u = g(y; \mu, \varepsilon),$$

where A is an $n \times n$ matrix, B and C are $n \times r$ and $l \times n$ matrices, all depending on a bifurcation parameter $\mu \in R$ and a vector $\varepsilon = (\varepsilon_1, \cdots, \varepsilon_k)$ of auxiliary parameters, and $g(\cdot)$ is a nonlinear function belonging to $C^{2(q+1)}$, in which q is the order of the harmonic balance approximation to be used later.

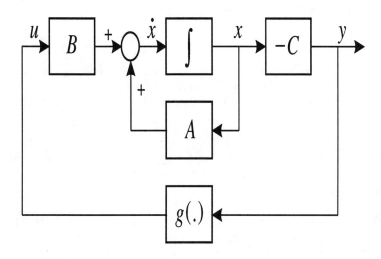

FIGURE 10.1
Nonlinear feedback control configuration.

It is easy to verify that this system has a linear feedforward transfer matrix, $G(s; \mu) = C[sI - A]^{-1}B$, obtained by taking Laplace transforms $\mathcal{L}\{\cdot\}$ with zero-initial conditions, and a nonlinear feedback function $\mathcal{L}\{g(y; \mu)\}$. After linearizing the nonlinear path about an equilibrium point, $\widehat{y}$, of the system, which is a solution of $G(0, \mu)g(\widehat{y}, \mu) = -\widehat{y}$, we obtain its Jacobian $J(\mu) = \frac{\partial g}{\partial y}\big|_{y=\widehat{y}}$. Assuming that the overall system transfer matrix $G(s; \mu)J(\mu)$ possesses an eigenvalue $\lambda = -1$, it is easy to verify that the matrix $G(\bar{s}; \mu)J(\mu)$ has the same eigenvalue. Thus, at the criticality $\mu = \mu_0$, we have $s = i\omega_0$ and $\bar{s} = -i\omega_0$, which are actually the two complex conjugate eigenvalues in the time-domain formulation [32].

We now consider the case where the system parameter μ is increasing and passing through the point μ_0, so that the complex variable s moves into the right-half plane and takes the value $\alpha + i\omega$, where α and ω are both positive real numbers. Suppose that for a value of μ larger than μ_0, the system has a periodic solution. In the following, $\mathbf{v}$ and $\mathbf{w}^\top$ denote respectively the right and left eigenvectors of the matrix $G(s;\mu)J(\mu)$ that correspond to the eigenvalue $\lambda = -1$, normalized by $|\mathbf{v}| = 1$ and $\mathbf{w}^\top \mathbf{v} = 1$.

For the system transfer matrix $G(s;\mu)J(\mu)$, the following expression can be used to detect periodic motions of the system output:

$$\mathbf{Y}^1 = \theta\mathbf{v} + \theta^3 \mathbf{V}_{13} + \theta^5 \mathbf{V}_{15} + \cdots , \qquad (10.2.1)$$

where $\mathbf{V}_{13}, \cdots, \mathbf{V}_{1,2q+1}$ are vectors orthogonal to $\mathbf{v}$, $q = 1, 2, \cdots$ and $\mathbf{Y}^1$ ($= \mathbf{E}^1$ in [32]) is the first harmonic of the output y.

We first observe that for a given value of $\hat{\omega}$ (approximation of the oscillatory frequency), we have

$$G(i\hat{\omega}) = G(s) + (-\alpha + i\delta\omega)G'(s) + \frac{1}{2}(-\alpha + i\delta\omega)^2 G''(s) + \cdots , \qquad (10.2.2)$$

where $\delta\omega = \hat{\omega} - \omega$, ω is the imaginary part of the bifurcating eigenvalues, and $G'(s)$ and $G''(s)$ are the first and second derivatives of $G(s)$, respectively, with respect to s. On the other hand, we recall the following equation of harmonic balance:

$$[G(i\omega)J + I] \sum_{j=0}^{q} \mathbf{V}_{1,2j+1} \theta^{2j+1} = -G(i\omega) \sum_{j=1}^{q} \mathbf{W}_{1,2j+1} \theta^{2j+1} , \qquad (10.2.3)$$

where $\mathbf{V}_{11} = \mathbf{v}$, with

$$\mathbf{W}_{1,2q+1} = \mathbf{p}_q \quad \text{and} \quad \mathbf{V}_{1,2q+1} \qquad q = 1, 2, \cdots , \qquad (10.2.4)$$

which are already known [34] and [32].

In a general situation, we need to solve the following equation:

$$[G(i\hat{\omega})J + I](\mathbf{v}\theta + \mathbf{V}_{13}\theta^3 + \mathbf{V}_{15}\theta^5 + \cdots) = -G(i\hat{\omega})[\mathbf{p}_1\theta^3 + \mathbf{p}_2\theta^5 + \cdots]. \qquad (10.2.5)$$

In so doing, by substituting (10.2.2) into (10.2.5) and then expanding the result in terms of the coefficients $\mathbf{V}_{1,2q+1}$ and $\mathbf{p}_q$, we obtain

$$\left[G(s)J - (\alpha - i\delta\omega)G'(s)J + \frac{1}{2}(\alpha - i\delta\omega)^2 G''(s)J + \cdots + I \right]$$
$$\times \left(\mathbf{v}\theta + \mathbf{V}_{13}\theta^3 + \mathbf{V}_{15}\theta^5 + \cdots - (\alpha - i\delta\omega)\mathbf{V}'_{13}\theta^3 - (\alpha - i\delta\omega)\mathbf{V}'_{15}\theta^5 \right)$$
$$= -[G(s) - (\alpha - i\delta\omega)G'(s) + \cdots]$$
$$\times \left[\mathbf{p}_1\theta^3 + \mathbf{p}_2\theta^5 - (\alpha - i\delta\omega)\mathbf{p}'_1\theta^3 - (\alpha - i\delta\omega)\mathbf{p}'_2\theta^5 + \cdots \right], \qquad (10.2.6)$$

where $\mathbf{V}'_{13} = \frac{d\mathbf{V}_{13}}{ds}, \mathbf{V}'_{15} = \frac{d\mathbf{V}_{15}}{ds}, \cdots$ and $\mathbf{p}'_1 = \frac{d\mathbf{p}_1}{ds}, \mathbf{p}'_2 = \frac{d\mathbf{p}_2}{ds}, \cdots$ are given in [35].

Then, by multiplying both sides of (10.2.6) with $\mathbf{w}^\top$ and noting that $\mathbf{w}^\top$ is the left eigenvector of $G(s)J$ corresponding to the eigenvalue -1, we can rewrite (10.2.6) in the following form:

$$-(\alpha - i\delta\omega)\mathbf{w}^\top G'(s)J\mathbf{v}\theta + \frac{1}{2}(\alpha - i\delta\omega)^2 \mathbf{w}^\top G''(s)J\mathbf{v}\theta$$
$$-(\alpha - i\delta\omega)\mathbf{w}^\top G'(s)J\mathbf{V}_{13}\theta^3 + (\alpha - i\delta\omega)^2 \mathbf{w}^\top G'(s)J\mathbf{V}'_{13}\theta^3 + \cdots$$
$$= -\mathbf{w}^\top G(s)\mathbf{p}_1\theta^3 + (\alpha - i\delta\omega)\mathbf{w}^\top G'(s)\mathbf{p}_1\theta^3 - \mathbf{w}^\top G(s)\mathbf{p}_2\theta^5$$
$$+(\alpha - i\delta\omega)\mathbf{w}^\top G'(s)\mathbf{p}_2\theta^5 + (\alpha - i\delta\omega)\mathbf{w}^\top G(s)\mathbf{p}'_1\theta^3 + \cdots. \quad (10.2.7)$$

Now, consider the first-order approximation of the above expression [32]:

$$(\alpha - i\delta\omega) = \frac{\mathbf{w}^\top G(i\omega)\mathbf{p}_1\theta^2}{\mathbf{w}^\top G'(i\omega)J\mathbf{v}} = \gamma_1\theta^2. \quad (10.2.8)$$

By substituting (10.2.8) into (10.2.7) and then grouping together all the coefficients of equal powers in θ, we arrive at

$$-(\alpha - i\delta\omega)\mathbf{w}^\top G'(s)J\mathbf{v}\theta + \frac{1}{2}\gamma_1^2 \mathbf{w}^\top G''(s)J\mathbf{v}\theta^5 - \gamma_1 \mathbf{w}^\top G'(s)J\mathbf{V}_{13}\theta^5 + \cdots$$
$$= -\mathbf{w}^\top \left[G(s)\mathbf{p}_1\theta^3 - \gamma_1 G'(s)\mathbf{p}_1\theta^5 + G(s)\mathbf{p}_2\theta^5 - \gamma_1 G(s)\mathbf{p}'_1\theta^5 \right.$$
$$\left. -\gamma_1 G'(s)\mathbf{p}_2\theta^7 + \cdots \right]. \quad (10.2.9)$$

In this expression, we can arrange the terms in θ as follows:

$$-(\alpha - i\delta\omega)\mathbf{w}^\top G'(s)J\mathbf{v}\theta = -\mathbf{w}^\top G(s)\mathbf{p}_1\theta^3$$
$$+\mathbf{w}^\top \begin{bmatrix} \gamma_1 G'(s)(J\mathbf{V}_{13} + \mathbf{p}_1) + \gamma_1 G(s)\mathbf{p}'_1 - \\ \frac{1}{2}\gamma_1^2 G''(s)J\mathbf{v} - G(s)\mathbf{p}_2 \end{bmatrix} \theta^5 + \cdots. \quad (10.2.10)$$

For $s = i\omega$, (10.2.10) can be rewritten as

$$(\alpha - i\delta\omega) = \frac{1}{\eta}\mathbf{w}^\top G(i\omega)\mathbf{p}_1\theta^2$$
$$+ \frac{1}{\eta}\mathbf{w}^\top \begin{bmatrix} -\gamma_1 G'(i\omega)(J\mathbf{V}_{13} + \mathbf{p}_1) - \gamma_1 G(i\omega)\mathbf{p}'_1 + \\ \frac{1}{2}\gamma_1^2 G''(i\omega)J\mathbf{v} + G(i\omega)\mathbf{p}_2 \end{bmatrix} \theta^4 + \cdots, \quad (10.2.11)$$

where $\eta = \mathbf{w}^\top G'(i\omega)J\mathbf{v}$.

In a simplified form, we can rewrite (10.2.11) as

$$(\alpha - i\delta\omega) = \gamma_1\theta^2 + \gamma_2\theta^4 + O(\theta^5). \quad (10.2.12)$$

Then, instead of using (10.2.8) as a first-order approximation of $(\alpha - i\delta\omega)$, we can use the more accurate expression (10.2.12).

Next, substituting (10.2.12) into (10.2.6) and then grouping together coefficients of the same power in θ, we obtain

$$(\alpha - i\delta\omega) = \gamma_1\theta^2 + \gamma_2\theta^4 + \gamma_3\theta^6 + O(\theta^7), \quad (10.2.13)$$

where

$$\gamma_3 = \frac{1}{\eta}\mathbf{w}^\top G(i\omega)\mathbf{p}_3 - \frac{\gamma_1}{\eta}\mathbf{w}^\top [G'(i\omega)\,[\mathbf{p}_2 + J\mathbf{V}_{15}] + G(i\omega)\mathbf{p}'_2]$$
$$-\frac{\gamma_2}{\eta}\mathbf{w}^\top [G'(i\omega)\,[\mathbf{p}_1 + J\mathbf{V}_{13}] + G(i\omega)\mathbf{p}'_1] + \frac{\gamma_1\gamma_2}{\eta}\mathbf{w}^\top G''(i\omega)J\mathbf{v}$$
$$+\frac{\gamma_1^2}{\eta}\mathbf{w}^\top \left[\frac{1}{2}G''(i\omega)[\mathbf{p}_1 + J\mathbf{V}_{13}] + G'(i\omega)[\mathbf{p}'_1 + J\mathbf{V}'_{13}] + \frac{1}{2}G(i\omega)\mathbf{p}''_1\right]$$
$$-\frac{\gamma_1^3}{6\eta}\mathbf{w}^\top G'''(i\omega)J\mathbf{v}\,. \tag{10.2.14}$$

Again, we reinsert the better approximation (10.2.13) for $(\alpha - i\delta\omega)$ into (10.2.6), so that the coefficients up to the 6th-power remain the same as before in the new formula. We can then calculate the next higher-order approximation. The result is

$$(\alpha - i\delta\omega) = \gamma_1\theta^2 + \gamma_2\theta^4 + \gamma_3\theta^6 + \gamma_4\theta^8 + O(\theta^9)\,, \tag{10.2.15}$$

where

$$\gamma_4 = \frac{1}{\eta}\mathbf{w}^\top G(i\omega)\mathbf{p}_4 - \frac{\gamma_3}{\eta}\mathbf{w}^\top [G'(i\omega)\,[\mathbf{p}_1 + J\mathbf{V}_{13}] + G(i\omega)\mathbf{p}'_1]$$
$$-\frac{\gamma_2}{\eta}\mathbf{w}^\top [G'(i\omega)\,[\mathbf{p}_2 + J\mathbf{V}_{15}] + G(i\omega)\mathbf{p}'_2]$$
$$-\frac{\gamma_1}{\eta}\mathbf{w}^\top [G'(i\omega)\,[\mathbf{p}_3 + J\mathbf{V}_{17}] + G(i\omega)\mathbf{p}'_3]$$
$$+\frac{\gamma_1^2}{\eta}\mathbf{w}^\top \left[\frac{1}{2}G''(i\omega)\,[\mathbf{p}_2 + J\mathbf{V}_{15}] + G'(i\omega)[\mathbf{p}'_2 + J\mathbf{V}'_{15}] + \frac{1}{2}G(i\omega)\mathbf{p}''_2\right]$$
$$+\frac{\gamma_2^2}{2\eta}\mathbf{w}^\top G''(i\omega)J\mathbf{v} + \frac{\gamma_1\gamma_2}{\eta}\mathbf{w}^\top \left[\begin{array}{l}G''(i\omega)\,[\mathbf{p}_1 + J\mathbf{V}_{13}] + 2G'(i\omega)[J\mathbf{V}'_{13} + \\ \mathbf{p}'_1] + G(i\omega)\mathbf{p}''_1\end{array}\right]$$
$$+\frac{\gamma_1\gamma_3}{\eta}\mathbf{w}^\top G''(i\omega)J\mathbf{v} - \frac{\gamma_1^2\gamma_2}{2\eta}\mathbf{w}^\top G'''(i\omega)J\mathbf{v}$$
$$-\frac{\gamma_1^3}{\eta}\mathbf{w}^\top \left[\begin{array}{l}\frac{G'''(i\omega)}{6}[\mathbf{p}_1 + J\mathbf{V}_{13}] + \frac{1}{2}G''(i\omega)[\mathbf{p}'_1 + \\ J\mathbf{V}'_{13}] + \frac{1}{2}G'(i\omega)\,[\mathbf{p}''_1 + J\mathbf{V}''_{13}] + \\ \frac{1}{6}G(i\omega)\mathbf{p}'''_1\end{array}\right]$$
$$+\frac{\gamma_1^4}{24\eta}\mathbf{w}^\top G''''(i\omega)J\mathbf{v}\,. \tag{10.2.16}$$

The values of γ_1, γ_2 and γ_3 were given above, and it is easy to verify that they remain the same in any higher-order approximations.

Taking only the real and imaginary parts of (10.2.15), we obtain

$$\alpha = \text{Re}\,(\gamma_1)\,\theta^2 + \text{Re}\,(\gamma_2)\,\theta^4 + \text{Re}\,(\gamma_3)\,\theta^6 + \text{Re}\,(\gamma_4)\,\theta^8 + \cdots\,. \tag{10.2.17}$$

When $\operatorname{Re}(\gamma_1) \neq 0$, we use the approximation

$$\alpha \approx \operatorname{Re}(\gamma_1)\theta^2. \tag{10.2.18}$$

To have a solution for small θ, we need to require both α and $\operatorname{Re}(\gamma_1)$ have the same sign, i.e., $\alpha\sigma_1 < 0$, where $\sigma_1 = -\operatorname{Re}(\gamma_1)$ is the *first* curvature coefficient:

$$\sigma_1 = -\operatorname{Re}\left(\frac{\mathbf{w}^\top G(i\omega)\mathbf{p}_1}{\mathbf{w}^\top G'(i\omega)J\mathbf{v}}\right).$$

Note that since α is positive, after increasing μ from the criticality, the equilibrium solution is unstable. But if the sign of σ_1 is negative, then the periodic solution surrounding the equilibrium state is stable.

Let us suppose that $\sigma_1 = 0$ at the criticality. In order to determine the stability of the emerging periodic solution, we need to consider the expansion of α up to the coefficient of θ^4, i.e.,

$$\alpha = \operatorname{Re}(\gamma_1)\theta^2 + \operatorname{Re}(\gamma_2)\theta^4. \tag{10.2.19}$$

Applying the same reasoning, $\sigma_2 = -\operatorname{Re}(\gamma_2)$ is the *second* curvature coefficient.

In a similar way, if both σ_1 and σ_2 are zero for certain critical combinations of the system parameters, then we need to explore the sign of the coefficient of the term θ^6, and so on.

In the terminology of degenerate Hopf bifurcations, the degeneracies are labeled as H_{km} for simplicity: H_{00} represents the classical Hopf bifurcation without any degeneracy (i.e., without any failure of the curvature coefficients or the transversality condition). The first subscript k indicates the order up to which the curvature coefficients vanish at criticality, while the second subscript m stands for the order up to which the derivatives of the formula in the failure of the transversality condition vanish. In this chapter, we only discuss the H_{k0}-degeneracies, i.e., failures on the curvature coefficients but not on the transversality condition. Note that multiple limit cycles emerge when perturbing the curvature coefficients near the value zero, after alternating the signs of the curvature coefficients in an increasing (or decreasing) order. For example, to have four limit cycles in the vicinity of a degenerate Hopf bifurcation – the H_{30}-degeneracy, (i.e., at the criticality we have $\sigma_1 = \sigma_2 = \sigma_3 = 0$ but $\sigma_4 \neq 0$), we need to perturb the system parameters in such a way that, for example, $\alpha > 0$, $\sigma_1 < 0$, $\sigma_2 > 0$, $\sigma_3 < 0$, and $\sigma_4 > 0$.

10.3 Normal Forms and Limit Cycles

In this section, bifurcation theory for planar systems is discussed by means of normal forms, for the case of limit cycles of small amplitudes. This method aims at finding a coordinate system in which the equations governing the dynamics of the system are simpler and more amenable to analysis. In these new coordinates, limit cycles look like circumferences and, therefore, the dynamics can be better analyzed in polar coordinates. In this setting, points satisfying $\dot{r} = 0$, $\dot{\phi} \neq 0$ and of constant sign, clearly indicates the presence of a limit cycle. However, this is a local method in the sense that the coordinate transformation is defined only in a neighborhood of the equilibrium point. From now on, and without loss of generality, we will assume that the equilibrium point is $\hat{x} = (0,0)$.

10.3.1 Normal Form Preliminaries

Consider the system

$$\dot{x} = f(x; \mu),$$

where $x \in R^2, \mu \in I \subset R^p$. Assume that $f(0; \mu) = 0$ for small values of $\|\mu\|$, and that $D_x f(0,0)$ has two purely imaginary eigenvalues $\lambda(0) = \pm i\omega(0)$. Then it is possible to find a linear transformation such that

$$D_x f(0; \mu) = \begin{pmatrix} \text{Re}\lambda(\mu) & -\text{Im}\lambda(\mu) \\ \text{Im}\lambda(\mu) & \text{Re}\lambda(\mu) \end{pmatrix} = \begin{pmatrix} \alpha(\mu) & -\omega(\mu) \\ \omega(\mu) & \alpha(\mu) \end{pmatrix}$$

to express the system as

$$\begin{pmatrix} \dot{x} \\ \dot{y} \end{pmatrix} = \begin{pmatrix} \alpha & -\omega \\ \omega & \alpha \end{pmatrix} \begin{pmatrix} x \\ y \end{pmatrix} + \begin{pmatrix} f_1(x,y;\mu) \\ f_2(x,y;\mu) \end{pmatrix}.$$

The computations for transforming the system to the normal form are easier if complex coordinates are used. Let us define $z = x + iy$, $\bar{z} = x - iy$, and then let $x = (z + \bar{z})/2$, $y = (z - \bar{z})/2i$. Then, system becomes

$$\begin{pmatrix} \dot{z} \\ \dot{\bar{z}} \end{pmatrix} = \begin{pmatrix} \lambda & 0 \\ 0 & \bar{\lambda} \end{pmatrix} \begin{pmatrix} z \\ \bar{z} \end{pmatrix} + \begin{pmatrix} F_1(z,\bar{z};\mu) \\ F_2(z,\bar{z};\mu) \end{pmatrix},$$

where

$$F_1(z,\bar{z};\mu) = f_1(x(z,\bar{z}), y(z,\bar{z}); \mu) + i f_2(x(z,\bar{z}), y(z,\bar{z}); \mu),$$
$$F_2(z,\bar{z};\mu) = f_1(x(z,\bar{z}), y(z,\bar{z}); \mu) - i f_2(x(z,\bar{z}), y(z,\bar{z}); \mu).$$

To analyze the behavior of the system, it only needs to consider the first component:

$$\dot{z} = \lambda z + F_1(z, \bar{z}; \mu),$$

since the second component is its conjugated. It is also simple to find the variation of the radius vector in polar coordinates:

$$r\dot{r} = \text{Re}(\bar{z}\,\dot{z}),$$
$$r^2\dot{\phi} = \text{Im}(\bar{z}\,\dot{z}). \qquad (10.3.20)$$

The goal is to find a coordinate transformation, such that the expression of the variation of the radius has the following *normal form*:

$$\dot{r} = r(\alpha + c_1 r^2 + c_2 r^4 + c_3 r^6 + \cdots + O(r^m)),$$

where $O(r^m)$ is an infinitesimum of order m when $r \to 0$, and c_i are the stability coefficients in the normal form. The nonzero roots of this equation correspond to limit cycles, whenever $\dot{\phi} > 0$ or $\dot{\phi} < 0$. In the new coordinates, the cycles are circles since $\dot{r} = 0$ implies r is a constant.

We work with systems where $F_1(z, \bar{z}; \mu)$ is a polynomial in z and $\bar{z}$, whose coefficients are functions of the parameter μ. If this is not the case, one could expand F_1 in a Taylor series and obtain a polynomial of degree N that is its good approximation near the origin. Thus, we assume that F_1 has the following structure:

$$F_1 = \sum_{i=2}^{N} \left(\sum_{j=0}^{i} a_{i,i-j}\, z^i\, \bar{z}^{i-j} \right).$$

10.3.2 Two Ways to Obtain a Normal Form

According to (10.3.20), to assure a negligible dependence of ϕ in the equation for $\dot{r}$, it is necessary to cancel the real part of all the terms in the product $\bar{z}\,\dot{z}$ that contain the monomials $\bar{z}^m z^n = r^{n+m} e^{i(n-m)\phi}$, with $m \neq n$. If $m = n$, then $\bar{z}^n z^n = r^{2n}$. This suggests two possibilities: (a) to eliminate the undesired terms; (b) to cancel the real part of the undesired monomials. It is possible to carry out a near-identity coordinate transformation, $z = Z + h(Z, \bar{Z})$, where h is a polynomial in the variables $Z, \bar{Z}$ of the same order as the monomials which we want to cancel. Therefore,

$$\dot{z} = \left(1 + \tfrac{\partial h}{\partial Z}\right) \dot{Z} + \tfrac{\partial h}{\partial \bar{Z}}\, \dot{\bar{Z}},$$
$$\dot{\bar{z}} = \tfrac{\partial \bar{h}}{\partial Z} \dot{Z} + \left(1 + \tfrac{\partial \bar{h}}{\partial \bar{Z}}\right) \dot{\bar{Z}},$$

and thus,

$$\Delta \dot{Z} = \dot{z}\left(1 + \frac{\partial \bar{h}}{\partial \bar{Z}}\right) - \dot{\bar{z}}\,\frac{\partial h}{\partial \bar{Z}},$$

where Δ is the Jacobian of the transformation

$$\Delta = \left(1 + \frac{\partial h}{\partial Z}\right)\left(1 + \frac{\partial \bar{h}}{\partial \bar{Z}}\right) - \frac{\partial \bar{h}}{\partial Z}\frac{\partial h}{\partial \bar{Z}}.$$

This is a real expression, because all the terms are products of conjugated expressions. In every simply connected region where $\Delta \neq 0$, the transformation is well-defined, and it is a bijection.

Method I

One way to obtain the normal form is to start with the complex normal form

$$\dot{Z} = \left(\dot{z} \left(1 + \frac{\partial \bar{h}}{\partial \bar{Z}}\right) - \dot{\bar{z}}\, \frac{\partial h}{\partial \bar{Z}} \right) \frac{1}{\Delta}.$$

In this case, since we are using a near-identity transformation, Δ^{-1} can be approximated by

$$\frac{1}{\Delta} = \frac{1}{(1 + \frac{\partial h}{\partial Z})(1 + \frac{\partial \bar{h}}{\partial \bar{Z}}) - \frac{\partial \bar{h}}{\partial Z} \frac{\partial h}{\partial \bar{Z}}} \cong 1 - \left(\frac{\partial h}{\partial Z} + \frac{\partial \bar{h}}{\partial \bar{Z}} + \frac{\partial h}{\partial Z} \frac{\partial \bar{h}}{\partial \bar{Z}} - \frac{\partial \bar{h}}{\partial Z} \frac{\partial h}{\partial \bar{Z}} \right).$$

Note that $\dot{Z}$ is a polynomial, whose coefficients can be adjusted by appropriately choosing the coefficients of h. In this way, it is possible to eliminate all the monomials in $\dot{Z}$, except those in the form of $z^n \bar{z}^{n-1}$:

$$\dot{Z} = \lambda Z + (a + i b) Z^2 \bar{Z} + \cdots + O(|Z|^{m+1}).$$

If the degree of the polynomial h is m, the coefficients of $\dot{Z}$ up to order m can be adjusted. For local analysis, higher-order terms can be neglected, and the truncated normal form for $m = 5$, for example, is

$$\dot{Z} = \lambda Z + (a + i b) Z^2 \bar{Z} + (c + i d) Z^3 \bar{Z}^2.$$

This can be written in polar coordinates as

$$\dot{r} = \alpha r + a r^3 + c r^5,$$
$$\dot{\phi} = \omega + b r^2 + d r^4.$$

Method II

Consider the polynomial $\Delta \dot{Z}$. In polar coordinates, since $\Delta \neq 0$ is real, we have

$$\Delta r \dot{r} = \Delta \operatorname{Re}(\bar{Z} \dot{Z}) = \operatorname{Re}(\bar{Z} \Delta \dot{Z}),$$
$$\Delta r^2 \dot{\phi} = \operatorname{Im}(\bar{Z} \Delta \dot{Z}).$$

The real part of the unwanted monomials of $\bar{Z} \triangle \dot{Z}$ can be cancelled if the coefficients of $\triangle \dot{Z}$ satisfy $a_{i,j} = -\bar{a}_{j+1,i-1}, \forall i > 0, j \neq i - 1$, and $a_{0,j} = 0$. If the coefficients of the transformation are chosen appropriately, every coefficient of $\triangle \dot{Z}$ up to the degree of h can be modified. For example, if $\triangle \dot{Z} = \alpha Z^3 + \beta Z^2 \bar{Z} - \bar{\alpha} Z \bar{Z}^2$, then $\mathrm{Re}\left(\bar{Z} \triangle \dot{Z}\right) = \mathrm{Re}(\beta)|Z|^4$. It is clear that the limit cycles are associated with the real positive roots of $\dot{r}$, or equivalently, with the roots of $\mathrm{Re}(\bar{Z} \triangle \dot{Z})$. Thus, it is necessary to study the roots of the polynomial

$$\alpha + c_1 r^2 + c_2 r^4 + c_3 r^6 + \cdots + O(r^m),$$

where the coefficients depend on the parameters of the system, as well as the transformation. Considering a local approach, higher-order terms can be ignored, leading to a polynomial in r only:

$$\alpha + c_1 r^2 + c_2 r^4 + c_3 r^6 + \cdots + c_n r^n. \tag{10.3.21}$$

The real positive roots r_i^* of this polynomial, for which $\dot{\phi}$ does not change sign and the Jacobian of the transformation does not vanish for any $r < \max(r_i^*)$, provide good approximations of the limit cycles of the system that enclose the equilibrium point.

To generate nested cycles, we can first locate the roots and then find the coefficients of the polynomial. Since the expressions of c_i are functions of the parameters of the system, we can find out in which range of parameter values the proposed cycles can be obtained.

10.4 Controlling the Multiplicity of Limit Cycles

In the pursuit of controlling the multiplicity of limit cycles, the results of curvature coefficients computed for n-dimensional systems can be applied. If only one limit cycle is desired, then the key of control methodology is to choose a suitable combination of parameters that can ensure the existence of a unique limit cycle. This means that a proper and suitable selection of the system parameters should be such that there is no alternation in sign in the consequent curvature coefficients (e.g., the first three curvature coefficients are all negative), so as to avoid the presence of multiple limit cycles, at least locally.

For a general nonlinear dynamical system, it is usually necessary to use a computer to implement the curvature coefficient formulas, so as to analyze different possibilities of valid regions, at least locally, such that only one stable limit cycle can appear. This methodology is also useful if we wish to avoid other degenerate Hopf bifurcations, such as those that have simultaneous failures of the transversality condition and the vanishing of curvature coefficients. In this case, the selection of system parameters for satisfying a specific sign-definition of the curvature coefficients leads to a well-known local bifurcation diagrams [22]. Hence, a control system designer can first choose a desired (local) region over which the corresponding conditions can be obtained in terms of the system parameters.

One advantage of this methodology is that in order to drive the system orbit to a desired region, there is no need to modify the feedback control path by adding any nonlinear terms: we can simply modify the system parameters as shown in the next section. This is a kind of parametric variation, according to the expressions of the curvature coefficients. Note that for some systems, it may be easier to modify the system parameters than fabricating a specific nonlinear feedback controller for the control of the appearance of limit cycles and bifurcations. For other systems, of course, it may be more convenient to modify the nonlinear feedback controller – sometimes it might even be necessary.

Under small variations in the main and auxiliary bifurcation parameters, distinct local bifurcation diagrams (i.e., plots of equilibrium and periodic solutions obtained via slight variations of these parameters) can be found, in the unfoldings of degenerate Hopf bifurcations involving the vanishing of several curvature coefficients (the H_{k0}-singularity). This singularity, after a suitable perturbation of system parameters, produces multiple (up to $k+1$) cycles in the unfoldings of the degeneracy. Generally, this type of degenerate Hopf bifurcations have large-amplitude limit cycles emerging from criticality, together with a hysteresis phenomenon. Therefore, it generally has multiple (small- and large-amplitude) limit cycles. For this reason, this degeneracy is not a good candidate for exploring control of small-amplitude limit cycles, and so it should be avoided.

The following scheme is developed based on the fact that in order to have multiple oscillations, an alternation of signs of the real part of the bifurcating eigenvalues and the ordered curvature coefficients is required. That is, each curvature coefficient gives a type of stable or unstable region in the phase space, depending on the values of the variables. As usual, we adopt the negative sign to indicate stabilizing effect; so one possible combination for a three-nested limit cycles (stable-unstable-stable, "SUS" for short) configuration from the equilibrium is to have a positive real part of the bifurcating eigenvalues α plus $\sigma_1(\mu,\varepsilon) < 0$, $\sigma_2(\mu,\varepsilon) > 0$ and $\sigma_3(\mu,\varepsilon) < 0$ ($c_1 < 0$, $c_2 > 0$, $c_3 < 0$, in normal forms), all at the criticality $\mu_0 = 0$, where ε is a vector of auxiliary parameters. The scheme is:

STEP 1: Locate an H_{10}-degeneracy in the parameter space, (μ, ε_1), i.e., where $\sigma_1(\mu^{(1)}, \varepsilon_1^{(1)}) = 0$. On this set, vary the auxiliary parameter ε_2, trace the Hopf bifurcation curve until the following conditions are satisfied: $\sigma_1(\mu^{(2)}, \varepsilon_1^{(2)}, \varepsilon_2^{(2)}) = \sigma_2(\mu^{(2)}, \varepsilon_1^{(2)}, \varepsilon_2^{(2)}) = 0$, i.e., the H_{20}-degeneracy. From this set, repeat the search, now by incorporating the variation of ε_3 and the formula of σ_3, to obtain the degenerate Hopf bifurcation curve (the H_{30}-degeneracy) that satisfies the Hopf bifurcating conditions, subject to the following additional constraints: $\sigma_1(\mu^{(3)}, \varepsilon_1^{(3)}, \varepsilon_2^{(3)}, \varepsilon_3^{(3)}) = \sigma_2(\mu^{(3)}, \varepsilon_1^{(3)}, \varepsilon_2^{(3)}, \varepsilon_3^{(3)}) = \sigma_3(\mu^{(3)}, \varepsilon_1^{(3)}, \varepsilon_2^{(3)}, \varepsilon_3^{(3)}) = 0$.

STEP 2: Perturb the values of the auxiliary parameters, $\varepsilon_1, \varepsilon_2$ and/or ε_3, in order to deal with one, two, or three periodic solutions from the degenerate point in the parameter space, $(\mu^{(3)}, \varepsilon_1^{(3)}, \varepsilon_2^{(3)}, \varepsilon_3^{(3)})$, by appropriately adjusting the sign alternation of the real parts of the complex eigenvalues and the ordered curvature coefficients, and then check with a continuation software package such as LOCBIF [27] and AUTO [17].

10.5 Two Illustrative Examples

In this section, we show two applications of the theoretical results developed above. The first example has been discussed, earlier in [38] and then later in [12, 30, 29], which is a planar cubic system. The second example is taken from the chemical engineering field (a three-dimensional dynamical system) for which the formulas derived in Section 10.2 are used.

10.5.1 A Cubic Polynomic System Example

The planar cubic system under consideration is described by

$$\dot{x} = \lambda x - y + (a - \omega - \theta) x^3 + (3\mu - \eta) x^2 y +$$
$$(3\theta + \xi - 3\omega - 2a) xy^2 + (\nu - \mu) y^3$$
$$\dot{y} = x + \lambda y + (\mu + \nu) x^3 + (3\omega + 3\theta + 2a) x^2 y +$$
$$(\eta - 3\mu) xy^2 + (\omega - \theta - a) y^3, \qquad (10.5.22)$$

where λ plays the role of the main bifurcation parameter, and $a, \omega, \theta, \mu, \eta, \xi$ and ν are all auxiliary parameters (for historical reasons we keep the original notation instead of μ and ε). If the system is linear, i.e.,

$$\begin{aligned} \dot{z}_1 &= \lambda z_1 - z_2, \\ \dot{z}_2 &= z_1 + \lambda z_2, \end{aligned} \qquad (10.5.23)$$

then a simple equilibrium at the origin is obtained. But it still has a nonlinear function (for example, a feedback), so that creation of multiple equilibrium solutions as well as limit cycles is possible by appropriately varying the main and auxiliary bifurcation control parameters.

After some algebraic computations (see [35] for details), the *first* curvature coefficient is obtained as

$$\sigma_1 = \frac{1}{16}\xi. \qquad (10.5.24)$$

As can be seen from (10.5.24), $\sigma_1 = 0$ if $\xi = 0$.

Considering the case where $\xi = 0$ and applying the formulas given in Section 10.2, we have

$$\sigma_2 = \frac{-5}{32} a\nu. \qquad (10.5.25)$$

In this case, we have two possibilities for σ_2 to be zero. One is $a = 0$; this possibility is ruled out because in this situation the equilibrium point is a center. Thus, the only possibility to have a weak focus of order 2 is $\nu = 0$. In this case, a computation of the third curvature coefficient yields

$$\sigma_3 = \frac{25}{128} a\omega\theta. \qquad (10.5.26)$$

This expression is equivalent to the result (up to a multiplication by a positive constant) obtained in [38] (see also [16]).

Following the normal-form method, the curvature coefficients can be obtained for system (10.5.22). We only consider the case of $\lambda = 0$ in order to simplify the expressions. Using the first method described in Section 10.3, we have

$$
\begin{aligned}
c_1 &= \tfrac{1}{8}\xi, \\
c_2 &= \tfrac{1}{16}(\nu\xi - 10a\nu), \\
c_3 &= \tfrac{1}{2048}\bigl(-320\,a\,\eta\,\nu + 6400\,a\,\mu\,\nu - 3104\,a^2\,\xi + 20\,\eta^2\,\xi - 1280\,a\,\theta\,\xi + \\
&\quad 320\,\theta^2\,\xi - 160\,\eta\,\mu\,\xi + 320\,\mu^2\,\xi + 32\,\eta\,\nu\,\xi - 640\,\mu\,\nu\,\xi + \\
&\quad 224\,\nu^2\,\xi + 640\,a\,\xi^2 + 208\,\theta\,\xi^2 - 19\,\xi^3 + 12800\,a\,\theta\,\omega - \\
&\quad 192\,a\,\xi\,\omega - 1280\,\theta\,\xi\,\omega - 160\,\xi^2\,\omega + 896\,\xi\,\omega^2\bigr).
\end{aligned}
$$
$$(10.5.27)$$

For the second method described in Section 10.3, the same values are obtained for c_1 and c_2, but a slightly different one is obtained for c_3:

$$
\begin{aligned}
c_{3,II} &= \tfrac{1}{2048}\bigl(-80\,a\,\eta\,\nu + 1600\,a\,\mu\,\nu - 448\,a^2\,\xi - 4\,\eta^2\,\xi - 320\,a\,\theta\,\xi - \\
&\quad 64\,\theta^2\,\xi + 32\,\eta\,\mu\,\xi - 64\,\mu^2\,\xi + 8\,\eta\,\nu\,\xi - 160\,\mu\,\nu\,\xi + \\
&\quad 128\,\nu^2\,\xi + 56\,a\,\xi^2 + 16\,\theta\,\xi^2 + 3\,\xi^3 + 3200\,a\,\theta\,\omega + \\
&\quad 336\,a\,\xi\,\omega - 320\,\theta\,\xi\,\omega - 136\,\xi^2\,\omega + 512\,\xi\,\omega^2\bigr).
\end{aligned}
$$

This is not surprising, because the near-identity transformations used to take the system to any of the two aforementioned normal forms are not the same. For the sake of clarity, the higher-order coefficients (c_4 and c_5) are given in the Appendix.

A simple test for verifying the correctness of the above formulas, we can vanish the lower-order coefficients, and then compare the results with those given in [38, 16]. These simplified results are entirely equivalent, up to positive constant multiple:

$$
\begin{aligned}
c_1 &= \tfrac{1}{8}\xi \propto \sigma_1, \\
c_2 &= -\tfrac{5}{8}a\nu \propto \sigma_2, \quad \text{if } \xi = 0 \ (c_1 = 0), \\
c_3 &= \tfrac{25}{16}a\theta\omega \propto \sigma_3, \quad \text{if } \xi = \nu = 0 \ (c_1 = c_2 = 0), \\
c_4 &= \tfrac{5}{64}a^2\theta\eta, \quad \text{if } \xi = \nu = \omega = 0, \ (c_1 = c_2 = c_3 = 0), \\
c_5 &= -\tfrac{5}{192}a^2\theta\left[4\left(\mu^2 + \theta^2\right) - a^2\right], \\
&\quad \text{if } \xi = \nu = \omega = \eta = 0 \ (c_1 = c_2 = c_3 = c_4 = 0).
\end{aligned}
$$

A simple way to obtain (or to predict) multiple limit cycles is to choose the roots of the normal-form polynomial, by means of determining circles (in the transformed plane) whose radii are the values of the chosen roots. These circles correspond to limit cycles in the original coordinates, whose amplitudes will be about the same as the radii of the circles, due to the near-identity transformations. The shape, however, could be highly distorted, since we did not take into account the phase of the transformation.

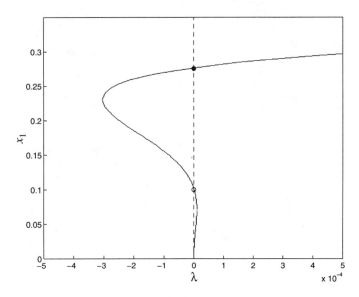

FIGURE 10.2
Continuation of the periodic solutions by varying λ.

To this end, let us consider the amplitude polynomial (10.3.21), with $\lambda = 0$, as our expressions for the computed curvature coefficients c_i, under this simplifying assumption. To obtain three nested limit cycles, we propose three roots $r_1^* = 0.1$, $r_2^* = 0.3$ and $r_3^* = 0.7$. Note that the amplitude polynomial (10.3.21) is not monic; thus we are able to choose an arbitrary

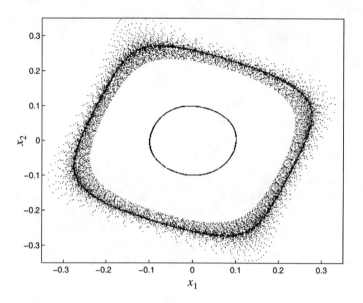

FIGURE 10.3
Two periodic solutions ($\lambda = 0$).

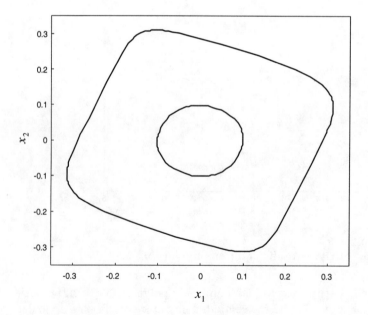

FIGURE 10.4
Periodic solutions predicted by the normal form.

value for the higher-order curvature coefficients in the approximation.
For a fourth-order expansion, the polynomial for $c_4 = 10$ is given by

$$-0.00441\, r^2 + 0.499\, r^4 - 5.9\, r^6 + 10\, r^8, \qquad (10.5.28)$$

that is, $c_1 = -0.00441$, $c_2 = 0.499$, and $c_3 = -5.9$. Given these values, it is possible to obtain the system parameters from (10.5.27). Due to the highly nonlinear nature of these expressions, there are many possible solutions, among which we choose $a = 1$, $\omega = 0.8654330$, $\theta = -4.318510$, $\mu = 0$, $\eta = -1$, $\xi = -1$ and $\nu = -0.795593$. With these parameter values, some simulations were performed to check the validity of the results. As shown in Figures 10.2 and 10.3, only two limit cycles are found. The one corresponding to the root $r_3^* = 0.7$ lies in a region where the coordinate transformation is not valid.

A prediction of the location and shape of the cycle can be performed via the coordinate transformation used to obtain the normal form $z = Z + h(Z, \overline{Z})$, thereby replacing $Z \to r_i^* e^{i\phi}$, $0 \le \phi \le 2\pi$, as shown in Figure 10.4. This clearly indicates the local nature of the normal form.

It seems that using higher-order normal forms may produce more accurate results. To verify this, we computed the curvature coefficients up to the fifth order, using the expressions (10.5.27) obtained with the first method in Section 10.3. With the values of the parameters given above, we obtain $c_1 = -0.0041$, $c_2 = 0.499$, $c_3 = -5.93052$, $c_4 = 13.742$, and $c_5 = -205.827$. The real positive roots of the polynomial are $r_1^* = 0.100018$ and $r_2^* = 0.269843$, indicating that only two cycles emerge. These values agree with the amplitudes of the cycles obtained by both simulation and the continuation method, as shown in Figure 10.2.

Another way to obtain the desired behavior is to take into account only the sign of the curvature coefficients, and choosing the parameters accordingly. This trial-and-error procedure is practical if it is possible to count on good software such as LOCBIF or AUTO that help "continue" the periodic solutions to validate the results. In this case, it is sometimes advisable to use the simplified expressions for the curvature coefficients, due to the empirical nature of this methodology.

For example, if we wish to obtain a weak focus of fourth-order (H_{40}) degeneracy and a stable limit cycle for $\lambda > 0$, it is required that $c_1 = 0$, $c_2 = 0$, $c_3 = 0$, $c_4 = 0$ and $c_5 < 0$. It is easy to verify that this can be attained by choosing the parameter values to be $a = 2.05$, $\omega = 0$, $\theta = -0.05$, $\mu = 1$, $\eta = 0$, $\xi = 0$ and $\nu = 0$. This is also confirmed by the results obtained from the continuation curves (see Figure 10.5), in which a large-amplitude periodic solution emerges from the Hopf bifurcation point.

Four nested cycles can be found for the parameter values $a = 1.92$, $\omega = 0.33 \times 10^{-1}$, $\theta = 1$, $\mu = 0$, $\eta = -0.1$, $\xi = 0.3 \times 10^{-1}$ and $\nu = 0.3 \times 10^{-1}$. The approximate curvature coefficients have signs $c_1 > 0$, $c_2 < 0$, $c_3 > 0$, $c_4 < 0$, $c_5 < 0$, and this seems to be appropriate in order to obtain four

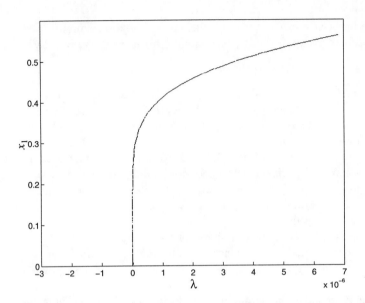

FIGURE 10.5
Continuation of the periodic solutions by varying λ ($c_1 = c_2 = c_3 = c_4 = 0$, and $c_5 < 0$).

cycles (named from the equilibrium labeled as "USUS"). Continuation of periodic solutions (see Figure 10.6) and simulation results (Figure 10.7) confirm this hypothesis.

Although the above curvature coefficients were computed for $\lambda = 0$, the amplitude polynomial (10.3.21) can be used to estimate the amplitude of the limit cycles for smaller λ values. For the set of parameter values given above, the curvature coefficients are $c_1 = 0.00375$, $c_2 = -0.0359437$, $c_3 = 0.0494555$, $c_4 = 0.314707$, and $c_5 = -1.16288$. For $\lambda = -2.5 \times 10^{-3}$, the real positive roots of the polynomial are $r_1^* = 0.0845689$ and $r_2^* = 0.36758$. These results agree quite well with those estimated from the continuation curve closest to the equilibrium point (the first two small cycles) which yield amplitudes 0.084 and 0.397. However, there are also cycles with amplitudes 0.8674 and 1.228, obtained by using continuation software that we did not detect with the normal-form approach. This approach fails to predict these two *large-amplitude* cycles, reflecting the local nature of the method. If the approximate coefficients are used to estimate the cycles, then we will obtain $c_1 = 0.00375$, $c_2 = -0.036$, $c_3 = 0.099$, $c_4 = -0.0288$, and $c_5 = -0.0301056$. For $\lambda = -2.5 \times 10^{-3}$, the real positive roots of the amplitude polynomial are $r_1^* = 0.0845428$ and $r_2^* = 1.05705$. Once again, the small cycle is predicted quite accurately, but we obtain a spurious estimate of a larger cycle that does not even exist in the original system.

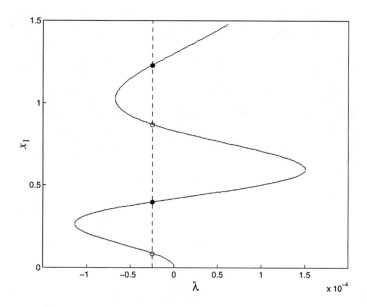

FIGURE 10.6
Continuation of the periodic solutions by varying λ
($c_1 > 0$, $c_2 < 0$, $c_3 > 0$, $c_4 < 0$, and $c_5 < 0$).

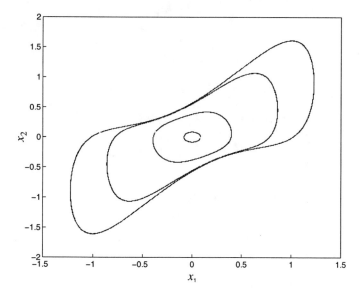

FIGURE 10.7
Four nested limit cycles.

The failure to predict larger-amplitude limit cycles leads to a consideration of using lower-order expansions. For example, a second-order expansion, using only c_1 and c_2, gives the estimates $r_1^* = 0.0846034$ and $r_2^* = 0.311724$, which are reasonable enough for such simple calculations.

Finally, four nested cycles, with opposite stabilities ("SUSU"), can be found for $a = 2.08$, $\omega = -0.33 \times 10^{-1}$, $\theta = 1$, $\mu = 0$, $\eta = 0.063$, $\xi = -0.3 \times 10^{-1}$ and $\nu = -0.3 \times 10^{-1}$, for which the simplified curvature coefficients have opposite signs as compared with the previous case. Its corresponding continuation curve is shown in Figure 10.8.

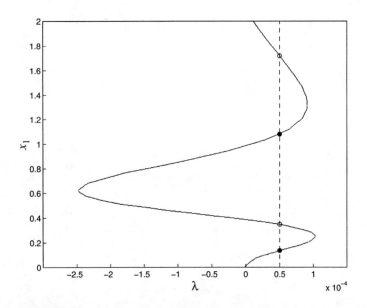

FIGURE 10.8
Continuation of the periodic solutions by varying λ ($c_1 < 0$, $c_2 > 0$, $c_3 < 0$, $c_4 > 0$, and $c_5 > 0$).

10.5.2 A Chemical Reactor Example

Consider the mathematical model of a continuous stirred tank reactor (CSTR) that has a single, irreversible, exothermic, first order reaction $A \to B$ with extraneous thermal capacitance (see [37]):

$$\dot{x}_1 = -x_1 + D(1 - x_1)\exp(x_2),$$
$$\dot{x}_2 = -(1 + \beta + \beta\rho)x_2 + \beta\rho x_3 + DB(1 - x_1)\exp(x_2),$$
$$\dot{x}_3 = -\beta\rho(x_2 - x_3)/\varepsilon, \qquad (10.5.29)$$

where D is the main bifurcation control parameter, and B, β, ρ, and ε are auxiliary parameters. Multiple oscillations, using the curvature coefficient formulas σ_1, σ_2, and σ_3 in the frequency domain and applying the procedure described in Section 10.4, have been obtained in [34] (and more recently, in [10]).

A nested configuration "USUS" of periodic solutions that enclose one equilibrium is shown in Figure 10.9, where the parameter values are: $D = 0.13689166$, $B = 23.825$, $\beta = 4.99$, $\rho = 1.0$, and $\varepsilon = 2.0$. The simulation results obtained using LOCBIF were double-checked by using the software AUTO. The parameter region in which four limit cycles coexist is very small, and the previous study on Sibirskii's cubic polynomic system example has drawn our attention to a very careful estimate of the size of this region. However, the location of the degenerate Hopf bifurcation point in a higher-codimensional case (such as the H_{k0}-singularity) enables us to find multiple oscillations by suitable perturbations of system parameters.

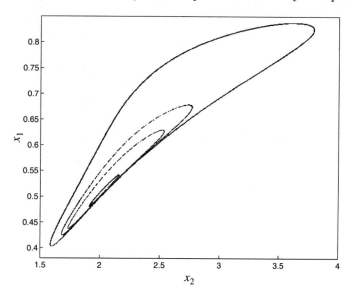

FIGURE 10.9
A nested configuration "USUS" of periodic solutions for the CSTR.

10.6 Conclusions

In this chapter, we have presented a simple procedure for detecting multiple oscillations existing in a nonlinear dynamical system, which is useful for the control objective of avoiding such dynamical behaviors. The detection procedure has been verified both analytically and numerically, and illustrated by a planar cubic system and a chemical reactor model. The planar cubic example provides simple relationships among the auxiliary parameters of the system, which can be used to evaluate the system curvature coefficients. In a general nonlinear system, the computational procedure described in Section 10.4 should be carried out along with some useful formulas provided in [34].

Acknowledgments

G. Calandrini and J. Moiola appreciate the financial support of CONICET (the National Council of Scientific Research of Argentina) and a grant of the Secretaría General de Ciencia y Tecnología U.N.S.; G. Chen is grateful to the support of the US Army Research Office under the Grant DAAG55-98-1-0198.

Appendix

The curvature coefficient obtained by using method I is:

$$
\begin{aligned}
c_4 = \frac{1}{8192} \big(& 640\,a^2\,\eta\,\theta + 15680\,a^3\,\nu - 40\,a\,\eta^2\,\nu - 9600\,a\,\theta^2\,\nu + 1280\,a\,\eta\,\mu\,\nu \\
& - 9600\,a\,\mu^2\,\nu - 2880\,a\,\nu^3 - 40\,a^2\,\eta\,\xi - 4\,\eta^3\,\xi + 576\,a\,\eta\,\theta\,\xi - 64\,\eta\,\theta^2\,\xi \\
& + 736\,a^2\,\mu\,\xi + 32\,\eta^2\,\mu\,\xi - 64\,\eta\,\mu^2\,\xi - 5088\,a^2\,\nu\,\xi + 4\,\eta^2\,\nu\,\xi - 864\,a\,\theta\,\nu\,\xi \\
& + 960\,\theta^2\,\nu\,\xi - 128\,\eta\,\mu\,\nu\,\xi + 960\,\mu^2\,\nu\,\xi + 32\,\eta\,\nu^2\,\xi - 384\,\mu\,\nu^2\,\xi + 288\,\nu^3\,\xi \\
& - 80\,\eta\,\theta\,\xi^2 + 160\,a\,\mu\,\xi^2 + 474\,a\,\nu\,\xi^2 + 240\,\theta\,\nu\,\xi^2 - 7\,\eta\,\xi^3 - 8\,\mu\,\xi^3 + 7\,\nu\,\xi^3 \\
& - 7040\,a\,\eta\,\theta\,\omega + 3840\,a^2\,\nu\,\omega + 80\,a\,\eta\,\xi\,\omega + 704\,\eta\,\theta\,\xi\,\omega - 3072\,a\,\mu\,\xi\,\omega \\
& + 1536\,a\,\nu\,\xi\,\omega - 1536\,\theta\,\nu\,\xi\,\omega + 120\,\eta\,\xi^2\,\omega - 384\,\nu\,\xi^2\,\omega - 11520\,a\,\nu\,\omega^2 \\
& - 640\,\eta\,\xi\,\omega^2 + 1536\,\mu\,\xi\,\omega^2 + 1152\,\nu\,\xi\,\omega^2 \big)
\end{aligned}
$$

$$
\begin{aligned}
c_5 = \frac{1}{75497472} \big(& 1966080\,a^4\,\theta - 4915200\,a^2\,\eta^2\,\theta - 7864320\,a^2\,\theta^3 + 3932160\,a^2\,\eta\,\theta\,\mu \\
& - 7864320\,a^2\,\theta\,\mu^2 + 19630080\,a^3\,\eta\,\nu + 264960\,a\,\eta^3\,\nu + 45527040\,a\,\eta\,\theta^2\,\nu \\
& - 215654400\,a^3\,\mu\,\nu - 2257920\,a\,\eta^2\,\mu\,\nu + 115752960\,a\,\theta^2\,\mu\,\nu - 12349440\,a\,\eta\,\mu^2\,\nu \\
& + 115752960\,a\,\mu^3\,\nu - 133693440\,a^2\,\theta\,\nu^2 - 7833600\,a\,\eta\,\nu^3 + 116858880\,a\,\mu\,\nu^3 \\
& - 49630080\,a^4\,\xi + 217472\,a^2\,\eta^2\,\xi + 41904\,\eta^4\,\xi + 36298752\,a^2\,\theta\,\xi \\
& - 3118080\,a\,\eta^2\,\theta\,\xi + 8480768\,a^2\,\theta^2\,\xi + 635904\,\eta^2\,\theta^2\,\xi - 21577728\,a\,\theta^3\,\xi \\
& - 552960\,\theta^4\,\xi - 8863744\,a^2\,\eta\,\mu\,\xi - 317952\,\eta^3\,\mu\,\xi + 10788864\,a\,\eta\,\theta\,\mu\,\xi \\
& + 552960\,\eta\,\theta^2\,\mu\,\xi + 10446848\,a^2\,\mu^2\,\xi + 497664\,\eta^2\,\mu^2\,\xi - 21577728\,a\,\theta\,\mu^2\,\xi \\
& - 1105920\,\theta^2\,\mu^2\,\xi + 552960\,\eta\,\mu^3\,\xi - 552960\,\mu^4\,\xi + 2982912\,a^2\,\nu\,\xi \\
& - 26496\,\eta^3\,\nu\,\xi + 9630720\,a\,\eta\,\theta\,\nu\,\xi - 4552704\,\eta\,\theta^2\,\nu\,\xi + 38645760\,a^2\,\mu\,\nu\,\xi \\
& + 225792\,\eta^2\,\mu\,\nu\,\xi + 28938240\,a\,\theta\,\mu\,\nu\,\xi - 11575296\,\theta^2\,\mu\,\nu\,\xi + 1234944\,\eta\,\mu^2\,\nu\,\xi \\
& - 11575296\,\mu^3\,\nu\,\xi - 34740480\,a^2\,\nu^2\,\xi - 577152\,\eta^2\,\nu^2\,\xi - 8134656\,a\,\theta\,\nu^2\,\xi \\
& + 19187712\,\theta^2\,\nu^2\,\xi - 1161216\,\eta\,\mu\,\nu^2\,\xi + 19187712\,\mu^2\,\nu^2\,\xi + 783360\,\eta\,\nu^3\,\xi \\
& - 11685888\,\mu\,\nu^3\,\xi + 397440\,\nu^4\,\xi + 21533952\,a^3\,\xi^2 - 140800\,a\,\eta^2\,\xi^2 \\
& - 10425856\,a^2\,\theta\,\xi^2 + 519936\,\eta^2\,\theta\,\xi^2 - 11149312\,a\,\theta^2\,\xi^2 + 1959936\,\theta^3\,\xi^2 \\
& + 3255296\,a\,\eta\,\mu\,\xi^2 - 979968\,\eta\,\theta\,\mu\,\xi^2 - 5754880\,a\,\mu^2\,\xi^2 + 1959936\,\theta\,\mu^2\,\xi^2 \\
& - 436800\,a\,\eta\,\nu\,\xi^2 - 4336128\,\eta\,\theta\,\nu\,\xi^2 - 372480\,a\,\mu\,\nu\,\xi^2 - 2893824\,\theta\,\mu\,\nu\,\xi^2 \\
& + 3459840\,a\,\nu^2\,\xi^2 + 8080896\,\theta\,\nu^2\,\xi^2 - 3670432\,a^2\,\xi^3 + 166040\,\eta^2\,\xi^3 \\
& - 565504\,a\,\theta\,\xi^3 + 1723904\,\theta^2\,\xi^3 - 938368\,\eta\,\mu\,\xi^3 + 1199360\,\mu^2\,\xi^3 \\
& - 540768\,\eta\,\nu\,\xi^3 + 319872\,\mu\,\nu\,\xi^3 + 345696\,\nu^2\,\xi^3 + 255488\,a\,\xi^4 + 261056\,\theta\,\xi^4 \\
& - 4093\,\xi^5 - 370851840\,a^3\,\theta\,\omega + 41011200\,a\,\eta^2\,\theta\,\omega + 231505920\,a\,\theta^3\,\omega \\
& - 115752960\,a\,\eta\,\theta\,\mu\,\omega + 231505920\,a\,\theta\,\mu^2\,\omega - 88719360\,a^2\,\eta\,\nu\,\omega \\
& + 260505600\,a^2\,\mu\,\nu\,\omega + 348733440\,a\,\theta\,\nu^2\,\omega - 16819200\,a^3\,\xi\,\omega + 973056\,a\,\eta^2\,\xi\,\omega \\
& + 114008064\,a^2\,\theta\,\xi\,\omega - 4101120\,\eta^2\,\theta\,\xi\,\omega + 94531584\,a\,\theta^2\,\xi\,\omega - 23150592\,\theta^3\,\xi\,\omega \\
& - 14825472\,a\,\eta\,\mu\,\xi\,\omega + 11575296\,\eta\,\theta\,\mu\,\xi\,\omega + 36655104\,a\,\mu^2\,\xi\,\omega - 23150592\,\theta\,\mu^2\,\xi\,\omega \\
& + 1007616\,a\,\eta\,\nu\,\xi\,\omega + 33730560\,\eta\,\theta\,\nu\,\xi\,\omega - 53575680\,a\,\mu\,\nu\,\xi\,\omega + 34882560\,a\,\nu^2\,\xi\,\omega + 13358592\,a\,\theta\,\xi^2\,\omega \\
& - 34873344\,\theta\,\nu^2\,\xi\,\omega + 7002624\,a^2\,\xi^2\,\omega - 3008640\,\eta^2\,\xi^2\,\omega + 13358592\,a\,\theta\,\xi^2\,\omega \\
& - 24803328\,\theta^2\,\xi^2\,\omega + 15903744\,\eta\,\mu\,\xi^2\,\omega - 19015680\,\mu^2\,\xi^2\,\omega + 6136320\,\eta\,\nu\,\xi^2\,\omega \\
& - 1781760\,\mu\,\nu\,\xi^2\,\omega - 4124160\,\nu^2\,\xi^2\,\omega - 94656\,a\,\xi^3\,\omega - 5035776\,\theta\,\xi^3\,\omega \\
& - 92448\,\xi^4\,\omega - 13762560\,a^2\,\theta\,\omega^2 + 83681280\,a\,\eta\,\nu\,\omega^2 + 7372800\,a\,\mu\,\nu\,\omega^2 \\
& - 9784320\,a^2\,\xi\,\omega^2 + 14556672\,\eta^2\,\xi\,\omega^2 - 139689984\,a\,\theta\,\xi\,\omega^2 + 76750848\,\theta^2\,\xi\,\omega^2 \\
& - 72105984\,\eta\,\mu\,\xi\,\omega^2 + 76750848\,\mu^2\,\xi\,\omega^2 - 8368128\,\eta\,\nu\,\xi\,\omega^2 - 737280\,\mu\,\nu\,\xi\,\omega^2 \\
& + 3179520\,\nu^2\,\xi\,\omega^2 - 5397504\,a\,\xi^2\,\omega^2 + 28760064\,\theta\,\xi^2\,\omega^2 + 1135488\,\xi^3\,\omega^2 \\
& + 474808320\,a\,\theta\,\omega^3 + 24514560\,a\,\xi\,\omega^3 - 47480832\,\theta\,\xi\,\omega^3 - 4995072\,\xi^2\,\omega^3 \\
& + 6359040\,\xi\,\omega^4 \big)
\end{aligned}
$$

The curvature coefficient obtained by using method II is:

$$
\begin{aligned}
c_4 = \frac{1}{16384} \big(& 1280\,a^2\,\eta\,\theta + 14720\,a^3\,\nu + 280\,a\,\eta^2\,\nu - 13440\,a\,\theta^2\,\nu - 320\,a\,\eta\,\mu\,\nu \\
& - 13440\,a\,\mu^2\,\nu - 11520\,a\,\nu^3 + 1120\,a^2\,\eta\,\xi - 8\,\eta^3\,\xi + 1152\,a\,\eta\,\theta\,\xi - 128\,\eta\,\theta^2\,\xi \\
& + 4352\,a^2\,\mu\,\xi + 64\,\eta^2\,\mu\,\xi - 128\,\eta\,\mu^2\,\xi - 5952\,a^2\,\nu\,\xi - 28\,\eta^2\,\nu\,\xi - 1248\,a\,\theta\,\nu\,\xi \\
& + 1344\,\theta^2\,\nu\,\xi + 32\,\eta\,\mu\,\nu\,\xi + 1344\,\mu^2\,\nu\,\xi + 64\,\eta\,\nu^2\,\xi - 768\,\mu\,\nu^2\,\xi + 1152\,\nu^3\,\xi \\
& - 160\,\eta\,\theta\,\xi^2 - 448\,a\,\mu\,\xi^2 + 822\,a\,\nu\,\xi^2 + 432\,\theta\,\nu\,\xi^2 - 26\,\eta\,\xi^3 + 32\,\mu\,\xi^3 + \nu\,\xi^3 \\
& - 14080\,a\,\eta\,\theta\,\omega + 15360\,a^2\,\nu\,\omega - 2240\,a\,\eta\,\xi\,\omega + 1408\,\eta\,\theta\,\xi\,\omega - 4224\,a\,\mu\,\xi\,\omega \\
& + 6144\,a\,\nu\,\xi\,\omega - 3072\,\theta\,\nu\,\xi\,\omega + 480\,\eta\,\xi^2\,\omega - 192\,\mu\,\xi^2\,\omega - 1152\,\nu\,\xi^2\,\omega - 46080\,a\,\nu\,\omega^2 \\
& - 1280\,\eta\,\xi\,\omega^2 + 3072\,\mu\,\xi\,\omega^2 + 4608\,\nu\,\xi\,\omega^2 \big)
\end{aligned}
$$

References

[1] E. H. Abed, "Bifurcation-theoretic issues in the control of voltage collapse," in *Proc. of IMA Workshop on Sys. and Contr. Theory for Power Systems*, J. H. Chow, P. V. Kokotovic and R. J. Thomas (Eds.), Springer-Verlag, New York, pp. 1-21, 1995.

[2] E. H. Abed and J. H. Fu, "Local feedback stabilization and bifurcation control, I. Hopf bifurcation," *Sys. Contr. Letts.*, vol. 7, pp. 11-17, 1986.

[3] E. H. Abed, and J. H. Fu, "Local feedback stabilization and bifurcation control, II. Stationary bifurcation," *Sys. Contr. Letts.*, vol. 8, pp. 467-473, 1987.

[4] E. H. Abed and H. O. Wang, "Feedback control of bifurcation and chaos in dynamical systems," in *Nonlinear Dynamics and Stochastic Mechanics*, N. Sri Namachchivaya and W. Kliemann (Eds.), CRC Press, Boca Raton, pp. 153-173, 1995.

[5] N. Aboud, A. Sathaye, and H. W. Stech, "BIFDE: Software for the investigation of the Hopf bifurcation problem in functional differential equations," *Proc. of the 27th Conf. on Decis. Contr.*, Austin, TX, pp. 821-824, 1988.

[6] D. J. Allwright, "Harmonic balance and the Hopf bifurcation theorem," *Math. Proc. of Cambridge Phil. Soc.*, vol. 82, pp. 453-467, 1977.

[7] N. Ananthkrishnan and K. Sudhakar, "Characterization of periodic motions in aircraft lateral dynamics," *J. of Guidance, Control and Dynamics*, vol. 19, pp. 680-685, 1996.

[8] A. A. Andronov, E. A. Leontovich, I. I. Gordon, and A. G. Maier, *Theory of Bifurcations of Dynamical Systems on a Plane*, Israel Prog. for Sci. Transl., Wiley, New York, 1973.

[9] N. N. Bautin, "On the number of limit cycles which appear with the variation of the coefficients from an equilibrium position of focus or center type," *Amer. Math. Soc. Transl. No. 100*, Providence, RI, 1954. Reprinted in *Stability and Dynamic Systems*, Amer. Math. Soc. Transl., Series 1, vol. 5, pp. 1962, 396-413 (original in Russian in *Mat. Sb. (N.S.)*, vol. 30, pp. 181-196, 1952).

[10] D. W. Berns and J. L. Moiola, "Multiple oscillations in a chemical reactor," *Latin American Applied Research*, vol. 28, pp. 49-56, 1998.

[11] D. W. Berns, J. L. Moiola, and G. Chen, "Feedback control of limit cycle amplitudes from a frequency domain approach," *Automatica*, vol. 34, pp. 1567-1573, 1998.

[12] T. R. Blows and N. G. Lloyd, "The number of limit cycles of certain polynomial differential equations," *Proc. of Royal Society of Edinburgh*, vol. 98A, pp. 215-239, 1984.

[13] G. Chen and X. Dong, "From chaos to order - Perspectives and methodologies in controlling nonlinear dynamical systems," *Int. J. of Bifur. Chaos*, vol. 3, pp. 1363-1409, 1993.

[14] G. Chen and X. Dong, *From Chaos to Order: Methodologies, Perspectives, and Applications*, World Scientific Pub. Co., Singapore, 1998.

[15] G. Chen and J. L. Moiola, "An overview on bifurcation, chaos and nonlinear dynamics in control systems," *J. of the Franklin Institute*, vol. 331B, pp. 819-858, 1994.

[16] S.-N. Chow, C. Li, and D. Wang, *Normal Forms and Bifurcation of Planar Vector Fields*, Cambridge University Press, Cambridge, UK, 1994.

[17] E. J. Doedel, "AUTO: Software for continuation and bifurcation problems in ordinary differential equations," *User Manual*, Applied Mathematics, Caltech, 1986.

[18] F. Dumortier, R. Roussarie, and C. Rousseau, "Hilbert's 16th problem for quadratic vector fields," *J. of Diff. Eqns.*, vol. 110, pp. 86-133, 1994.

[19] W. W. Farr, C. Li, I. S. Labouriau, and W. F. Langford, "Degenerate Hopf bifurcation formulas and Hilbert's 16th Problem," *SIAM J. of Math. Anal.*, vol. 20, pp. 13-30, 1989.

[20] J. M. Franke and H. W. Stech, "Extensions of an algorithm for the analysis of nongeneric Hopf bifurcations, with applications to delay-difference equations," in *Delay Differential Equations and Dynamical Systems*, S. Busenberg and M. Martelli (Eds.), Springer-Verlag, pp. 161-175, 1991.

[21] F. Göbber and K. -D. Willamowski, "Lyapunov approach to multiple Hopf bifurcation," *J. Math. Anal. Appl.*, vol. 71, pp. 333-350, 1979.

[22] M. Golubitsky and W. F. Langford, "Classification and unfoldings of degenerate Hopf bifurcations," *J. of Diff. Eqns.*, vol. 41, pp. 375-415, 1981.

[23] B. D. Hassard, N. D. Kazarinoff, and Y. H. Wan, *Theory and Applications of Hopf Bifurcations*, London Math. Soc. Lecture Note Series, vol. 41, Cambridge University Press, Cambridge, UK, 1981.

[24] B. D. Hassard and Y. H. Wan, "Bifurcation formulae derived from center manifold theory," *J. of Math. Anal. Appl.*, vol. 63, pp. 297-312, 1978.

[25] E. M. James and N. G. Lloyd, "A cubic system with eight small-amplitude limit cycles," *IMA J. of Appl. Math.*, vol. 47, pp. 163-171, 1991.

[26] V. Kertész and R. E. Kooij, "Degenerate Hopf bifurcation in two dimensions, " *Nonlinear Analysis, Theory, Methods and Applications*, vol. 17, pp. 267-283, 1991.

[27] A. I. Khibnik, Yu. A. Kuznetsov, V. V. Levitin, and E. V. Nikolaev, "Continuation techniques and interactive software for bifurcation analysis of ODE's and iterated maps," *Physica D*, vol. 62, pp. 360-371, 1993.

[28] C. Li, "Two problems of planar quadratic systems," *Scientia Sinica - Series A*, vol. 26, pp. 471-481, 1983.

[29] N. G. Lloyd, "Limit cycles of polynomial systems – Some recent developments," in *New Directions in Dynamical Systems*, T. Bedford and J. Swift (Eds), London Math. Soc. Lect. Note Series, vol. 127, pp. 192-234, 1988.

[30] N. G. Lloyd, T. R. Blows, and M. C. Kalenge, "Some cubic systems with several limit cycles," *Nonlinearity*, vol. 1, pp. 653-669, 1988.

[31] N. G. Lloyd and J. M. Pearson, "Conditions for a center and the bifurcation of limit cycles in a class of cubic systems," in *Bifurcations of Planar Vector Fields*, J.P. Françoise and R. Roussarie (Eds.), Lecture Notes in Mathematics, vol. 1455, Springer-Verlag, pp. 230-242, 1990.

[32] A. I. Mees and L. O. Chua, "The Hopf bifurcation theorem and its applications to nonlinear oscillations in circuits and systems," *IEEE Trans. on Circ. Sys.*, vol. 26, pp. 235-254, 1979.

[33] A. M. Mohamed and F. P. Emad, "Nonlinear oscillations in magnetic bearing systems," *IEEE Trans. on Auto. Contr.*, vol. 38, pp. 1242-1245, 1993.

[34] J. L. Moiola and G. Chen, *Hopf Bifurcation Analysis: A Frequency Domain Approach*, World Scientific Pub. Co., Singapore, 1996.

[35] J. L. Moiola and G. Chen, "On the birth of multiple limit cycles in nonlinear systems," *Int. J. of Bifur. Chaos*, vol. 6, pp. 2587-2603, 1996.

[36] A. H. Nayfeh, A. M. Harb, and C. M. Chin, "Bifurcations in a power system model," *Int. J. of Bifur. Chaos*, vol. 6, pp. 497-512, 1996.

[37] J. B. Planeaux, "Bifurcation Phenomena in CSTR Dynamics," *Ph.D. Thesis*, University of Minnesota, 1993.

[38] K. S. Sibirskii, "On the number of limit cycles in a neighborhood of singular points," *J. of Diff. Eqns.*, vol. 1, pp. 36-47, 1965 (translated from Differentsial'nye Uvavneniya, vol. 1, pp. 53-66, 1965).

[39] A. Tesi, E. H. Abed, R. Genesio, and H. O. Wang, "Harmonic balance analysis of period-doubling bifurcations with implications for control of nonlinear dynamics," *Automatica*, vol. 32, pp. 1255-1271, 1996.

[40] Y. -Q. Ye, S. Cai, L. Chen, K. Huang, D. Luo, Z. Ma, E. Wang, M. Wang and X. Yang, *Theory of Limit Cycles*, Eds: Gould, S. H. and Hale, J. K., Transl. of Math. Monographs, vol. 66, Amer. Math. Soc., 1986.

[41] P. K. Yuen and H. H. Bau, "Rendering a subcritical Hopf bifurcation supercritical," *J. of Fluid Mech.*, vol. 317, pp. 91-109, 1996.

[42] H. Żołądek, "On a certain generalization of Bautin's theorem," *Nonlinearity*, vol. 7, pp. 273-279, 1994.

[43] H. Żołądek, "Eleven small limit cycles in a cubic vector field," *Nonlinearity*, vol. 8, pp. 843-860, 1995.

11

Theory and Experiments on Nonlinear Time-Delayed Feedback Systems with Application to Chaos Control

Patrick Celka

Department of Electrical Engineering
Swiss Federal Institute of Technology
1015 Lausanne, Switzerland
patrick.celka@epfl.ch

Abstract

This chapter is devoted to show basic theoretical study on delay-differential equations. Connection between a singular approximation of the delay-differential equation and a 1D map is shown to provide information for the development of a chaos control design technique. Experimental results on control of time-delayed feedback systems are presented. Finally, the noise influence on the control effect together with an adaptive noise cancellation are reported.

11.1 Introduction

It has been reported that nonlinear time-delayed systems can produce a wide variety of behaviours: stable equilibrium points, periodic solutions [5] as well as chaotic solutions [8, 12]. Many experimental analyses of complex dynamical regimes have been conducted and especially in the contex of interferometer based optical systems [7, 18, 2]. The fundamental periodic solution which appears after a Hopf bifurcation has a period of about $T_0 \approx 2(\tau + T)$ where τ is the delay introduced in the feedback path and T is the time response of the device. While the model is infinite-dimensional and evolves on an infinite-dimensional manifold, the attractor dimension is finite due to the contraction effect of the nonlinear mapping [8]. Ikeda and Matsumoto [11] have given an estimate of the attractor Lyapunov dimension D_L for the Ikeda map, and it ranges approximately from 2 to 13 when some bifurcation parameter are varied. Ikeda found high-dimensional attractors for large delay τ and feedback gain m.

It has been mentioned that several co-existing attractors, belonging to different bifurcation branches, can be found in these systems [12]. The fundamental solution evolves trough a period doubling $T_0 \to 2T_0 \ldots \to 2^N T_0$, as one parameter is varied, and we will call this the *fundamental branch*. These solutions are called $2^N \tau$-periodic and the cascade accumulates at the Feigenbaum point. Ikeda and Matsumoto [12] have discovered the existence of other co-existing periodic solutions and called them *harmonics* and *isomers* of these harmonics. Several type of isomers exists and they are all different in shape. Each branch follows a period doubling cascade. These solutions are mostly periodic below the Feigenbaum point, but when the bifurcation parameter is varied towards the fully developed chaotic region, each branch evolves independently within a given parameter region, and gives rise to different stable or unstable periodic or chaotic solutions. A merging process occurs for higher values of the parameters, yielding a final merger and a unique chaotic attractor [12, 7]. Within this complex paradigm, control is a great challenging objective and we try to provide some interesting issues towards this goal.

11.2 Delay-Differential Equations as Models

11.2.1 Delay-Differential Equation

The nonlinear element in an optical ring cavity has a sinusoidal transfer characteristic $f(x) = (1 - sin(x - \varphi_0))/2$. In the case of the Mach-Zehnder interferometer [2], the state variable x is either the intensity of the electromagnetic wave vector or equivalently the phase induced by an electrooptical effect via the electrode of the device. The dynamical system is usually described by the nonlinear map $f(x)$, a nth order linear low-pass filter modeled by the transfer function $H(p)$ in the Laplace domain

$$H(p) = \frac{1}{1 + P_n(p)} \qquad (11.2.1)$$

with $P_n(p) = \sum_{k=1}^{n} a_k p^k$ and $a_k = \alpha_k T^k$, and a set of bifurcation parameters $\{m, \varphi_0, \tau, \ldots\}$. In optical systems, the time constant T of the low-pass filter is determined by the smallest bandwidth value of all the devices in the system and the order n is usually one. The gain m is produced either by an electronic circuit or by the laser power. In the case of an electronic circuit, the time constant T is proportional to the RC characteristic. Let x be the state variable of the system and X its Laplace transform, the dynamical equation is given by

$$X(p) = mH(p)Y(p)e^{-\tau p} \qquad (11.2.2)$$

where $Y(p) = \mathcal{L}[y(t)]$ is the Laplace transform of the output variable $y = f(x)$ of the nonlinear element. The nonlinear function is usually a unimodal map [16] such as the Ikeda, logistic or tent map. If the linear part of the system is a second order ($n = 2$) low-pass filter, the state equation of the sytem is given by

$$T^2 \alpha_2 \frac{d^2 x(t)}{dt^2} + T\alpha_1 \frac{dx(t)}{dt} + x(t) = mf(x(t - \tau)) \qquad (11.2.3)$$

The next section shows two different realizations of the delay-differential equation (DDE) (11.2.3)

11.2.2 Realization with an Electronic Circuit

Figure 11.1 shows a schematic circuit realization of (11.2.3) together with the nonlinear subsystem which implement a tent map.

The parameters of the circuit are related to those of the normalized equation (11.2.3) by: $T^2 \alpha_2 = LC$ and $T\alpha_1 = RC$. The time constant T is controllable by the values of the inductance L, capacitance C and linear positive resistance R. The delay τ is obtained with an analog delay line RD107, which can produce delays within the range 1ms to more than 2s, and $\tau = 1024/\nu_c$ where ν_c is the clock frequency of the RD107 circuit. The actual values for the circuit and normalized parameter of (11.2.3) are summarized in Table 11.1.

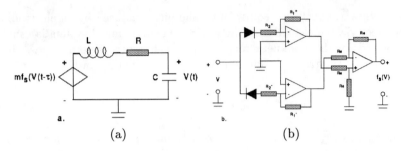

FIGURE 11.1
(a) Circuit realization of (11.2.3) with a voltage controlled voltage source, and (b) the nonlinear subsystem with $R_e = 100k\Omega$.

TABLE 11.1
Parameter values

$R[\Omega]$	$L[mH]$	$C[nF]$	$T[\mu s]$	$\tau[ms]$	α_2	α_1
303	18	330	76.8	1.28	1	1.3

The actual values of τ and T gives us $\tau/T \approx 16.6$. The bifurcation parameter m is setted by an ajustable operational amplifier (LM741) based gain circuit for which $m = 1 + R_1/R_2$. This simple circuit is not shown in Figure 11.1. Due to the low-pass effect of the operational amplifiers, the hat of the tent map $f(x) = 1 - 0.5|x|$, which we will refer to as the *standard tent map*, is smoothed. Actually, the hardware implementation of the tent map can be modeled by

$$f_s(V) = \left(V_{DC} - \alpha A^+ \left| \frac{3(V/\delta)}{\beta(3 - |V|) + |V|} \right|^\nu \right) \text{ if } V > 0$$

$$= \left(V_{DC} + \alpha A^- \left| \frac{3(V/\delta)}{\beta(3 - |V|) + |V|} \right|^\nu \right) \text{ if } V < 0 \quad (11.2.4)$$

$$(11.2.5)$$

where A^+ and A^- are the absolute values of the gain factors of the operational amplifier (LM741) gain circuits: $A^+ = R_1^+/R_2^+$ and $A^- = R_1^-/R_2^-$. These gain factors are not ajustable and $A^+ \approx A^- \approx 0.5$. The DC bias V_{DC} is set to 1.25V. The α, β, δ and ν factors take into account the fact that the circuit realization introduce a smoothing of the hat resulting in f_s instead of f as the nonlinear function. A SPICE simulation of the nonlinearity subsystem have enabled us to determine $\alpha = 0.78399$, $\beta = 0.56876$, $\delta = 1.3979$ and $\nu = 1.5005$ with a linear regression coefficient of 0.99994.

The shape of the map (11.2.4) is quite different from the standard tent map, but it has been shown that our model (11.2.4) is in agreement with our measurements and SPICE simulation on the nonlinearity [4].

11.2.3 Realization with an Electrooptic Circuit

Equation (11.2.3) model quite accurately an electrooptic dynamical system composed of a Mach-Zehnder (MZ) type interferometer circuit, optical fiber, optical power combiner/splitter and electronic circuits. Indeed, the dynamical equation of such systems is given by $T\dot{x}(t) + x(t) = mf(I, x(t - \tau))$ where $f(x, y) = x(1 - Sin(y))/2$ and I is the input light power injected in the system by the laser. Some recent works have proposed other possible optical realization [14, 18].

In our MZ based system, the state variable x is the phase of the output light of the MZ, i.e., $x = KV$ (V is the voltage applied across the electrodes of the MZ) the m factor stands for the bifurcation parameter and is given by $m = GK$ where G is the feedback loop gain (unit Volt/W). Figure 11.2 shows the physical realization of the electrooptic dynamical system.

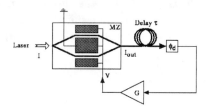

FIGURE 11.2
The electrooptical dynamical system composed of a MZ, a photoreceiver and a single mode fiber for the delay τ.

The K factor depends on the laser source, the geometry and the material used in the integrated optical circuit realizing the MZ: the laser wavelength l, the distance between the electrodes d, the length of the electrodes L, the refraction index n and the electrooptical coefficient r_{33} relative to the polarization of the input lightwave and the crystal cut. Finally this factor is expressed as $K = 2\pi r_{33} n^3 L/(d\lambda)$ if the input light is polarized in the TE mode and the material is Ti:LiNbO3. The transfer characteristic of the MZ is sinusoidal and represented by the function f. This function can be considered as an unimodal map in the second argument when restricted to a given interval $[a, b]$: $f(.,.) : R \times [a, b] \to [a, b]$ ($a, b \in \mathcal{R}$ the set of real numbers). Instead of considering the state variable as the phase factor $x = KV$, we can use an equivalent state variable $y = I_{out}$ where I_{out} is the output lightwave intensity, and obtain a similar equation to (11.2.3): $T\dot{y}(t) + y(t) = f(I, my(t - \tau))$.

It has been shown by Ikeda et al. and Hopf et al. that an optical ring cavity with and electronical time-delayed feedback could produce stable equilibrium points, periodical and chaotic solutions.

The delay τ could be realized electronically or optically. We have chosen the optical delay realization because it introduces much less noise and a greater stability then its electronic counterpart. The typical single mode fiber length we have used where 700m and 300m so we were able to achieve optical delays of τ=3.476ms and τ =1.7ms respectively. The MZ we have used were 1300nm dual output Y-Fed Balanced Bridge Modulators optimized for the TE propagation with an electrical bandwidth of about 1GHz and a DC half-wave voltage V_π=4.68V (this determine $K = \pi/V_\pi$). Insertion losses were about 3dB for each outputs. The wide dynamics acceptable at both DC and RF electrodes enable us to reach a large variety of behaviors in the system.

It is well established that depending on the ratio τ/T, several situations may occur. If τ/T is much greater than one, then a discrete time approximation of Equation (11.2.3) leads to the so called Ikeda map which is well known to produce N-Cycles (N=1,2,3,...) and chaotic solutions. On the other hand, if τ/T is about one, the discrete time approximation is no more valid and we have to consider the DDE (11.2.3). The following section presents some aspect of the singular 1D-map approximation of the DDE (11.2.3).

11.3 From Continuous to Discrete-Time Models

When the time constant T is much lower than the delay τ, it has been shown [12] that the DDE (11.2.3) can be approximated by the discrete-time equation $x \mapsto mf(x)$ in the stable regimes, and the dynamics of the discrete-time equation reproduce quite accurately the behaviour of the continuous-time system. Singular approximations (when $\tau/T \gg 1$) delay-differential equations have been studied by numerous authors from which [5, 15, 16] give interesting results.

In the standard tent map f, there is no period-doubling route to chaos [23]. The standard tent map is well known to produce stable fixed point solutions if $m < 2$, and chaotic solutions for $m \geq 2$. Correspondingly, the continuous-time Equation (11.2.3) exhibits stable equilibrium points and chaotic solutions in the respective regions. The discrete-time system with the smooth tent map $x \mapsto mf_s(x)$ is characterized by a limited period-doubling route to chaos giving rise to 2^N-cycle periodic solutions $\mathbf{X}^N = \{\bar{x}_1, \bar{x}_2, \ldots, \bar{x}_{2^N}\}$. In the continuous-time system (11.2.3) with the nonlinearity being f_s, these cycles correspond to $2^N\tau$-periodic solutions

$\mathbf{X}_\tau^N$ [4]. It is not surprising to find the same appearance of harmonics, isomers and also of different types of chaotic attractors in the electronic circuit as well as in interferometer optical systems because they are modeled by the same kind of DDE (with unimodal map). When $m \in [m_f, m_*[$, m_f standing for the value of m that ends the period doubling in the map $x \mapsto m f_s(x)$ (in the standard tent map f, $m_f = 2$), the chaotic attractors of the smoothed or standard tent map belonging to the fundamental branch are composed of two sets of unconnected chaotic intervals and the solutions of (11.2.3) are square-wave like [5, 13]. The value of m_* can be computed with the so-called critical lines associated with the map f_s [9, 3]. If $m \geq m_{Max}$ where $m_{Max} \geq m_*$ is the solution of $f_s(m) + f_s^{-1}(-m) = 0$ (for the standard tent map $m_{Max} = 4$), the solution becomes unbounded. Throughout this section, we will use f as a generic unimodal map.

Mean Floquet multipliers associated with stable solutions of equation (11.2.3) have been introduced in [4] in order to make the comparison with the map $x \mapsto mf(x)$. Furthermore, we show in [4] that within a given $[m_f, m_*[$ bifurcation interval where chaotic solutions appear, the discrete-time approximation can also be used if we introduce a novel definition for the Lyapunov exponent (LE) associated with the DDE (11.2.3). This generalized LE is very close to the LE of the discrete map and thus both approaches give similar chaotic-like solutions from a statistical point of view. This point is very important for the meaning of our control scheme. This comparison between the two LE is motivated by two facts: first, controlling chaotic trajectories with high LE is more difficult than with low LE; second, controlling two trajectories which have the same LE, even from two different systems, with the same control method, is of the same order of difficulty. We thus conclude that if the control scheme we present works for the discrete-time system, then it will give good results for the continuous-time system in the same range of the parameter set values. Let us first recast Equation (11.2.3), recalling that $\alpha_2 = 1$, in the following way:

$$\frac{d\mathbf{x}}{dt} + A\mathbf{x} = \frac{m}{T^2}\mathbf{h}(t); \; A = \begin{pmatrix} 0 & -1 \\ 1/T^2 & \alpha_1/T \end{pmatrix} \quad (11.3.6)$$

with $\mathbf{x} = (x, dx/dt)^T$ and $\mathbf{h}(t) = (0, f(x(t-\tau)))^T$. Equation (11.3.6) can be rewritten as $\dot{\mathbf{x}} = -A\mathbf{x} + F_m(\mathbf{x}_\tau)$ with $\mathbf{x}_\tau(t) = \mathbf{x}(t-\tau)$, and the related variational equation is given by

$$\delta\dot{\mathbf{x}} = -A\delta\mathbf{x} + \frac{\partial F_m(\mathbf{x}_\tau)}{\partial \mathbf{x}_\tau}\delta\mathbf{x}_\tau \quad (11.3.7)$$

In order to have an idea of the complexity of these types of DDE, it is interesting to estimate the Lyapunov dimension, which gives a good approximation of the fractal dimension of the attacting set. Using numerical integration (4th order Runge-Kutta, $\Delta t = \tau/(N_b-1)$) of Equations (11.3.6) and (11.3.7) together with the Gram-Schmidt procedure as described by Farmer and by Ikeda and Matsumoto [8, 11], we were able to compute the Lyapunov spectrum of Equation (11.3.6), and the Lyapunov dimension D_L. The Lyapunov dimension D_L range from 4 to more than 19 when m and τ/T are varied. The linear relation between D_L and $m\tau/T$ observed experimentally by Ikeda [11, 8], seems to be effective for the m part but not for the τ/T one [4].

Equation (11.3.6) can be solved on each interval $[(n+1)\tau, (n+2)\tau]$ for ($\mathcal{N}$ is the set of natural numbers) $n \in \mathcal{N} \cup \{-1\}$, giving rise to the solution $\mathbf{x}(.) = \{\mathbf{x}_n(.)\}_{-1}^{\infty}$ with the initial condition $\mathbf{x}_{-1}(t) = (\varphi(t), d\varphi(t)/dt)^T = \mathbf{\Phi}(t)$ on $[0, \tau]$. The vector valued function $\mathbf{x}(.)$ belongs to the space $C^2([0, +\infty], I) \times C^1([0, +\infty], I)$ where I is a length m_{Max} interval. The solution $\mathbf{x}(.)$ with respect to the initial condition $\mathbf{\Phi}(.)$ is computed iteratively on each τ-time interval via the *evolution operator* $\mathcal{F}_m$. Within the given range $m \in [0, m_{Max}[$, the nonlinear operator $\mathcal{F}_m$ is bounded, and due to the low-pass behaviour of Equation (11.3.6), $\mathcal{F}_m$ will be a continuous functional space valued operator. Introducing the time normalization $t \to (n+1)\tau + \hat{t}$ with $\hat{t} \in [0, \tau]$ and $n \in \mathcal{N} \cup \{-1\}$, we introduce $\mathcal{F}_m$ in the following definition

DEFINITION 11.1 *We will call* $\mathcal{F}_m$ *the* evolution operator *of the equation (11.3.6) if*
$\mathcal{F}_m : C^2([0,\tau], I) \times C^1([0,\tau], I) \to C^2([0,\tau], I) \times C^1([0,\tau], I) : \mathbf{x}_{n-1}(\hat{t}) \to \mathcal{F}_m(\mathbf{x}_{n-1}(\hat{t})) = \mathbf{x}_n(\hat{t})$ *is a continuously differentiable map.* $(\hat{t}, \{\mathbf{x}_n(\hat{t})\}_{n=0}^{\infty})$ *is the trajectory associated to one solution of Equation (11.3.6) for which* $\mathbf{x}_n(\hat{t}) = \Xi_n(\hat{t}) + e^{-A\hat{t}}\mathbf{x}_{n-1}(\tau)$, *with* $\mathbf{x}_{-1}(\hat{t}) = \mathbf{\Phi}(\hat{t})$ *the initial condition and* $\Xi_n(\hat{t}) = \frac{m}{T^2} \int_0^{\hat{t}} dt'\, e^{-A(\hat{t}-t')} \mathbf{h}_n(t')$, *with* $\mathbf{h}_n(t') = (0, f(x_{n-1}(t')))^T$ *for* $\mathbf{x}_n(0) = \mathbf{x}_{n-1}(\tau)$ *the continuity condition.*

We now give a Lemma that will be useful in the proofs of the main theorem. This Lemma states that under the assumption that $\tau/T \gg 1$ the kernel $e^{-A(\hat{t}-t')}$ in the evolution operator can be approximated by a Dirac distribution up to a matrix multiplication factor. This observation was already mentioned by Ikeda et al. [12].

LEMMA 11.1
Let $\delta(t-t_0)$ *be the scalar valued Dirac distribution centered at* t_0, $d\xi(t, t_0) = dt\, e^{A(t-t_0)}$ *a matrix valued measure defined on* $[0, \tau]$ *with A defined in Equation (11.3.6). If we impose* $d\xi(t, t_0) = B(t, t_0)\delta(t - t_0)dt$ *then* $B(t, t_0) \approx A^{-1}e^{A(t-t_0)}$ *if* $\tau/T \gg 1$ *and* $T > 0$.

From this lemma, we have the following trivial but important result on the matrix $B(t, t_0)$

COROLLARY 11.2
With the same assumption as in Lemma 11.1, $\lim_{t \to t_0} B(t, t_0) = A^{-1}$

Let us now derive the discrete-time approximation of Equation (11.3.6).

THEOREM 11.3
In the case $\tau/T \to \infty$ the Equation (11.3.6) reduces to the difference equation $x(t) \approx mf(x(t-\tau))$ with $\dot{x}(t) \approx 0 \ \forall t \geq 0$.

PROOF If again $\hat{t} \in [0, \tau]$, the solution of the Equation (11.3.6) is of the form

$$\mathbf{x}_n(\hat{t}) = \mathcal{F}_m(\mathbf{x}_{n-1}(\hat{t}))$$

where $\mathcal{F}_m$ is defined in Definition 11.1. Let us write the right-hand side of the previous equation as

$$\mathbf{g}_n(\hat{t}) = \frac{m}{T^2} \int_0^{\hat{t}} d\xi(t, \hat{t}) \mathbf{h}_n(t)$$

Using Lemma 11.1, we can write

$$\begin{aligned} \mathbf{g}_n(\hat{t}) &\approx \tfrac{m}{T^2} \int_0^{\hat{t}} dt B(t, \hat{t}) \delta(t - \hat{t}) \mathbf{h}_n(t) \\ &\approx \tfrac{m}{T^2} B(\hat{t}, \hat{t}) \mathbf{h}_n(\hat{t}) \end{aligned}$$

with Corollary 11.2 we get

$$\mathbf{g}_n(\hat{t}) \approx \frac{mA^{-1}}{T^2} \mathbf{h}_n(\hat{t})$$

and with the definition of the vector $\mathbf{h}_n(\hat{t})$ and the matrix

$$A^{-1} = \begin{pmatrix} \alpha_1/T & 1 \\ -1/T^2 & 0 \end{pmatrix} T^2$$

we obtain

$$\begin{pmatrix} x_n(\hat{t}) \\ y_n(\hat{t}) \end{pmatrix} \approx \begin{pmatrix} mf(x_{n-1}(\hat{t})) \\ 0 \end{pmatrix}$$

completing the proof of the theorem. □

Theorem 11.3 implies that using the Lemma 11.1, Equation (11.3.6) reduces to a discrete-time equation when $\tau/T \gg 1$, as the solution must be constant on each time interval $[n\tau, (n+1)\tau]$. Nevertheless, the difference equation is a continuous time equation and it is difficult to establish clearly the frontier between discrete and continuous models. It is interesting to go further in the analysis of the relationship between the continuous and discrete time equation in defining Mean Floquet Multipliers (MFM) $\mu_{\tau,p}$ and Mean Lyapunov Exponent (MLE) $\Lambda_{\tau,p}$ related to Equation (11.3.6) (see [4]). These are scalar quantities that must be compared to the corresponding scalar ones for the discrete map.

In the limiting case $\tau/T \to \infty$ the Floquet multipliers and the Lyapunov exponents of the discrete map $x \mapsto mf(x)$ are close to the MFM and MLE of the continuous-time Equation (11.3.6), respectively. In practice, for a finite ratio $\tau/T \approx 20$ there exists a value of p such that $\Lambda_{\tau,p}$ is as close as we want to the Lyapunov exponent of the map $x \mapsto mf(x)$, and the same holds for the Floquet multiplier.

In the practical case where τ/T is finite and in the range [15, 80], we have observed that there exists a p value such that $\Lambda_{\tau,p}$ is very closed to the LE and $\mu_{\tau,p}$ very close to the Floquet multiplier of the discrete map for given values of m.

If $m \in [m_f, m_*[$ and $\tau/T \gg 1$, the mean trajectory match as a discrete-time trajectory because the discrete-time attractor lies on two unconnected sets. If $m \geq m_*$, the continuous time trajectory fills the entire interval I and the connection between the mean trajectory and the discrete-time approximation is completely lost.

We have thus established the different *continuous − discrete* connections with MFM, MLE and mean trajectories in DDE [4].

These important observations allow us to design a control scheme for the discrete map and expect corresponding results for the DDE.

11.4 Control of DDE Using Delayed Self-Feedback

11.4.1 Control towards Stable Periodic Orbits of the Fundamental Branch

Pyragas has introduced a very simple time-delay error feedback control method in order to stabilize unstable periodic orbits (UPO) embedded in a chaotic attractor [19]. Some experimental and theoretical approches based on this method have been the subject of numerous recent papers [20, 21, 22]. Applying this scheme to an optical bistable system running in chaotic mode has been done previously [1]. However, the stabilized orbits where not UPO, but ones close to stable uncontrolled system orbits. The first possible way of controlling the DDE (11.2.3) with the map (11.2.4) is to stabilize chaotic trajectories toward some periodic orbits whose behaviours are close to stable $2^N\tau$-periodic solutions of the original system. In the following sections, we will use the electronic system of section 11.2.2 and the parameter values presented in this paper.

The control will be done in real time assuming the state x is observable. Let us suppose that the original system is in a chaotic region with $m = m_c \in [m_f, m_*[$ and that $\tau/T \gg 1$ (in our case 16.6). Notice that the interval $[m_f, m_*[$ corresponds to chaotic trajectories that are square-wave like solutions of Equation (11.2.3). We will force the system to be close to one of its $2^N \tau$-periodic solutions by using an error feedack scheme with a control signal of the form $c(t) = K(x(t - nT_0) - x(t))$ with $K \in \Re$, $n \in \{1/2, 1\}$ and $T_0 = 2\tau$. The controlled system can be modeled by

$$T^2 \alpha_2 \frac{d^2 x(t)}{dt^2} + T\alpha_1 \frac{dx(t)}{dt} + x(t) = mf(x(t - \tau)) + c(t) \quad (11.4.8)$$

The use of other values of n will be the subject of the next section. The evolution operator of Equation (11.4.8) will be noted $\mathcal{F}_{m,K}$ as it depends on K. It is easy to derive the discrete map that corresponds to Equation (11.4.8) with Theorem 11.3 and in the case where $n = 1/2$

$$x_k = (mf(x_{k-1}) + Kx_{k-1})/(1 + K) = F_{m,K}(x_{k-1}) \quad (11.4.9)$$

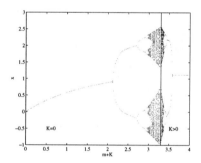

FIGURE 11.3
Comparison between numerical simulations of controlled ($K > 0$) and uncontrolled ($K = 0$) systems modeled by $F_{m,K}$ with f_s.

Figure 11.3 shows the bifurcation diagram corresponding to the map f_s. This bifurcation diagram has been computed with Equation (11.4.9), and the corresponding map, varying m while $K = 0$ until $m = m_c$, and then varying K while $m = m_c$ was kept unchanged. Keeping $m = m_c$ fixed and varying K in the map (11.4.9) results in a reverse period-doubling ending with fixed point solutions independent of the K values. We observe that the smooth map gives effectively a limited period doubling before entering the chaotic region. Right after m_f, the solution evolves to chaotic intervals as in the standard tent map where $m_f = 2$. The ν factor determines the value of m_f and also the number N of different stable cycles. The K factor determines stability intervals one wants to access, called in [1] the *Target Intervals*. When K is chosen in such a way that it corresponds to a 2^N-cycle, the continuous-time system modeled by Equation (11.4.8) will evolve to a $2^N\tau$-periodic solution that is close to the original $2^N\tau$-periodic solution.

In order to confirm the correspondance between controlled discrete- and continuous-time solutions, we have computed the Lyapunov exponent of Equation (11.4.9) and the MLE of Equation (11.4.8) using a time interval length $|\Omega| = T_0/2$. Parameters for (11.4.8) are given in Table 1. The evolution operator $\mathcal{F}_{m,K}$ of Equation (11.4.8) depends on the vector $\mathbf{h}_n$

$$\mathbf{h}_n(\hat{t}) = \begin{pmatrix} 0 \\ \frac{mf(x_{n-1}(\hat{t}))+Kx_{n-1}(\hat{t})}{m(1+K)} \end{pmatrix} \qquad (11.4.10)$$

From a trajectory $(\hat{t}, \{\mathbf{x}_n(\hat{t})\}_{n=-1}^N)$ solution of Equation (11.4.8), we have computed the MLE for $m_c = 3.2$ and $\tau/T = 100/6$, varying K for three values of p. Numerical simulations to compute MFM and MLE were done for 1000 Ω intervals.

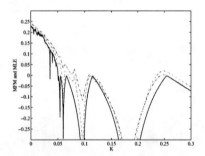

FIGURE 11.4
Computation of MLE and MFM. Finite ratio $\tau/T = 100/6$ for two values of p: $p = 0.1$ (dashed) and $p = 1$ (dotted-dashed). Solid line indicate result from the discrete map.

Figure 11.4 shows that the MLE and MFM are close to their discrete equivalent values computed with $F_{m,K}$, and that there are several Target Intervals for the controlled system. Each region where the MLE is negative corresponds to a stable equilibrium point or periodic solution. These solutions of Equation (11.4.8) are close to original stable solutions of Equation (11.3.6) or (11.2.3) on the fundamental branch.

The bifurcation diagrams of Figure 11.3 show that there is a reverse period-doubling cascade $2^N T_0 \to 2^{N-1} T_0 \ldots \to T_0$ as parameter K is varied. Instead of using a control signal of the form $c(t) = K(x(t-nT_0)-x(t))$

with $K \in \Re$, $n \in \{1/2, 1\}$, we can also consider a weighted linear combination of $x(t - n\tau)$ such as $c(t) = K((1 - R)\sum_{n=1}^{\infty} R^{n-1}x(t - n\tau) - x(t))$. This method has been used by Socolar et al. [21] in the case of control towards unstable periodic orbits. In certain cases, they have shown that using $R \neq 0$ could lead to the stabilization of UPO that were uncontrollable in the case $R = 0$ as in Pyragas's method. The large range of possible values for n shows a first limitation of the present method, because we cannot predict what kind of solution we will control with such a control signal.

Remark A: All time series and phase plane pictures have been digitally acquired on a computer. The K values are approximate because of a 5% tolerance on the circuit elements R, R_1, R_2, $R_1^{\pm}$, $R_2^{\pm}$, R_e, L and C. The m gain factor was $1 + R_1/R_2$ and also subject to 5% tolerance.

The 4τ-periodic orbit was controllable for $K = 0.080 \pm 0.004$. From Figure 11.4, it is far from being obvious that it was possible because the MLE has a sudden jump around this K value for all p values. Figure 11.5 shows a controlled chaotic trajectory towards some 8τ-periodic solution as predicted by Figure 11.4 for $m = 3.20 \pm 0.16$

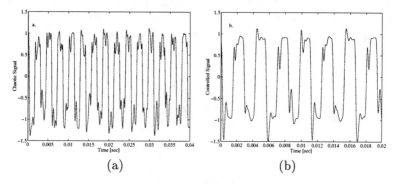

FIGURE 11.5
(a) Experimental time series of a chaotic solution and (b) control towards a 8τ-periodic trajectory for $K = 0.05 \pm 0.0025$.

Another drawback of this method is that the control may generate new periodic solutions that are not at all related to stable periodic solutions of the original system [4].

11.4.2 Control towards Stable Periodic Harmonics and Isomers

Already stated in the introduction, Ikeda et al. have predicted the existence of harmonics and isomers of the fundamental branch [13, 12]. Derstine et al. [6] and Celka [2] have verified experimentally these predictions with a hybrid optically bistable system. One problem in observing these solutions is their relative random appearence as a function of the parameter m and the speed at which we vary it because they usually co-exist for a given m. Moreover, some of these isomers are unstable and unobservable unless we use some control on them. Dertsine et al. [6] have provided a simple and efficient locking method in order to *stabilize* the waveforms. Our approach is quite different since we start from a chaotic trajectory, N^p which is a chaotic square-wave like solution, or L_C^p which is a chaotic trajectory with a predominant power spectral peak located at $f = p/T_0$, and control it towards periodic isomers. Using our time-delay control scheme, we were able to control such isomers and observe more efficiently the bifurcation route to fully developed chaos. We have used the formulation of Derstine [6] for the waveforms: L^1 is the fundamental branch, L^n is an nth order harmonic branch while $L_{m/n}^n$ for $n = 3, 5, 7, \ldots$ and $m = 2, 4, 8, \ldots$ are its various isomers with period mT_0/n. Figure 11.6 shows a chaotic harmonic trajectory. We were able to control $L_{4/3}^3$ and $L_{8/3}^3$ using a control signal of the form $c(t) = K(x(t - T_{(m,n)}) - x(t))$ where $T_{(4,3)} \approx 1.8 \ ms$ and $T_{(8,3)} \approx 3.8 \ ms$ [4]. The control delay $T_{(4,3)}$ corresponds to $2T_0/3$ and $T_{(8,3)}$ to $4T_0/3$ with less then 5% error. This error could be due to either the inaccuracy on the time constant T (precision over R, L, and C) and on τ (precison on ν_c in the analog delay line RD107 set-up), either the mismatch between the effective period of the square-wave and our approximation $T_0 \approx 2(\tau + T)$.

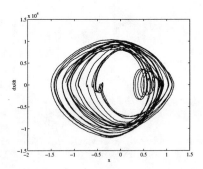

FIGURE 11.6
Experimental fissured chaotic attractor L_C^3.

The number of existing harmonics depends on the ratio τ/T: the greater this ratio, the greater the harmonic and isomers number. We have mentioned above that the control was made on a L_C^p attractor. We can access it by applying the control with a high value of K and a control delay $T_{(m,n)}$ with $n = 3, 5, 7, \ldots$ and $m = 2, 4, 8, \ldots$, that results in "fissuring" N^p into L_C^n. If we proceed directly from N^p the control signal is large and the eventual resulting periodic orbit does not correspond to either one harmonic or isomer solution. This could introduce new periodic solutions that are only related to the controlled system as already mentioned in the previous section. Figures 11.7 and 11.8 show stable periodic orbits that are not standard isomers. We can check this by comparing the shape of the chaotic attractor and those of the stabilized orbits.

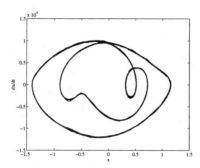

FIGURE 11.7
Experimental control towards stable periodic orbits.

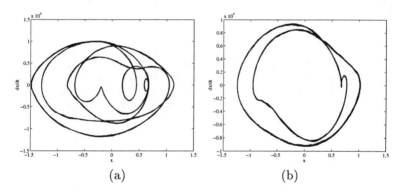

FIGURE 11.8
Experimental control towards stable periodic orbits.

11.4.3 Noise Influence on Control Effect

While the electronical setup we have used does not introduce much noise, in many practical situations (biology, mechanics, chemistry), noise is of major importance. In this section, we examine numerically the influence of additive Gaussian noise $n(t)$ on the controllability of chaotic trajectories within our continuous time control scheme. In order to assess this influence, we have monitored the controlled trajectories for various signal to noise ratios (SNR)($SNR = 10\ Log(P_x/P_n)$ where P_x is the power of the state $x(t)$ and P_n is the power of the additive noise $n(t)$). We cannot, on theoretical grounds infer about additive noise influence, and we have therefore assessed it by using numerical simulations. Figure 11.9 displays both the state $x(t)$ and the biased control signal $c(t) - 1.5$ for clarity purposes.

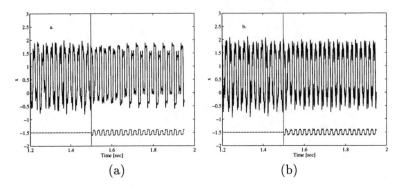

FIGURE 11.9
Effect of additive Gaussian noise on the control trajectories for $K = 0.05$, $\tau/T = 100/6$ and $m = 3.2$: (a) with SNR=40dB, (b) SNR=20dB.

Depending on definition of controllability, up to SNR=30dB the resulting controlled trajectories are still close to the noiseless ones. But obviously, for SNR approximately smaller then 20dB, stabilization is highly perturbed by the noise. The control signal $c(t)$ does not go to zero but remains periodic (in noise free environments). This clearly establishes evidence of the fact that our controlled trajectories are not UPO's of the original system, which in that case should result in $c(t) \to 0$ as $t \to \infty$.

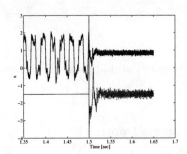

FIGURE 11.10
Effect of additive Gaussian noise on the control trajectories for $K = 1$, $\tau/T = 100/6$ and $m = 3.2$ with SNR=20dB.

When considering the large feedback gain situation $K = 1$, the controlled orbit should be a stable equilibrium state (see Figure 11.3), but in the presence of noise, the control results in some way in the extraction of the noise component as seen in Figure 11.10.

What is somewhat less evident is that the probability density function (pdf) of $c(t)$ and of the controlled trajectories have different shapes. Indeed, as Figure 11.11 shows, $c(t)$ as a Gaussian pdf shape, but $x(t)$ as a double-Gaussian pdf shape.

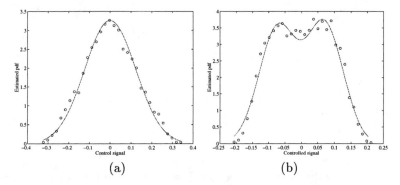

FIGURE 11.11
Estimation of the probability density function of $c(t)$ (a) and $x(t)$ with removed mean value (b).

The control variable $c(t)$ and the controlled state $x(t)$ are obviously correlated. One can thus use an adaptive noise cancelling scheme to clean the controlled trajectories using $c(t)$ as a noise reference signal [10]. Figure 11.12 shows the control in presence of noise (SNR=12dB) without noise cancellation while Figure 11.13 shows the result of adaptive noise cancelling. The adaptive system is turned on at the same time as the control one. The adaptive noise reduction is effective but still does not allow to recover a periodic orbit. The controlled trajectory is filtered and thus distorted.

The filter used in the adaptation scheme was a 60 taps transversal filter, and a normalized LMS adaptive algorithm was used to learn the filter weights with an adaptive step size $\mu = 0.01$.

11.5 Conclusions

This chapter has considered delay-differential equations as possible models for electronic circuits and electrooptic systems. The complexity of such models has been shown in terms of fractal dimensions (the state space geometrical structure) and dynamical properties (multiple attractors, Lya-

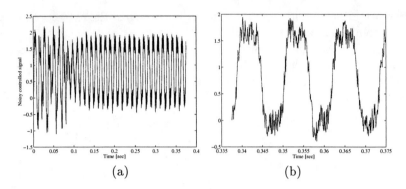

FIGURE 11.12
Control of a noisy chaotic trajectory (a) and a piece of controlled trajectory (b).

punov exponents). A discrete-time approximation of the DDE has lead us to consider relationships between DDE and disctrete-time quantities and to introduce generalized mean Floquet multipliers and mean Lyapunov exponents for nonlinear delay-differential equations. A smoothed nonlinear tent map was implemented with operational amplifiers in an electronical circuit, and experimental evidence of the existence of different solutions belonging to different bifurcation branches has been performed.

We have applied a continuous time control method to a high-dimensional system modeled by a second order nonlinear delay-differential equation in order to stabilize the system onto different stable periodic solutions. The stability regions for controlled 2^N-cycles computed from the discrete map have proved to be in good agreement with experimentally measured stability regions for controlled $2^N \tau$-periodic solutions. Several isomers of harmonic solutions have been successfully controlled with our method. Controlled solutions were not unstable periodic orbits but rather some closely related stable or unstable periodic solutions of the original system.

In the case where the system behaves in a chaotic way, due to noise, the experimental control signal $c(t) = K(x(t - T_{(m,n)}) - x(t))$ could never be zero and the controlled periodic orbits are limited to be as close as possible to the original ones. A first drawback of this control method is the impossibility to determine in advance a periodic orbit to which we want the controlled system to go, and a second one is the possible generation of new stable periodic solutions that are neither UPO nor close to stable solutions of the uncontrolled system (see also discussions given in the chapter by Yu et al. in this book).

Several propositions have been made to use delay-differential equations for information storage in periodic solutions [12, 17], but, due to noise considerations, the device must be under control. As we have emphasized,

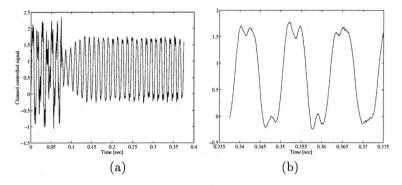

FIGURE 11.13
Adaptive noise cancellation together with control of a noisy chaotic trajectory (a) and a piece of cleaned controlled trajectory (b).

this implies to consider new solutions generated by the controlled system and to take into account the previously stated drawback when using our control method.

Finally, due to the high correlation between the control and the controlled signals, we have proposed to use an adaptive noise cancellation method. Numerical experiments have proven to be effective in the noise suppression guaranteeing a "quasi-periodic" behaviour of the controlled system.

References

[1] P. Celka, "Control of time-delayed feedback systems with application to optic," *Int. J. Electr.*, vol. 79, pp. 787-795, 1995.

[2] P. Celka, *Nonlinear Integrated Optical Circuits and Applications*, PhD thesis, Swiss Federal Institute of Technology, 1995.

[3] P. Celka, "A simple way to compute the existence region of 1D chaotic attactors in 2D-map," *Physica D*, vol. 90, pp. 235-241, 1996.

[4] P. Celka, "Delay-differential equation versus 1D-map: application to chaos control," *Physica D*, vol. 104, pp. 127-148, 1997.

[5] S.-N. Chow and D. Green, *Some Results on Singular Delay-Differential Equations*, Kluwer Academic Publishers, Netherlands, 1993.

[6] M. W. Derstine, H. M. Gibbs, F. A. Hopf and D. L. Kaplan, "Alternate paths to chaos in optical bistability," *Phys. Rev. A*, vol. 27, pp. 3200-3208, 1983.

[7] F. A. Hopf, D. L. Kaplan, H. M. Gibbs and R. L. Shoemaker, "Bifurcations to chaos in optical bistability," *Phys. Rev. A*, vol. 25, pp. 2172-2182, 1982.

[8] J. D. Farmer, "Chaotic attractors of an infinite-dimensional dynamical system," *Physica D*, vol. 4, pp. 366-393, 1982.

[9] I. Gumovski and C. Mira, *Dynamique Chaotique*, Cambridge University Press, 1990.

[10] S. Haykin, *Adaptive Filter Theory*, Prentice-Hall, 1991.

[11] K. Ikeda and K. Matsumoto, "Study of a high-dimensional chaotic attractor," *J. Stat. Phys.*, vol. 44, pp. 955-983, 1986.

[12] K. Ikeda and K. Matsumoto, "High-dimensional chaotic behaviour in systems with time-delayed feedback," *Physica D*, vol. 29, pp. 223-235, 1987.

[13] K. Ikeda, K. Kondo and O. Akimoto, "Successive higher-harmonic bifurcations in systems with delayed feedback," *Phys. Rev. Lett.*, vol. 49, pp. 1467-1470, 1982.

[14] L. Larger, J.-P. Goedgebuer and J.-M. Merolla, "Chaotic oscillator in wavelength: a new setup for investigating differential difference equations describing nonlinear dynamics," *IEEE J. Q. Electr.*, vol. 34, pp. 594-601, 1998.

[15] S. Lepri, G. Giacomelli, A. Politi and F.T. Arecchi, "High-dimensional chaos in delayed dynamical systems," *Physica D*, vol. 70, pp. 235-249, 1993.

[16] Yu. L. Maistrenko, A. N. Sharkovsky and E.Yu. Romanenko, *Difference Equations and Their Applications*, Kluwer Academic Publishers, Netherlands, 1993.

[17] B. Mensour and A. Longtin, "Controlling chaos to store information in delay-differential equations," *Phys. Lett. A*, vol. 205, pp. 18-24, 1995.

[18] J. Mork, B. Tromborg and J. Mark, "Chaos in semiconductor lasers with optical feedback: theory and experiment," *IEEE J. Q. Electr.*, vol. 28, pp. 93-108, 1992.

[19] K. Pyragas, "Continuous control of chaos by self-controlling system," *Phys. Lett. A*, vol. 170, pp. 421-428, 1992.

[20] K. Pyragas and A. Tamasevicius, "Experimental control of chaos by delayed self-controlling feedback," *Phys. Lett. A*, vol. 180, pp. 99-102, 1993.

[21] J. E. S. Socolar, D. W. Sukow and D. J. Gauthier, "Stabilizing unstable periodic orbits in fast dynamical system," *Phys. Rev. E*, vol. 50, pp. 3245-3248, 1994.

[22] Y. C. Tian and F. R. Gao, "Adaptive control of chaotic continuous-time systems with delay," *Physica D*, vol. 117, pp. 1-12, 1998.

[23] T. Yoshida, H. Mori and H. Shigematsu, "Analytic study of chaos of the tent map: band structure, power spectra and critical behaviors," *J. Stat. Phys.*, vol. 31, pp. 279-308, 1983.

12

Time Delayed Feedback Control of Chaos

Xinghuo Yu[1], Yuping Tian[2], and Guanrong Chen[3]

[1]Faculty of Informatics and Communication
Central Queensland University
Rockhampton, QLD 4702, Australia
x.yu@cqu.edu.au

[2]Department of Automatic Control
Southeast University, Nanjing, 210096, P.R.China

[3]Department of Electrical and Computer Engineering
University of Houston, Houston, Texas, 77204 USA

Abstract

> This chapter presents an overview of several recently developed chaos control design methodologies based on the time delayed feedback control (TDFC) concept. The linear TDFC for chaos control is first discussed, with some of its peculiar properties explored. It is noted that linear TDFC based chaos control suffers from the so-called "odd number multiplier limitation," which limits the application scope of this control strategy. Two different alternative chaos control design methodologies are then discussed: one is the sliding mode based TDFC and the other, an optimal principle based TDFC. The sliding mode based TDFC has the advantage of small magnitude control using only partial system state(s), while the optimal principle based TDFC overcomes the "odd number multiplier limitation." The two methodologies are tested using the Rössler system as an example and their effectiveness is demonstrated.

12.1 Introduction

Recently, stabilizing unstable periodic orbits (UPOs) of chaotic systems has become an active and focusing direction in the field of chaos control [4]. This problem can be formulated as a (target) tracking problem in classical control theory. Therefore, the rich literature of conventional tracking control theory is readily applicable for the tasks of stabilizing UPOs, provided that the UPOs as reference signals are available for use. In practice, it is very difficult to obtain exact and analytic formulas for UPOs, except the degenerate case of unstable equilibria, and is extremely difficult (if not impossible) to implement UPOs by physical means such as circuitry due to the instability nature of such orbits. As an alternative, various approximation approaches were proposed for simulating and implementing UPOs (e.g., [8, 10]).

There is a time delayed feedback control (TDFC) method in classical control theory [2], which receives a renewal of great interest spurred by Pyragas' paper [19] for stabilization of UPOs in chaotic systems. The novel idea in this methodology is to use the current as well as past system states in the feedback, thereby avoiding a direct use of the target UPO in the controller. This linear TDFC method has lately been extended and applied to various systems [22, 30].

The TDFC concept has posed a number of new problems about chaos control to the research communities of nonlinear dynamics and control systems. In traditional control theory, for example, control strategies are designed so that the time delayed terms, which are usually considered as disturbances or undesirable system variations, can be suppressed [2]. On the contrary, the TDFC concept in chaos control makes use of the time delayed terms for good purposes. This conceptual difference makes the TDFC a distinct research topic in the field of chaos control, which deserves more thorough studies.

Research on the TDFC for chaos has been only for a relatively short period of time, but it has already had some interesting theoretical and practical results [1, 3, 4, 5, 6, 5, 10, 11, 12, 16, 17, 25, 29, 30]. The first TDFC strategy of Pyragas, which is a linear TDFC method [19], was found to incur some difficulties in analyzing the controlled system stability, which may result in impossible prediction of its success (see, e.g., [13]). Also, there has been no systematic procedures for determining the linear feedback gain matrix in this approach, although experimental adjustments have been reported for lower-dimensional chaotic systems [23]. Furthermore, the inherent "odd number multiplier limitation" of the linear TDFC, pointed out by Ushio [26] and then by Nakajima et al. [18], has further restricted the application scope of all linear TDFC strategies. This limitation suggests that there does not exist any linear TDFC for stabilizing a UPO if the Jacobian of the chaotic system has an odd number of real characteristic multipliers whose values are greater than unity. To overcome this limitation, other classes of TDFC strategies ought to be, and in fact have been developed. Examples include the sliding mode based control (to be discussed in this chapter) and the optimal principle based control methods [25]. The effectiveness of the TDFC methods generally requires a good knowledge of the time delay constant (i.e., the period of the desired UPO), which is crucial to the success of these control methods. An adaptive algorithm has recently been developed [30, 19], which solved this problem so that the TDFC strategies can be widely applied.

In this chapter, we first offer a brief introduction to some recent developments of the authors' research on linear TDFC for chaotic systems. We then discuss a sliding mode based TDFC method and an optimal principle based TDFC technique for chaos control. Some simulations are finally given to show the effectiveness of the new approaches, followed by the conclusions.

12.2 Chaos Control with Linear TDFC

Consider a general continuous–time chaotic dynamical system:

$$\dot{x} = f(x), \qquad (12.2.1)$$

where $x \in R^n$ is the system state, $f(x)$ a differentiable vector-valued function. Here, we discuss autonomous systems only for convenience; most results developed in this chapter can be easily extended to nonautonomous systems.

Let $\bar{x}(t)$ be a τ-periodic solution of (12.2.1), assumed to be unstable, which satisfies

$$\dot{\bar{x}}(t) = f(\bar{x}) \quad \text{and} \quad \bar{x}(t) = \bar{x}(t - \tau). \qquad (12.2.2)$$

The control task is to design a (simple) controller for the given system, so as to track this inherent UPO.

The linear TDFC mechanism proposed in [19] has the following form:

$$\dot{x} = f(x) + K\big(x(t) - x(t - \tau)\big). \qquad (12.2.3)$$

When the system trajectory converges to a τ-period orbit, the feedback term in (12.2.3) vanishes automatically. Therefore, it is guaranteed that the stabilized orbit is a solution of Equation (12.2.1). However, the converse assertion needs some additional condition which is indicated in the following lemma.

LEMMA 12.1
If the controlled orbit $x(t)$ tracks the inherent UPO $\bar{x}(t)$ as $t \to \infty$ and if the matrix K is nonsingular, then the delay-time τ is (an integer multiple of) the period of $\bar{x}(t)$.

PROOF If $x(t)$ tracks $\bar{x}(t)$ as $t \to \infty$, then we have both $f(x(t)) - f(\bar{x}(t)) \to 0$ and $x(t) - \bar{x}(t) \to 0$, so $\frac{d}{dt}(x(t) - \bar{x}(t)) \to 0$, as $t \to \infty$. It follows from (12.2.2) and (12.2.3) that

$$K(x(t) - x(t-\tau)) \to 0,$$

implying that $x(t) - x(t-\tau) \to 0$ since K is nonsingular. Note that $x(t) = \bar{x}(t)$ by periodicity, τ must be (an integer multiple of) the period of the inherent UPO. □

Remark A. This property demands a special attention when dealing with chaos control by using the linear TDFC approach:

1. the period $t_p = \tau$ has to be known a *prior*;
2. if K is singular, some component(s) of $x(t)$ may be periodic but some others may not; therefore, τ may not be the common period for all components.

In the linear TDFC, delayed states are used as reference signals for some control purposes. In so doing, the analytic solution of the target UPO is not necessarily to be known and controller design becomes much easier. Experimental results have shown its effectiveness in stablizing UPOs [23, 22]. However, a main problem with this linear TDFC method is that the user usually does not know beforehand to which particular UPO the controlled system output will converge, if there is more than one UPO of different periods.

The linear TDFC idea can be further extended, for instance, to [20]

$$\dot{x} = f(x) + K\left(x(t) - R_M \sum_{m=1}^{M} r_{m-1} x(t - m\tau)\right),$$

where $R_M = \left(\sum_{m=1}^{M} r_{m-1}\right)^{-1}$, with M being finite or infinite [3, 4], or to some other nonlinear forms [5, 6, 5]. In general, the linear TDFC technique is subjected to the following "odd number multiplier limitation" [18, 20].

LEMMA 12.2
If the Jacobian of system (12.2.1) at the target UPO $\bar{x}(t)$, $J(x) = \partial f(x)/\partial x$, has an odd number of real characteristic multipliers that are greater than

unity, then the UPO cannot be tracked by the linear TDFC design with any constant feedback gain matrix K.

Remark B. This limitation holds for any linear feedback controller which is a function of $x(t) - x(t - \tau)$.

We now reformulate the above chaos control system in a convenient format for further discussion. Because of (12.2.2), the error dynamics become

$$\dot{e} = f(x, \bar{x}) + K(x(t) - x(t - \tau)), \qquad (12.2.4)$$

where $e(t) = x(t) - \bar{x}(t)$ and $f(x, \bar{x}) = f(x) - f(\bar{x})$. The Jacobian about the given UPO is denoted as $J(\bar{x}) = \partial f(x, \bar{x})/\partial x$ below. To quantitatively determine the gain matrix K, we provide the following local results.

THEOREM 12.3
For the error dynamical system (12.2.4), if there exist two positive definite and symmetric constant matrices, P and Q, and a constant gain matrix, K, such that the Riccati polynomial matrix

$$J^\top(x)P + PJ(x) + PKQ^{-1}K^\top P + PK + K^\top P + Q \qquad (12.2.5)$$

is either zero or (semi)-negative definite ($= 0$, ≤ 0, or < 0), then when $\|e(t)\|$ is small enough, it will always approach zero: $\|e(t)\| \to 0$ as $t \to \infty$.

PROOF   Construct a Lyapunov function of the form

$$V(e) = e^\top P e + \int_{t-\tau}^{t} e^\top(s) Q e(s) ds.$$

where $e = x(t) - x(t - \tau)$. Since $\bar{x}(t)$ is a fixed point of $f(x, \bar{x})$, we have a Taylor expansion $f(x, \bar{x}) = J(x)e(t) + [H.O.T.]$, where $[H.O.T.]$ are higher-order terms in $e(t)$. Thus,

$$\begin{aligned}\dot{V}(e(t)) &= \dot{e}^\top(t)Pe(t) + e^\top(t)P\dot{e}(t) + e^\top(t)Qe(t) - e^\top(t-\tau)Qe(t-\tau) \\ &= [f(x,\bar{x}) + Ke(t))]^\top Pe(t) + e^\top(t)P[f(x,\bar{x}) + Ke(t))] \\ &\quad + e^\top(t)Qe(t) - e^\top(t-\tau)Qe(t-\tau) \\ &= [J(x)e(t) + [H.O.T.] + Ke(t))]^\top Pe(t) \\ &\quad + e^\top(t)P[J(x)e(t) + [H.O.T.] + Ke(t))] \\ &\quad + e^\top(t)Qe(t) - e^\top(t-\tau)Qe(t-\tau) \\ &= -\left[Q^{1/2}e(t-\tau) + Q^{-1/2}K^\top Pe(t)\right]^\top \\ &\quad \times \left[Q^{1/2}e(t-\tau) + Q^{-1/2}K^\top Pe(t)\right]\end{aligned}$$

$$+e^\top(t)\left[J^\top(x)P + PJ(x) + PKQ^{-1}K^\top P + PK\right.$$
$$\left.+K^\top P + Q\right]e(t) + [H.O.T.]^\top Pe + e^\top P[H.O.T.]$$
$$< 0, \qquad \forall \text{ small } \|x\|.$$

To this end, a standard verification using class-$\mathcal{K}$ functions for this nonautonomous system [14] completes the proof of the theorem. □

COROLLARY 12.4
Theorem 12.3 still holds if condition (12.2.5) is replaced by the following condition: the Riccati equation

$$\dot{P}(t) = J^\top(x)P(t) + P(t)J(x) + P(t)KQ^{-1}K^\top P(t) + PK + K^\top P + Q \tag{12.2.6}$$

has a positive definite and symmetric solution $P(t) > 0$, $t \in [t_0, \infty)$.

12.3 Chaos Control by Sliding Mode Based TDFC

A new approach to chaos control is the sliding mode based TDFC method. The main idea of this methodology is to perturb some parameter(s) in the chaotic system, to create a local attraction region (usually, a neighborhood of the target UPO) for the controlled system. Then, due to the ergodicity of chaotic dynamics, sooner or later the controlled system orbit will enter this region. Once the orbit enters this region, it will be further attracted to the target UPO.

In this approach, the controlled chaotic system takes the following form:

$$\dot{x} = f(x) + g(x)\gamma, \tag{12.3.7}$$

where $f, g \in R^n$ but parameter γ is specifically chosen from among the chaotic system parameters to be perturbed for control purpose.

Only partial information about the system states is needed in this approach. Assume that this partial state information is given via a one-dimensional manifold:

$$y = h(x), \tag{12.3.8}$$

where h is a smooth scalar field on R^n. Using the input-output linearization approach [24], we can reformulate (12.3.8) as

$$\dot{\mu}_i = \mu_{i+1}, \quad i = 1, \cdots, l-1, \tag{12.3.9}$$

$$\dot{\mu}_r = a(\mu, \phi) + b(\mu, \phi)\gamma, \tag{12.3.10}$$

$$\dot{\phi} = w(\mu, \phi), \tag{12.3.11}$$

where

$$\mu_i = \mathcal{L}_f^{i-1} h(x), \quad i = 1, \cdots, l,$$
$$a(\mu, \phi) = \mathcal{L}_f^n h(x), \quad b(\mu, \phi) = \mathcal{L}_g \mathcal{L}_f^{l-1} h(x),$$

in which l is the relative degree of $h(x)$, and the output is defined as $y = \mu_1$, with the state vectors

$$\mu = (\mu_1, \cdots, \mu_r)^\mathsf{T} \quad \text{and} \quad \phi = (\phi_1, \cdots, \phi_{n-r})^\mathsf{T}.$$

Here, $\mathcal{L}_f h$ represents the Lie derivative of h with respect to f, i.e., $\mathcal{L}_f h = \nabla h$, where $\nabla h = (\partial h / \partial x)$ representing the gradient of h. Higher order Lie derivatives can be defined recursively, as $\mathcal{L}_f^i h = \nabla (\mathcal{L}_f^{i-1} h) f$ for $i = 1, 2, \cdots$, with $\mathcal{L}_f^0 h = h$.

Since the perturbation is permitted only within a small range, we propose to use the sliding mode control (SMC), known also as the variable structure control (VSC) approach [27].

THEOREM 12.5
In system (12.3.7), if the control structure is chosen as

$$\gamma(\xi) = \begin{cases} \gamma_u & s(t, \tau, \mu) > 0 \\ \gamma_l & s(t, \tau, \mu) < 0, \end{cases} \quad (12.3.12)$$

where $s(t, \tau, \mu)$ is an asymptotically stable switching manifold, then there always exists an attraction region defined by

$$\gamma_l < -\left(\frac{\partial s}{\partial \mu} b(\mu, \phi)\right)^{-1} \frac{\partial s}{\partial \mu} a(\mu, \phi) < \gamma_u. \quad (12.3.13)$$

The controlled system orbits are always bounded outside the attraction region. Once the system trajectory enters this attraction region, stabilizing the target UPO will be realized.

PROOF It is well known from the SMC design principle [27] that to ensure the sliding on the manifold $s(t, \tau, \mu) = 0$, the following condition is needed within a neighborhood of $s(t, \tau, \mu) = 0$:

$$s\dot{s} < 0. \quad (12.3.14)$$

When in sliding,

$$s(t, \tau, \mu) = 0 \quad \text{and} \quad \dot{s}(t, \tau, \mu) = 0. \quad (12.3.15)$$

The average ideal value of the smooth feedback control, known as the *equivalent control* and denoted by $\gamma_{eq}(t, \tau)$, can confine the system states to the switching manifold $s(t, \tau, \mu) = 0$.

Note that γ_{eq} can be calculated by solving the following equation:

$$\dot{s} = \frac{\partial}{\partial \mu}\dot{\mu} = \big\langle \nabla s, a(\mu,\phi) + b(\mu,\phi)\gamma_{eq} \big\rangle = 0, \tag{12.3.16}$$

which means that

$$a(\mu,\phi) + b(\mu,\phi)\gamma_{eq} \in \mathrm{Ker}(\nabla s),$$

where $\mathrm{Ker}(\nabla s)$ represents the kernel of the gradient, which is the tangent subspace to $s = 0$. Solving (12.3.16) yields

$$\gamma_{eq} = -\left(\frac{\partial s}{\partial \mu}b(\mu,\phi)\right)^{-1}\frac{\partial s}{\partial \mu}a(\mu,\phi).$$

The attraction region can then be constructed as [27]

$$\gamma_l < \gamma_{eq} < \gamma_u. \tag{12.3.17}$$

The boundedness of the chaotic dynamics is maintained, even outside the attraction region, owning to the SMC mechanism. Indeed, since we have assumed $\gamma \in [\gamma_l, \gamma_u]$ and that small values of γ_l and γ_u do not change the chaotic nature of the system (which may only change slightly the orientation of the chaotic flows), there exist a Lyapunov function $V(\mu,\phi,\gamma)$ and a positive constant M such that

$$V(\mu,\phi,\gamma_l) < M \quad \text{and} \quad V(\mu,\phi,\gamma_u) < M.$$

When the sliding mode is not reached, the perturbation switches between γ_l and γ_u, so that

$$V(\mu,\phi,\gamma) < M.$$

When the sliding mode is reached, $s(t,\tau,\mu) = 0$, so that γ as an "equivalent control" satisfies $\gamma_l \leq \gamma_{eq} \leq \gamma_u$. Hence,

$$V(\mu,\phi,\gamma_{eq}) < M.$$

This implies that the SMC strategy will not cause instability even if the controlled system orbit is located outside the attraction region. □

Remark C. Even in the case that the time delay constant is not equal to the period of the target UPO, the boundedness of the controlled trajectory is still maintained under the SMC strategy. This demonstrates the robustness of the SMC based chaos control method.

12.4 TDFC Design Based on an Optimal Principle

In this section, we introduce another type of time-delayed control method [25]. Differing from the commonly used linear TDFC, we adopt an optimal control principle for the design of the time delayed feedback controller. In this approach, we explore the inherent properties of chaotic systems and use the system states and time delayed states to form a performance index. When the performance index is minimized, the resulting controller enables stabilization of the desired UPO. Unlike linear TDFC methods, this new controller is not subject to the "odd number multiplier limitation."

Consider the controlled chaotic system

$$\dot{x} = f(x) + u, \tag{12.4.18}$$

where $u \in R^n$ is the control input to be determined. Observe that when the target UPO $\bar{x}(t)$ is reached, we have the following:

$$\bar{x}(t) - \bar{x}(t-\tau) = \int_{t-\tau}^{t} f(\bar{x}(s))\,\mathrm{d}s = 0,$$

$$\frac{\mathrm{d}}{\mathrm{d}t}\bar{x}(t) - \frac{\mathrm{d}}{\mathrm{d}t}\bar{x}(t-\tau) = f(\bar{x}(t)) - f(\bar{x}(t-\tau)) = 0,$$

$$\frac{\mathrm{d}^2}{\mathrm{d}t^2}\bar{x}(t) - \frac{\mathrm{d}^2}{\mathrm{d}t^2}\bar{x}(t-\tau) = \frac{\mathrm{d}}{\mathrm{d}t}f(\bar{x}(t)) - \frac{\mathrm{d}}{\mathrm{d}t}f(\bar{x}(t-\tau)) = 0,$$

$$\cdots\cdots$$

Therefore, the process of stabilizing a chaotic system onto one of its inherent UPOs can be regarded as a process of minimizing $\int_{t-\tau}^{t} f(x(s))\,\mathrm{d}s$ and its derivatives.

Consider only the first and second order derivatives. Let

$$\psi(x(t)) = \int_{t-\tau}^{t} f(x(s))\mathrm{d}s, \tag{12.4.19}$$

$$\psi'(x(t)) = f(x(t)) - f(x(t-\tau)), \tag{12.4.20}$$

$$\psi''(x(t)) = \frac{\mathrm{d}}{\mathrm{d}t}f(x(t)) - \frac{\mathrm{d}}{\mathrm{d}t}f(x(t-\tau)). \tag{12.4.21}$$

Introduce the following performance index:

$$J = \int_{t_0}^{\infty} \left[\psi^\mathrm{T}(x(t))\Gamma\psi(x(t)) + {\psi'}^\mathrm{T}(x(t))\Lambda\psi'(x(t)) + {\psi''}^\mathrm{T}(x(t))\psi''(x(t))\right]\mathrm{d}t,$$

$$\tag{12.4.22}$$

where $\Gamma = \Gamma^\top \in R^{n \times n}$ and $\Lambda = \Lambda^\top \in R^{n \times n}$ are positive definite weight matrices.

THEOREM 12.6
If the weight matrices Λ and Γ are taken as
$$2\Gamma^{1/2} - \Lambda > 0 \quad \text{and} \quad \Lambda\Gamma = \Gamma\Lambda,$$
then the following controller will minimize index (12.4.22):
$$u^{\mathrm{op}}(t) = -\left[\frac{\partial f}{\partial x}(x(t))\right]^{-1}\left[-\frac{\partial f}{\partial x}(x(t-\tau))\big(f(x(t-\tau)) + u^{\mathrm{op}}(t-\tau)\big)\right.$$
$$+\frac{\partial f}{\partial x}(x(t))f(x(t)) + L\big(f(x(t)) - f(x(t-\tau))\big)$$
$$\left.+G\int_{t-\tau}^{t} f(x(s))\,\mathrm{d}s\right], \qquad (12.4.23)$$

where
$$G = \Gamma^{1/2}, \qquad (12.4.24)$$
$$L = (2\Gamma^{1/2} - \Lambda)^{1/2}. \qquad (12.4.25)$$

Moreover, if the matrix
$$M = \begin{bmatrix} 0 & I \\ -G & -L \end{bmatrix}$$
has all its eigenvalues located on the left-hand side of the complex plane, then
$$\lim_{t\to\infty} \psi(x(t)) = 0, \quad \lim_{t\to\infty} \frac{\mathrm{d}}{\mathrm{d}t}\psi(x(t)) = 0, \quad \text{and} \quad \lim_{t\to\infty} \frac{\mathrm{d}^2}{\mathrm{d}t^2}\psi(x(t)) = 0.$$

PROOF We first discuss how to minimize the performance index (12.4.22). The classical variational principle [15] shows that the optimal solution given by the performance index (12.4.22) must satisfy the following second order Euler equation:
$$\frac{\partial F}{\partial \psi} - \frac{\mathrm{d}}{\mathrm{d}t}\frac{\partial F}{\partial \psi'} + \frac{\mathrm{d}^2}{\mathrm{d}t^2}\frac{\partial F}{\partial \psi''} = 0,$$
where
$$F = \psi^\top(x(t))\Gamma\Psi(x(t)) + \psi'^\top(x(t))\Lambda\psi'(x(t)) + \psi''^\top(x(t))\psi''(x(t)).$$

So we have
$$\frac{\mathrm{d}^4}{\mathrm{d}t^4}\psi - \Lambda\frac{\mathrm{d}^2}{\mathrm{d}t^2}\psi + \Gamma\psi = 0. \qquad (12.4.26)$$

A special solution of (12.4.26) can be easily derived, as

$$\frac{\mathrm{d}^2}{\mathrm{d}t^2}\psi(x(t)) + L\frac{\mathrm{d}}{\mathrm{d}t}\psi(x(t)) + G\psi(x(t)) = 0. \tag{12.4.27}$$

Substituting (12.4.19), (12.4.20) and (12.4.21) into (12.4.27) yields

$$\frac{\partial f}{\partial x}(x(t))\left[f(x(t)) + u(t)\right] - \frac{\partial f}{\partial x}(x(t-\tau))\left[f(x(t-\tau)) + u(t-\tau)\right]$$

$$L[f(x(t)) - f(x(t-\tau))] + G\int_{t-\tau}^{t} f(x(s))\,\mathrm{d}s = 0,$$

which gives the optimal controller (12.4.23) when $\frac{\partial f}{\partial x}(x(t)) \neq 0$ for all $x \in \Omega \subset R^n$, where Ω denotes a neighborhood of the target UPO.

Now, we verify the asymptotic property of the system under the feedback control $u^{\mathrm{op}}(t)$. Denote $z = [z_1^\mathsf{T}, z_2^\mathsf{T}]^\mathsf{T}$, where $z_1 = \psi$ and $z_2 = \frac{\mathrm{d}}{\mathrm{d}t}\psi$. Note that (12.4.27) can be rewritten as

$$\begin{cases} \dot{z}_1 = z_2, \\ \dot{z}_2 = -Gz_1 - Lz_2. \end{cases}$$

The asymptotic stability of z with respect to the equilibrium point $z = 0$ is determined by the matrices G and L. It is well known that if the matrix

$$M = \begin{bmatrix} 0 & I \\ -G & -L \end{bmatrix}$$

has all its eigenvalues located on the left-hand side of the complex plane, then the stability of z is guaranteed, that is,

$$\lim_{t\to\infty} \psi(x(t)) = 0, \quad \lim_{t\to\infty} \frac{\mathrm{d}}{\mathrm{d}t}\psi(x(t)) = 0, \quad \text{and} \quad \lim_{t\to\infty} \frac{\mathrm{d}^2}{\mathrm{d}t^2}\psi(x(t)) = 0,$$

as claimed. □

Remark D. Small-magnitude chaos control actions are preferable in chaos control so that the existing chaos is not overkilled. For this purpose, we use the following saturated control law for stabilizing UPOs:

$$u_k^{\mathrm{s}}(t) = \begin{cases} \epsilon_k, & \text{if } u_k^{\mathrm{op}}(t) \geq \epsilon_k \\ u_k^{\mathrm{op}}(t), & \text{if } -\epsilon_k < u_k^{\mathrm{op}}(t) < \epsilon_k \\ -\epsilon_k, & \text{if } u_k^{\mathrm{op}}(t) \leq -\epsilon_k, \end{cases} \tag{12.4.28}$$

where subscripts $k = 1, \cdots, n$ denote the element indexes of the input vector, and ϵ_k are small positive numbers.

Due to the limited control magnitudes, the target UPO can be effectively controlled only within a local region. However, the reachability of the attraction region can be guaranteed in the global sense due to the ergodicity of the chaotic dynamics. This is also the main reason to use small-magnitude controls for stabilization of chaotic systems.

Remark E. When a τ-periodic UPO is stabilized, we have

$$x(t) - x(t-\tau) = \int_{t-\tau}^{t} f(x(s))\,\mathrm{d}s + \int_{t-\tau}^{t} u(s)\,\mathrm{d}s \to 0, \quad \text{as } t \to \infty.$$

Since $\int_{t-\tau}^{t} f(x(s))\mathrm{d}s \to 0$ when $t \to \infty$, we have

$$\int_{t-\tau}^{t} u(s)\,\mathrm{d}s \to 0 \quad \text{when } t \to \infty,$$

which means that the average effect of the control (12.4.28) over a period of τ will vanish as $t \to \infty$. If ϵ_k is sufficiently small, according to the average theory [28], the τ-period UPO reached can be regarded as an inherent UPO of the original chaotic system. This is another reason why small-magnitude control is preferable when a τ–periodic orbit is stablized.

Remark F. Since an equilibrium point can be regarded as a trivial UPO with period zero, we also have

$$\psi(x) = \lim_{\tau \to 0} \int_{t}^{t-\tau} f(x(s))\,\mathrm{d}s = f(x(t)).$$

In this case, the optimal performance index (12.4.22) reduces to

$$J = \int_{t_0}^{\infty} [f^{\mathrm{T}}(x(t))\Gamma f(x(t)) + f'^{\mathrm{T}}(x(t))f'(x(t))]\,\mathrm{d}t, \qquad (12.4.29)$$

where $\Gamma = G^{\mathrm{T}}G \in R^{n \times n}$. The optimal stabilizing controller in accordance with this criterion is

$$u^{\mathrm{op}} = -\left[\frac{\partial f}{\partial x}\right]^{-1}\left[\frac{\partial f}{\partial x}f(x(t)) + Gf(x(t))\right].$$

To guarantee the asymptotic stability of this control system, the matrix G should have all its eigenvalues located on the left-hand side of the complex plane. Then, the controller u^{op} generates an attraction region in the set

$$E = \{\,x \in R^n : f(x) = 0\,\},$$

which contains all system equilibria.

12.5 Estimation of Delay Time τ

One common assumption about the TDFC strategies for chaos control is that the delay time τ has to be known *a priori*, in order to deliver an effective control performance. The acquisition of τ is equally difficult compared to acquiring the analytic solution of UPOs. In this section, we introduce an adaptive algorithm based on the gradient descent approach for searching an accurate estimate of τ.

The performance index for search is defined as

$$E = \int_{t_0}^{T+t_0} \| (y(t) - y(t-\tau)) \| \, dt, \qquad (12.5.30)$$

where $y(t)$ represents either the full system states or some manifold(s) of partial system states, and T is a large enough instant that can cover a sufficiently long length of time for estimation.

The adaptive search of τ is carried out as follows:

1. Set the tolerance error ϵ and a large T. Simulation starts from $t_0 = 0$. For a given initial condition, $\tau = \tau_0$, let the chaotic system run freely for a period of time, τ_0.

2. Enable the control $u(t)$ and let the system run for a period of T. Set $i = 1$ and let $\tau_1 = \tau_0$.

3. Compute the adjustment to τ_i:

$$\tau_{i+1} = \tau_i - \beta \frac{\partial E}{\partial \tau_i}, \qquad (12.5.31)$$

where

$$\frac{\partial E}{\partial \tau_i} = \int_{iT+t_0}^{(i+1)T+t_0} (y(t) - y(t-\tau_i))\dot{y}(t-\tau_i) \, dt \qquad (12.5.32)$$

with β being a proper adaptation parameter. Set $i = i + 1$.

4. If

$$E > \epsilon,$$

go to Step 3; otherwise, stop.

The formulas (12.5.31) and (12.5.32) are not convenient for computation. But we can scale it and use the average of E and $\partial E/\partial \tau$. Its discrete version is

$$E = \frac{1}{N} \sum_{k=1}^{N} \Delta t \, \| y(t_{k-1}) - y(t_{k-1} - \tau) \|^2 \qquad (12.5.33)$$

and its recursive version becomes

$$E_k = \frac{k-1}{k} E_{k-1} + \frac{\Delta t \, \| y(t_{k-1}) - y(t_{k-1} - \tau) \|^2}{k}. \quad (12.5.34)$$

Similarly,

$$\frac{\partial E}{\partial \tau_k} = \frac{k-1}{k} \frac{\partial E}{\partial \tau_{k-1}} + \frac{\Delta t \, (y(t_{k-1}) - y(t_{k-1} - \tau)) \dot{y}(t_{k-1} - \tau)}{k}. \quad (12.5.35)$$

When $\dot{y}(t_{k-1} - \tau)$ is approximated by

$$\dot{y}(t_{k-1} - \tau) = (y(t_{k-1} - \tau) - y(t_{k-2} - \tau))/\Delta t,$$

Equation (12.5.35) becomes

$$\frac{\partial E}{\partial \tau_k} = \frac{k-1}{k} \frac{\partial E}{\partial \tau_{k-1}}$$
$$+ \frac{(y(t_{k-1}) - y(t_{k-1} - \tau))(y(t_{k-1} - \tau) - y(t_{k-2} - \tau))}{k} \quad (12.5.36)$$

12.6 Simulation Studies

In this section, simulation results on the Rössler system are presented to show the effectiveness of the two new TDFC methods discussed above. The Rössler system is described by

$$\begin{bmatrix} \dot{x}_1 \\ \dot{x}_2 \\ \dot{x}_3 \end{bmatrix} = \begin{bmatrix} 0 & -1 & -1 \\ 1 & a & 0 \\ x_3 & 0 & -c \end{bmatrix} \begin{bmatrix} x_1 \\ x_2 \\ x_3 \end{bmatrix} + \begin{bmatrix} 0 \\ 0 \\ b \end{bmatrix},$$

where the parameters were taken as $a = b = 0.2$ and $c = 5.7$, respectively. A typical chaotic behavior of the Rössler system ($u \equiv 0$) is shown in Figure 12.1.

We first used the sliding mode based TDFC for chaos control. We used the x_2-component to form the one-dimensional manifold was chosen as

$$y(t) = x_2(t) - x_2(t - \tau).$$

The control signal

$$u(t) = -k_0 \, \text{sign}\big(x_2(t) \cdot (x_2(t) - x_2(t - \tau))\big)$$

was added to the second equation (x_2) of the Rössler system, which as discussed above, can form an local attraction region.

In this simulation, we first set the intial condition to be $x(0) = (0\ 0\ 0)^\top$, $\tau_0 = 4$, $\beta = 5$, and $k_0 = 0.2$. Figure 12.2 shows the convergence of the

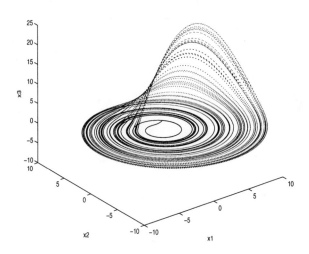

FIGURE 12.1
The chaotic attractor of the Rössler system.

time delay constant τ. Clearly, τ converges to 5.861, which is the period for the target UPO. Figure 12.3 depicts the convergence of function E, showing that the estimation error converges to zero. The system behavior when $\tau = 5.860$ is shown in Figure 12.4, in which the stabilizing period one UPO is achieved.

We now apply the optimal principle based TDFC to stabilize the inherent UPO in the Rössler system, with the derived delay time $\tau = 5.861$. The parameters were $\Gamma = \text{diag}(0.25,\ 0.25,\ 0.81)$ and $\Lambda = \text{diag}(0.84,\ 0.84,\ 1.16)$. From (12.4.24) and (12.4.25), we have $G = \Gamma^{1/2} = \text{diag}(0.5,\ 0.5,\ 0.9)$ and $L = (2\Gamma^{1/2} - \Lambda)^{1/2} = \text{diag}(0.4,\ 0.4,\ 0.8)$. It is easy to verify that the matrix M has all its eigenvalues located on the left hand side of the complex plane. The saturation level was taken as $\epsilon_k = 0.08$. The controlled system behavior is shown in Figure 12.5. The final converged orbit is also shown in the same figure, where the inherent period one orbit is depicted by the dotted curve and the converged orbit as by the solid one. They are very close to each other. Figure 12.6 shows that the average effect of the control input over a period of τ, i.e., $\frac{1}{\tau} \int\limits_{t-\tau}^{t} u(s)\,ds$, is very small.

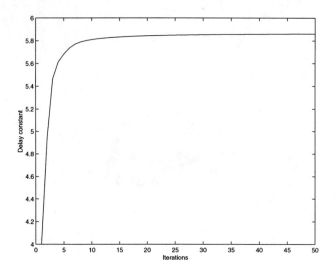

FIGURE 12.2
The convergence of the time delay constant τ.

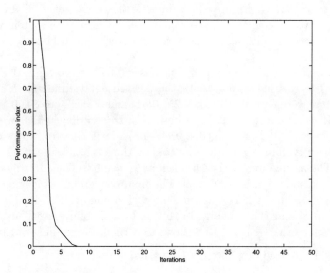

FIGURE 12.3
The convergence of function E (the estimation error).

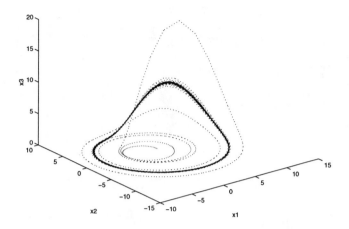

FIGURE 12.4
The period one UPO is stabilized.

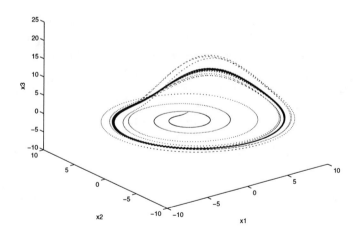

FIGURE 12.5
The controlled system behavior at saturation level $\epsilon_k = 0.08$.

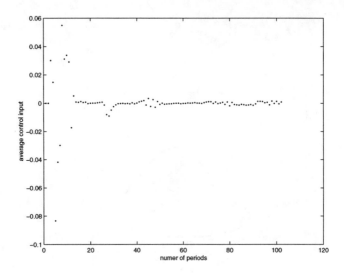

FIGURE 12.6
The average effect of the control input over a period of τ.

12.7 Conclusions

We have reviewed three recently developed chaos control methodologies that are based on the concept of time delayed feedback control (TDFC). Some inherent properties of chaos control systems using the linear TDFC have been discussed, where one drawback is its "odd number multiplier limitation." Two alternatives have then been introduced: one is the sliding mode based TDFC method and the other, an optimal principle based TDFC technique. An adaptive algorithm has been introduced to estimate an accurate period constant for UPOs. The sliding mode based TDFC has the advantage of small-magnitude controls and uses only partial system states information, while the optimal principle based TDFC overcomes completely the "odd number multiplier limitation." These two new TDFC methodologies have been tested by using the Rössler system as an example.

Acknowledgment

The authors would like to thank the Australian Research Council and the National Key Project of China for the financial supports of this research work.

References

[1] M. Basso, R. Genesio, and A. Tesi, "Stabilizing periodic orbits of forced systems via generalized Pyragas controllers," *IEEE Trans. on Circ. Sys.*, I, vol. 44, pp. 1023-1027, 1997.

[2] R. Bellman and K. L. Cooke, *Differential-Difference Equations*, Academic Press, New York, 1963.

[3] M. E. Bleich and J. E. S. Socolar, "Stability of periodic orbits controlled by time-delayed feedback," *Phys. Lett. A*, vol. 210, pp. 87-94, 1996.

[4] M. E. Bleich and J. E. S. Socolar, "Controlling spatiotemporal dynamics with time-delay feedback," *Phy. Rev. E*, vol. 54, pp. R17-R20, 1996.

[5] M. E. Brandt and G. Chen, "Controlling the dynamical behavior of a circle map model of the human heart," *Biol. Cybern.*, vol. 74, pp. 1-8, 1996.

[6] M. E. Brandt and G. Chen, "Feedback control of a quadratic map model of cardiac chaos," *Int. J. of Bifur. Chaos*, vol. 6, pp. 715-723, 1996.

[7] M. E. Brandt and G. Chen, "Bifurcation control of two nonlinear models of cardiac activity," *IEEE Trans. on Circ. Sys.*, I, vol. 44, pp. 1031-1034, 1997.

[8] G. Chen and X. Dong, "On Feedback control of chaotic continuous-time systems," *IEEE Trans. on Circ. Sys.*, I, vol.40, pp. 591-600, 1993.

[9] G. Chen and X. Dong, *From Chaos to Order: Methodologies, Perspectives and Applications*, World Scientific Pub. Co., Singapore, 1998.

[10] G. Chen and X. Yu, "On time delayed feedback control of chaos," *IEEE Trans. on Circ. Sys.*, I, 1999, in press.

[11] T. Hikihara, "Controlling chaos for magneto-elastic beam by delayed feedback control," *Proc. of Int. Conf. on Nonlin. Bifur. Chaos*, Sept. 16-18, Lodz-Dobieszkow, Poland, 1996.

[12] W. Just, T. Bernard, M. Ostheimer, E. Reibold, and H. Benner, "Mechanism of time-delayed feedback control," *Phys. Rev. Lett.*, vol. 78, pp. 203-206, 1997.

[13] W. Just, D. Reckwerth, J. Mockel, E. Reibold, and H. Benner, "Delayed feedback control of periodic orbits in autonomous systems," *Phys. Rev. lett.*, vol. 81, pp.562-565, 1998.

[14] H. K. Khalil, *Nonlinear Systems*, 2nd Ed., Prenticed-Hall, New Jersey, 1996.

[15] D. E. Kirk, *Optimal Control Theory: An Introduction*, Prentice-Hall, New Jersey, 1970.

[16] A. Kittel, J. Parisi, and K. Pyragas, "Delayed feedback control of chaos by self-adapted delay time," *Phys. Lett. A*, vol. 198, pp. 433-436, 1995.

[17] K. Konishi, M. Ishii, and H. Kokame, "Stabilizing unstable periodic points of one-dimensional nonlinear systems using delayed-feedback signals," *Phys. Rev. E*, vol. 54, pp. 3455-3460, 1996.

[18] H. Nakajima, "On analytical properties of delayed feedback control of chaos," *Phys. Lett. A*, vol. 232, pp. 207-210, 1997.

[19] H. Nakajima, H. Ito and Y. Ueda, "Automatic adjustment of delay time and feedback gain in delayed feedback control of chaos," *IEICE Transactions Fundamentals*, vol. E80-A, pp. 1554-1559, 1997.

[20] H. Nakajima and Y. Ueda, "Limitation of generalized delayed feedback control," *Physica D*, vol. 111, pp.143-150, 1998.

[21] K. Pyragas, "Continuous control of chaos by self-controlling feedback," *Phys. Lett. A.*, vol. 170, pp. 421-428, 1992.

[22] K. Pyragas, "Control of chaos via extended delay feedback," *Phys. Lett. A*, vol. 206, pp. 323-330, 1995.

[23] K. Pyragas and A. Tamasevicius, "Experimental control of chaos by delayed self-controlling feedback," *Phys. Lett. A*, vol. 180, pp. 99-102, 1993.

[24] J.-J. Slotine and W. Li, *Applied Nonlinear Control*, Prentice-Hall, New Jersey, 1991.

[25] Y.-P. Tian and X. Yu, "Stabilizing unstable periodic orbits of chaotic systems via an optimal principle," *Physica D*, 1998, submitted.

[26] T. Ushio, "Limitation of delayed feedback control in nonlinear discrete-time systems," *IEEE Trans. on Circ. Sys.*, I, vol. 43, pp. 815-816, 1996.

[27] V. I. Utkin, *Sliding Modes in Control Optimization*, Springer-Verlag, New York, 1992.

[28] H. Verhulst, *Nonlinear Differential Equations and Dynamical Systems*, Second Ed., Springer-Verlag, Berlin, 1996.

[29] M. de S. Vieira and A. J. Lichtenberg, "Controlling chaos using nonlinear feedback with delay," *Phys. Rev. E*, vol. 54, pp. 1200-1207, 1996.

[30] X. Yu, "Tracking inherent periodic orbits in chaotic systems via adaptive variable structure time delayed self–control," *IEEE Trans. on Circ. Sys.*, I, 1999, in press.

13

Impulsive Control and Synchronization of Chaos

J.A.K. Suykens,[1] T. Yang,[2] J. Vandewalle,[1] L.O. Chua[2]

[1]Dept. of Electrical Engineering
Katholieke Universiteit Leuven
ESAT-SISTA, Kardinaal Mercierlaan 94
B-3001 Leuven (Heverlee), Belgium
johan.suykens,joos.vandewalle@esat.kuleuven.ac.be

[2]Department of Electrical Engineering and Computer Sciences
University of California at Berkeley, Berkeley, CA 94720, USA
taoyang,chua@fred.eecs.berkeley.edu

Abstract

This chapter presents methods for impulsive control and synchronization of chaos. Basic theory on asymptotic stability of impulsive differential equations in the sense of Lakshmikantham is given and applied to impulsive stabilization of Chua's circuit. Based on this theory, a systematic design procedure for impulsive synchronization of chaotic Lur'e systems by measurement feedback is derived and illustrated on a hyperchaotic system of coupled Chua's circuits. Existence of periodic solutions in impulsive differential equations is discussed. An application to impulsive control of the Rössler system to periodic motion is shown. Finally, an experimental confirmation of impulsively synchronized Chua's circuits is given together with applications to secure communications and chaotic digital code-division multiple access (CDMA) systems.

13.1 Introduction

Many methods have been proposed for control and synchronization of chaotic systems [1, 4, 9, 18, 20, 22, 25, 27, 30, 31, 38], including the OGY method, optimal control techniques, adaptive control, time-delay feedback control, predictive Poincaré control, occasional proportional feedback control, etc. A recent overview of existing methods is given in [5]. Among these approaches, predictive Poincaré control and occasional proportional feedback control are examples of impulsive control schemes with varying impulse intervals. Impulsive control is an attractive method in the sense that it allows stabilizing of chaotic systems using small control impulses. It also offers a direct method for modulating digital information onto a chaotic carrier signal for spread spectrum applications in secure communications [42, 44] with applications in CDMA [43]. However, due to a lack of effective tools for analyzing impulsive differential equations [14], most impulse control schemes were initially designed by trial-and-error.

Recently, methods for impulsive control and synchronization of nonlinear systems were developed in [44] based upon the theoretical work of [14]. At discrete time instants, jumps in the system's state are caused by a control input. Global asymptotic stability of a system is proven by means of a Lyapunov function and is characterized by a set of conditions related to the time instants, the time intervals in between, and a coupling condition. Initially, the method has been demonstrated ad hoc to the special cases of Chua's circuit [42, 44] and the Lorenz system [48, 49], with an experimental confirmation on impulsively synchronized Chua's circuits [21]. In these examples full state information was assumed, which means that knowledge of the full state vector of the system is needed in order to synchronize the systems by impulses. Besides global asymptotic stability, practical stability [15, 19] has also been investigated in [47]. Impulsive control of chaos to periodic motions has been studied in [50] and illustrated on the Rössler attractor.

A general and systematic design procedure for master-slave synchronization of identical Lur'e systems was proposed in [35, 36]. Examples of chaotic and hyperchaotic Lur'e systems [13, 34, 37] are Chua's circuit [8, 6, 7, 17], generalized Chua's circuits that exhibit n-scroll attractors [33] and arrays that consist of such chaotic cells [12, 29]. In [35] the case of measurement feedback has been studied with sufficient conditions for global asymptotic stability of the error system. This error is defined between the outputs (instead of states) of the master and slave system. In practice the full state vector is indeed often not available, not measurable, or too expensive to measure. The conditions for synchronization have been expressed in terms of matrix inequalities [3], which also occur in the context of nonlinear H_∞ synchronization methods for secure communication applications [32, 31].

Controller design is done then by solving a nonlinear optimization problem which involves the matrix inequality.

This chapter is organized as follows. In Section 13.2 basic theory on asymptotic stability of impulsive differential equations is presented. In Section 13.3 impulsive synchronization of Lur'e systems is discussed. Section 13.4 presents work on impulsive control to periodic motion. In Section 13.5 an experimental confirmation and applications in secure communications and CDMA are discussed.

13.2 Basic Theory of Impulsive Differential Equations

In this Section we present basic theory on asymptotic stability of impulsive differential equations. The method is illustrated on stabilization of Chua's oscillator. For further details and additional examples we refer to [14, 48, 42, 44, 49, 41].

13.2.1 Impulsive Differential Equation, Comparison System and Asymptotic Stability

Consider the nonlinear system

$$\dot{\mathbf{x}} = \mathbf{f}(t, \mathbf{x}) \qquad (13.2.1)$$

where $\mathbf{f} : \mathbf{R}_+ \times \mathbf{R}^n \mapsto \mathbf{R}^n$ is continuous, $\mathbf{x} \in \mathbf{R}^n$ is the state variable, and $\dot{\mathbf{x}} \triangleq \frac{d\mathbf{x}}{dt}$. Consider a discrete set $\{\tau_k\}$ of time instants, where $0 < \tau_1 < \tau_2 < ... < \tau_k < \tau_{k+1} < ...$. Let

$$U(k, \mathbf{x}) = \Delta \mathbf{x}|_{t=\tau_k} \triangleq \mathbf{x}(\tau_k^+) - \mathbf{x}(\tau_k^-) \qquad (13.2.2)$$

be the "jump" in the state vector at the time instant τ_k. Then the impulsive system is described by

$$\begin{cases} \dot{\mathbf{x}} = \mathbf{f}(t, \mathbf{x}), & t \neq \tau_k \\ \Delta \mathbf{x} = U(k, \mathbf{x}), & t = \tau_k \\ \mathbf{x}(t_0^+) = \mathbf{x}_0, & t_0 \geq 0, \ i = 1, 2, ... \end{cases} \qquad (13.2.3)$$

This is called an impulsive differential equation [14]. If f is locally Lipschitz continuous in x uniformly in t and $\tau_k - \tau_{k-1}$ does not vanish with k (or bounded from below), then the Cauchy problem is well posed (at least locally). In order to study the stability of the impulsive differential equation (13.2.3) we make use of the following definitions and theorems [14].

DEFINITION 13.1 Let $V : \mathbf{R}_+ \times \mathbf{R}^n \mapsto \mathbf{R}_+$, then V is said to belong to class $\mathcal{V}_0$ if

1. V is continuous in $(\tau_{k-1}, \tau_k] \times \mathbf{R}^n$ and for each $\mathbf{x} \in \mathbf{R}^n$, $i = 1, 2, ...,$

$$\lim_{(t,\mathbf{y}) \to (\tau_k^+, \mathbf{x})} V(t, \mathbf{y}) = V(\tau_k^+, \mathbf{x}) \quad (13.2.4)$$

   exists;

2. V is locally Lipschitzian in $\mathbf{x}$.

DEFINITION 13.2 For $(t, \mathbf{x}) \in (\tau_{k-1}, \tau_k] \times \mathbf{R}^n$, we define

$$D^+ V(t, \mathbf{x}) \triangleq \limsup_{h \to 0} \frac{1}{h}[V(t+h, \mathbf{x} + h\mathbf{f}(t, \mathbf{x})) - V(t, \mathbf{x})] \quad (13.2.5)$$

DEFINITION 13.3 [Comparison system] Let $V \in \mathcal{V}_0$ and assume that

$$\begin{cases} D^+ V(t, \mathbf{x}) \leq g(t, V(t, \mathbf{x})), & t \neq \tau_k \\ V(t, \mathbf{x} + U(k, \mathbf{x})) \leq \psi_k(V(t, \mathbf{x})), & t = \tau_k \end{cases} \quad (13.2.6)$$

where $g : \mathbf{R}_+ \times \mathbf{R}_+ \mapsto \mathbf{R}$ is continuous and $\psi_k : \mathbf{R}_+ \mapsto \mathbf{R}_+$ is nondecreasing. Then the system

$$\begin{cases} \dot{w} = g(t, w), & t \neq \tau_k \\ w(\tau_k^+) = \psi_k(w(\tau_k)) \\ w(t_0^+) = w_0 \geq 0 \end{cases} \quad (13.2.7)$$

is called the comparison system of (13.2.3).

DEFINITION 13.4

$$S_\rho = \{\mathbf{x} \in \mathbf{R}^n | \, \|\mathbf{x}\|_2 < \rho\} \quad (13.2.8)$$

where $\|\cdot\|$ denotes the Euclidean norm on $\mathbf{R}^n$.

DEFINITION 13.5 A function α is said to belong to class $\mathcal{K}$ if $\alpha \in C[\mathbf{R}_+, \mathbf{R}_+]$, $\alpha(0) = 0$ and $\alpha(x)$ is strictly increasing in x.

Assumption. $\mathbf{f}(t, 0) = 0$, $U(k, 0) = 0$, and $g(t, 0) = 0$ for all k.

Remark A. With the above assumptions we find that the trivial solutions of (13.2.3) and (13.2.7) are identical for all times except at the discrete set $\{\tau_k\}$.

Basic Theory of Impulsive Differential Equations

THEOREM 13.1 (Theorem 3.2.1, page 139, [14])
Assume that the following three conditions are satisfied:

1. $V : \mathbf{R}_+ \times S_\rho \mapsto \mathbf{R}_+$, $\rho > 0$, $V \in \mathcal{V}_0$, $D^+V(t, \mathbf{x}) \leq g(t, V(t, \mathbf{x}))$, $t \neq \tau_k$.
2. *there exists a* $\rho_0 > 0$ *such that* $\mathbf{x} \in S_{\rho_0}$ *implies that* $\mathbf{x} + U(k, \mathbf{x}) \in S_{\rho_0}$ *for all* k *and* $V(t, \mathbf{x} + U(k, \mathbf{x})) \leq \psi_k(V(t, \mathbf{x}))$, $t = \tau_k$, $\mathbf{x} \in S_{\rho_0}$.
3. $\beta(\|\mathbf{x}\|_2) \leq V(t, \mathbf{x}) \leq \alpha(\|\mathbf{x}\|_2)$ *on* $\mathbf{R}_+ \times S_\rho$, *where* $\alpha(\cdot), \beta(\cdot) \in \mathcal{K}$.

Then the stability properties of the trivial solution of the comparison system (13.2.7) *imply the corresponding stability properties of the trivial solution of* (13.2.3).

THEOREM 13.2 (Corollary 3.2.1, page 142, [14])
Let $g(t, w) = \lambda(t)w$, $\lambda \in C^1[\mathbf{R}_+, \mathbf{R}_+]$, $\psi_k(w) = d_k w$, $d_k \geq 0$ *for all* k. *Then the origin of system* (13.2.3) *is asymptotically stable if*

$$\lambda(\tau_{k+1}) + \log(\gamma d_k) \leq \lambda(\tau_k), \text{ for all } k, \text{ where } \gamma > 1 \qquad (13.2.9)$$

and

$$\dot{\lambda}(t) \geq 0. \qquad (13.2.10)$$

13.2.2 Impulsive Stabilization of Chua's Oscillator

We illustrate the theory here on impulsive stabilization of Chua's oscillator. In dimensionless form Chua's oscillator [6, 17] is given by

$$\begin{cases} \dot{x} = \alpha_0[y - x - f(x)] \\ \dot{y} = x - y + z \\ \dot{z} = -\beta_0 y - \gamma_0 z \end{cases} \qquad (13.2.11)$$

where $f(x)$ is the piecewise-linear characteristic of the Chua's diode, which is given by

$$f(x) = b_0 x + \frac{1}{2}(a_0 - b_0)(|x + 1| - |x - 1|) \qquad (13.2.12)$$

where $a_0 < b_0 < 0$ are constants.

Let $\mathbf{x} = [x; y; z]$, then Chua's oscillator equation can be written as

$$\dot{\mathbf{x}} = A\mathbf{x} + \Phi(\mathbf{x}) \qquad (13.2.13)$$

where

$$A = \begin{bmatrix} -\alpha_0 & \alpha_0 & 0 \\ 1 & -1 & 1 \\ 0 & -\beta_0 & -\gamma_0 \end{bmatrix}, \Phi(\mathbf{x}) = \begin{bmatrix} -\alpha_0 f(x) \\ 0 \\ 0 \end{bmatrix}. \qquad (13.2.14)$$

The impulsive stabilization of Chua's oscillator corresponds to

$$\begin{cases} \dot{\mathbf{x}} = A\mathbf{x} + \Phi(\mathbf{x}), \ t \neq \tau_k \\ \Delta\mathbf{x}|_{t=\tau_k} = B\mathbf{x}. \end{cases} \quad (13.2.15)$$

COROLLARY 13.3
Let d_1 denote the largest eigenvalue of $(I+B^T)(I+B)$, B be a symmetric matrix, and $\rho_R(I+B) \leq 1$, where $\rho_R(\cdot)$ denotes the spectral radius of $I+B$. Let q be the largest eigenvalue of $(A+A^T)$ and let the impulses be equidistant with time interval Δ_τ. If

$$0 \leq q + 2|\alpha_0 a_0| \leq -\frac{1}{\Delta_\tau}\log(\xi d_1), \ where \ \xi > 1 \quad (13.2.16)$$

is satisfied, then the origin of the impulsively controlled Chua's oscillator is globally asymptotically stable.

PROOF Let us employ the Lyapunov function $V(t,\mathbf{x}) = \mathbf{x}^T\mathbf{x}$. For $t \neq \tau_k$, we have

$$\begin{aligned} D^+V(t,\mathbf{x}) &= \mathbf{x}^T A\mathbf{x} + \mathbf{x}^T A^T \mathbf{x} + \mathbf{x}^T \Phi(\mathbf{x}) + \Phi^T(\mathbf{x})\mathbf{x} \\ &\leq q\mathbf{x}^T\mathbf{x} + 2|\alpha_0 a_0|\mathbf{x}^T\mathbf{x} \\ &= (q + 2|\alpha_0 a_0|)V(t,\mathbf{x}). \end{aligned} \quad (13.2.17)$$

Hence, condition 1 of Theorem 13.1 is satisfied with $g(t,w) = (q+2|\alpha_0 a_0|)w$.

Since B is symmetric, $(I+B)$ is also symmetric and one has $\rho_R(I+B) = \|I+B\|_2$. Given any $\rho_0 > 0$ and $\mathbf{x} \in S_{\rho_0}$, one obtains

$$\|\mathbf{x} + B\mathbf{x}\|_2 \leq \rho_R(I+B)\|\mathbf{x}\|_2 \leq \|\mathbf{x}\|_2. \quad (13.2.18)$$

The last inequality follows from $\rho_R(I+B) \leq 1$. Hence $\mathbf{x} + B\mathbf{x} \in S_{\rho_0}$.
For $t = \tau_k$, one obtains

$$\begin{aligned} V(\tau_k, \mathbf{x} + B\mathbf{x}) &= (\mathbf{x} + B\mathbf{x})^T(\mathbf{x} + B\mathbf{x}) \quad (13.2.19) \\ &= \mathbf{x}^T(I+B^T)(I+B)\mathbf{x} \\ &\leq d_1 V(\tau_k, \mathbf{x}). \end{aligned}$$

Hence condition 2 of Theorem 13.1 is satisfied with $\psi_k(w) = d_1 w$. One sees that condition 3 of Theorem 13.1 is also satisfied. It follows from Theorem 13.1 that the asymptotic stability of the impulsively controlled Chua's oscillator in (13.2.15) is implied by the following comparison system:

$$\begin{cases} \dot{\omega} = (q + 2|\alpha_0 a_0|)\omega, \ t \neq \tau_k \\ \omega(\tau_k) = d_1 \omega(\tau_k) \\ \omega(t_0) = \omega_0 \geq 0. \end{cases} \quad (13.2.20)$$

From (13.2.16) we have

$$\int_{\tau_k}^{\tau_{k+1}} (q + 2|\alpha_0 a_0|)dt + \log(\xi d_1) \leq 0, \ \xi > 1 \qquad (13.2.21)$$

and $\dot{\lambda}(t) = q + 2|\alpha a| \geq 0$. It follows from Theorem 13.2 and the Lyapunov function that the trivial solution of (13.2.15) is globally asymptotically stable.

The result from Corollary 13.3 gives an estimate for the upper bound $\Delta_{\tau,max}$ on the time interval Δ_τ:

$$\Delta_{\tau,max} = \left|\frac{\log(\xi d_1)}{q + 2|\alpha_0 a_0|}\right|, \ \xi \to 1^+. \qquad (13.2.22)$$

Note that this upper bound follows from a sufficient condition for asymptotic stability.

For the simulations shown in Figure 13.1 the parameters $\alpha_0 = 15$, $\beta_0 = 20$, $\gamma_0 = 0.5$, $a_0 = -120/7$, $b_0 = -75/7$ were chosen. A fourth-order Runge-Kutta method with step size 10^{-5} has been used. The initial condition is given by $\mathbf{x}(0) = [-2.121304; -0.066170; 2.881090]$. The uncontrolled trajectories are shown in Figure 13.1, which correspond to the double scroll attractor. In the simulations, $B = \text{diag}\{\nu, -1, -1\}$ has been chosen, which corresponds to a "strong" impulsive control law. It follows from Corollary 13.3 that $\rho_R(I+B) \leq 1$ should be satisfied, which implies that $-2 \leq \nu \leq 0$. By choosing this matrix B, it is easy to see that $d_1 = (\nu + 1)^2$. Hence an estimate for the boundaries of the stable region is given by

$$0 \leq \Delta_\tau \leq -\frac{\log \xi + \log(\nu + 1)^2}{q + 2|\alpha_0 a_0|}, \ -2 \leq \nu \leq 0. \qquad (13.2.23)$$

Figure 13.1 shows the stable region for different values of ξ. The entire region below the curve corresponding to $\xi = 1$ is the estimated stable region. When $\xi \to \infty$, the stable region shrinks to a line $\nu = -1$.

13.3 Impulsive Synchronization of Lur'e Systems

In this section, we apply the theory of impulsive differential equations to the synchronization of Lur'e systems. For further details and additional examples, see [35, 36].

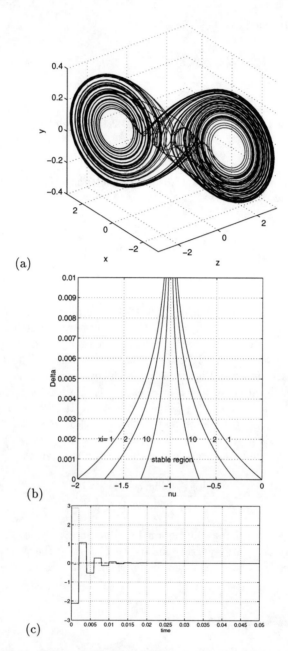

FIGURE 13.1
Impulsive stabilization of Chua's oscillator: (a) double scroll attractor; (b) estimation of boundaries of stable regions for different values of ξ; (c) impulsive stabilization, shown is x(t).

13.3.1 Synchronization Scheme and Error System

We consider the following master-slave synchronization scheme:

$$\mathcal{M}: \begin{cases} \dot{\mathbf{x}} = A\mathbf{x} + B\sigma(C\mathbf{x}) \\ \mathbf{p} = L\mathbf{x} \end{cases}$$

$$\mathcal{S}: \begin{cases} \dot{\mathbf{z}} = A\mathbf{z} + B\sigma(C\mathbf{z}), \\ \mathbf{q} = L\mathbf{z} \end{cases} \quad t \ne \tau_k$$

$$\mathcal{C}: \begin{cases} \dot{\xi} = E\xi + F(\mathbf{p}-\mathbf{q}) + W_F \sigma(V_{F_1}\xi + V_{F_2}(\mathbf{p}-\mathbf{q})), & t \ne \tau_k \\ \Delta \mathbf{z} = D_1 \mathbf{u}, & t = \tau_k \\ \Delta \xi = D_2 \mathbf{v}, & t = \tau_k \\ \mathbf{u} = G_1 \xi + H_1(\mathbf{p}-\mathbf{q}) \\ \mathbf{v} = G_2 \xi + H_2(\mathbf{p}-\mathbf{q}) \end{cases}$$

(13.3.24)

which consists of a master system $\mathcal{M}$, a slave system $\mathcal{S}$, and a controller $\mathcal{C}$. $\mathcal{M}$ and $\mathcal{S}$ are two identical Lur'e systems with state vectors $\mathbf{x}, \mathbf{z} \in \mathbf{R}^n$ and matrices $A \in \mathbf{R}^{n \times n}$, $B \in \mathbf{R}^{n \times n_h}$, $C \in \mathbf{R}^{n_h \times n}$. A Lur'e system is a linear dynamical system, feedback interconnected to a static nonlinearity $\sigma(.)$ that satisfies a sector condition [13, 37] (here it has been represented as a recurrent neural network with one hidden layer, activation function $\sigma(\cdot)$ and n_h hidden units [34]). We assume that $\sigma(\cdot) : \mathbf{R}^{n_h} \mapsto \mathbf{R}^{n_h}$ is a diagonal nonlinearity with $\sigma_i(\cdot)$ belonging to sector $[0, K]$ for $i = 1, ..., n_h$. The output (or measurement) vectors of $\mathcal{M}$ and $\mathcal{S}$ are $\mathbf{p}, \mathbf{q} \in \mathbf{R}^l$ with $l \le n$ and $L \in \mathbf{R}^{l \times n}$. For the sake of generality, a nonlinear dynamic output feedback controller of Lur'e form is taken for the state equation with state vector $\xi \in \mathbf{R}^{n_\xi}$. By means of the matrices D_1 and D_2, it is decided on which state equations the impulsive controls $\mathbf{u} \in \mathbf{R}^{m_z}$ and $\mathbf{v} \in \mathbf{R}^{m_\xi}$ are applied. The output difference $\mathbf{p} - \mathbf{q}$ is taken as input of the controller $\mathcal{C}$. The matrices of the controller are of dimension $E \in \mathbf{R}^{n_\xi \times n_\xi}$, $F \in \mathbf{R}^{n_\xi \times l}$, $W_F \in \mathbf{R}^{n_\xi \times n_{h_\xi}}$, $V_{F_1} \in \mathbf{R}^{n_{h_\xi} \times n_\xi}$, $V_{F_2} \in \mathbf{R}^{n_{h_\xi} \times l}$, $D_1 \in \mathbf{R}^{n_z \times m_z}$, $D_2 \in \mathbf{R}^{n_\xi \times m_\xi}$, $G_1 \in \mathbf{R}^{m_z \times n_\xi}$, $G_2 \in \mathbf{R}^{m_\xi \times n_\xi}$, $H_1 \in \mathbf{R}^{m_z \times l}$, $H_2 \in \mathbf{R}^{m_\xi \times l}$, where n_{h_ξ} is the number of hidden units in the Lur'e system of $\mathcal{C}$. The control law includes the cases of static output feedback ($G_1 = 0$, $G_2 = 0$) and linear dynamic output feedback ($W_F = 0$, $V_{F_1} = 0$, $V_{F_2} = 0$) [30].

The output error system for $\mathbf{e}_L$ becomes

$$\mathcal{E}: \begin{cases} \dot{\mathbf{e}}_L = LA\mathbf{e} + LB\,\eta(C\mathbf{e}; \mathbf{z}), & t \ne \tau_k \\ \dot{\xi} = E\xi + F(\mathbf{p}-\mathbf{q}) + W_F \sigma(V_{F_1}\xi + V_{F_2}(\mathbf{p}-\mathbf{q})), & t \ne \tau_k \\ \Delta \mathbf{e}_L = -LD_1 \mathbf{u}, & t = \tau_k \\ \Delta \xi = D_2 \mathbf{v}, & t = \tau_k \\ \mathbf{u} = G_1 \xi + H_1(\mathbf{p}-\mathbf{q}) \\ \mathbf{v} = G_2 \xi + H_2(\mathbf{p}-\mathbf{q}) \end{cases}$$

(13.3.25)

where the synchronization error is defined as $\mathbf{e} = \mathbf{x} - \mathbf{z}$ for the state vectors and $\mathbf{e}_L = \mathbf{p} - \mathbf{q} = L\mathbf{e}$ for the outputs. Furthermore $\Delta \mathbf{e}_L = -L\Delta \mathbf{z}$ and $\eta(C\mathbf{e}; \mathbf{z}) = \sigma(C\mathbf{e} + C\mathbf{z}) - \sigma(C\mathbf{z})$.

13.3.2 Matrix Inequalities and Controller Synthesis

In order to derive a sufficient condition for global asymptotic stability of the error system $\mathcal{E}$, we take the Lyapunov function

$$V(\mathbf{e}_L, \xi) = \zeta^T P \zeta = [\mathbf{e}_L^T \ \xi^T] \begin{bmatrix} P_{11} & P_{12} \\ P_{21} & P_{22} \end{bmatrix} \begin{bmatrix} \mathbf{e}_L \\ \xi \end{bmatrix}, \ P = P^T > 0. \tag{13.3.26}$$

According to the previous section, it is sufficient to prove that

$$\begin{cases} \dot{V} \leq \alpha_* V, & \alpha_* > 0, & t \neq \tau_k \\ V(\zeta + \Delta\zeta) < \beta_* V, & \beta_* > 0, & t = \tau_k \\ \|\zeta + \Delta\zeta\|_2 < \|\zeta\|_2, & & t = \tau_k \\ \alpha_*(\tau_{k+1} - \tau_k) + \log \beta_* < 0. \end{cases} \tag{13.3.27}$$

From (13.3.27) we find that $\beta_* < 1$ should be satisfied. We express the conditions now as matrix inequalities. In the derivation we exploit the inequalities

$$\begin{cases} \eta(C\mathbf{e})^T \Lambda \left[\eta(C\mathbf{e}) - C\mathbf{e} \right] \leq 0, & \forall \mathbf{e} \in \mathbf{R}^n \\ \sigma(\varphi)^T \Gamma \left[\sigma(\varphi) - V_{F_1}\xi - V_{F_2}L\mathbf{e} \right] \leq 0, & \forall \mathbf{e} \in \mathbf{R}^n, \xi \in \mathbf{R}^{n_\xi}. \end{cases} \tag{13.3.28}$$

These are related to the sector conditions on the nonlinearities $\eta(\cdot)$ and $\sigma(\cdot)$, which are assumed to belong to sector $[0, 1]$. Λ and Γ are diagonal matrices with positive diagonal elements and $\varphi = V_{F_1}\xi + V_{F_2}L\mathbf{e}$. By employing (13.3.28) in an application of the S-procedure [3, 11, 39] a matrix inequality is obtained by writing

$$\dot{V} - \alpha_* V - 2\eta(C\mathbf{e})^T \Lambda \left[\eta(C\mathbf{e}) - C\mathbf{e} \right] - 2\sigma(\varphi)^T \Gamma \left[\sigma(\varphi) - V_{F_1}\xi - V_{F_2}L\mathbf{e} \right] \leq 0 \tag{13.3.29}$$

as a quadratic form $\mathbf{w}^T Z \mathbf{w} \leq 0$ in $\mathbf{w} = [\mathbf{e}; \xi; \eta; \sigma]$. Imposing this quadratic form to be negative semidefinite for all $\mathbf{w}$, one obtains

$$Z = Z^T = \begin{bmatrix} Z_{11} & Z_{12} & Z_{13} & Z_{14} \\ Z_{12}^T & Z_{22} & Z_{23} & Z_{24} \\ Z_{13}^T & Z_{23}^T & Z_{33} & 0 \\ Z_{14}^T & Z_{24}^T & 0 & Z_{44} \end{bmatrix} \leq 0 \tag{13.3.30}$$

with

$$Z_{11} = A^T P_{11} L + L^T P_{11} A + L^T P_{12} F L \\ \qquad + L^T F^T P_{21} L - \alpha_* L^T P_{11} L$$

$$Z_{22} = E^T P_{22} + P_{22} E - \alpha_* P_{22}$$
$$Z_{33} = -2\Lambda$$
$$Z_{44} = -2\Gamma$$

$$Z_{12} = A^T P_{12} + L^T P_{12} E \\ \qquad + L^T F^T P_{22} - \alpha_* L^T P_{12}$$
$$Z_{13} = L^T P_{11} LB + C^T \Lambda$$
$$Z_{14} = L^T P_{12} W_F + L^T V_{F_2}^T \Gamma$$
$$Z_{23} = P_{21} LB$$
$$Z_{24} = P_{22} W_F + V_{F_1}^T \Gamma.$$

Note that for matrices the notation $> 0, < 0$ means positive and negative definite matrices, respectively. In order to express the other conditions (13.3.27) as matrix inequalities, we write

$$\zeta + \Delta\zeta = \begin{bmatrix} e_L \\ \xi \end{bmatrix} + \begin{bmatrix} \Delta e_L \\ \Delta \xi \end{bmatrix} = M \begin{bmatrix} e_L \\ \xi \end{bmatrix} \qquad (13.3.31)$$

with

$$M = \begin{bmatrix} I - LD_1 H_1 & -LD_1 G_1 \\ D_2 H_2 & I + D_2 G_2 \end{bmatrix}.$$

This yields the matrix inequality $M^T P M < \beta_* P$ and $M^T M < I$.

The controller is designed by solving the feasibility problem:

Find $\quad \theta_c, Q, \Lambda, \Gamma, \alpha_*, \beta_*$

such that
$$\begin{cases} Z \leq 0 \\ M^T P M < \beta_* P \\ M^T M < I \\ \alpha_* (\tau_{k+1} - \tau_k) + \log \beta_* < 0 \end{cases} \qquad (13.3.32)$$

with $P = Q^T Q > 0$ and the controller parameter vector θ_c containing the elements of the matrices $E, F, W_F, V_{F_1}, V_{F_2}, G_1, H_1, G_2, H_2$. This problem has to be solved for given matrices A, B, C, L, D_1, D_2 and the choice of a fixed time interval $\Delta_\tau = \tau_{k+1} - \tau_k$.

13.3.3 Example: Synchronization of Coupled Chua's Circuits

We illustrate impulsive master-slave synchronization of Lur'e systems on a hyperchaotic system which consists of two unidirectionally coupled Chua circuits [12]:

$$\begin{cases} \dot{x}_1 &= a\,[x_2 - h(x_1)] \\ \dot{x}_2 &= x_1 - x_2 + x_3 \\ \dot{x}_3 &= -b\,x_2 \\ \dot{x}_4 &= a\,[x_5 - h(x_4)] + K_c\,(x_4 - x_1) \\ \dot{x}_5 &= x_4 - x_5 + x_6 \\ \dot{x}_6 &= -b\,x_5 \end{cases} \qquad (13.3.33)$$

with $h(x_i) = m_1 x_i + \frac{1}{2}(m_0 - m_1)(|x_i + c| - |x_i - c|)$ (i=1,4). For $m_0 = -1/7$, $m_1 = 2/7$, $a = 9$, $b = 14.286$, $c = 1$, $K_c = 0.01$, the system exhibits hyperchaotic behavior with a double-double scroll attractor. The system can be represented in Lur'e form with $n = 6$, $n_h = 2$, and

$$A = \left[\begin{array}{ccc|ccc} -a\,m_1 & a & 0 & 0 & 0 & 0 \\ 1 & -1 & 1 & 0 & 0 & 0 \\ 0 & -b & 0 & 0 & 0 & 0 \\ \hline 0 & 0 & 0 & -a\,m_1 + K_c & a & 0 \\ 0 & -K_c & 0 & 1 & -1 & 1 \\ 0 & 0 & 0 & 0 & -b & 0 \end{array}\right],$$

$$B = \left[\begin{array}{c|c} -a\,(m_0 - m_1) & 0 \\ 0 & 0 \\ 0 & 0 \\ \hline 0 & -a\,(m_0 - m_1) \\ 0 & 0 \\ 0 & 0 \end{array}\right], \quad C = \left[\begin{array}{ccc|ccc} 1 & 0 & 0 & 0 & 0 & 0 \\ 0 & 0 & 0 & 1 & 0 & 0 \end{array}\right].$$

(13.3.34)

In Figure 13.2 results for the linear dynamic output feedback case are shown. Two inputs and outputs are defined for the system by taking $\mathbf{p} = [x_2; x_5]$, $\mathbf{q} = [z_2; z_5]$, and $D_1 = \mathrm{diag}\{0, 1, 0, 0, 1, 0\}$. Furthermore, $n_\xi = 2$, $m_\xi = 2$, $W_{EF} = 0$, $V_{F_1} = 0$, $V_{F_2} = 0$ was chosen. Sequential quadratic programming [10] by means of the function *constr* of Matlab has been applied. The first constraint in (13.3.32) was used as objective function $\lambda_{max}(Z)$ (where $\lambda_{max}(.)$ denotes the maximal eigenvalue of a symmetric matrix) while the remaining three constraints have been imposed as hard constraints. The following starting points have been chosen for the optimization: θ_c random according to a Gaussian distribution with zero mean and standard deviation 0.1; $Q = I$; $\Lambda = I$; $\alpha_* = 10$; $\beta_* = 0.1$. The fixed time interval $\tau_{k+1} - \tau_k$ was chosen equal to 0.1. Instead of β_*, the parameter $1/[1 + \exp(-\beta_*)]$ (which belongs to $(0, 1)$) was taken as unknown from the optimization problem. While synchronization is only proven for $[x_2 - z_2; x_5 - z_5]$ by the Lyapunov function (13.3.26), the synchronization error $\mathbf{e} = \mathbf{x} - \mathbf{z}$ for the complete state vector is also converging to the origin for this example.

13.4 Impulsive Control to Periodic Motions

In this section we discuss the existence of periodic solutions of impulsive differential equations and impulsive control of chaos to periodic behavior. For further details, see [50].

13.4.1 Existence of Periodic Solutions of Impulsive Differential Equations

In order to control the system (13.2.3) to a T-periodic trajectory by means of impulsive control methods, we study the existence of periodic solutions of the following impulsive differential equation:

$$\begin{cases} \dot{\mathbf{x}} = \mathbf{f}(t,\mathbf{x}), & t \neq \tau_k, \, t \in J \\ \mathbf{x}(\tau_k^+) = \Psi_k(\mathbf{x}(\tau_k)), & k = 1,...,q \\ \mathbf{x}(0) = \mathbf{x}(T) \end{cases} \quad (13.4.35)$$

where $J \in [0,T]$, $\tau_k \in (0,T)(k=1,...,q)$ and $\Psi_k : \mathbf{R}^n \mapsto \mathbf{R}^n$, $k=1,...,q$. Given two vectors $\mathbf{u} = (u_1, u_2, ..., u_n)^T \in \mathbf{R}^n$ and $\mathbf{v} = (v_1, v_2, ..., v_n)^T \in \mathbf{R}^n$, $\mathbf{u} \preceq \mathbf{v}$ denotes that $u_i \leq v_i$ for each $i = 1, 2, ..., n$ and $\mathbf{u} \succeq \mathbf{v}$ denotes that $u_i \geq v_i$ for each $i = 1, 2, ..., n$. We have the following from [2].

DEFINITION 13.6 *Let $PC(J, \mathbf{R}^n)$ denote the set of functions $\phi : J \mapsto \mathbf{R}^n$ which are continuous for $t \in J, t \neq \tau_k$ and are continuous from the left for $t \in J$. At points $\tau_k \in J$, ϕ has a discontinuity of the first kind. $PC^1(J, \mathbf{R}^n)$ denotes the set of functions $\psi : J \mapsto \mathbf{R}^n$ which have derivatives $\dot{\psi} \in PC(J, \mathbf{R}^n)$.*

DEFINITION 13.7 *The function $\mathbf{v} \in PC^1(J, \mathbf{R}^n)$ is said to be a lower solution of system (13.4.35) if*

$$\begin{cases} \dot{\mathbf{v}}(t) \preceq \mathbf{f}(t, \mathbf{v}(t)), & t \neq \tau_k, t \in J \\ \mathbf{v}(\tau_k^+) \preceq \Psi_k(\mathbf{v}(\tau_k)), & k = 1,...,q \\ \mathbf{v}(0) \preceq \mathbf{v}(T) \end{cases} \quad (13.4.36)$$

and $\mathbf{v}$ is not a solution of system (13.4.35).

DEFINITION 13.8 *The function $\mathbf{v} \in PC^1(J, \mathbf{R}^n)$ is said to be an upper solution of system (13.4.35) if*

$$\begin{cases} \dot{\mathbf{v}}(t) \succeq \mathbf{f}(t, \mathbf{v}(t)), & t \neq \tau_k, t \in J \\ \mathbf{v}(\tau_k^+) \succeq \Psi_k(\mathbf{v}(\tau_k)), & k = 1,...,q \\ \mathbf{v}(0) \succeq \mathbf{v}(T) \end{cases} \quad (13.4.37)$$

and $\mathbf{v}$ is not a solution of system (13.4.35).

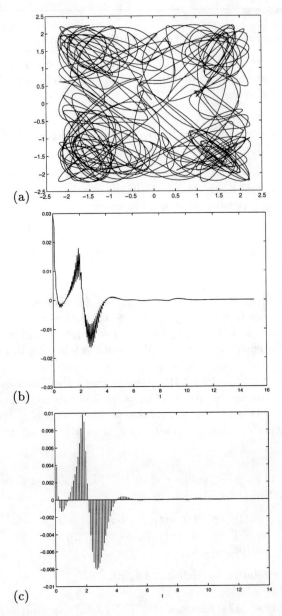

FIGURE 13.2
Synchronization of two identical hyperchaotic systems (coupled Chua's circuits) by impulsive linear dynamic output feedback with two outputs and two control inputs: (a) double-double scroll attractor according to Kapitaniak & Chua, shown is (x_1, x_4) for the system with 6 state variables; (b) output synchronization error $e_5(t) = x_5 - z_5$; (c) impulsive control $u_5(t)$ applied to the slave system.

Remark B. In Definitions 13.7 and 13.8, **v** not being a solution of the system (13.4.35) means that at least one inequality should be satisfied.

DEFINITION 13.9 *The function* $\mathbf{f}: J \times \mathbf{R}^n \mapsto \mathbf{R}^n$ *is said to be quasi-monotone nondecreasing if* $\mathbf{u}, \mathbf{v} \in \mathbf{R}^n$, $\mathbf{u} \preceq \mathbf{v}$, $u_i = v_i$ *for some* $1 \leq i \leq n$, *implies* $f_i(t, \mathbf{u}) \leq f_i(t, \mathbf{v})$. *Here* f_i *indicates the i-th component of the vector-valued function* $\mathbf{f}$.

DEFINITION 13.10 *A function* $\mathbf{g}: \mathbf{R}^n \mapsto \mathbf{R}^n$ *is said to be nondecreasing in* $\mathbf{R}^n$ *if for* $\mathbf{u}, \mathbf{v} \in \mathbf{R}^n$, $\mathbf{u} \preceq \mathbf{v}$ *implies* $\mathbf{g}(\mathbf{u}) \preceq \mathbf{g}(\mathbf{v})$.

Let $\Delta_{\tau,0} = [0, \tau_1], \Delta_{\tau,k} = (\tau_k, \tau_{k+1}]$ for $k = 1, ..., q-1$ and $\Delta_{\tau,q} = (\tau_q, T]$, then the following theorem gives sufficient conditions for the existence of periodic solutions of the system (13.4.35).

THEOREM 13.4 (Theorem 13.1, page 162, [2])
Let the following conditions hold:

1. *The functions* $\mathbf{v}, \mathbf{w}$ *are lower and upper solutions of system* (13.4.35) *such that* $\mathbf{v}(t) \preceq \mathbf{w}(t)$ *in* J.

2. *The function* $\mathbf{f}: J \times \mathbf{R}^n \mapsto \mathbf{R}^n$ *is quasi-monotone nondecreasing in* $J \times \mathbf{R}^n$, *continuous in the sets* $\Delta_{\tau,k} \times \mathbf{R}^n (k = 0, ..., q)$, *and for each* $k = 1, ..., q$ *and* $\mathbf{x} \in \mathbf{R}^n$ *there exists the finite limit of* $\mathbf{f}(t, \mathbf{y})$ *as* $(t, \mathbf{y}) \to (\tau_k, \mathbf{x}), t > \tau_k$.

3. *There exists a function* $\lambda \in \mathcal{L}^1(J, \mathbf{R}_+)$ *such that*

$$\sup_{\mathbf{v}(t) \preceq \mathbf{x}(t) \preceq \mathbf{w}(t)} |f_i(t, \mathbf{x})| \leq \lambda(t) \qquad (13.4.38)$$

almost everywhere on J *(i=1, ..., n)*.

4. *The functions* $\Psi_k : \mathbf{R}^n \mapsto \mathbf{R}^n$, $(k = 1, ..., q)$ *are continuous and nondecreasing with respect to* $\mathbf{x} \in \mathbf{R}^n$.

Then the system (13.4.35) *has a periodic solution* $\mathbf{x}(t)$ *such that* $\mathbf{v}(t) \preceq \mathbf{x}(t) \preceq \mathbf{w}(t)$ *in* J.

13.4.2 Impulsive Control of Rössler Attractor to Periodic Motion

We illustrate impulsive control of chaos to periodic behavior on the Rössler system. The impulsively controlled Rössler system is given by

$$\begin{cases} \dot{x} = -y - z \\ \dot{y} = x + a_r y \\ \dot{z} = zx + b_r - c_r z, \quad t \neq \tau_k, t \in J \\ \mathbf{x}(\tau_k^+) = \Psi_k(\mathbf{x}(\tau_k)), \quad k = 1, ..., q, \end{cases} \quad (13.4.39)$$

where $\mathbf{x} = [x; y; z]$, Ψ_k is a continuous and nondecreasing function and a_r, b_r, c_r are parameters of the Rössler system.

COROLLARY 13.5
Assume that $a_r, b_r, c_r > 0$, the free Rössler system is chaotic and $z > 0$, then (13.4.39) can be impulsively suppressed into periodic trajectories.

Figure 13.3 shows simulation results for impulsive control to periodic motion for the Rössler system with $a_r = 0.398$, $b_r = 2$, $c_r = 4$, and $\mathbf{x}(0) = [-2.277838; -2.696438; 0.304911]$. A fourth-order Runge-Kutta algorithm with step size of 0.005 is used. A Poincaré section $x = 0$ is chosen and the proportional impulses are generated by

$$\begin{cases} x(\tau_k^+) = x(\tau_k) \\ y(\tau_k^+) = \psi_2(\mathbf{x}(\tau_k)) = [1 - \lambda \operatorname{sign}(y(\tau_k))] y(\tau_k), \quad |\lambda| < 1 \\ z(\tau_k^+) = z(\tau_k) \end{cases} \quad (13.4.40)$$

which is continuous and nondecreasing. Figure 13.3(a) shows the Rössler attractor of the free system. One can see that $z > 0$ is satisfied. Figure 13.3(b) shows the bifurcation diagram with respect to the control parameter λ, from which one can find the period 1, 2, and 4 windows. Figure 13.3(c) shows the controlled period 1 trajectory when $\lambda = 0.65$. Figure 13.3(d) shows the controlled period 2 trajectory when $\lambda = 0.45$. Figure 13.3(e) shows the controlled period 4 trajectory when $\lambda = 0.35$. Figure 13.3(f) shows the controlled period 8 trajectory when $\lambda = 0.305$. We also found that if λ is much larger than 0.77, the controlled Rössler system becomes unstable. Another possible choice for Ψ_k is the additive impulse generating law:

$$\begin{cases} x(\tau_k^+) = x(\tau_k) \\ y(\tau_k^+) = \lambda + y(\tau_k) \\ z(\tau_k^+) = z(\tau_k). \end{cases} \quad (13.4.41)$$

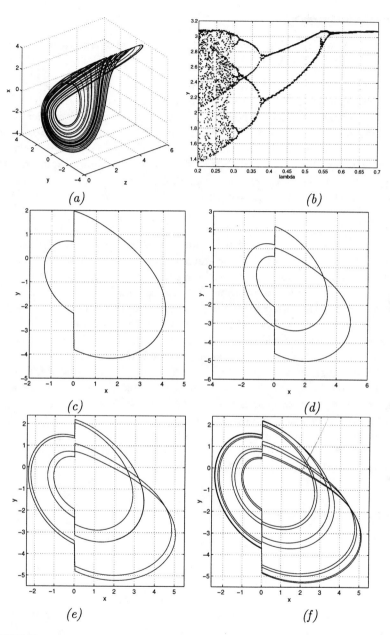

FIGURE 13.3
Impulsive control of the Rössler system to periodic trajectories using proportional impulses at Poincaré section $x = 0$: (a) attractor of the free Rössler system; (b) bifurcation diagram of the impulsively controlled Rössler system; (c) controlled period 1 trajectory; (d) controlled period 2 trajectory; (e) controlled period 4 trajectory; (f) controlled period 8 trajectory.

13.5 Experimental Confirmation, Secure Communications, and $(CD)^2MA$

The impulsive synchronization of Chua's circuits has been experimentally confirmed in [21]. In Figure 13.4 the implementation is shown together with experimental results. For its application to secure communication exploiting chaos, the transmitted signals are divided into small time frames [42, 44]. In each time frame, the synchronization impulses and the scrambled message signal are embedded. Conventional cryptographic methods are used to scramble the message signal. In this scheme the security is enhanced with respect to more classical approaches for which chaotic cryptanalysis has shown a number of deficiencies [26, 40, 51, 52, 53, 54]. This approach forms the basis for chaotic digital code-division multiple access $(CD)^2MA$ communication systems, introduced in [43, 45, 46]. Unlike existing CDMA systems, $(CD)^2MA$ systems use continuous pseudo-random time series to spread the spectrum of the message signal and the signal is then directly sent through the channel to the receiver. In this sense the carrier used in $(CD)^2MA$ is a continuous pseudo-random signal instead of a single tone as used in CDMA. In a $(CD)^2MA$ system every mobile station has the same structure and parameters, only different initial conditions are assigned to different mobile stations. Instead of synchronizing two binary pseudo-random sequences as in CDMA systems, an impulsive control scheme is used in order to synchronize two chaotic systems in $(CD)^2MA$. Simulation results show that the channel capacity of $(CD)^2MA$ is twice as large as for CDMA in wireless environments. A pending patent of $(CD)^2MA$ had been documented in the University of California at Berkeley under the title: "Chaotic digital code-division multiple access for wireless communication systems" with case No. B97-080.

13.6 Conclusions

In this chapter we presented theory on the asymptotic stability and the existence of period solutions of impulsive differential equations. The theory has been illustrated on Chua's circuit and the Rössler system. A systematic design procedure for impulsive synchronization of chaotic Lur'e systems by measurement feedback is derived and illustrated on a hyperchaotic system of coupled Chua's circuits. Impulsive synchronization of Chua's circuits has been experimentally confirmed. Its use for secure communications exploiting chaos has been demonstrated. The method has been applied to chaotic CDMA communication systems.

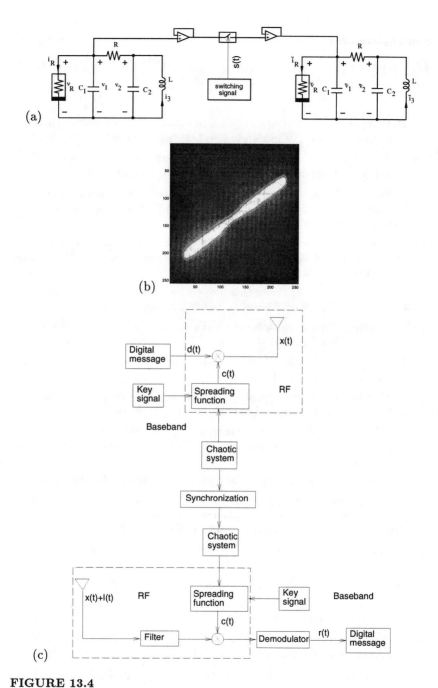

FIGURE 13.4
(a) Implementation of impulsive synchronization of two Chua circuits; (b) oscilloscope experiments; (c) $(CD)^2MA$ communication system.

References

[1] R. E. Amritkar and N. Gupte, "Synchronization of chaotic orbits: The effect of a finite time step," *Phys. Rev. E*, vol. 47, pp. 3889-3895, 1993.

[2] D. D. Bainov and P. S. Simeonov, *Impulsive Differential Equations: Periodic Solutions and Applications*, Longman Group UK Limited, 1993.

[3] S. Boyd, L. El Ghaoui, E. Feron, and V. Balakrishnan, *Linear Matrix Inequalities in System and Control Theory*, SIAM (Studies in Applied Mathematics), vol. 15., 1994.

[4] G. Chen and X. Dong, "From chaos to order – Perspectives and methodologies in controlling nonlinear dynamical systems," *Int. J. of Bifur. Chaos*, vol. 3, pp. 1363-1409, 1993.

[5] G. Chen and X. Dong, *From Chaos to Order – Methodologies, Perspectives and Applications*, World Scientific Pub. Co., Singapore, 1998.

[6] L. O. Chua, "Global unfolding of Chua's circuit," *IEICE Trans. Fundamentals.*, vol. E76-A, pp.704-734, 1993.

[7] L. O. Chua, "Chua's circuit 10 years later," *Int. J. Circuit Theory and Appl.*, vol. 22, pp. 279-305, 1994.

[8] L. O. Chua, M. Komuro, and T. Matsumoto, "The double scroll family," *IEEE Trans. on Circ. Sys.*, I, vol. 33, pp. 1072-1118, 1986.

[9] L. O. Chua, T. Yang, G. Q. Zhong, and C. W. Wu, "Adaptive synchronization of Chua's Oscillators," *Int. J. of Bifur. Chaos*, vol. 6, pp.189-201, 1996.

[10] R. Fletcher, *Practical Methods of Optimization*, Wiley, New York, 1987.

[11] A. L. Fradkov and V. A. Yakubovich, "The S-procedure and duality relations in nonconvex problems of quadratic programming," *Vestnik Leningrad. Univ. Math.*, vol. 6, pp.101-109, 1979 (in Russian, 1973).

[12] T. Kapitaniak and L. O. Chua, "Hyperchaotic attractors of unidirectionally-coupled Chua's circuits," *Int. J. Bifur. Chaos*, vol. 4, pp. 477-482, 1994.

[13] H. Khalil, *Nonlinear Systems*, Macmillan, New York, 1992.

[14] V. Lakshmikantham, D. D. Bainov, and P. S. Simeonov, *Theory of Impulsive Differential Equations*, World Scientific Pub. Co., Signapore, 1989.

[15] V. Lakshmikantham, S. Leela, and A. A. Martynyuk, *Practical Stability of Nonlinear Systems*, World Scientific Pub. Co., Singapore, 1990.

[16] E. N. Lorenz, "Deterministic nonperiodic flow," *J. of Atmos. Sci.*, vol. 20, pp. 130-141, 1963.

[17] R. N. Madan (Ed.), *Chua's Circuit: A Paradigm for Chaos*, World Scientific Pub. Co., Singapore, 1993.

[18] M. A. Matias and J. Guemez, "Stabilization of chaos by proportional pulses in the system variable", *Phys. Rev. Lett.*, vol. 72, pp. 1455-1458, 1994.

[19] F. A. McRae, "Practical stability of impulsive control systems," *J. of Math. Anal. Appl.*, vol. 181, pp. 656–672, 1994.

[20] E. Ott, C. Grebogi, and J. A. Yorke, "Controlling chaos," *Phys. Rev. Lett.*, vol. 64, pp. 1196-1199, 1990.

[21] A. I. Panas, T. Yang, and L. O. Chua, "Experimental results of impulsive synchronization between two Chua's circuits," *Int. J. of Bifur. Chaos*, vol. 8, pp.639-644, 1998.

[22] K. Pyragas, "Continuous control of chaos by self-controlling feedback," *Phys. Lett. A.*, vol. 170, pp. 421-428, 1992.

[23] O. E. Rössler, "An equation for continuous chaos," *Phys. Lett. A*, vol. 57, p. 397, 1976.

[24] A. M. Samoilenko and N. A. Perestyuk, *Impulsive Differential Equations*, World Scientific Pub. Co., Singapore, 1995.

[25] J. Schweizer and M. P. Kennedy, "Predictive Poincaré control: A control theory for chaotic systems, " *Phys. Rev. E*, vol. 52, pp. 4865-4876, 1995.

[26] K. M. Short, "Steps toward unmasking secure communications," *Int. J. of Bifur. Chaos*, vol.4, pp. 957-977, 1994.

[27] T. Stojanovski, L. Kocarev, and U. Parlitz, "Driving and synchronizing by chaotic impulses," *Phys. Rev. E*, vol. 54, pp. 2128-2138, 1996.

[28] T. Stojanovski, L. Kocarev, and U. Parlitz, "Digital coding via chaotic systems," *IEEE Trans. on Circ. Sys.*, I, vol. 44, pp. 562-565, 1997.

[29] J. A. K. Suykens and L. O. Chua, "n-Double scroll hypercubes in 1D-CNNs," *Int. J. of Bifur. Chaos*, vol. 7, pp. 1873-1885, 1997.

[30] J. A. K. Suykens, P. F. Curran, and L. O. Chua, "Master-slave synchronization using dynamic output feedback," *Int. J. of Bifur. Chaos*, vol. 7, pp. 671-679, 1997.

[31] J. A. K. Suykens, P. F. Curran, J. Vandewalle, and L. O. Chua, "Robust nonlinear H_∞ synchronization of chaotic Lur'e systems," *IEEE Trans. on Circ. Sys.*, I, vol. 44, pp. 891-904, 1997.

[32] J. A. K. Suykens, P. F. Curran, T. Yang, J. Vandewalle, and L. O. Chua, "Nonlinear H_∞ synchronization of Lur'e systems: dynamic output feedback case," *IEEE Trans. on Circ. Sys.*, I, vol. 44, pp. 891-904, 1997.

[33] J. A. K. Suykens, A. Huang, and L. O. Chua, "A family of n-scroll attractors from a generalized Chua's circuit," *Archiv für Elektronik und Ubertragungstechnik (Int. J. of Electronics and Communications)*, vol. 51, pp. 131-138, 1997.

[34] J. A. K. Suykens, J. Vandewalle, and B. L. R. De Moor, *Artificial Neural Networks for Modeling and Control of Non-Linear systems*, Kluwer Academic Pub., Boston, 1996.

[35] J. A. K. Suykens, T. Yang, and L. O. Chua, "Impulsive synchronization of chaotic Lur'e systems by measurement feedback," *Int. J. of Bifur. Chaos*, vol. 8, 1371-1381, 1998.

[36] J. A. K. Suykens, T. Yang, J. Vandewalle, and L. O. Chua, "Impulsive synchronization of chaotic Lur'e systems: state feedback case," *IEEE Int. Conf. on Decision and Control*, Tampa, FL, USA, 1998.

[37] M. Vidyasagar, *Nonlinear Systems Analysis*, Prentice-Hall, New Jersey, 1993.

[38] C. W. Wu and L. O. Chua, "A unified framework for synchronization and control of dynamical systems," *Int. J. Bifur. Chaos*, vol. 4, pp. 979-989, 1994.

[39] V. A. Yakubovich, "The S-procedure in nonlinear control theory," *Vestnik Leningrad. Univ. Math.*, vol. 4, pp.73-93, 1977 (in Russian, 1971).

[40] T. Yang, "Recovery of digital signals from chaotic switching," *Int. J. of Circ. Theory and Appl.*, vol. 23, pp. 611-615, 1995.

[41] T. Yang, "Impulsive control," *IEEE Trans. on Auto. Control*, 1999, in press.

[42] T. Yang and L. O. Chua, "Impulsive stabilization for control and synchronization of chaotic systems: Theory and application to secure communication," *IEEE Trans. on Circ. Sys.*, I, vol. 44, pp. 976-988, 1997.

[43] T. Yang and L.O. Chua, "Chaotic digital code-division multiple access (CDMA) systems," *Int. J. of Bifur. Chaos*, vol. 7, pp. 2789-2805, 1997.

[44] T. Yang and L. O. Chua, "Impulsive control and synchronization of nonlinear dynamical systems and application to secure communication," *Int. J. of Bifur. Chaos*, vol. 7, pp. 645-664, 1997.

[45] T. Yang and L. O. Chua, "Applications of chaotic digital code-division multiple access (CDMA) to cable communication systems," *Int. J. of Bifur. Chaos*, vol. 8, pp. 1657-1669, 1998.

[46] T. Yang and L. O. Chua, "Error performance for chaotic digital code-division multiple access (CDMA)," *Int. J. of Bifur. Chaos*, vol. 8, pp. 2047-2059, 1998.

[47] T. Yang, J. A. K. Suykens, and L. O. Chua, "Impulsive control of nonautonomous chaotic systems using practical stabilization," *Int. J. of Bifur. Chaos*, vol. 8, pp. 1557-1564, 1998.

[48] T. Yang, L.-B. Yang, and C.-M. Yang, "Impulsive synchronization of Lorenz systems," *Phys. Lett. A*, vol. 226, pp. 349-354, 1997.

[49] T. Yang, L.-B. Yang, and C.-M. Yang, "Impulsive control of Lorenz system," *Physica D*, vol. 110, pp. 18-24, 1997.

[50] T. Yang, L.-B. Yang, and C.-M. Yang, "Control of Rössler system to periodic motions using impulsive control method," *Phys. Lett. A*, vol. 232, pp. 356-361, 1997.

[51] T. Yang, L.-B. Yang, and C.-M. Yang, "Breaking chaotic secure communication using a spectrogram," *Phys. Lett. A*, vol. 247, pp. 105-111, 1998.

[52] T. Yang, L.-B. Yang, and C.-M. Yang, "Application of neural networks to unmasking chaotic secure communication," *Physica D*, vol. 124, pp. 248-257, 1998.

[53] T. Yang, L.-B. Yang, and C.-M. Yang, "Cryptanalyzing chaotic secure communication using return maps," *Phys. Lett. A*, vol. 245, pp. 495-510, 1998.

[54] T. Yang, L.-B. Yang, and C.-M. Yang, "Break chaotic switching using generalized synchronization: Examples," *IEEE Trans. on Circ. Sys.*, I, vol. 45, pp. 1062-1067, 1998.

Acknowledgments

Part of this research work was carried out at the ESAT laboratory and the Interdisciplinary Center of Neural Networks ICNN of the Katholieke Universiteit Leuven, in the framework of the Belgian Programme on Interuniversity Poles of Attraction, initiated by the Belgian State, Prime Minister's Office for Science, Technology and Culture (IUAP P4-02) and in the framework of a Concerted Action Project MIPS (Modelbased Information Processing Systems) of the Flemish Community. J. Suykens is a postdoctoral researcher with the National Fund for Scientific Research FWO - Flanders. T. Yang and L.O. Chua were supported by the Office of Naval Research under grant numbers N00014-97-1-0463 and N00014-96-1-0753.

14

Control and Anticontrol of Bifurcations with Application to Active Control of Rayleigh-Bénard Convection

Hua O. Wang and Dong S. Chen

Department of Electrical and Computer Engineering
Duke University, Durham, North Carolina, 27708 USA
hua@ee.duke.edu

Abstract

Bifurcation control deals with the modification of the bifurcation characteristics of a parameterized nonlinear system by a judiciously designed control input. This chapter summarizes some of the recent advances in this area, including both theory and applications. First, the problem of anticontrol of bifurcations is considered. Anticontrol of bifurcations refers to the situation where a certain type of bifurcation is created at a preferred location with certain desired properties by appropriate controls. Second, the results on amplitude control of bifurcations are summarized. It is shown that the amplitude of bifurcated solutions is directly related to the so-called bifurcation stability coefficients. Finally, the problem of active control of Rayleigh-Bénard is tackled using the techniques of amplitude control of bifurcations. Both theoretical and experimental results are presented.

14.1 Introduction

In recent years, there has been rapidly growing interest in control of nonlinear dynamical systems exhibiting bifurcation phenomena. There are a wide variety of promising potential applications of bifurcation controls, including possible resolution for stall of compression system in jet engines, high incidence flight, voltage collapse in power systems, oscillatory behavior of tethered satellites, magnetic bearing systems, rotating chains, thermal convection loop, and cardiac alternans as well as other pathological heart rhythms (see [2] and references therein). In general, the goal of bifurcation control is to design a controller that can modify the bifurcation characteristics of a bifurcating nonlinear system, thereby achieving some desirable dynamical behaviors [30, 2]. This goal is often achieved by delaying the onset of an inherent bifurcation and/or stabilizing ("softening") an existing bifurcation. Other bifurcation control objectives can entail optimization of a performance index near the bifurcation, re-shaping of a bifurcation diagram, or a combination of some of these [2]. In this chapter, we investigate two bifurcation control problems.

First, we consider the "inverse" problem, i.e., the problem of anticontrol of bifurcations, analogous to the anticontrol problem for chaos. Anticontrol of chaos means that chaos is created or enhanced when it is healthy and beneficial. Similarly, anticontrol of bifurcations refers to the situation where a certain type of bifurcation is created at a preferred location with certain desired properties by appropriate controls. Our work on anticontrol of bifurcations is motivated by observations that in some applications, it may be advantageous to introduce new bifurcations to the nominal branch of system output. These new bifurcated solutions may serve as new and more desirable operating conditions; they may also serve as warning signals of impending collapse or catastrophe. Alternatively, they may be judiciously combined with existing dynamical features of the system to extend the operating region via an enlargement of the system parameter ranges which otherwise cannot be accomplished by conventional control methods.

More precisely, we present some results on the introduction of new Hopf bifurcations into a given system via feedback control. Hopf bifurcations are common in many nonlinear oscillations and limit cycle behaviors, in systems of biological, ecological, social and economic sciences and engineering. Within this context, anticontrol of Hopf bifurcations can be viewed as one approach to designing limit cycles with specified oscillatory behaviors into a system by appropriate feedback controls. In the controller design, we employ dynamic feedback control laws incorporating washout filters. The use of washout filters ensures that lower-frequency orbits of the system are retained in the closed-loop system, while the transient dynamics and higher-frequency orbits are modified. Washout filter-aided design

Introduction

techniques have been used to control the location and stability of Hopf bifurcations [26, 28, 36], which are employed in this chapter to derive a systematic procedure for the design of some effective control laws to create Hopf bifurcations in a given system.

The second part of this chapter deals with amplitude control of bifurcations. In this problem, the control objective is to control the amplitude of bifurcated solutions in bifurcating and/or chaotic nonlinear systems. Both pitchfork and Hopf bifurcations have been addressed. In the case of pitchfork bifurcation, the control seeks to modify (e.g., reduce) the amplitude (or rather magnitude) of the bifurcated equilibrium branches. In the case of Hopf bifurcation, the control aims at controlling the amplitude of the limit cycles born through the Hopf bifurcation. There are a number of control objectives that can be achieved via the bifurcation amplitude control. For instance, chaotic behavior in a nonlinear system can be converted to limit cycle behavior with desired amplitude. There are circumstances in which it is undesirable or impossible to operate the system on a steady state equilibrium. Amplitude control would enable the system to operate on a small amplitude limit cycle surrounding the equilibrium. Also in applications such as Rayleigh-Bénard convection (RBC), it is important to control the amplitude of the bifurcated solutions (convection).

Specifically, we report theoretical results on amplitude control of bifurcated solutions in pitchfork and Hopf bifurcations. Control of limit cycle amplitude associated with Hopf bifurcation and period-doubling bifurcation was pursued in [19, 23] using a harmonic balance approach. In this chapter, results are presented in state space for general n-dimensional nonlinear systems. We show that the amplitude of the bifurcated solutions is directly related to the so-called *bifurcation stability coefficients*.

The third part of this chapter contains an application of bifurcation control theory to the problem of active control of Rayleigh-Bénard convection. The Rayleigh-Bénard convection is a classical bifurcation phenomena that has been investigated in detail [9]. Control of Rayleigh-Bénard convection is a problem of importance to theoretical research as well as industrial applications. For instance, in Czochralski crystal growth, convection causes defects and dopant inhomogeneities and degrades crystal properties [20]. Based on techniques of amplitude control of bifurcations, two control strategies are developed, one in the pseudo-spectral coordinate and the other one in the physical spatial coordinate. Theoretical as well as experimental results confirm the effectiveness of the control strategies.

14.2 Anticontrol of Bifurcations

Bifurcation theory deals with the qualitative change of a system under parametric variations. Consider a general one-parameter finite dimensional system:

$$\dot{\mathbf{x}} = \mathbf{F}(\mathbf{x}; \mu). \tag{14.2.1}$$

Here $\mathbf{x} \in \mathbb{R}^n$ is the state vector, $\mu \in \mathbb{R}$ is the bifurcation parameter, the vector field $\mathbf{F}$ is smooth in $\mathbf{x}$ and μ, and $\mathbf{F}(0; \mu) = \mathbf{0}$, i.e., $\mathbf{x} = \mathbf{0}$ is always a fixed point of the system. Suppose the Jacobian $L_0 = \left. \frac{\partial \mathbf{F}}{\partial \mathbf{x}} \right|_{(\mathbf{x}, \mu) = (\mathbf{0}, 0)}$ is singular at $\mu = 0$, and the fixed point $\mathbf{x} = \mathbf{0}$ loses stability as μ changes its sign, then a bifurcation happens at $\mu = 0$. When the fixed point loses stability, the system gives rise to either a new branch of fixed points or periodic oscillations. One important issue of bifurcation is the *direction*, or stability, of the bifurcation. Superciritcal bifurcations permit smooth transition of system states, while subcritical, transcritical, and saddle-node bifurcations normally lead to hysteresis and "jump" behaviors which are undesirable. Therefore a typical bifurcation control objective is to delay and/or stabilize an existing bifurcation [30, 2]. In the framework of bifurcation control, we consider the following general n-dimensional control system:

$$\dot{\mathbf{x}} = \mathbf{F}(\mathbf{x}, u; \mu), \tag{14.2.2}$$

where u represents the control input.

In this section, we present results on the problem of anticontrol of bifurcations which refers to the situation where a certain type of bifurcation is created at a preferred location with certain desired properties by appropriate controls. Specifically, we present results on the introduction of new Hopf bifurcations into a given system via feedback control. The control objective is to design a controller, u, that is capable of creating a Hopf bifurcation at a desired point $(\mathbf{x}, \mu)$ and moreover endowing the Hopf bifurcation certain preferable characteristics. To achieve this control objective, we start from the simple cases of one-dimensional and two-dimensional systems, and then generalize the results to the general n-dimensional case.

14.2.1 Hopf Bifurcation Theorem

Before presenting the main results, the classical criteria for Hopf bifurcation in continuous-time systems are briefly reviewed in this subsection. This background material is needed for the development of the bifurcation analysis and controller design in the sequel.

Rewrite the system (14.2.1) in the linearized form

$$\dot{\mathbf{x}} = A(\mu)\mathbf{x} + \mathbf{F}(\mathbf{x}; \mu) \tag{14.2.3}$$

where $\mathbf{F} \in C^k(\mathbb{R}^n)$ with $k \geq 3$, $\mathbf{F}(\mathbf{x}_0; \mu) = \mathbf{0}$ and $\frac{\partial \mathbf{F}}{\partial \mathbf{x}}(\mathbf{x}_0; \mu) = \mathbf{0}$ for all sufficiently small $|\mu - \mu_0|$, where μ_0 is the value of μ at the bifurcation point. Assume that the linear part $A(\mu)$ at the origin has a pair of eigenvalues $\lambda_{1,2}(\mu) = \alpha(\mu) \pm i\omega(\mu)$ with $\alpha(\mu_0) = 0$ and $\omega(\mu_0) \neq 0$. Furthermore, suppose that the eigenvalue pair cross the imaginary axis with nonzero speed, i.e.,

$$\frac{\partial \alpha}{\partial \mu}(\mu_0) \neq 0, \qquad (14.2.4)$$

which is known as the *transversality* condition for the crossing of the eigen-loci at the imaginary axis. Then in any neighborhood U of the point $\mathbf{x}_0$ and for any given $\varepsilon > 0$, there is a $\bar{\mu}$ with $|\bar{\mu} - \mu_0| < \varepsilon$ such that the differential equation (14.2.3) has a nontrivial periodic orbit in U. In this case, the system is said to undergo a *Hopf bifurcation* at the bifurcation point $(\mathbf{x}_0, \mu_0)$ [17].

More precisely, if $\lambda_{1,2}$ cross the imaginary axis from the left-half plane to the right-half plane as μ increases, then in any small left-neighborhood of μ ($\mu < \mu_0$), $\mathbf{x}_0$ is a stable focus, while in any small right-neighborhood of μ_0 ($\mu > \mu_0$), this focus becomes unstable, surrounded by a limit cycle p_ε of amplitude $O(\sqrt{|\mu - \mu_0|})$. The asymptotic stability of p_ε is governed by one characteristic exponent given by a real smooth even function

$$\beta(\varepsilon) = \beta_2 \, \varepsilon^2 + \beta_4 \, \varepsilon^4 + \ldots . \qquad (14.2.5)$$

It is known that p_ε is orbitally asymptotically stable if $\beta(\varepsilon) < 0$ but is unstable if $\beta(\varepsilon) > 0$. Typically, the local stability of the bifurcated periodic solution p_ε, i.e., the *direction* of the bifurcation, is determined by the sign of β_2, which is called the *bifurcation stability coefficient*. The computation of β_2 is relatively complicated, but a scheme is available for its evaluation (see [30] for details).

14.2.2 Bifurcation Creation in 1-D Systems

Consider a one-dimensional system described by a parameterized ordinary differential equation,

$$\dot{x} = f(x; \mu) + u, \qquad (14.2.6)$$

where μ is a parameter and u is the control input. To achieve the control objective, we design the controller with the aid of a washout filter. The washout-filter-aided controller assumes the following structure:

$$\dot{w} = x - dw \stackrel{\Delta}{=} y, \qquad (14.2.7)$$

$$u = g(y; \mathbf{K}), \qquad (14.2.8)$$

where **K** is the control gain vector, considered as a parameter vector of the nonlinear control function g, and d is the washout filter time constant. Due to the nature of washout filter, $u = 0$ when $\dot{w} = y = 0$, so all the equilibria remain unchanged when control actions are applied. The following constraints should be fulfilled:

1. $d > 0$, which guarantees the stability of the washout filter.
2. $g(0, \mathbf{K}) = 0$, which preserves the original equilibrium points.

In most problems, the bifurcation is required to be supercritical to obtain stable bifurcated periodic orbits (limit cycles), but it can rarely be accomplished with a linear feedback alone. For this simple system, our controller can also be used to control the direction of the bifurcation. For this purpose, consider a specialized nonlinear control function of the form

$$g(y; \mathbf{K}) = k_1(y + k_2 y^2 + k_3 y^3), \qquad (14.2.9)$$

where the control gain vector $\mathbf{K} = (k_1, k_2, k_3)^\top$. Use the transformation $x_1 = x - x_0$, $w_1 = w - w_0 = w - x_0/d$ for localization. Then, Taylor-expand the system function as

$$f(x_1) = f(x - x_0) = f_1 x_1 + \frac{1}{2} f_2 x_1^2 + \frac{1}{6} f_3 x_1^3 + O(x_1^4), \qquad (14.2.10)$$

where $f_1 = \frac{\partial f}{\partial x}(x_0)$, $f_2 = \frac{\partial^2 f}{\partial x^2}(x_0)$, $f_3 = \frac{\partial^3 f}{\partial x^3}(x_0)$, etc. From the elementary bifurcation theory [30], we know that, generically, only up to the third order derivatives are necessary for bifurcation analysis. Therefore, the higher-order terms are dropped.

With linear stability analysis, conditions for a Hopf bifurcation at $\mu = \mu_0$ are found to be

$$k_1 = d - f_1(\mu_0), \quad f_1(\mu_0) < 0, \quad \text{and} \quad \frac{\partial \alpha}{\partial \mu}(\mu_0) = \frac{\partial f_1}{\partial \mu}(\mu_0) \neq 0. \qquad (14.2.11)$$

Note that $f_1(\mu_0) < 0$ simply means that the bifurcation point μ_0 should be stable in the original system. The bifurcation stability coefficient can be computed according to the schemes described in [30],

$$\beta_2 = \frac{f_3}{8} - \frac{f_2^2}{8 f_1} + \frac{f_2}{4} k_2 - \frac{3 f_1}{4} k_3. \qquad (14.2.12)$$

With this formula, we can design nonlinear control terms in (14.2.9) to change the direction of the bifurcation directly.

14.2.3 Bifurcation Creation in 2-D Systems

The problem of creating a Hopf bifurcation in a two-dimensional system is much more complicated than the one-dimensional setting. For simplicity,

Anticontrol of Bifurcations

we only consider the case with a pair of conjugated complex eigenvalues. In Subsection 14.2.4, we will see that the case with two real eigenvalues can be decomposed into two one-dimensional subsystems and then solved by the procedure proposed in Subsection 14.2.2.

Assume the controlled system equation to be

$$\dot{x}_1 = f_1(x_1, x_2; \mu) + u, \qquad (14.2.13)$$

$$\dot{x}_2 = f_2(x_1, x_2; \mu). \qquad (14.2.14)$$

As before, a washout filter in x_2 is used,

$$\dot{w} = x_2 - dw \triangleq y, \qquad (14.2.15)$$

$$u = g(y; \mathbf{K}). \qquad (14.2.16)$$

The linearized system matrix at the equilibrium (x_{10}, x_{20}) is

$$A(\mu) = \begin{pmatrix} f_{11}(\mu) & f_{12}(\mu) & -dg_1 \\ f_{21}(\mu) & f_{22}(\mu) & 0 \\ 0 & 1 & -d \end{pmatrix}, \qquad (14.2.17)$$

where $f_{mn}(\mu) = \frac{\partial f_m}{\partial x_n}(x_{10}, x_{20}; \mu)$, $(m, n = \{1, 2\})$, and $g_1 = \frac{\partial g}{\partial y}(0)$.

Conditions for $A(\mu)$ to have a pair of imaginary eigenvalues with the third one located on the left-half plane are

$$\Big(\Delta - d(f_{11} + f_{22}) - f_{21}g_1\Big)(d - f_{11} - f_{22}) - d\Delta = 0, \quad (14.2.18)$$

$$\Delta - d(f_{11} + f_{22}) - f_{21}g_1 > 0, \; d \cdot \Delta > 0, \; d - f_{11} - f_{22} > 0. \quad (14.2.19)$$

where $\Delta \triangleq f_{11}f_{22} - f_{12}f_{21}$. Since the original system has a pair of conjugated complex eigenvalues, Δ should be positive and never be zero, with both f_{12} and f_{21} nonzero. The value of g_1 can be obtained by solving the equation (14.2.18). The result is

$$g_1 = -\frac{(f_{11} + f_{22})(d^2 - d(f_{11} + f_{22}) + f_{11}f_{22} - f_{12}f_{21})}{(d - f_{11} - f_{22})f_{21}} \qquad (14.2.20)$$

All three inequalities (14.2.19) are satisfied if $d > 0$ and $d - f_{11} - f_{22} > 0$. If the designated bifurcation point in the original system is stable, then $f_{11} + f_{22} < 0$. Thus, the condition is automatically satisfied. If the bifurcation point is unstable, then we have to choose d large enough ($d > f_{11} + f_{22} = 2\mathbf{Re}[\lambda_0]$) to overcome this instability, where λ_0 is the unstable eigenvalue of the openloop system (14.2.13)–(14.2.14).

14.2.4 Bifurcation Creation in n-D Systems

With the preliminary results in one- and two-dimensional cases, we can proceed to deal with n-dimensional systems. From the results in Subsection

14.2.2, we assume that the designated bifurcation point $(\mathbf{x}_0, \mu_0)$ is stable, or the system Jacobian has only two unstable conjugated complex eigenvalues (all the others are stable). In order to analyze this n-dimensional system, we employ the real canonical Jordan form to decompose the whole system into interconnected subsystems.

First, rewrite the control system equation (14.2.2) in linearized form

$$\dot{\mathbf{x}} = A(\mu)\mathbf{x} + B(\mu)u + \mathbf{F}(\mathbf{x}, u; \mu), \qquad (14.2.21)$$

where $\mathbf{F}(\mathbf{x}, u; \mu)$ denotes the second and higher-order terms in $\mathbf{x}$. Note that A, B, and $\mathbf{F}$ all depend on μ. In the following discussion, we only use their values at $\mu = \mu_0$ and drop all the parameter dependency for notational simplicity.

From the matrix theory [11], we know that if the eigenvalues are $\lambda_1, \ldots, \lambda_p$, $\alpha_{p+1} \pm i\omega_{p+1}, \ldots, \alpha_r \pm i\omega_r$, where $\lambda_i \in \mathbb{R}$, $\alpha_i \in \mathbb{R}$, $\omega_i \in \mathbb{R}$, and n_i is the multiplicity of these eigenvalues, then a real matrix P can be found to transform the system matrix A into the real Jordan canonical form:

$$\hat{A} = \begin{pmatrix} J_{n_1}(\lambda_1) & & & & & \\ & \ddots & & & 0 & \\ & & J_{n_p}(\lambda_p) & & & \\ & & & C_{n_{p+1}}(\alpha_{p+1}, \omega_{p+1}) & & \\ & 0 & & & \ddots & \\ & & & & & C_{n_r}(\alpha_r, \omega_r) \end{pmatrix}, \qquad (14.2.22)$$

where $J_{n_k}(\lambda_k)$ and $C_{n_k}(\alpha_k, \omega_k)$ are real and complex Jordan blocks. Using the coordinate transformation

$$\mathbf{x} = P\mathbf{z}, \qquad (14.2.23)$$

the system (14.2.21) is transformed into its Jordan canonical form,

$$\dot{\mathbf{z}} = P^{-1}AP\mathbf{z} + P^{-1}Bu + P^{-1}\mathbf{F}(P\mathbf{z}, u) = \hat{A}\mathbf{z} + \hat{B}u + \hat{\mathbf{F}}(\mathbf{z}, u), \qquad (14.2.24)$$

where $\hat{A} = P^{-1}AP$, $\hat{B} = P^{-1}B$, and $\hat{\mathbf{F}}(\mathbf{z}, u) = P^{-1}\mathbf{F}(P\mathbf{z}, u)$.

For each Jordan block of real eigenvalues, if one state is controllable (i.e., $\hat{b}_k \neq 0$), and if the subsystem is stable at the designated bifurcation point $\mu = \mu_0$ (i.e., $\lambda_k(\mu_0) < 0$), we can introduce a Hopf bifurcation at this point via feedback control through the washout filter with respect to this state by the procedure given at Subsection 14.2.2. The subsystem is of the form

$$\dot{z}_k = \hat{f}(z_k; \mathbf{z}_0, \mu) + \hat{b}_k u, \qquad (14.2.25)$$

with $\mathbf{z}_0$, the equilibrium of $\mathbf{z}$, as a parameter. The washout filter is introduced as

$$\dot{w} = z_k - dw \stackrel{\Delta}{=} y, \qquad (14.2.26)$$

$$u = \frac{1}{\hat{b}_k} g(y; \mathbf{K}). \tag{14.2.27}$$

Since the system matrix is in the real Jordan canonical form, only this state is destabilized while all the others are untouched. So the original system does have a Hopf bifurcation at this point. The criticality of this bifurcation can also be designed analytically, with the formula (14.2.12) for β_2 under the transformation (14.2.23).

For Jordan blocks with complex eigenpairs, we can also introduce a Hopf bifurcation for any controllable state. Let the subsystem for this state be written as

$$\dot{z}_1 = \hat{f}_1(z_1, z_2; \mathbf{z}_0, \mu) + \hat{b}_1 u, \tag{14.2.28}$$

$$\dot{z}_2 = \hat{f}_2(z_1, z_2; \mathbf{z}_0, \mu) + \hat{b}_2 u. \tag{14.2.29}$$

As above, we view all the state variables of $\mathbf{z}$, except z_1 and z_2, as parameters, and use their equilibrium values in the bifurcation analysis.

Next, suppose $\hat{b}_1 \neq 0$. Substituting z_1 and z_2 with $\nu = \frac{z_1}{\hat{b}_1}$ and $\xi = z_2 - \frac{\hat{b}_2}{\hat{b}_1} z_1$ gives

$$\dot{\nu} = \tilde{f}_1(\nu, \xi; \mathbf{z}_0, \mu) + u, \tag{14.2.30}$$

$$\dot{\xi} = \tilde{f}_2(\nu, \xi; \mathbf{z}_0, \mu), \tag{14.2.31}$$

where $\tilde{f}_1(\nu, \xi) = \frac{1}{\hat{b}_1}\hat{f}_1(\hat{b}_1\nu, \xi + \hat{b}_2\nu)$ and $\tilde{f}_2(\nu, \xi) = \hat{f}_2(\hat{b}_1\nu, \xi + \hat{b}_2\nu) - \frac{\hat{b}_2}{\hat{b}_1}\hat{f}_1(\hat{b}_1\nu, \xi + \hat{b}_2\nu)$. This equation is in the form of (14.2.13)–(14.2.14), so we can use the procedure described therein to design the controller as

$$\dot{w} = \xi - dw \stackrel{\Delta}{=} y, \tag{14.2.32}$$

$$u = k_1 y + k_3 y^3, \tag{14.2.33}$$

where k_1 is used to relocate the bifurcation point and k_3 is used to change its criticality.

Finally, we should determine the effects of the controller on the original system. In view of the inverse of the transformation (14.2.23), where $P = [P_{ij}]$, we know that if a bifurcation is introduced to the state z_j, there will also be a bifurcation at the state x_i if and only if $P_{ij} \neq 0$. If we want to induce a bifurcation to x_i, we have to find a j such that both P_{ij} and $\hat{b}_j$ are nonzero. A cross reference table can then be constructed with this condition, so as to facilitate the controller design.

14.2.5 Design Example

Consider a simple 1-D system

$$\dot{x} = -\mu - x^2. \tag{14.2.34}$$

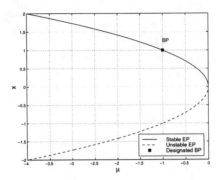

FIGURE 14.1
Bifurcation diagram of the system $\dot{x} = -\mu - x^2$, where BP is the designated bifurcation point.

The bifurcation diagram of this system is shown in Figure 14.1. There is one saddle-node bifurcation occurring at $\mu = 0$ in this system. The occurrence of dynamic bifurcations such as saddle node bifurcations have been linked to a wide variety of incipient instabilities in engineering and physical systems, including voltage collapse phenomenon in electric power systems [6, 24] and thermal runaway of continuous stirred tank reactors [8].

Our control or anticontrol objective, rather, is to introduce a supercritical Hopf bifurcation at the point BP: $\mu = -1, x = 1$ in this system, thereby creating a stable limit cycle near this bifurcation point. The small amplitude stable limit cycle may serve as a warning signal of impending collapse or catastrophe or it may also serve as a new and more desirable operating condition depending on the application.

To generate the intended Hopf bifurcation at the point BP, a controller with both linear and cubic terms is used:

$$u = g(y) = k_1 y + k_3 y^3. \quad (14.2.35)$$

From (14.2.11), the gain k_1 should be

$$k_1 = d - f_x(x_0; \mu_0) = d + 2x_0 = d + 2. \quad (14.2.36)$$

The value of d can be selected arbitrarily, as long as it is positive. Here, we use $d = 0.2$, so $k_1 = 2.2$. If only the linear term is used, the bifurcation is subcritical (see Figure 14.2), so the nonlinear term has to be added. According to (14.2.12), this bifurcation can be rendered supercritical if $k_3 < -\frac{1}{6}k_1$. Here we use $k_3 = -10$ to control the criticality. The location of the bifurcation point is unchanged but the Hopf bifurcation becomes supercritical (Figure 14.3).

Anticontrol of Bifurcations

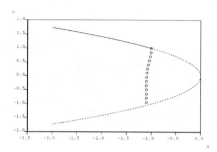

FIGURE 14.2
Bifurcation diagram for $d = 0.2, k_1 = 2.2, k_3 = 0$. Empty circles denote the minimum amplitudes of unstable period orbits.

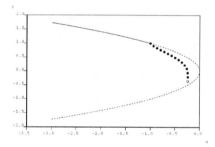

FIGURE 14.3
Bifurcation diagram for $d = 0.2, k_1 = 2.2, k_3 = -10$. Empty circles denote the minimum amplitudes of unstable periodic orbits, with filled circles for the minimum amplitudes of stable periodic orbits.

14.3 Amplitude Control of Bifurcations

In this section, the problem of amplitude control of bifurcations is considered. The objective is to control the amplitude of bifurcated solutions in bifurcating and/or chaotic nonlinear systems. We consider two type of bifurcations, namely, the pitchfork bifurcation and Hopf bifurcation. In the case of pitchfork bifurcation, the control seeks to modify (e.g., reduce) the amplitude (or rather magnitude) of the bifurcated equilibrium branches. In the case of Hopf bifurcation, the control aims at controlling the amplitude of the limit cycles born through the Hopf bifurcation.

To begin with, rewrite system (14.2.2) in the series form

$$\dot{\mathbf{x}} = L_0\mathbf{x} + u\gamma + uL_1\mathbf{x} + Q_0(\mathbf{x},\mathbf{x}) + u^2 L_2\mathbf{x} + \qquad (14.3.37)$$

$$+ uQ_1(\mathbf{x},\mathbf{x}) + C_0(\mathbf{x},\mathbf{x},\mathbf{x}) + \text{H.O.T.} \qquad (14.3.38)$$

With the feedback control, the characteristic parameters associated with stability as well as the amplitude of the bifurcated solutions will be affected. A superscript * is employed in the remainder of this section to indicate that the corresponding expression pertains to the controlled system.

14.3.1 Pitchfork Bifurcation Case

Pitchfork bifurcation occurs in systems with symmetry. In many situations, the vector field is an odd function of $\mathbf{x}$, i.e., $\mathbf{F}(-\mathbf{x};\mu) = -\mathbf{F}(\mathbf{x};\mu)$. From basic bifurcation theory [17, 30], in a neighborhood of the origin in $(\mathbf{x},\mu)$ space there always exist a one-parameter family $(\mathbf{x}(\varepsilon),\mu(\varepsilon))$ of stationary solutions of (14.2.1) (with $\mathbf{x}(\varepsilon) \to 0$, $\mu \to 0$ as $\varepsilon \to 0$). When $F(\mathbf{x};\mu)$ is analytic, ε can be chosen so that $\mathbf{x}$, μ are also analytic, and

$$\mathbf{x}(\varepsilon) = \mathbf{x}_1\varepsilon + \mathbf{x}_2\varepsilon^2 \ldots \qquad (14.3.39)$$

$$\mu(\varepsilon) = \mu_2\varepsilon^2 + \mu_4\varepsilon^4 \ldots \qquad (14.3.40)$$

Generically, $\mu_2 \neq 0$. When ε is small enough, $\varepsilon \approx \sqrt{\frac{\mu}{\mu_2}}$, so $\mathbf{x} \approx \frac{\mathbf{x}_1}{\sqrt{|\mu_2|}}\sqrt{\mu}$, i.e., $\mathbf{x} \propto 1/\sqrt{\mu_2}$. Therefore, μ_2 can be selected as the characteristic parameter of the bifurcation amplitude. The desired bifurcation amplitude can be achieved when μ_2 is set to an appropriate value.

The characteristic parameter μ_2 is given by

$$\mu_2 = -(\mathbf{l}Q_0(\mathbf{r},\mathbf{x}_2) + \mathbf{l}C_0(\mathbf{r},\mathbf{r},\mathbf{r}))/\lambda'(0). \qquad (14.3.41)$$

where $\mathbf{l}$ and $\mathbf{r}$ are the normalized left and right eigenvectors of the linearized system corresponding to the zero eigenvalue, i.e., $\mathbf{l}L_0 = L_0\mathbf{r} = 0$. $\lambda'(0) = \frac{\partial\lambda(0)}{\partial\mu} \neq 0$ is the crossing speed of the zero eigenvalue through the imaginary axis. From Factorization Theorem, μ_2 is directly related to the *bifurcation stability coefficient* β_2 with

$$\beta_2 = -2\mu_2 \lambda'(0). \qquad (14.3.42)$$

The bifurcated solution is asymptotically stable if $\beta_2 < 0$ but is unstable if $\beta_2 > 0$.

To preserve the symmetry, the control should also be an odd function of **x**, i.e., $u(-\mathbf{x}) = -u(\mathbf{x})$; therefore, the controller cannot have constant or quadratic terms. In addition, to avoid the relocation of the bifurcation point, only the cubic terms are used in our controller:

$$u(\mathbf{x}) = C_u(\mathbf{x}, \mathbf{x}, \mathbf{x}). \qquad (14.3.43)$$

Now substituting this control function into the control system equation (14.2.2) and calculating Q_0 and C_0, from (14.3.41), we get

$$\mu_2^* = \mu_2 - \mathbf{l}\gamma C_u(\mathbf{r}, \mathbf{r}, \mathbf{r})/\lambda'(0). \qquad (14.3.44)$$

If the system is linearly controllable (i.e., $\mathbf{l}\gamma \neq 0$), and $C_u(\mathbf{r}, \mathbf{r}, \mathbf{r}) \neq 0$, the amplitude of the bifurcation can be controlled to any desired value.

14.3.2 Hopf Bifurcation Case

Recall that the hypothesis associated with the occurrence of a Hopf bifurcation in system (14.2.1) can be stated as follows.

(1) L_0 has a pair of simple, complex conjugate eigenvalues $\lambda_1 = i\omega_c$, $\lambda_2 = -i\omega_c$ on the imaginary axis, where $\omega_c \neq 0$. Moreover, all the other eigenvalues of L_0 have a negative real part. Let **l** and **r** be the normalized left and right eigenvectors of L_0 with respect to eigenvalue $i\omega_c$, i.e. $\mathbf{l}L_0 = i\omega_c \mathbf{l}$, $L_0 \mathbf{r} = i\omega_c \mathbf{r}$, and $\mathbf{lr} = 1$.

(2) The continuous extension to $\mu \neq 0$ of the eigenvalue $i\omega_0$, say $\lambda(\mu)$, should transversely cross the imaginary axis at $\mu = 0$, i.e.,

$$\mathbf{Re}\left(\frac{d\lambda}{d\mu}\right)\bigg|_{\mu=0} \neq 0 \quad \text{or} \quad \mathbf{Re}(\mathbf{l}L_1\mathbf{r}) \neq 0. \qquad (14.3.45)$$

From Hopf bifurcation theorem, in a neighborhood of the origin in $(\mathbf{x}, \mu)$ space, there always exists a one-parameter family $(\mathbf{x}(t, \varepsilon), \mu(\varepsilon))$ of periodic solutions of (14.2.1) (with $\mathbf{x}(t, \varepsilon) \to \mathbf{0}$, $\mu \to 0$ as $\varepsilon \to 0$), having period $T(\varepsilon)$ (with $T \to \frac{2\pi}{\omega_0}$ as $\varepsilon \to 0$). When $\mathbf{F}_\mu(\mathbf{x})$ is analytic, ε can be chosen so that x, μ, and T are all analytic, and

$$\mathbf{x}(t, \varepsilon) = \varepsilon\, \mathbf{x}_1(t) + \varepsilon^2\, \mathbf{x}_2(t) + \dots, \qquad (14.3.46)$$

$$\mu(\varepsilon) = \mu_2\, \varepsilon^2 + \mu_4\, \varepsilon^4 + \dots, \qquad (14.3.47)$$

$$T = (2\pi/\omega_c)(1 + T_2\, \varepsilon^2 + T_4\, \varepsilon^4 \dots), \qquad (14.3.48)$$

where $\mathbf{x}_1(t) = \mathbf{Re}(\mathbf{r}e^{i2\pi t/T})$. When ε is small enough, $\varepsilon \approx \sqrt{\frac{\mu}{\mu_2}}$, so

$$\mathbf{x} \approx \sqrt{|\mu|}\mathbf{Re}(\frac{\mathbf{r}}{\sqrt{|\mu_2|}}e^{i2\pi t/T}) \tag{14.3.49}$$

In the vicinity of the bifurcation point, the amplitude of limit cycle $A \propto \frac{1}{\sqrt{|\mu_2|}}$. The value of A becomes smaller when the absolute value of μ_2 gets larger. Therefore, μ_2 can be selected as the characteristic parameter of the bifurcation amplitude. The desired amplitude of oscillation can be achieved when μ_2 is set to an appropriate value.

From the factorization theorem [17],

$$\beta_2 = -2\mu_2 \mathbf{Re}\lambda'(0), \tag{14.3.50}$$

where β_2 is the *bifurcation stability coefficient* of the Hopf bifurcation. When $\beta_2 > 0$, the limit cycle is unstable and the bifurcation is subcritical. When $\beta_2 < 0$, the limit cycle is stable and the system undergoes a supercritical bifurcation. When the bifurcation is supercritical, the amplitude of bifurcated limit cycle becomes smaller and smaller as β_2 becomes more and more negative.

Assume the feedback control to be of the following form:

$$u(\mathbf{x}) = C_u(\mathbf{x}, \mathbf{x}, \mathbf{x}), \tag{14.3.51}$$

where C_u is a cubic symmetric multilinear function. Note that $u(\mathbf{x})$ contains no terms linear in $\mathbf{x}$. This ensures that the bifurcation point is unaffected by the feedback control. The bifurcation coefficient of the controlled system is

$$\mu_2^* = \mu_2 - \frac{3}{4}\mathbf{Re}C_u(\mathbf{r},\mathbf{r},\bar{\mathbf{r}})\mathbf{l}\gamma/\lambda'(0). \tag{14.3.52}$$

When the system is linearly controllable, or $\mathbf{l}\gamma \neq 0$, the amplitude of the bifurcation can be changed arbitrarily. If the system is linearly uncontrollable, this cubic control law has no effect on bifurcation amplitude; however, it has been shown that a controller with quadratic terms can be used to attain the desired control objective.

14.4 Bifurcation Control of Rayleigh-Bénard Convection

In this section, we consider the problem of active control of Rayleigh-Bénard convection. In particular, techniques from amplitude control of bifurcations are employed to design controllers to relocate the onset of convection and/or suppress the convection amplitude. Active control of

Rayleigh-Bénard convection is a problem of importance to both theoretical research and industrial applications.

14.4.1 Rayleigh-Bénard Convection

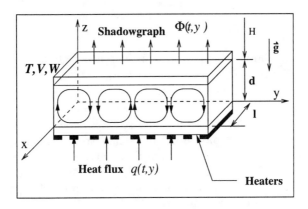

FIGURE 14.4
Schematic plot of Rayleigh-Bénard convection experiment apparatus.

Consider a horizontal, homogeneous layer of fluid between two infinite, parallel plates positioned distance d apart. The plates are normal to the Cartesian coordinate z that is parallel to the gravity vector. Plane (x, y) define the lower plate. The origin of the coordinate system is located in the lower plate. The bottom plate is heated with a constant heat flux q and the top plate is maintained at a uniform temperature T_1. We use the height (d) as the length scale, the thermal diffusion time d^2/κ (where κ is the thermal diffusivity of the fluid) as a time scale. In the dimensionless coordinate system, the bottom and top plates are located at $z = 0$ and $z = 1$. With the assumption of Boussinesq approximation, the motion of the fluid is described by the Oberbeck-Boussinesq's (OB) equations [9, 18] which is a set of dimensionless equations written as

$$\nabla \cdot \mathbf{u} = 0, \qquad (14.4.53)$$

$$\mathrm{Pr}^{-1}(\partial_t \mathbf{u} + (\mathbf{u} \cdot \nabla)\mathbf{u}) = -\nabla p + RT\hat{e}_z + \Delta \mathbf{u}, \qquad (14.4.54)$$

$$\partial_t T + \mathbf{u}\nabla T = \Delta T. \qquad (14.4.55)$$

In the above equations, $\mathbf{u}^\top = u_x, u_y, u_z$ is the divergence free, dimensionless velocity vector which satisfies zero normal velocity and the non-slip condition on all solid surfaces; p is the pressure field; T is the temperature field; $\mathrm{Pr} = \nu/\kappa$ is the Prandtl number; $R = \frac{g\alpha\bar{q}d^4}{\kappa\nu k}$ is the Rayleigh number; g is the gravitational acceleration; α is the thermal expansion coefficient; κ is the thermal diffusivity; ν is the kinematic viscosity; and $\bar{q}$ is the spatially-averaged heat flux in the bottom plate.

The equations of motion (14.4.53)–(14.4.55) admit the no-motion (conduction) solution,

$$\mathbf{u}(x,y,z,t) = 0, \qquad (14.4.56)$$

$$T(x,y,z,t) = T_2 - z(T_2 - T_1), \qquad (14.4.57)$$

where T_2 is the uniform temperature in the bottom plate.

With linear stability analysis, the no-motion solution loses stability at $R = R_c \approx 1708$ via a supercritical bifurcation into time-independent, cellular motion. For $R < R_c$, the no-motion solution is globally stable.

To facilitate the analysis, we first use some standard transformation to simplify the equations. Considering the symmetry in the geometry of the experimental apparatus and the arrangement of heaters, we can ignore the motion along the heater and only consider two-dimensional flows in the system. Taking the curl of the momentum equation (14.4.54) to eliminate pressure, the resulting equations are

$$\frac{\partial v}{\partial y} + \frac{\partial w}{\partial z} = 0, \qquad (14.4.58)$$

$$\frac{\partial T}{\partial t} + v\frac{\partial T}{\partial y} + w\frac{\partial T}{\partial z} = \frac{\partial^2 T}{\partial y^2} + \frac{\partial^2 T}{\partial z^2}, \qquad (14.4.59)$$

$$\frac{1}{\mathrm{Pr}}\left(\frac{\partial \Omega}{\partial t} + v\frac{\partial \Omega}{\partial y} + w\frac{\partial \Omega}{\partial z}\right) = \nabla^2 \Omega + \mathrm{R}\frac{\partial T}{\partial y}, \qquad (14.4.60)$$

where $v(t,y,z)$ and $w(t,y,z)$ are the vertical and horizontal velocity fields, $T(t,y,z)$ is the temperature field in the fluid, and Ω is the vorticity defined by

$$\Omega = \frac{\partial w}{\partial y} - \frac{\partial v}{\partial z}. \qquad (14.4.61)$$

In the experiment, we use the shadowgraph to measure the vertical average of the variations in density field of the fluid. Shadowgraph method is a widely used visualization procedure for flow with density variations. Because the refractive index of the fluid is directly related to the fluid density, two light rays entering parallel exit the layer at a slightly different place. This changes the distribution of the light ray and generates an image which reflects the density variations in the fluid. The relationship of the shadowgraph signal with the fluid temperature field is readily derived as [24]

$$\Phi(t,y) = \frac{\delta\rho}{\rho_0} = \frac{2H}{n_0}\left(\frac{\mathrm{d}n_0}{\mathrm{d}T}\right)\int_0^l \mathrm{d}x\,(\partial_{xx} + \partial_{yy})\int_0^d \mathrm{d}z\,T(t,x,y,z). \qquad (14.4.62)$$

The control action is put into the system through perturbation of the thermal boundary conditions. In the experimental apparatus, we have twenty heaters at the bottom boundary whose heating power can be controlled individually. The boundary condition on the bottom side is therefore

$$k\frac{\partial T(t,y,z)}{\partial z}\bigg|_{z=z_0} = q(t,y). \quad (14.4.63)$$

When the control is activated, the Rayleigh number R should be maintained. Since $R = \frac{g\alpha\bar{q}d^4}{\kappa\nu k}$ is related only to the spatial mean of the heat flux, we can keep this constant but use the spatial variation around this mean as the control input.

In this problem, the control objective is to keep the system in no-motion mode or make the convection amplitude as small as possible. In earlier analysis and control experiments [12, 13], it was found that simply by using linear control one can delay the pitchfork bifurcation, but will meet a limitation past which the bifurcation point cannot be moved. Beyond this point, amplitude control methods become more attractive to accomplish the control objective. In higher Rayleigh numbers and high control gains, Hopf bifurcations also come up and make the bifurcation diagram complicated. This also makes the search for control strategies other than simple linear control law necessary.

14.4.2 Analysis and Control Design

To analyze the system, we simplify the equations with the pseudo-spectral method. Because of the problem's symmetry and expected steady roll patterns at the moderate Rayleigh numbers used in this chapter, considering the boundary conditions, we represent all the fields by Fourier series of the spatial coordinates. In the following, we use subscripts to denote the Fourier coefficients. Based upon the "generalized Lorenz truncation" [10], which retains only the two lowest vertical modes, a simple Lorenz-like model is derived. It can be confirmed that the system is linearly controllable at the bifurcation point, therefore, the linear control law can be used to delay the bifurcation to a higher Rayleigh number. However, in our former analysis, it is shown that there is a limit for moving the bifurcation point with linear control law [14]. In this case, the amplitude control theory of Section 14.3 tells us that cubic control law can be used to suppress the amplitude of convection in the linearly uncontrollable region. The control law used here assumes a linear term and a cubic term:

$$\mathbf{q} = \mathbf{G}_l\mathbf{\Phi} + \mathbf{C_u}(\mathbf{\Phi},\mathbf{\Phi},\mathbf{\Phi}), \quad (14.4.64)$$

where $\mathbf{G}_l$ is the linear gain matrix and $\mathbf{C_u}$ is a set of trilinear symmetric functions. From intuition and confirmed by the simulations, we assume that the transfer function matrix of $\mathbf{q}$ to $\mathbf{\Phi}$ is diagonally dominant. Thus, a diagonal control matrix is expected to give a good control result. The control law then becomes

$$q_k(t) = g_l \Phi_k(t) + g_{cs} \Phi_k^3(t), \qquad k = 1, \ldots, N, \qquad (14.4.65)$$

where g_l and g_{cs} are linear and cubic control gains. For simplicity, here we use the same control gain for all modes, which should be adjusted further by experiments. When implementing this control law, the pseudo-spectral coefficients q_k and Φ_k should be transformed back into spatial coordinates. A linear operation in pseudo-spectral domain corresponds to a linear operation in spatial coordinates, while a multiplication operation in pseudo-spectral domain corresponds to a convolution integral in spatial coordinates. So in spatial coordinates,

$$q(t,y) = g_l \Phi(t,y) + g_{cs} \int_0^\lambda dy_1 \int_0^\lambda dy_2\, \Phi(t, y - y_1 - y_2)\Phi(t, y_1)\Phi(t, y_2). \qquad (14.4.66)$$

The corresponding discrete form is

$$q_m = g_l \Phi_m + g_{cs} \sum_{m_1=1}^{M} \sum_{m_2=1}^{M} \Phi_{((m-m_1-m_2+1))_M+1} \Phi_{m_1} \Phi_{m_2}, \qquad m = 1, \ldots, M, \qquad (14.4.67)$$

where Φ_m and q_m are spatially sampling of $\Phi(t,y)$ and $q(t,y)$, respectively, and $((m))_M$ is used to denote (m modulo M).

With the help of Fast Fourier Transform, this control law can be implemented efficiently. However, another simpler control law may be more realistic and easier to implement:

$$q_m = g_l \Phi_m + g_{cp}\left(\Phi_m^3 - \frac{1}{M}\sum_{n=1}^{M} \Phi_n^3\right). \qquad (14.4.68)$$

The corresponding control law in pseudo-spectral space is

$$q_k = g_l \Phi_k + \frac{g_{cp}}{N^2} \sum_{k_1=1}^{N} \sum_{k_2=1}^{N} \Phi_{((k-k_1-k_2))_N} \Phi_{k_1} \Phi_{k_2}, \qquad k = 1, \ldots, N. \qquad (14.4.69)$$

Since the nonlinear terms are still in trilinear symmetric form, the former theory of amplitude control is still applicable. In the following section, we will use simulation to compare the effectiveness of these two control strategies.

14.4.3 Control Results

In our experiments, a rectangular container with a 1.6 × 8 aspect ratio filled with ethylene glycol (Pr = 198) is used. The height of the container

$d = 8.1$mm. Under the bottom of the container, twenty individual heater strips provide ohmic heating to the 0.20-mm-thick mirrored silica glass lower boundary. These heaters can be controlled individually, so that the spatial distribution of heat flux can be used to control the fluid motion. On top of the container, cooling water with constant temperature maintains an appropriate constant temperature boundary condition on the 1.0mm-thick upper boundary. Shadowgraph visualization is used to measure the patterns in the fluid and the convection amplitude. The shadowgraph image is captured by a CCD camera which sends the image signal into a computer. The computer then uses the shadowgraph signal to calculate the proper control output. These results are converted into voltage signals and applied to the heaters. The whole configuration is delineated in Figure 14.5.

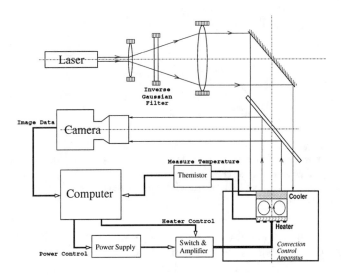

FIGURE 14.5
Configuration of the Raleigh-Bénard convection control experiment setup. The entire system includes optical measurement instruments, convection cell, and computer-based controller.

Simulation

Before applying the controller into real experiments, we solve the partial differential equations numerically to simulate the convection system. A mixed finite difference pseudo-spectral technique is employed for numerical computation [15]. The states are initially set to an arbitrary random value for a certain Rayleigh number. Then the evolution of system state variables are computed and the time history is recorded. After the convection amplitude becomes invariant, the Rayleigh number is changed to a new value to start another simulation. We use the standard deviation of the shadowgraph image to represent the convection amplitude [12, 13]. With sufficient simulation points, the bifurcation diagram is drawn and bifurcation points are found.

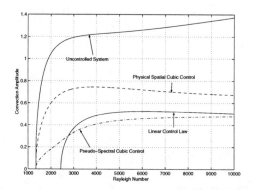

FIGURE 14.6
Bifurcation diagram of Rayleigh-Bénard system with linear control law($g_l = 2.0$), pseudo-spectral cubic control law($g_{cs} = 30.0$), spatial cubic control law($g_{cp} = 1.0$), and without control generated by numerical simulation.

Bifurcation diagrams of different control laws as well as the original uncontrolled one are delineated in Figure 14.6. It is shown that the bifurcation can be delayed by linear control law, while both cubic control laws can change the amplitude of bifurcation without moving the bifurcation point. Note that linear control law also suppresses the bifurcation amplitude.

From the simulations, we find that both cubic control algorithms are effective in controlling the bifurcation amplitude. However, the result of physical spatial control law seems not as effective as the pseudo-spectral control law. This will be investigated further in future analysis and experiments.

Experimental Results

In the experiments, we employed only the physical spatial cubic control law for ease of implementation and execution speed. We first implemented

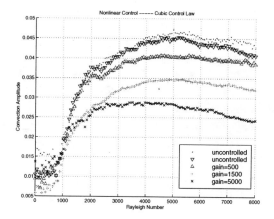

FIGURE 14.7
Experimental results of convection amplitude control with spatial cubic control law. Note that the Hopf bifurcation point remains unchanged.

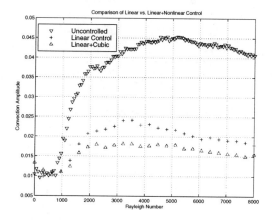

FIGURE 14.8
Experimental results using the combination of linear and nonlinear controller.

purely cubic control law to control the convection. It is shown in Figure 14.7 that the cubic control law suppresses the amplitude of the bifurcation but does not change the location of bifurcation point. When the control gain is increased, the convection amplitude is suppressed. This is in agreement with our simulation studies. In the second set of experiments, we compared the combination of linear and cubic control law with linear control law. The results (Figure 14.8) confirm that the combination of linear and cubic control law yields better results in suppressing the bifurcation amplitude than the linear control law alone.

In some recent experiments we found the occurrence of Hopf bifurcations when the linear control gain was varied. Traveling waves were observed in the experimental apparatus. Investigations are underway to suppress the Hopf bifurcations with nonlinear and differential controllers.

14.5   Conclusions

This chapter presents some of the recent advances in the area of bifurcation control, including both theory and applications. We have discussed the problem of anticontrol of Hopf bifurcations, that is, a Hopf bifurcation is created with desired location and preferable properties by an appropriate dynamic feedback control. In particular, washout-filter-aided dynamic feedback control laws are developed for creation of the intended Hopf bifurcation. As Hopf bifurcations give rise to limit cycles, anticontrol of Hopf bifurcations suggests a new approach to designing limit cycles and oscillations into a system via feedback control when such dynamical behaviors are desirable. The proposed approach also provides a new way of designing warning signals of impending collapse or catastrophe for monitoring and control purposes.

We also have dealt with the problem of controlling the amplitude of bifurcated solutions in bifurcating and/or chaotic systems. It is shown that the amplitude of the bifurcated solutions is directly related to the so-called bifurcation stability coefficients. As an application of the theory, we have investigated the problem of active control of Rayleigh-Bénard convection. It has been shown that linear control laws can delay the onset of the convection, a stationary bifurcation, and nonlinear control laws can suppress the bifurcation amplitude without relocating the bifurcation point. A combination of linear and cubic control law is considered to be most effective and flexible for this problem. For the nonlinear part, two (spatially) distributed nonlinear control laws, one in pseudo-spectral coordinates and one in physical spatial coordinates, are presented to control the amplitude of the bifurcation. The performance of the controllers are evaluated by numerically solving the controlled partial differential equations and then are applied to the experimental system. Simulation results along with experimental results demonstrate the feasibility and effectiveness of the proposed control approach. Ongoing research will focus on exploring the control of Hopf bifurcations (time-dependent motion) occurring with some control setting.

Acknowledgment

The authors would like to express their thanks to Drs. Guanrong Chen and Laurens Howle who have collaborated with the authors in many aspects of the research leading to this chapter. This research is supported in part by the Lord Foundation of North Carolina.

References

[1] E. H. Abed and J.-H. Fu, "Local feedback stabilization and bifurcation control," *Systems Control Lett.*, Part I. Hopf bifurcation, vol. 7, pp. 11-17, 1986, and Part II. Stationary bifurcation, vol. 8, pp. 467-473, 1987.

[2] E. H. Abed, H. O. Wang, and A. Tesi, "Control of bifurcations and chaos," *The Control Handbook*, W.S. Levine (Ed.), CRC Press & IEEE Press, Boca Raton, FL, 1995, pp. 951-966.

[3] J. Baillieul, S. Dahlgren, and B. Lehman, "Nonlinear control design for systems with bifurcations with applications to stabilization and control of compressors," *Proc. of IEEE Conf. on Decis. Contr.*, 1995, pp. 3063-3067.

[4] D. W. Berns, J. L. Moiola, and G. Chen, "Feedback control of limit cycle amplitudes from a frequency domain approach," *Automatica*, vol. 34, pp. 1567-1573, 1998.

[5] M. E. Brandt and G. Chen, "Bifurcation control of two nonlinear models of cardiac activity," *IEEE Trans. on Circ. Sys.*, I, vol. 44, pp. 1031-1034, 1997.

[6] H.-D. Chiang, I. Dobson, R. J. Thomas, J. S. Thorp, and L. Fekih-Ahmed, "On voltage collapse in electric power systems," *IEEE Trans. on Power Systems*, vol. 5, 1990, pp. 601-611.

[7] D. S. Chen, H. O. Wang, L. E. Howle, M. R. Gustafson, and T. Meressi, "Amplitude control of bifurcations and application to Rayleigh-Bénard Convection," *Proc. IEEE Conference on Decision and Control*, Tampa, FL, Dec. 16-18, 1998, pp. 1951-1956.

[8] M. Cibrario and J. Lévine, "Saddle-node bifurcation control with application to thermal runaway of continuous stirred tank reactors," *Proc. IEEE Conference on Decision and Control*, Brighton, England, December 1991, pp. 1551-1552.

[9] S. Chandrasekhar, *Hydrodynamic and Hydromagnetic Stability*, Dover, New York, 1981.

[10] P. C. Hohenberg and J. B. Swift, "Hexagons and rolls in periodically modulated Rayleigh-Bénard convection," *Phys. Rev. A*, vol. 35, n. 9, pp. 3855-3873, 1987.

[11] R. A. Horn and C. R. Johnson, *Matrix Analysis*, Cambridge University Press, Cambridge, 1993.

[12] L. E. Howle, "Control of Rayleigh-Bénard convection in a small aspect ratio container," *Int. J. Heat Mass Transfer.*, vol. 40, pp. 817-822, 1997.

[13] L. E. Howle, "Active control of Rayleigh-Bénard convection," *Phys. Fluids*, vol. 9, pp. 1861-1863, 1997.

[14] L. E. Howle, "Linear stability analysis of controlled Rayleigh-Bénard convection using shadowgraphic measurement," *Phys. Fluids*, vol. 9, pp. 3111-3113, 1997.

[15] L. E. Howle, "A comparison of the reduced Galerkin and pseudo-spectral methods for simulation of steady Rayleigh-Bénard convection," *Int. J. Heat Mass Transfer.*, vol. 39, pp. 2401-2407, 1996.

[16] L. N. Howard, "Nonlinear Oscillations," in F. C. Hoppensteadt (Ed.), *Nonlinear Oscillations in Biology*, Amer. Math. Society, Providence, RI, 1979, pp. 1-68.

[17] G. Iooss and D. D. Joseph, *Elementary Stability and Bifurcation Theory*, Springer, New York, 1980.

[18] D. D. Joseph, *Stability of Fluid Motions*, vol. II, Springer-Verlag.

[19] J. L. Moiola, D. W. Berns, and G. Chen, "Feedback control of limit circle amplitudes," *Proc. IEEE Conference on Decision and Control*, San Diego, 1997, pp. 1479-1485.

[20] G. Muller, *Crystals, Growth, Properties and Applications*, Springer, Berlin, 1988.

[21] J. Singer, Y. Z. Wang, and H. H. Bau, "Controlling a chaotic system," *Phys. Rev. Lett.*, vol. 66, pp. 1123-1125.

[22] J. Singer and H. H. Bau, "Active control of convection," *Phys. Fluids A*, vol. 3, No. 12, pp. 2859-2865, 1991.

[23] A. Tesi, E. H. Abed, R. Genesio, and H. O. Wang, "Harmonic balance analysis of period-doubling bifurcations with implications for control of nonlinear dynamics," *Automatica*, vol. 32, pp. 1255-1271, 1996.

[24] A. Thess and S. A. Orszag, "Surface-tension-driven Bénard convection at infinite Prandtl number," *J. Fluid Mech.*, vol. 283, pp. 201-230, 1995.

[25] H. O. Wang, E. H. Abed, and A. M. A. Hamdan, "Bifurcations, chaos and crises in voltage collapse of a model power system," *IEEE Trans. on Circ. Sys.*, I, vol. 41, pp. 294-302, 1994.

[26] H. O. Wang and E. H. Abed, "Bifurcation control of a chaotic system," *Automatica*, vol. 31, pp. 1213-1226, 1995.

[27] H. O. Wang, R. A. Adomaitis, and E. H. Abed, "Nonlinear analysis and control of rotating stall in axial flow compressors," *Proc. Amer. Control Conf.*, Baltimore, Maryland, June 1994.

[28] H. O. Wang, D. S. Chen, and L. Bushnell, "Control of bifurcations and chaos in heart rhythms," *Proc. IEEE Conference on Decision and Control*, San Diego, 1997, pp. 395-400.

[29] H. O. Wang, D. S. Chen, and G. Chen, "Bifurcation control of pathological heart rhythms," *Proc. IEEE Conf. on Contr. Appl.*, Trieste, Italy, Sept 1-4, 1998, pp. 858-862.

[30] S. Weibel, and J. Baillieul, "Oscillatory control of bifurcations in rotating chains," *Proc. Amer. Contr. Conf.*, Albuquerque, NM, June 1997, pp. 2713-2718.

15

Delay Feedback Control of Cardiac Activity Models

Michael E. Brandt[1] and Guanrong Chen[2]

[1]Neurosignal Analysis Laboratory
University of Texas-Houston Medical School
Houston, Texas 77030 USA
mbrandt@ped1.med.uth.tmc.edu

[2]Department of Electrical and Computer Engineering
University of Houston, Houston, Texas 77204 USA
gchen@uh.edu

Abstract

This chapter describes methods based on time-delay feedback (TDF) for controlling several nonlinear models of cardiac electrophysiologic activity in the presence of additive Gaussian noise. We discuss use of self-tuning controllers that incorporate a feedback reference signal, which need not be the target, and a linear autoregressive formulation for the control gain. Such controllers are effective at stabilizing three different simulated maps to a variety of fixed-point or period-2 orbits. We contrast our approach with the OGY method which has been used to control some chaotic biological processes as well as some nonchaotic, stochastic ones.

15.1 Introduction and Background

Throughout recorded history, nonlinear phenomena displaying complex, possibly chaotic behavior have long been viewed with fear. In recent times however, there has been a reversal of attitude with regard to these phenomena. They are now generally thought of as "... dynamics freed from the shackles of order and predictability." Furthermore, they "permit systems to randomly explore their every dynamical possibility. It is exciting variety, richness of choice, a cornucopia of opportunities" [11]. It is now known that under the proper conditions, such phenomena can be stabilized and/or controlled.

Within the context of biomedical engineering, healthy dynamics have been traditionally regarded as regular and predictable, whereas disease states such as fatal arrhythmias, aging, and drug toxicity arise from disorder and possibly even chaos. In the last decade however, several studies have shown the opposite: the complex variability of healthy dynamics in a variety of physiologic systems has features reminiscent of nonlinear dynamical systems and a wide class of disease processes (including drug toxicities and aging) may actually decrease (but not eliminate) the amount of complexity in physiologic systems [15]. An example of a system in which healthy dynamics may be chaotic in the technical sense of the word is the timing of the normal heartbeat. In contrast to the common belief that healthy heartbeats are seemingly regular, if one carefully examines the interbeat interval variations in healthy individuals, it is observed that the normal heart rate apparently fluctuates in an erratic fashion even at rest, which turns out to be consistent with deterministic chaos. These notions have important implications both for basic mechanisms in physiology as well as for clinical monitoring, including the problem of anticipating sudden cardiac death [15]. The term "dynamical disease" has been used to describe the relationship between complex and orderly behaviors in clinical medicine [23].

15.1.1 Complexity and "Dynamical Disease"

Examples of dynamical diseases include cell counts in hematological disorders, stimulant drug-induced abnormalities in temporal patterns of brain enzyme and receptor behavior, the resting record in a variety of signal sensitive biological systems following desensitization, epilepsy, hormone release patterns correlated with the spontaneous mutation of a neuroendocrine cell to a neoplastic tumor, the prediction of immunological rejection of heart transplants, the electroencephalographic behavior of the human brain in the presence of neurodegenerative disorders, physiological changes with aging, imminent ventricular fibrillation in humans, and interbeat interval patterns in a variety of cardiac disorders.

15.1.2 Complex Dynamics and the Human Heart

Many human heart rhythms can be classified, described and analyzed through the theories and methodologies of nonlinear dynamics. It has been demonstrated that considering the human heart as an excitable medium driven by limit-cycle oscillators (e.g., the heart's natural pacemaker) has enabled cardiologists to gain important insights into the prevention and control of deadly arrhythmias [13]. One reason for this is that a strategy for the clinical application of nonlinear dynamics is based on the hypothesis that cardiac arrhythmias may be associated with complex dynamics in comparatively simple mathematical models. For example, laboratory experiments have demonstrated complex nonlinear dynamics (e.g., chaos) in the activity of portions of animal hearts. Because it is likely that some cardiac arrhythmias can be modeled by deterministic equations for wave generation and propagation in the heart, understanding the dynamics may well yield fruitful new proposals for forestalling or controlling dangerous arrhythmias through the application of imposed stimuli [13].

15.1.3 Controlling Complex Cardiac Rhythms

In recent years, much attention has been given to the notion of controlling nonlinear dynamical systems [4]. In cardiology, devices such as pacemakers, defibrillators, and antitachycardia devices have been developed mainly by engineers and cardiologists using empirical methods. Though these devices have proven themselves highly reliable, further research in control theory will likely lead to new breakthroughs [13].

The first attempt at controlling complex cardiac rhythms can be traced back to Garfinkel et al. [12], where the goal was to stabilize drug-induced cardiac arrhythmia in the *in vitro* rabbit ventricle. It was shown that administering electrical stimuli to the septum at irregular times determined by the OGY chaos control method [18] could convert the arrhythmia to periodic beating. Despite some success using this method, there remain difficulties in applying the OGY technique in cardiac control, particularly when systemwide parameters are not available for manipulation. Recently, this difficulty was addressed and somewhat resolved via a modified OGY approach [10]. Some other approaches are discussed in the chapter by Wang and Chen in this book.

Chaos control has found some potential applications in some other biomedical areas as well. For instance, the hypothesis that chaotic dynamics may serve as "the essential ground state for the neural-perceptual apparatus" [20] has led to the study of nonlinear dynamical control and anticontrol in neurophysiological systems [21].

15.2 TDF Control of a Quadratic Map Model of Cardiac Chaos

A quadratic map has been suggested by Glass and Zeng [14] as a model of irregular dynamics of the heart such as that which occurs in ventricular fibrillation. The map is given as

$$x_{i+1} = f(\mu, x_i) = \mu - x_i^2 + \xi_i, \qquad (15.2.1)$$

where x_i represents the cardiac interbeat interval, μ is a parameter, and ξ_i models both intrinsic and measurement noise distributed as $N(0, \sigma^2)$.

It was demonstrated in [14] that introducing a flattening of the chaotic quadratic map leads to a simple control method. We showed in [1] that self-tuning feedback (STF) control can also be used to stabilize this map in its chaotic mode (for $\mu = 2$) to a targeted periodic orbit. This can be done by adding the TDF controller $g_i(x_{i-d} - \bar{x}_i)$ to (15.2.1) where $\bar{x}_i$ is a target trajectory of period-$(d+1)$ satisfying the original map (15.2.1). The gain g_i is of the form

$$g_i = C + \left(\frac{-\tilde{f}(\mu, x_i) + x_{i-d}}{x_{i-d} - \bar{x}_i} \right), \qquad (15.2.2)$$

where $\tilde{f}(\mu, x_i)$ is an estimate of x_{i+1} based on the value of x_i, and C is a constant in the range $(-2 < C < 0)$ determined by stability analysis. This is a type of anticipatory controller whose success depends on the amount of estimation error. Even a moderate amount of error (up to 5% as we showed in [1]) allows successful control. In practice, however, it may not be possible to perform adequate one-step-ahead prediction of the controlled map prior to actually effecting the control strategy. A better approach may be to use TDF control where only previous system state values are explicitly required. This is discussed next.

15.2.1 The TDF Control Method

The controlled form of the map (15.2.1) is

$$x_{i+1} = \mu - x_i^2 + \xi_i + g_i\, e_i, \qquad (15.2.3)$$

where g_i is a self-tuning control gain to be automatically determined at each step, and $e_i = x_i - v_{ref}$ is the tracking error with respect to an auxiliary constant amplitude reference signal, v_{ref}. Note that v_{ref} takes the place of the target trajectory $\bar{x}_i$ since the latter may not be available.

A traditional self-tuning feedback (STF) controller is one in which the control gain is modified by the feedback error on each iteration. The objective of the design is to find a simple and implementable g_i to achieve the goal of automatic control, i.e.,

$$x_i \to \bar{x} \quad \text{as} \quad i \to \infty,$$

where $\bar{x}$ is a desired target state which is usually (but not necessarily) an unstable fixed point (UFP), x^*, of the original map, perhaps of large period. When the control objective is finally realized, $x_i - \bar{x}$ will have been minimized (to a small, ideally zero, constant) at the time iteration halts. A simple design for g_i is one that uses a linear combination of the supplied reference signal and a few available previous system states. It is given by the following linear relation:

$$g_i = k v_{ref} + \sum_{j=0}^{n} a_j x_{i-j}, \qquad (15.2.4)$$

where n is a small integer, independent of the system dimension, and the k and a_j are to be determined for stable tracking to the target $\bar{x}$.

The mathematical justification is similar to the one we used to control the Hénon map in [3]. In the simplest form of the controller in which $n = 0$ (also we let $\xi_i = 0$ in this analysis), Equation (15.2.3) becomes

$$x_{i+1} = \alpha + \beta x_i^2 + \eta x_i, \qquad (15.2.5)$$

where

$$\alpha = \mu - k v_{ref}^2, \quad \beta = a_0 - 1, \quad \text{and} \quad \eta = k v_{ref} - a_0 v_{ref}.$$

If we wish to control the orbit x_i specifically to the target x^* of the map (15.2.1), it may turn out to be very difficult in practice to use the UFP x^* itself as a term in the feedback controller due to its instability and sensitivity to noise. Instead we make use of the auxiliary reference signal v_{ref} to achieve the goal $x_i \to \bar{x}$ as $i \to \infty$.

Let's assume that the control objective has been successfully achieved. It then follows from (15.2.5) that

$$\beta \bar{x}^2 + (\eta - 1) \bar{x} + \alpha = 0. \qquad (15.2.6)$$

This indicates that the target $\bar{x}$, originally a UFP of the uncontrolled system (15.2.1), is now a stable fixed point of the controlled system (15.2.6) whose solutions are given by

$$\bar{x} = \frac{(1 - \eta) \pm \sqrt{(\eta - 1)^2 - 4\alpha \beta}}{2\beta}. \qquad (15.2.7)$$

Necessary conditions for the existence of two real, finite fixed points of (15.2.6) are

$$\beta \neq 0 \quad \text{or} \quad a_0 \neq 1, \qquad (15.2.8)$$

and $((\eta - 1)^2 - 4\alpha\beta) > 0$, or

$$\left((v_{ref}(k - a_0) - 1\right)^2 - 4\left(\mu - kv_{ref}^2\right)(a_0 - 1)) > 0. \tag{15.2.9}$$

This also provides bounds on v_{ref} as a function of k, a_0, and the parameter μ to ensure stable feedback control. For example, for $\mu = 2$ (the quadratic map in its chaotic mode) and $\xi_i = 0$, if we set $k = a_0 = 0.4$, then (15.2.9) reduces to the inequality $|v_{ref}| < 2.458$. Notice that this range is greater than that produced by the original map $x_{i+1} = 2 - x_i{}^2$ (-2 to $+2$). Thus we can confine our selection of v_{ref} to the range of the data.

If we now insist that $\bar{x} = x^*$ as $i \to \infty$, then we must simultaneously satisfy

$$x^{*2} + x^* - \mu = 0, \quad \text{and} \tag{15.2.10}$$

$$\beta x^{*2} + (\eta - 1)x^* + \alpha = 0, \tag{15.2.11}$$

in which x^* is an unstable solution of (15.2.10) but a stable solution of (15.2.11) as mentioned above. For the quadratic map (15.2.10),

$$x^* = \frac{-1 + \sqrt{1 + 4\mu}}{2}. \tag{15.2.12}$$

Hence, to control the system trajectory to this target, we substitute it into (15.2.11) and obtain

$$(\alpha + \beta)\left(\frac{-1 + \sqrt{1 + 4\mu}}{2}\right)^2 + (\eta - 1)\left(\frac{-1 + \sqrt{1 + 4\mu}}{2}\right) = 0. \tag{15.2.13}$$

This simply indicates that v_{ref} must be chosen to be x^* in this case (substitute x^* for v_{ref} in (15.2.13) to recover (15.2.10)), which is consistent with the classical feedback control theory where the target is used in the feedback.

To complete the analysis, we determine a sufficient condition for the control of x_i to the target x^* for the case where the reference signal $v_{ref} = x^*$. Substituting $v_{ref} = x^*$ in (15.2.5) gives

$$x_{i+1} = \mu - x_i^2 + (kx^* + a_0 x_i)(x_i - x^*).$$

Note that x^* satisfies the original system, i.e.,

$$x^* = \mu - x^{*2}.$$

A subtraction of these last two equations gives

$$e_{i+1} = (a_0 - 1)e_i^2 + (k + a_0 - 2)(x^* e_i), \tag{15.2.14}$$

where $e_{i+1} = x_{i+1} - x^*$. For stability of the controlled system it is sufficient to require the eigenvalue of the Jacobian of the above autonomous error dynamical system be strictly less than one:

$$|\lambda| = |(k + a_0 - 2) x^*| < 1. \quad (15.2.15)$$

For the model under inspection, both theory and simulation confirm that condition (15.2.15) guarantees $e_i \to 0$; therefore, $x_i \to x^*$, as $i \to \infty$.

The above analysis indicates that we can control the system orbit either to the UFP, or we can make use of (15.2.8) or (15.2.9) to specify one of a range of possible values for v_{ref}, and thus tune the control outcome to one of a set of $\bar{x}$ depending on the specific values chosen for the parameters v_{ref}, k, and a_0.

Note that while (15.2.8) and (15.2.9) do not require explicit knowledge of x^*, they do require that we know the values of the system parameters (in this case, μ). However, it turns out in practice that we need only make some observations of the system over a short time window to determine the data range to come up with reasonable choices for the gain parameters.

15.2.2 Simulation Results

Here we present examples of control under four different modes of the map (15.2.1), namely, 1) nonchaotic, nonstochastic, 2) nonchaotic, stochastic, 3) chaotic, nonstochastic, and 4) chaotic, stochastic. We give examples for these cases since they may apply to the particular biological processes under discussion in this chapter. The gain parameters are chosen to be the same for all examples discussed here, and we specify that $a_0 = k = C$ (for $n = 0$). With this condition, (15.2.9) becomes

$$|v_{ref}| < \sqrt{\frac{1 - 4\mu(C - 1)}{4(C - C^2)}}. \quad (15.2.16)$$

This relation gives the stable range of values for v_{ref} as a function of the system (μ) and gain (C) parameters. Solving for C, (15.2.16) yields

$$0 < C < \frac{(v^2_{ref} + \mu) \pm 4\sqrt{(v^2_{ref} + \mu)^2 - v^2_{ref} - 4v^2_{ref}\,\mu}}{2v^2_{ref}}. \quad (15.2.17)$$

For example, for $\mu = 2$ and $v_{ref} = \pm 2$ (covering the data range), then $0 < C < 0.75$. For $v_{ref} = 1$ (a UFP), then $0 < C < 1.5$, and for $v_{ref} = 0$, then $0 < C < \infty$.

Case 1. Nonchaotic, nonstochastic: Here we use (15.2.4) to attempt to control (15.2.1) with $\xi_i = 0$ and the nonchaotic case of $\mu = 1.3$ (period-4). Figure 15.1 shows control solutions ($\bar{x}_i$ vs. v_{ref}) for values of $C \in (0.1, 0.9)$. Figure 15.2(a) is an example of period-1 control for $C = 0.4$, and Figure 15.2(b) shows period-2 control for $C = 0.2$ (control is initiated at $i = 500$). Note that v_{ref} is set to 1 in Figures 15.2 and 15.4.

Case 2. Nonchaotic, stochastic: For this example we again select $\mu = 1.3$ and $\sigma_\xi \approx 0.03$ (ξ_i nonzero). Figure 15.2(c) is an example of noisy period-1 control for $C = 0.4$, and Figure 15.2(d) noisy period-2 control for $C = 0.2$ (both with $v_{ref} = 1$).

Case 3. Chaotic, nonstochastic: For this case we specify $\mu = 2$ (chaotic) and $\xi_i = 0$. Figure 15.3 shows control results ($\bar{x}_i$ vs. v_{ref}) for values of $C \in (0.1, 0.9)$. Figure 15.4(a) is an example of period-1 control for $C = 0.8$, and Figure 15.4(b) shows period-2 control for $C = 0.4$.

Case 4. Chaotic, stochastic: Here, $\mu = 1.9$ and $\sigma_\xi \approx 0.03$. Figure 15.4(c) is an example of noisy period-1 control for $C = 0.8$, and Figure 15.4(d) noisy period-2 control for $C = 0.4$.

15.3 TDF Control of a Circle-Map Cardiac Model

In [2] we demonstrated STF control of a model of human modulated parasystole – a condition characterized by competition between the normal sinus pacemaker of the heart and an abnormal ventricular (ectopic) focus [8]. The model is given by the following one-dimensional, two-parameter nonlinear circle map [9, 24]:

$$\phi_{i+1} = F(\phi_i, p) = \begin{cases} \phi_i + t_s/t_e + \xi_i \pmod 1 \\ \quad \text{if } 0 \leq \phi_i < t_s/t_e - \theta/t_e, \\ \phi_i + t_s/t_e + f(\phi_i, a) + \xi_i \pmod 1 \\ \quad \text{if } t_s/t_e - \theta/t_e \leq \phi_i < 1, \end{cases} \qquad (15.3.18)$$

where $i = 0, 1, 2, \cdots$, and p is a list of parameters as follows:

t_e — period of a ventricular ectopic pacemaker
t_s — period of the normal sinus pacemaker
θ — refractory period associated with the normal sinus beats
ϕ_i — phase of the ith sinus beat in the ectopic cycle

ξ_i is random noise distributed as $N(0, \sigma^2)$, and $f(\cdot, \cdot)$ is an experimentally determined nonlinear function: for the embryonic chicken heart cell aggregate it has the form

$$f(\phi, a) = -C \exp\{-(\phi - \phi_{\max})^2/\sigma^2\} - S(\phi - 1)\phi^m/(\phi^m + B^m), \qquad (15.3.19)$$

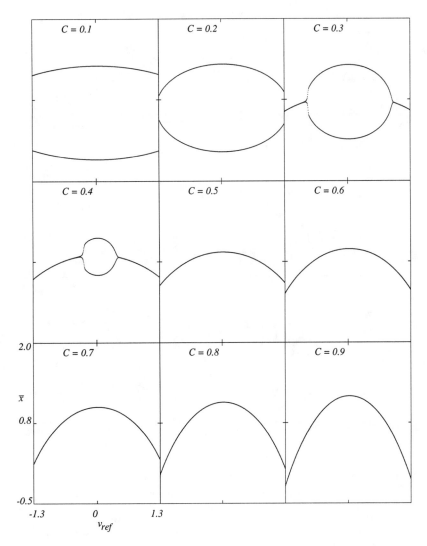

FIGURE 15.1
Control solutions for the nonchaotic quadratic map ($\mu = 1.3$) for the TDF gain parameter $C \in (0.1, 0.9)$ as a function of v_{ref}, showing regions of period-1 and period-2 control.

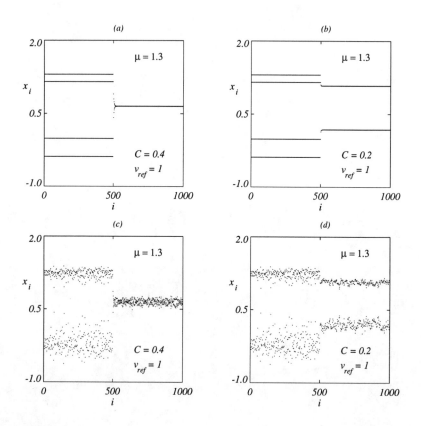

FIGURE 15.2
Time series examples for period-1 control (a and c) and period-2 control (b and d) of the nonchaotic quadratic map ($\mu = 1.3$). Top row (a and b) is without noise; bottom row (c and d) is with noise ($\sigma_\xi \approx 0.03$). TDF control is initiated at time $i = 500$ with parameters given in the figure.

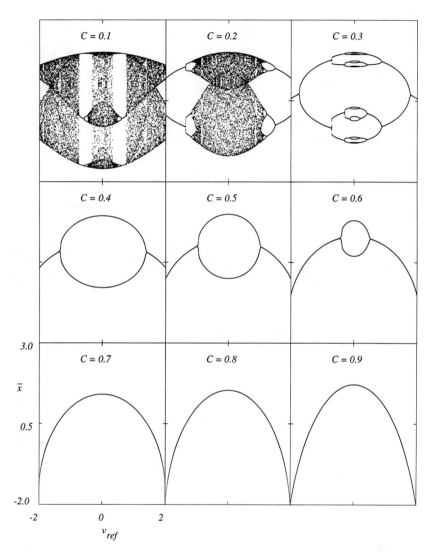

FIGURE 15.3
Control solutions for the chaotic quadratic map ($\mu = 2$) for the TDF gain parameter $C \in (0.1, 0.9)$ as a function of v_{ref}, showing regions of period-1 to period-n control, in addition to chaotic regions for $C = 0.1, \ 0.2$.

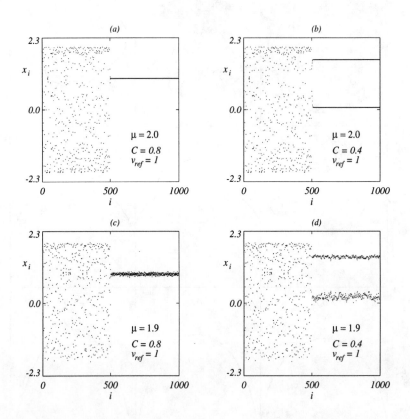

FIGURE 15.4
Time series examples for period-1 control (a and c) and period-2 control (b and d) of the chaotic quadratic map ($\mu = 2$). Top row (a and b) is without noise; bottom row (c and d) is with noise ($\sigma_\xi \approx 0.03$). TDF control is initiated at time $i = 500$ with parameters given in the figure.

in which a and t_s/t_e are parameters, and

$$\begin{aligned}
C &= 0.125 + 1.25a \\
\phi_{max} &= 0.34 + 0.12 \times 2^{-50a} \\
\sigma^2 &= 0.04 \times 2^{-50a} \\
B &= 0.34 + 0.48 \times 2^{-50a} \\
m &= 1.875 \times 2^{50a} \\
S &= 0.92 \\
\theta/t_s &= 0.4
\end{aligned}$$

Depending on the settings of the two parameters a and t_s/t_e, the dynamical behavior of ϕ_i in the circle map (15.3.18) can be periodic, quasiperiodic or chaotic [17]. It has a stable fixed point for $t_s/t_e \in [0.514156, 1.0)$.

In [2] we demonstrated that an STF controller of the form

$$g_i = K_i e_i^{(-d)} = K_i(\phi_{i-d} - \bar{\phi}_i), \qquad i \geq d \geq 0, \tag{15.3.20}$$

where d is an integer delay and $\bar{\phi}_i$ is the period-$(d+1)$ target trajectory, can be used to stabilize the system given a suitable choice for K_i determined via stability analysis. We designed a K_i similar in form to (15.2.2) that depends on predicted estimates of ϕ_{i+1}.

As was stated previously in regard to the quadratic map, it may not be possible to obtain reasonable one-step-ahead predictions while at the same time applying the control. We therefore applied TDF control to this map as well. The controlled map is

$$\phi_{i+1} = F(\phi_i, p) + g_i e_i, \tag{15.3.21}$$

where $e_i = \phi_i - \phi_{ref}$ and ϕ_{ref} is an external reference signal in the range 0 to 1. The gain is given by

$$g_i = k\,\phi_{ref} + \sum_{j=0}^{n} a_j\,\phi_{i-j}. \tag{15.3.22}$$

As in the case of the quadratic map we examined using the specific gain

$$g_i = C(\phi_{ref} + \phi_i), \tag{15.3.23}$$

for $k = a_0 = C$ with $-1 < C < 0$. We operate the circle map in its chaotic regime with the parameters $a = 0.072$ and $t_s/t_e = 0.465$.

Figure 15.5 shows the controlled trajectories $\bar{\phi}_i$ of (15.3.21) for values of C from -0.1 to -0.9 and $\xi_i = 0$. It is seen that in addition to some regions of chaos, there are many combinations of values of C and ϕ_{ref} which stabilize (15.3.18) to period-1, -2, or -3 trajectories.

Figure 15.6(a) is an example of period-1 control with $\xi_i = 0$, and Figure 15.6(c) shows "noisy" period-1 control with ξ_i nonzero and $\sigma_\xi \approx 0.05$ (in both cases $C = -0.6$ and $\phi_{ref} = 0.8$). The controller is turned on at

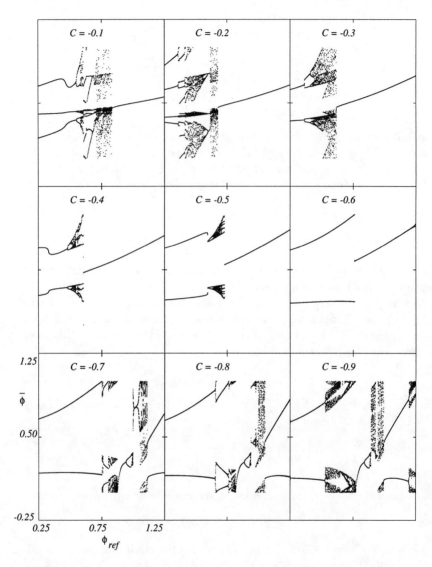

FIGURE 15.5
Control solutions for the chaotic circle map cardiac model for the TDF gain parameter $C \in (-0.1, -0.9)$ as a function of ϕ_{ref}, showing regions of period-1 to period-n control, in addition to some chaotic regions.

time $i = 500$. Likewise, Figure 15.6(b) is an example of period-2 control with $\xi_i = 0$, and Figure 15.6(d) shows "noisy" period-2 control with ξ_i nonzero and $\sigma_\xi \approx 0.05$ (again, in both cases, $C = -0.6$ and $\phi_{ref} = 0.4$).

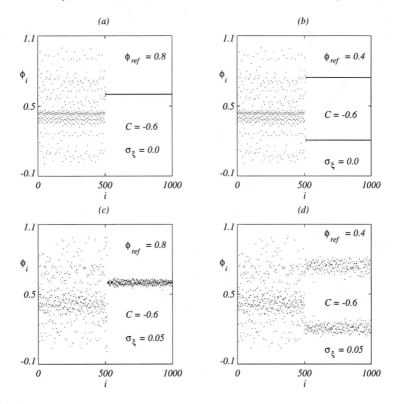

FIGURE 15.6
Time series examples for period-1 control (a and c) and period-2 control (b and d) of the chaotic circle map cardiac model. Top row (a and b) is without noise; bottom row (c and d) is with noise ($\sigma_\xi \approx 0.05$). TDF control is initiated at time $i = 500$ with parameters given in the figure.

15.4 Linear TDF Control of a Cardiac Conduction Model

In a recent study [7] it was demonstrated that the OGY method could be used to stabilize a pathological period-2 rhythm (alternans) in an atrioventricular (AV) nodal conduction model of the heart. The model used was the following nonlinear discrete-time map [22]:

$$A_{i+1} = f(A_i, H_i) = A_{min} + R_{i+1} + \beta_i \exp(-H_i/\tau_{rec}), \qquad (15.4.24)$$

where A_i represents the time interval between cardiac impulse excitation of the low interatrial septum and the bundle of His (the atrial-His interval), with $A_{min} = 33$ ms and $\tau_{rec} = 70$ ms, H is the interval between bundle of His activation and the next atrial activation (the nodal recovery time), with initial interval H_0, and

$$R_0 = \gamma \exp(-H_0/\tau_{fat}),$$
$$R_{i+1} = S_i \exp[-(A_i + H_i)/\tau_{fat}] + \gamma \exp(-H_i/\tau_{fat}), \quad \text{and}$$
$$\beta_i = \begin{cases} 201 \text{ ms} - 0.7 A_i, & \text{for } A_i < 130 \text{ ms} \\ 500 \text{ ms} - 3.0 A_i, & \text{for } A_i \geq 130 \text{ ms}, \end{cases}$$

in which $\gamma = 0.3$ ms, and $\tau_{fat} = 30$ s.

When rabbit hearts are electrically stimulated near the sinoatrial node with a fixed time interval following Bundle of His activation, the A_i may alternate in time characteristic of reentrant tachycardia [22]. This can be simulated in the model by setting the H_i to a constant interval < 57 ms. Doing so, A_i starts out as a period-one rhythm, but bifurcates into a period-two rhythm (alternans) at about $i = 200$ eventually alternating between values of 113 and 148 ms; R_i eventually reaches the steady-state value ≈ 51.

The controlled form of (15.4.24) is

$$A_{i+1} = f(A_i, H_i) + u_i, \qquad (15.4.25)$$

where u_i is a TDF control input to be automatically determined at each cycle i. The control term u_i actually represents a timing interval (with respect to time $f(.)$) when an electrical pulse stimulus of empirically-determined amplitude would be delivered to the cardiac preparation during the ith cycle. Experimental implementation details concerning dynamical control of the rabbit heart can be found in the studies by Garfinkel et al. [12], Sun et al. [22], and a recent study by Hall et al. [16]. The latter used a simple variant of the OGY method to suppress induced alternans in five rabbit heart preparations with good reported success.

A very simple design for u_i is

$$u_i = k\hat{A}_{i+1} \qquad (15.4.26)$$

where k is constant and $\hat{A}_{i+1}$ is an estimated value of A_{i+1}. From (15.4.24), autoregressive modeling of the A_i leads to the approximation

$$A_{i+1} = f(A_i, H_i) \approx A_{i-1} \qquad (15.4.27)$$

a short time after the occurrence of the period-two bifurcation. Equation (15.4.26) then becomes

$$u_i = kA_{i-1} \tag{15.4.28}$$

and in the controlled model the additive control term depends only on previous values of the system variable. To re-establish period-1 control of map (15.4.24), we need only select the constant parameter k.

In a real heart, the variable H_i may be more readily accessible than A_i. Therefore, we explored use of a related controller given by

$$u_i = kH_{i-1}. \tag{15.4.29}$$

This turns out in practice to be more stable and useful than the controller (15.4.28). We use the linear TDF controller (15.4.29) to stabilize (15.4.24) with H_i chosen as the constant interval $\bar{H} = 45$ ms.

Figure 15.7 is a plot of $(-2 < k < 2)$ vs. $\bar{A}$ (the final target trajectory). It can be observed that period-1 control can be effected by varying the value of k from slightly less than zero to large negative values ($k < -5$) without instability setting in. Use of $k < 0$ for period-1 control means that a perturbation proportional to a negative time quantity is added to the system. Physiologically, this is important since the interval A (or H) must be shortened rather than lengthened in a real cardiac system. For $k = -0.3$, the final period-1 control to $\bar{A} \approx 129.6$ is achieved as shown in Figure 15.8(a). Figure 15.8(b) shows an example of control of (15.4.24) with $k = -0.3$, and $H_i = \bar{H} + \xi_i$, where the Gaussian noise ξ_i has a variance $\sigma_\xi \approx 1$ ms. Thus, use of (15.4.29) turns out to be a much simpler control method for this conduction model than the OGY approach.

To implement this controller practically, we note that Equations (15.4.25) and (15.4.29) can be equivalently expressed as

$$A_i - kH_{i-2} = f(A_{i-1}, H_{i-1}), \quad \text{for } k < 0 \tag{15.4.30}$$

with a one cycle delay. Thus, during each cycle, the preparation is stimulated at time $T_i = A_i - kH_{i-2}$ under the conditions that $k < 0$ (stimulate after measurement of A_i) and that $T_i < A_{i+1}$. If the latter of these two conditions are not met, then do not stimulate during the current cycle i.

Sun et al. [22] further showed that when the rabbit hearts were stimulated at a fixed interstimulus interval S, a more complex A_i time series characteristic of fibrillation may be produced depending on the specific value of S chosen (see Figure 15.6 in [22]). A constant interstimulus interval is simulated in the model (15.2.1) by constraining $S = H_i + A_i$ via the substitution:

$$H_i = S - A_i \tag{15.4.31}$$

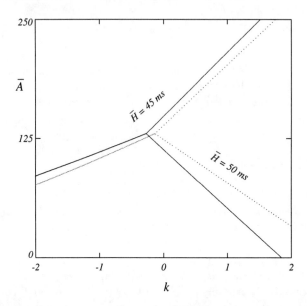

FIGURE 15.7
Control solutions for the cardiac conduction model and the linear TDF gain parameter $k \in (-2, 2)$ for $\bar{H} = 45$ (solid) and 50 (dotted) ms showing regions of period-1 and period-2 control.

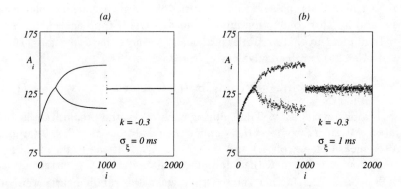

FIGURE 15.8
Time series examples for nonstochastic period-1 control (a) and stochastic ($\sigma_\xi \approx 1$ ms) period-2 control (b) of the cardiac conduction model. TDF control is initiated at time $i = 1000$ with parameters given in the figure.

with S constant. Results of LTDF control of this variation of the model are described in further detail in [5].

15.5 Discussion

In summary, we presented an effective method based on nonlinear TDF control for stabilizing a quadratic map model of cardiac fibrillation and a chaotic circle map model of modulated parasystole. We also showed that a simple linear TDF controller can be used to stabilize complex pathological rhythms in an atrioventricular cardiac conduction model. Examples of control of both stochastic and nonstochastic versions of each map were also presented.

Pyragas and colleagues [19] were among the first researchers to achieve success in using delayed output feedback mainly to control nonlinear continuous time chaotic systems. However, systematic procedures for determining the feedback gain have been problematical. There are three key features of our approach:

> By not explicitly using a UFP of the system in the design of the controller (in some cases using instead an easily tuned auxiliary reference signal), a range of target trajectories can be stabilized using our method including UFP of the original system.

> The method can easily be used to control nonstochastic as well as stochastic forms of the chaotic and nonchaotic maps.

> Some guidelines and procedures for determining the feedback gains can be derived under our framework based on the Lyapunov first and second methods.

Finally, we point out that the TDF control method described here relies on just a few easily tuned parameters which can be selected from short term observation of the system under study. The associated controller is simpler to implement than the OGY method, for example, which requires that the locations of saddle-type UFPs be found using the method of delay-coordinate embedding. There is no guarantee that finding these UFPs will require less samples of the process than the present method, or even that its success is ensured for arbitrary nonlinear systems such as those of higher dimension, or those with nonsaddle-type UFPs.

References

[1] M. E. Brandt and G. Chen, "Controlling the dynamical behavior of a circle map model of the human heart," *Biol. Cybern.*, vol. 74, pp. 1-8, 1996.

[2] M. E. Brandt and G. Chen, "Feedback control of a quadratic map model of cardiac chaos," *Int. J. of Bifur. Chaos*, vol. 6, pp. 715-723, 1996.

[3] M. E. Brandt, A. Ademoglu, D. Lai, and G. Chen, "Autoregressive self-tuning feedback control of the Henon map," *Phys. Rev. E*, vol. 54, pp. 6201-6206, 1996.

[4] M. E. Brandt, H. T. Shih, and G. Chen, "Linear time-delay feedback control of a pathological rhythm in a cardiac conduction model," *Phys. Rev. E*, vol. 56, pp. R1334-R1337, 1997.

[5] M. E. Brandt and G. Chen, "Time-delay feedback control of complex pathological rhythms in an atrioventricular conduction model," preprint, 1999.

[6] G. Chen and X. Dong, *From Chaos to Order: Methodologies, Perspectives and Applications*, World Scientific Pub. Co., Singapore, 1998.

[7] D. J. Christini and J. J. Collins, "Using chaos control to suppress a pathological nonchaotic rhythm in a cardiac model," *Phys. Rev. E*, vol. 53, pp. R49-52, 1996.

[8] M. Courtemanche, L. Glass, M. D. Rosengarten, and A. L. Goldberger, "Beyond pure parasystole: promises and problems in modeling complex arrythmias," *Amer. J. of Physiol.*, vol. 257, pp. H693-706, 1989.

[9] M. Courtemanche, L. Glass, J. Belair, D. Scagliotti, and D. Gordon, "A circle map in a human heart," *Physica D*, vol. 40, pp. 299-310, 1989.

[10] W. L. Ditto, "Applications of chaos in biology and medicine," in D. E. Herbert (Ed.), *Chaos and the Changing Nature of Science and Medicine: An Introduction*, AIP Press, New York, 1996, pp. 175-201.

[11] J. Ford, "What is chaos, that we should be mindful of it?" in P. Davies (Ed.), *The New Physics*, Cambridge University Press, New York, 1989, pp. 348-372.

[12] A. Garfinkel, M. L. Spano, W. L. Ditto, and J. N. Weiss, "Controlling cardiac chaos," *Science*, vol. 257, pp. 1230-1235, 1992.

[13] L. Glass, "Dynamics of cardiac arrhythmias," *Physics Today*, August 1996, pp. 40-45.

[14] L. Glass and W. Zeng, "Bifurcations in flat-topped maps and the control of cardiac chaos," *Int. J. of Bifur. Chaos*, vol. 4, pp. 1061-1067, 1994.

[15] A. L. Goldberger, "Applications of chaos to physiology and medicine," in J. H. Kim and J. Stringer (Eds.), *Applied Chaos*, Academic Press, New York, 1992, pp. 321-331.

[16] K. Hall, D. J. Christini, M. Tremblay, J. J. Collins, L. Glass, and J. Billette, "Dynamic control of cardiac alternans," *Phys. Rev. Lett.*, vol. 78, pp. 4518-4521, 1997.

[17] Y. He, R. Chi, and N. He, "Lyapunov exponents of the circle map in human hearts," *Phys. Lett. A*, vol. 170, pp. 29-32, 1992.

[18] E. Ott, C. Grebogi, and J. A. Yorke, "Controlling chaos," *Phy. Rev. Lett.*, vol. 64, pp. 1196-1199, 1990.

[19] K. Pyragas, "Continuous control of chaos by self-controlling feedback," *Phys. Lett. A*, vol. 170, pp. 421-428, 1992.

[20] C. A. Skarda and W. J. Freeman, "How brains make chaos in order to make sense of the world," *Behav. Brain Sci.*, vol. 10, pp. 161-195, 1987.

[21] S. J. Schiff, K. Jerger, D. H. Duong, T. Chang, M. L. Spano, and W. L. Ditto, "Controlling chaos in the brain," *Nature*, vol. 370, pp. 615-620, 1994.

[22] J. Sun, F. Amellal, L. Glass, and J. Billette, "Alternans and period-doubling bifurcations in atrioventricular nodal conduction," *J. Theor. Biol.*, vol. 173, pp. 79-91, 1995.

[23] W. Yang, M. Ding, A. J. Mandell, and E. Ott, "Preserving chaos: Control strategies to preserve complex dynamics with potential relevance to biological disorders," *Phys. Rev. E.*, vol. 51, pp. 102-110, 1995.

[24] W. Z. Zeng, M. Courtemanche, L. Sehn, A. Shrier, and L. Glass, "Theoretical computation of phase locking in embryonic atrial heart cell aggregates," *J. of Theor. Biol.*, vol. 145, pp. 225-244, 1990.

16

Bifurcation Stabilization with Applications in Jet Engine Control

Guoxiang Gu[1] and **Andrew Sparks**[2]

[1]Department of Electrical and Computer Engineering
Louisiana State University
Baton Rouge, LA 70803-5901 USA
ggu@ee.lsu.edu
[2]Flight Dynamics Directorate

Wright Laboratory
Wright-Patterson Air Force Base
Dayton, OH 45433-7531 USA

Abstract

Local output feedback stabilization with smooth nonlinear controllers is studied for parameterized nonlinear systems of which the linearized system possesses multiple pairs of imaginary eigenvalues, and the bifurcated solution is unstable at the critical value of the parameter. Necessary and sufficient conditions are sought for stabilization of such nonlinear bifurcated systems based on the projection method [16] and Lyapunov method [11]. The result was applied to axial flow compressors which are the essential part of jet engines. It is shown that the feedback control law proposed in [13] admits stabilizing property for axial flow compressors at the critical operating point.

16.1 Introduction

Stabilization of nonlinear control systems with smooth state feedback control has been studied by a number of people [5, 1, 2, 8]. An interesting situation for nonlinear stabilization is when the linearized system has uncontrollable modes on the imaginary axis with the rest of modes stable. This is so-called *critical cases*, for which the linearization theory is inadequate. It becomes more intricate if the underlying nonlinear system involves a real-valued parameter. At critical values of the parameter, the linearized system has unstable modes corresponding eigenvalues on imaginary axis, and additional equilibrium solutions will be born. The bifurcated solutions may or may not be stable. The instability of the bifurcated solution may cause "hysteresis loop" in bifurcation diagrams for both subcritical pitchfork bifurcation and Hopf bifurcation [16], and induce undesirable physical phenomena. This is manifested by a rotating stall in axial flow compressors which are the essential part of jet engines [4, 21, 22]. Thus, bifurcation stabilization is an important topic in nonlinear control.

Abed and Fu studied bifurcation stabilization using smooth local state feedback control [1, 2]. For Hopf bifurcation, stabilization conditions were obtained for both cases where the critical modes of the linearized system are controllable and uncontrollable. For stationary bifurcation, stabilization conditions were derived for the case where the critical mode of the linearized system is controllable. The uncontrollable case was investigated in [17] where normal forms of the nonlinear system are used, and bifurcation stabilization is characterized with invariance. In [15], these results are extended to output feedback systems where the critical modes of the linearized system are not controllable. It is shown in [15] that under some mild conditions, nonlinear controllers do not offer any advantage over the linear ones for bifurcation stabilization, if the critical modes of the linearized system are not observable, and that quadratic feedback controllers are adequate for bifurcation stabilization, if the critical modes of the linearized system are observable. In this chapter, nonlinear systems with multiple pairs of critical modes on the imaginary axis will be investigated. Two different approaches will be pursued: the projection method in [16] and the Lyapunov method in [11]. Conditions will be sought under which stabilization can be characterized for bifurcated systems with multiple pairs of critical modes.

An application platform for bifurcation stabilization is rotating stall control in axial flow compressors. Rotating stall is a severely non-axisymmetric distribution of axial flow velocity, taking the form of a wave or "stall cell," that propagates steadily in the circumferential direction at a fraction of the rotor speed. It is shown in [4, 3, 21] that rotating stall is associated with subcritical Hopf bifurcations born at the peak pressure rise. Once a compressor enters rotating stall, there is a substantial loss in both the pressure rise and average mass flow, and it is very difficult for the compressor to return to normal operating condition due to the hysteresis loop associated with subcritical Hopf bifurcations. Prolonged operation under stall may damage the jet engine, and bring catastrophic consequences. Feedback control was proposed in [10] to improve the compressor performance, and has since received great attention in recent years. The existing results include linear control law [23], bifurcation stabilization [19, 24], and backstepping method [18, 7]. However most past research has focused on the simple third order model involving only a single critical mode that is a crude approximation to the full PDE (partial differential equation) model of Moore and Greitzer, derived for the axial flow compression system. This chapter focuses on rotating stall control based on the multi-mode Moore-Greitzer model [20], which in the limiting case converges to the full PDE model in [22]. A feedback control law proposed in [13] for the simple third order model will be investigated because of its less restrictive requirement on the actuator and sensor. It will be shown that this same feedback control law stabilizes the critical operating condition at the peak pressure rise for the multi-mode Moore-Greitzer model, changes Hopf bifurcations from subcritical into supercritical, and thus eliminates the hysteresis loop associated with rotating stall. Detailed analysis for the feedback control law in [13] will be presented with simulation results.

16.2 Local Stability and Stabilization for Hopf Bifurcations

The nonlinear system under consideration is the following nth order parameterized nonlinear system:

$$\dot{x} = f(\gamma, x), \quad f(\gamma, x_0) = 0 \ \forall \ \gamma \in \mathbf{R}, \ x_0 \in \mathcal{S} \subset \mathbf{R}^n, \qquad (16.2.1)$$

where $x \in \mathbf{R}^n$, γ is a real-valued parameter, and $\mathcal{S}$ a linear subspace of $\mathbf{R}^n$, to be clarified later. It is assumed that $f(\cdot, \cdot)$ is sufficiently smooth such that the equilibrium solution x_e, satisfying $f(\gamma, x_e) = 0$, is a smooth function of γ. The smoothness of $f(\cdot, \cdot)$ implies the existence of a Taylor series expansion near the origin of $\mathbf{R}^n$ in the form of

$$\dot{x} = f(\gamma, x) = L(\gamma)x + Q(\gamma)[x, x] + C(\gamma)[x, x, x] + \cdots \qquad (16.2.2)$$

where $L(\gamma)x$, $Q(\gamma)[x, x]$, and $C(\gamma)[x, x, x]$ can each be expanded into

$$L(\gamma)x = L_0 x + \delta\gamma L_1 x + \delta\gamma^2 L_2 x + \cdots,$$
$$Q(\gamma)[x, x] = Q_0[x, x] + \delta\gamma Q_1[x, x] + \cdots,$$
$$C(\gamma)[x, x, x] = C_0[x, x, x] + \delta\gamma C_1[x, x, x] + \cdots,$$

with $\delta\gamma = \gamma - \gamma_c$ and L_0, L_1, L_2 constant matrices of size $n \times n$. Suppose that $L(\gamma)$ possesses m ($\leq n/2$) pairs of complex eigenvalues $\lambda_k(\gamma) = \alpha_k(\gamma) \pm j\beta_k(\gamma)$, dependent smoothly on γ. It is assumed that for $1 \leq k \leq m$,

$$\alpha_k(\gamma_c) = 0, \quad \beta(\gamma_c) = \omega_k \neq 0, \quad \alpha'_k(\gamma_c) = \frac{d\alpha_k}{d\gamma}(\gamma_c) \neq 0, \quad (16.2.3)$$

while all other eigenvalues of $L(\gamma)$ are stable at, and in a neighborhood of $\gamma = \gamma_c$. That is, m pairs of eigenvalues cross the imaginary axis *simultaneously*. Then each $\lambda_k(\gamma)$ is a critical eigenvalue, and so is its conjugate. The center space, or the eigen-space for the m pairs of critical eigenvalues is denoted by $\mathcal{S}_c$, and is assumed to be orthogonally complement to $\mathcal{S}$ in the sense that $\mathcal{S} \oplus \mathcal{S}_c = \mathbf{R}^n$. Since projection of $x_0 \in \mathcal{S}$ to $\mathcal{S}_c$ is zero, the equilibria x_0 satisfying (16.2.1) will be called zero solution. The strict crossing assumption implies that the zero solution changes its stability as γ crosses γ_c. For instance, $\alpha'_k(\gamma_c) > 0$ implies that the zero solution is locally stable for $\gamma < \gamma_c$ and becomes unstable for $\gamma > \gamma_c$. The crucial problem is the determination of local stability near the critical parameter γ_c at which Hopf bifurcations occur and the periodic solutions are born. Roughly speaking, local stability of Hopf bifurcations refers to stability property of such periodic solutions near the inception point.

The Projection Method

The center space $\mathcal{S}_c$, spanned by all the critical eigenvectors, is complete. If all the critical eigenvalues are distinct,

$$\mathcal{S}_c = \mathcal{S}_{c_1} \oplus \mathcal{S}_{c_2} \oplus \cdots \oplus \mathcal{S}_{c_m},$$

where $\mathcal{S}_{c_k}$ denotes the eigen-space spanned by the kth pair of the critical eigenvectors. The projection method in [16] projects the nonlinear dynamics into the subspace of $\mathcal{S}_{c_k}$ so that stability of the projected dynamics can be analyzed. Associated with each pair of the critical eigenvalues $(\lambda_k, \bar{\lambda}_k)$, a SCV (stability characteristic value) $\tilde{\lambda}_2^{(k)}$ can be defined, whose sign determines local stability of the projected dynamics. See [16, 2] for details. Let the Taylor series of $f(\gamma, x)$ be of the form in (16.2.2) where $L_0 = L(0)$. An algorithm to compute $\tilde{\lambda}_2^{(k)}$ is outlined next.

Step 1: Compute left row eigenvector ℓ_k and right column eigenvector r_k of L_0 corresponding to the kth critical eigenvalue of $\lambda_k(0) = j\omega_k$. Normalize by setting $\ell_k r_k = 1$.

Step 2: Solve column vectors μ_k and ν_k from the equations

$$-L_0\mu_k = \frac{1}{2}Q_0[r_k, \bar{r}_k], \quad (2j\omega_k I - L_0)\nu_k = \frac{1}{2}Q_0[r_k, r_k]. \quad (16.2.4)$$

Step 3: The coefficient $\tilde{\lambda}_2^{(k)}$ is given by

$$\tilde{\lambda}_2^{(k)} = 2\mathrm{Re}\left\{2\ell_k Q_0[r_k, \mu_k] + \ell_k Q_0[\bar{r}_k, \nu_k] + \frac{3}{4}\ell_k C_0[r_k, r_k, \bar{r}_k]\right\}. \quad (16.2.5)$$

THEOREM 16.1
Suppose that L_0 has m pairs of nonzero critical eigenvalues on the imaginary axis with the rest on open left half plane. Then the projected dynamics along the kth pair of the eigenvectors is stable if $\tilde{\lambda}_2 < 0$, and unstable if $\tilde{\lambda}_2 > 0$. For the case of $m = 1$, the associated Hopf bifurcation is supercritical or stable if $\tilde{\lambda}_2 < 0$, and subcritical or unstable if $\tilde{\lambda}_2 > 0$.

It should be clear that if L_0 admits only a single pair of imaginary eigenvalues, then local stability of the bifurcated system is equivalent to $\tilde{\lambda}_2^{(k)} < 0$. However if L_0 admits more than one pair of imaginary eigenvalues, local stability of each projected dynamics does not guarantee local stability of the bifurcated systems. In this case, local stability of the bifurcated systems can be tackled by the Lyapunov method [11].

The Lyapunov Method

The Lyapunov method for multiple critical modes is developed by Fu [11]. It gives sufficient conditions on the existence of the Lyapunov function that guarantees local stability of the bifurcated system. The following notion is needed.

DEFINITION 16.1 *Under the condition that L_0 admits $m > 1$ pairs of eigenvalues on the imaginary axis with the rest on open left half plane, the nonlinear system (16.2.2) is said to be locally center-symmetric in the sense of Lyapunov if the following conditions*

$$\begin{aligned}
&\omega_i \neq \omega_j, &&\omega_i \neq 2\omega_j, &&\omega_i \neq 3\omega_j, &&\text{if } m \geq 2,\\
&\omega_i \neq \omega_j + \omega_k, &&\omega_i \neq 2\omega_j + \omega_k, &&2\omega_i \neq \omega_j + \omega_k, &&\text{if } m \geq 3,\\
&\omega_i \neq \omega_j + \omega_k + \omega_l, &&\omega_i + \omega_j \neq \omega_k + \omega_l, &&&&\text{if } m \geq 4,
\end{aligned}$$

hold where all indices are distinct.

THEOREM 16.2
Suppose that L_0 admits $m > 1$ pairs of eigenvalues on the imaginary axis with the rest on open left half plane and the nonlinear system (16.2.2) is locally center-symmetric in the sense of Lyapunov. Then there exists a Lyapunov function that guarantees local stability of (16.2.2) at the criticality,

if

$$\chi_{kk} = 16\text{Re}\left\{\ell_k\left(2Q_0[r_k,\mu_k] + Q_0[\bar{r}_k,\nu_k] + \frac{3}{4}C_0[r_k,r_k,\bar{r}_k]\right)\right\} < 0,$$

for $k = 1, \cdots, m$, and

$$\chi_{kl} = 16\text{Re}\left\{\ell_k\left(Q_0[r_k,\mu_l] + Q_0[r_l,\mu_{k_l}] + Q_0[\bar{r}_l,\nu_{k_l}] + \frac{3}{4}C_0[r_k,r_l,\bar{r}_l]\right)\right\}$$

is nonpositive for $k,l = 1, \cdots, m$ with $k \neq l$, where μ_k, ν_k are the solutions to (16.2.4), and μ_{k_l}, ν_{k_l} are given by

$$\mu_{k_l} = -\frac{1}{2}(L_0 - j(\omega_k - \omega_l))^{-1}Q_0[r_k,\bar{r}_l], \tag{16.2.6}$$

$$\nu_{k_l} = -\frac{1}{2}(L_0 - j(\omega_k + \omega_l))^{-1}Q_0[r_k,r_l]. \tag{16.2.7}$$

It is noted that $\chi_{kk} = 8\tilde{\lambda}_2^{(k)}$. Hence local stability of Hopf bifurcation with multiple pairs of critical modes requires more than local stability of each projected dynamics. Roughly speaking, the nonpositivity of χ_{kl} with $k \neq l$ takes the coupling of different pairs of critical modes into account for local stability of the bifurcated system in (16.2.2).

Bifurcation Stability and Stabilization

Unstable or subcritical Hopf bifurcations are undesirable for nonlinear systems which induce a hysteresis loop [16]. Bifurcation control was proposed in [1] to change Hopf bifurcations from subcritical into supercritical thereby eliminating the associated hysteresis loop. Sufficient conditions for local stability of Hopf bifurcations were derived in [11] for the case of multiple pairs of critical modes. Theorem 16.2 is quoted from [11]. An interesting question is how close the sufficient condition in Theorem 16.2 is to necessity.

THEOREM 16.3
Suppose that hypotheses of Theorem 16.2 hold, and

$$\chi_{kl} = f_1(k)f_2(l), \tag{16.2.8}$$

for each k and l where either $f_1(k)$ or $f_2(l)$ is sign invariant. Then Hopf bifurcations involving multiple pairs of critical modes are locally stable, if and only if $\chi_{kk} = 8\tilde{\lambda}_2^{(k)} < 0$ for $k = 1, 2, \cdots, m$.

PROOF We need only show that $\chi_{kk} < 0$ implies that $\chi_{kl} \leq 0$. It is noted that $\chi_{kk} = f_1(k)f_2(k)$ by comparing the expressions of χ_{kk} and χ_{kl}. Thus if $f_1(k)$ has the same sign, so does $f_2(k)$ which, in turn, implies that χ_{kl}

has the same sign as χ_{kk} for each k and l. The proof for the case that $f_2(l)$ is sign invariant is similar, and is thus omitted.

Although the condition in (16.2.8) is rather restrictive, it gives an effective way to solve the jet engine control problem posed in [10] which will be presented in the next two sections.

16.3 Multi-Mode Moore-Greitzer Model

An interesting application platform for nonlinear bifurcation stability and stabilization is rotating stall control for axial flow compressors. Rotating stall is induced by subcritical Hopf bifurcations born at the peak pressure rise resulting in hysteresis loop. It is difficult to get the compressor out of stall which may bring about catastrophic consequences for jet engines. For this reason, large stall margins are used in practice that effectively limit compressor performance (see [12] and the references therein). A bifurcation approach was proposed in [19, 24] to change subcritical Hopf bifurcation into a supercritical one, thereby stabilizing the critical operating condition at the peak pressure rise and eliminating the hysteresis loop. However, the results in [19, 24] are established for the simple third order model, involving only a single critical mode of rotating stall; its effectiveness to the full PDE model remains unknown. This chapter will adopt a control law proposed in [13] for the multi-mode Moore-Greitzer Model which has N pairs of critical modes for rotating stall and converges to the full PDE model as $N \to \infty$. For completeness, this section will give a derivation on the multi-mode Moore-Greitzer model, and analyze its equilibria and linearized system.

A schematic axial flow compressor is shown in Figure 16.1.

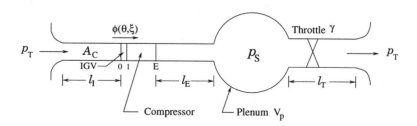

FIGURE 16.1
Schematic of compressor showing nondimensionalized lengths.

The equation governing the flow in compression systems is given by [22]:

$$\Psi + l_c \frac{d\Phi}{d\xi} = \psi_c(\phi) - m\left(\frac{\partial}{\partial \xi}\int_{-l_F}^{0} \varphi\, d\eta\right) - \left(\frac{1}{a}\frac{\partial \varphi}{\partial \xi} + \frac{1}{b}\frac{\partial \varphi}{\partial \theta}\right)\bigg|_{\eta=0} \quad (16.3.9)$$

where $\phi = \Phi + \varphi$ is the local flow coefficient at station 0 ($\eta = 0$) with Φ the mean flow, and Ψ the total pressure rise, and l_c the total aerodynamic length of the compressor. The incompressible and irrotational assumption on the gas flow implies the existence of disturbance flow potential that satisfies Laplace's equation with zero boundary condition at $\eta = -l_F \leq -l_c$, which in turn implies that the disturbance flow at station 0 has the form:

$$\varphi|_{\eta=0} = \sum_{n=1}^{N}(A_n \cos(n\theta) + B_n \sin(n\theta)),\quad N\to\infty, \quad (16.3.10)$$

Hence, for any uniformly distributed $2N+1$ points $\{\theta_k\}_{k=1}^{2N+1}$ on circumference of the compressor duct,

$$A_n = \frac{2}{2N+1}\sum_{k=1}^{2N+1}\phi(\theta_k)\cos n\theta_k,\quad B_n = \frac{2}{2N+1}\sum_{k=1}^{2N+1}\phi(\theta_k)\sin n\theta_k,$$

and Φ is the average value of the $2N+1$ local flow rates at station 0. Substituting (16.3.10) into (16.3.9) and taking each of the moments gives

$$\Psi + l_c \frac{d\Phi}{d\xi} = \frac{1}{2\pi}\int_{-\pi}^{\pi}\psi_c(\phi)\,d\theta,$$

$$\left(\frac{m\cosh(nl_F)}{n\sinh(nl_F)} + \frac{1}{a}\right)\frac{dA_n}{d\xi} = \frac{1}{\pi}\int_{-\pi}^{\pi}\psi_c(\phi)\cos n\theta\,d\theta - \frac{n}{b}B_n,$$

$$\left(\frac{m\cosh(nl_F)}{n\sinh(nl_F)} + \frac{1}{a}\right)\frac{dB_n}{d\xi} = \frac{1}{\pi}\int_{-\pi}^{\pi}\psi_c(\phi)\sin n\theta\,d\theta + \frac{n}{b}A_n,$$

for $n = 1, \ldots, N$. Although the integral involving $\psi_c(\cdot)$ is difficult to evaluate, it has an approximate Fourier series form

$$\psi_c(\phi) \approx \overline{\psi}_{cA_0}(\phi) + \sum_{n=1}^{N}(\psi_{cA_n}\cos n\theta + \psi_{cB_n}\sin n\theta). \quad (16.3.11)$$

The above approximation converges uniformly under a rather mild assumption. Thus, a multi-mode equation is now obtained for the local flow rate:

$$\Psi + l_c\frac{d\Phi}{d\xi} = \overline{\psi}_{cA_0}(\phi), \quad (16.3.12)$$

$$\left(\frac{m\cosh(nl_F)}{n\sinh(nl_F)} + \frac{1}{a}\right)\frac{dA_n}{d\xi} = \psi_{cA_n} - \frac{n}{b}B_n, \quad (16.3.13)$$

$$\left(\frac{m\cosh(nl_F)}{n\sinh(nl_F)} + \frac{1}{a}\right)\frac{dB_n}{d\xi} = \psi_{cB_n} + \frac{n}{b}A_n, \quad (16.3.14)$$

where $n = 1, \ldots, N$. Define the discrete Fourier coefficients $\hat{A}_n$, $\hat{B}_n$, $\hat{\psi}_{cA_n}$, and $\hat{\psi}_{cB_n}$ through the relation

$$\Phi = \frac{1}{\sqrt{2N+1}}\hat{A}_0, \quad A_n = \sqrt{\frac{2}{2N+1}}\hat{A}_n, \quad B_n = \sqrt{\frac{2}{2N+1}}\hat{B}_n,$$
$$\overline{\psi}_c = \frac{1}{\sqrt{2N+1}}\hat{\psi}_{cA_0}, \quad \psi_{cA_n} = \sqrt{\frac{2}{2N+1}}\hat{\psi}_{cA_n}, \quad \psi_{cB_n} = \sqrt{\frac{2}{2N+1}}\hat{\psi}_{cB_n}. \quad (16.3.15)$$

Define vectors of size $2N+1$ as

$$\hat{\phi}_N = \begin{bmatrix} \hat{A}_0 & \hat{A}_1 & \hat{B}_1 & \ldots & \hat{A}_N & \hat{B}_N \end{bmatrix}^T, \quad (16.3.16)$$

$$\phi_N = \begin{bmatrix} \phi(\theta_1) & \phi(\theta_2) & \ldots & \phi(\theta_{2N+1}) \end{bmatrix}^T, \quad (16.3.17)$$

$$\hat{\psi}_{cN} = \begin{bmatrix} \hat{\psi}_{cA_0} & \hat{\psi}_{cA_1} & \hat{\psi}_{cB_1} & \ldots & \hat{\psi}_{cA_N} & \hat{\psi}_{cB_N} \end{bmatrix}^T, \quad (16.3.18)$$

$$\psi_{cN}(\phi) = \begin{bmatrix} \psi_c(\phi(\theta_1)) & \psi_c(\phi(\theta_2)) & \ldots & \psi_c(\phi(\theta_{2N+1})) \end{bmatrix}^T. \quad (16.3.19)$$

Then $T\phi_N = \hat{\phi}_N$ and $T\psi_{cN} = \hat{\psi}_{cN}$, where T is a DFT (discrete Fourier transform) matrix:

$$T = \sqrt{\frac{2}{2N+1}} \begin{bmatrix} \frac{1}{\sqrt{2}} & \frac{1}{\sqrt{2}} & \cdots & \frac{1}{\sqrt{2}} \\ \cos\theta_1 & \cos\theta_2 & \ldots & \cos\theta_{2N+1} \\ \sin\theta_1 & \sin\theta_2 & \ldots & \sin\theta_{2N+1} \\ \vdots & \vdots & \ddots & \vdots \\ \cos N\theta_1 & \cos N\theta_2 & \ldots & \cos N\theta_{2N+1} \\ \sin N\theta_1 & \sin N\theta_2 & \ldots & \sin N\theta_{2N+1} \end{bmatrix} \quad (16.3.20)$$

so that $T^{-1} = T^T$. Substituting (16.3.15) into (16.3.12) – (16.3.14) gives

$$\Psi\sqrt{2N+1} + l_c\frac{d\hat{A}_0}{d\xi} = \hat{\psi}_{cA_0}(\phi) \quad (16.3.21)$$

$$\left(\frac{m\cosh(nl_F)}{n\sinh(nl_F)} + \frac{1}{a}\right)\frac{d\hat{A}_n}{d\xi} = \hat{\psi}_{cA_n} - \frac{n}{b}\hat{B}_n \quad (16.3.22)$$

$$\left(\frac{m\cosh(nl_F)}{n\sinh(nl_F)} + \frac{1}{a}\right)\frac{d\hat{B}_n}{d\xi} = \hat{\psi}_{cB_n} + \frac{n}{b}\hat{A}_n \quad (16.3.23)$$

for $n = 1, \ldots, N$. The above equations can be written in matrix form

$$E\dot{\hat{\phi}}_N = F\hat{\phi}_N + \hat{\psi}_{cN} - \sqrt{2N+1}e_N\Psi, \quad (16.3.24)$$

where $e_N^T = \begin{bmatrix} 1 & 0 & \cdots & 0 \end{bmatrix}$, and

$$E = \text{diag}(l_c, m_1, m_1, \cdots, m_N, m_N), \quad m_n = \frac{m\cosh(nl_F)}{n\sinh(nl_F)} + \frac{1}{a}, \quad (16.3.25)$$

$$F = \text{diag}\left(0, \begin{bmatrix} 0 & -\frac{1}{b} \\ \frac{1}{b} & 0 \end{bmatrix}, \begin{bmatrix} 0 & -\frac{2}{b} \\ \frac{2}{b} & 0 \end{bmatrix}, \cdots, \begin{bmatrix} 0 & -\frac{N}{b} \\ \frac{N}{b} & 0 \end{bmatrix}\right). \quad (16.3.26)$$

Using discrete Fourier transform $\hat{\phi}_N = T\phi_N$ gives vector-valued ODE:

$$ET\dot{\phi}_N = FT\phi_N + T\psi_{cN} - \sqrt{2N+1}e_N\Psi.$$

Multiplying both sides by T^T, and noting that $T^T e_N = \bar{e}_N/\sqrt{2N+1}$, with $\bar{e}_N$ a size of $2N+1$ vector of all entries 1, yield

$$T^T ET\dot{\phi}_N = T^T FT\phi_N + \psi_{cN} - \bar{e}_N\Psi.$$

Combined with pressure rise equation [22], an ODE model of

$$\dot{\phi}_N = T^T E^{-1} FT\phi_N + T^T E^{-1} T\psi_{cN} - T^T E^{-1} T\bar{e}_N\Psi, \quad (16.3.27)$$

$$\dot{\Psi} = \frac{1}{4B^2 l_c}\left(\Phi - f_T^{-1}(\psi)\right) = \frac{1}{4B^2 l_c}\left(\frac{\bar{e}_N^T \phi_N}{2N+1} - \gamma\sqrt{\Psi}\right) \quad (16.3.28)$$

is obtained. Since ϕ_N has size $2N+1$, the above equations are of $(2N+2)$th order.

The compressor performance characteristic equation is assumed to be of the form [22]:

$$\psi_c(\phi) = H\left[c_0 + c_1\left(\frac{\phi}{W} - 1\right) + c_3\left(\frac{\phi}{W} - 1\right)^3\right],$$

$$\psi_{cN} = H\left[c_0\bar{e}_N + c_1\Delta\phi_N + c_3\Delta\phi_N^{\cdot 3}\right], \quad \Delta\phi_N := \phi_N/W - \bar{e}_N,$$

where $x^{\cdot 2} = x \cdot x$ denotes element-wise product and $x^{\cdot k}$ raises each element of x to the power of k. To simplify the model in (16.3.27), the following substitutions

$$M_1 = T^T E^{-1} FT, \quad M_2 = T^T E^{-1} T,$$

are used. Therefore, the final compressor model has the form:

$$\dot{\phi}_N = M_1\phi_N + M_2 H(c_0\bar{e}_N + c_1\Delta\phi_N + c_3\Delta\phi_N^{\cdot 3}) - M_2\bar{e}_N\Psi, \quad (16.3.29)$$

$$\dot{\Psi} = \frac{1}{4B^2 l_c}\left(\frac{\bar{e}_N^T \phi_N}{2N+1} - \gamma\sqrt{\Psi}\right). \quad (16.3.30)$$

Suppose that the flow is perturbed near the equilibria of the uniform flow. Then

$$\phi_N = \phi_e\bar{e}_N + \delta\phi_N, \quad \Delta\phi_N = \frac{1}{W}\delta\phi_N + a\left(\frac{\phi_e}{W} - 1\right)\bar{e}_N, \quad (16.3.31)$$

where $\phi_e \neq 0$ is a scalar factor, representing the flow rate intensity. It can be easily verified that

$$\Delta\phi_N^3 = \frac{1}{W^3}\delta\phi_N^3 + \frac{3}{W^2}\left(\frac{\phi_e}{W} - 1\right)\delta\phi_N^2 + \frac{3}{W}\left(\frac{\phi_e}{W} - 1\right)^2 \delta\phi_N + \left(\frac{\phi_e}{W} - 1\right)^3 \bar{e}_N.$$

Substituting the above into (16.3.29) gives

$$\delta\dot{\phi}_N = (M_1 + M_2\alpha)\,\delta\phi_N + \frac{3c_3}{W^2}\left(\frac{\phi_e}{W} - 1\right)M_2\delta\phi_N^2 + \frac{c_3}{W^3}M_2\delta\phi_N^3$$

$$+ (M_1\phi_e + M_2\psi_c(\Delta\phi_N)) - M_2\bar{e}_N\Psi, \quad \alpha = \left.\frac{d\psi_c}{d\phi}\right|_{\phi=\phi_e}.$$

The following result concerns stationary equilibria of the multi-mode model. Since the proof is straightforward, it is omitted.

PROPOSITION 16.4
The uniform flow $\phi_N = \phi_e \bar{e}_N$ is a stationary equilibrium of (16.3.29), if and only if

$$\Psi_e = \psi_c(\phi_e) = H\left[c_0 + c_1\left(\frac{\phi_e}{W} - 1\right) + c_3\left(\frac{\phi_e}{W} - 1\right)^3\right] = \frac{\phi_e^2}{\gamma^2}. \quad (16.3.32)$$

The condition in (16.3.32) determines equilibria of the compressor model for uniform mass flow. It is somewhat surprising to note that this is exactly the same as the Moore-Greitzer model for the case of $N = 1$. Denote state-variable as $x = \delta\phi_N \oplus \delta\Psi$ where $\oplus$ denotes direct sum and $\delta\Psi = \Psi - \Psi_e$. The linearized model near the equilibria $\phi_N = \phi_e\bar{e}_N$ and $\Psi = \Psi_e$ is now given by

$$\dot{x} = \begin{bmatrix} \delta\dot{\phi}_N \\ \delta\dot{\Psi} \end{bmatrix} = \begin{bmatrix} L_{11} & L_{12} \\ L_{21} & L_{22} \end{bmatrix} x, \quad (16.3.33)$$

$$L_{11} = M_1 + M_2\alpha, \quad \alpha = \left.\frac{d\psi_c}{d\phi}\right|_{\phi=\phi_e} = \frac{H}{W}\left[c_1 + 3c_3\left(\frac{\phi_e}{W} - 1\right)^2\right],$$

$$L_{12} = -M_2\bar{e}_N, \quad L_{21} = \frac{\bar{\beta}^{-2}\bar{e}_N^T}{2N+1}, \quad L_{22} = -\frac{\gamma\bar{\beta}^{-2}}{2\sqrt{\Psi_e}}, \quad \bar{\beta} = 2B\sqrt{l_c}.$$

Substituting the expressions for M_1, and M_2 gives

$$L = S^{-1}\begin{bmatrix} \tau & \kappa\bar{e}_N^T \\ -\kappa\bar{e}_N & E^{-1/2}(F+\alpha I)E^{-1/2} \end{bmatrix}S, \quad \kappa = \frac{1}{2Bl_c}, \quad (16.3.34)$$

$$S = \begin{bmatrix} 0 & \bar{\beta}\sqrt{(2N+1)} \\ E^{1/2}T & 0 \end{bmatrix}, \quad \tau = -\frac{\gamma}{8B^2l_c\sqrt{\Psi_e}}, \quad (16.3.35)$$

where $T\bar{e}_N = e_N\sqrt{2N+1}$ and $E^{-1/2}e_N = e_N/\sqrt{l_c}$ are used. Since S is a matrix of similarity transformation, an eigenvalue λ of L satisfies

$$\det \begin{bmatrix} \lambda - \tau & -\kappa e_N^T \\ \kappa e_N & \lambda I - E^{-1/2}(F + \alpha I)E^{-1/2} \end{bmatrix} = 0.$$

PROPOSITION 16.5
Consider the compressor model as in (16.3.29) and (16.3.30). Its linearized system near $(\phi_e \bar{e}_N, \Psi_e)$ as in (16.3.32) has $(N+1)$ pairs of eigenvalues that are given as follows:

$$\lambda_{1,2} = 0.5(\tau + \alpha/l_c) \pm 0.5j\sqrt{4(\kappa^2 + \tau\alpha/l_c) - (\tau + \alpha/l_c)^2}, \quad (16.3.36)$$

$$\lambda_{2n+1,2n+2} = \left(\alpha \pm j\frac{n}{b}\right)m_n^{-1}, \quad m_n = \frac{m\cosh(nl_F)}{n\sinh(nl_F)} + \frac{1}{a}, \quad (16.3.37)$$

where $n = 1, 2, \cdots, N$, and $j = \sqrt{-1}$.

PROOF Careful observation of SLS^{-1} with L in (16.3.34) and S in (16.3.35) yields

$$D = SLS^{-1} = \text{diag}(D_0, D_1, \cdots, D_N), \quad D_0 = \begin{bmatrix} \tau & \kappa \\ -\kappa & \alpha/l_c \end{bmatrix}, \quad (16.3.38)$$

$$D_n = \begin{bmatrix} \alpha & -n/b \\ n/b & \alpha \end{bmatrix} m_n^{-1}, \quad n = 1, 2, \cdots, N. \quad (16.3.39)$$

Direct calculation for eigenvalues of L gives

$$\det(\lambda I - L) = \det(\lambda I - D) = \prod_{n=0}^{N} \det(\lambda I - D_n) = 0, \iff$$

$$\det(\lambda I - D_0) = \lambda^2 - (\tau + \alpha/l_c)\lambda + \kappa^2 + \tau\alpha/l_c = 0,$$

$$\det(\lambda I - D_n) = (m_n\lambda - \alpha)^2 + n^2/b^2 = 0, \quad n = 1, 2, \cdots, N.$$

Hence the expressions for $2(N+1)$ eigenvalues of L are those shown in (16.3.36) and (16.3.37).

Assume S-shape for the performance characteristic curve $\psi_c(\cdot)$ and $c_1 + 3c_3 = 0$ as in [22]. A schematic compressor characteristic is shown in Figure 16.2. The designed operating point is uniquely determined by the intersection of the throttle line (dashed lines A–B or C–D for two different values of the throttle position) with the compressor performance curve $\psi_c(\cdot)$. The maximum pressure rise takes place at $\alpha = 0$ (point A) because that is where the derivative of $\psi_c(\cdot)$ equals zero. Since $\alpha < 0$ on the right side of the maximum pressure rise, the N pairs of eigenvalues of L matrix in (16.3.37) are stable. By $\tau < 0$, the pair of eigenvalues of (16.3.36) are also stable. Hence, any point on right side of the peak of $\psi_c(\cdot)$ is a stable operating point and the uniform flow is a stable equilibrium, as highlighted by the solid line. However, if the throttle value γ decreases, then the flow rate intensity ϕ_e decreases, and the derivative of $\psi_c(\cdot)$ eventually changes its sign into positive. Thus left side of the peak point A on $\psi_c(\cdot)$ corresponds to unstable operating points, as indicated by dotted line. It is noted that the N pairs of eigenvalues of L in (16.3.37) become imaginary precisely at $\alpha = 0$. The critical mass flow rate and pressure rise are given by

$$\phi_e = \phi_c = 2W, \quad \Psi_e = \Psi_c = H(c_0 + c_1 + c_3), \qquad (16.3.40)$$

due to the assumption $c_1 + 3c_3 = 0$. At the criticality, Hopf bifurcations occur for the multi-mode model of the compression system in (16.3.29) and (16.3.30) that induce rotating stall. Furthermore, if the underlying bifurcation is subcritical or unstable, the A–C portion of the stall curve is unstable which is sketched for $N = 1$. Stall cells will be born at point A and grow, which will throttle the operating point from A to B quickly which is a stable operating point, although undesirable. There is a tremendous drop in both the pressure rise and flow rate. Moreover, increasing the throttle position and flow rate at point B does not increase the pressure rise. Rather, the pressure rise becomes even lower before it comes to point C at which it jumps back to (due to again loss of stability) the performance curve $\psi_c(\cdot)$. The hysteresis loop A-B-C-D is the main cause for loss of compressor performance and the potential damage to jet engines.

The above discussion indicates that stabilization of the critical operating point and A–C portion of the stall curve is a key issue for rotating stall control which will eliminate the hysteresis loop and allow smooth change of the operating point between axisymmetric flow and stalled flow for the compression system.

16.4 Rotating Stall Control

This section will adopt a control law proposed in [13] for the multi-mode Moore-Greitzer Model, which, in the limit, converges to the full PDE model. It will be shown that the feedback control law proposed in [13] achieves stabilization of the critical operating point for the multi-mode compression system and has the potential for industrial applications. Explicit condition will be derived for the feedback gain. The feedback system employs throttle position as actuator and pressure rise as sensored output, where

$$\gamma = \gamma_o + u, \quad u = \frac{K}{\sqrt{\Psi}}, \qquad (16.4.41)$$

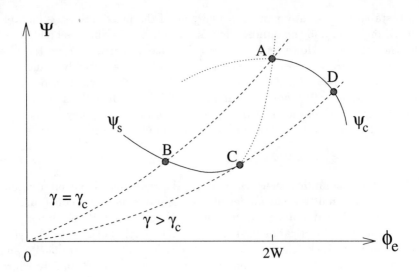

FIGURE 16.2
Schematic compressor characteristic, showing rotating stall.

with γ_o synthesized disturbance and u feedback control input. The critical value of γ_o is given by

$$\gamma_o = \gamma_c = \frac{\phi_c - K}{\sqrt{\Psi_c}}, \qquad (16.4.42)$$

with ϕ_c and Ψ_c at the peak of the compressor performance curve $\psi_c(\cdot)$. Now the multi-mode Moore-Greitzer Model in (16.3.29)–(16.3.30) admits a Taylor series expansion

$$\dot{x} = f(\gamma, x) = Lx + Q[x,x] + C[x,x,x] + \cdots, \qquad (16.4.43)$$

where $x = \delta\phi_N \oplus \delta\Psi$ with $\delta\phi_N = \phi_N - \phi_e \bar{e}_N$, $\delta\Psi - \Psi_e$, and

$$L = S^{-1} \begin{bmatrix} \tau & \kappa e_N^T \\ -\kappa e_N & E^{-1/2}(F+\alpha I)E^{-1/2} \end{bmatrix} S, \qquad (16.4.44)$$

$$Q[x,x] = \begin{bmatrix} \frac{3Hc_3}{W^2}\left(\frac{\phi_e}{W}-1\right) M_2 \; 0_{2N+1} \\ 0_{2N+1}^T & -\frac{\tau}{4\Psi_e} \end{bmatrix} x^{\cdot 2}, \qquad (16.4.45)$$

$$C[x,x,x] = \begin{bmatrix} \frac{Hc_3}{W^3} M_2 \; 0_{2N+1} \\ 0_{2N+1}^T & \frac{\tau}{8\Psi_e^2} \end{bmatrix} x^{\cdot 3}. \qquad (16.4.46)$$

It is noted that the parameter τ is a function of γ_o, instead of γ, and it is also interesting to note that this local model is exactly the same as the one without feedback control except that γ_o is replaced by γ.

In order to apply the projection and the Lyapunov method to rotating stall control, left and right critical eigenvectors corresponding to rotating stall need be computed. Denote Θ_n as a column vector of size $2N+1$ as

$$\Theta_n = \begin{bmatrix} e^{-jn\theta_1} & e^{-jn\theta_2} & \cdots & e^{-jn\theta_{2N+1}} \end{bmatrix}^{\mathrm{T}} \quad (16.4.47)$$

where $e^{j\theta_k}$'s are equally distributed on the unit circle as assumed. Denote also

$$p_n = \sqrt{\frac{2m_n^{-1}}{2N+1}}.$$

Then it can be shown that the right and left eigenvectors corresponding to the nth pair of the critical eigenvalues associated with rotating stall are given by

$$r_n = p_n \begin{bmatrix} 1 & j \end{bmatrix} u_n \begin{bmatrix} \Theta_n \\ 0 \end{bmatrix}, \quad \ell_n = m_n p_n u_n^{\mathrm{T}} \begin{bmatrix} 1 \\ -j \end{bmatrix} \begin{bmatrix} \Theta_n^{\mathrm{H}} & 0 \end{bmatrix},$$

respectively, for $1 \leq n \leq N$, where u_n is an arbitrary nonzero column vector of size 2. Moreover, $\ell_n r_n = 1$, if and only if $\|u_n\| = \sqrt{u_n^{\mathrm{T}} u_n} = 1/\sqrt{2}$. The solutions of μ_n and ν_n to (16.2.4) can also be computed, and are given by

$$\mu_n = \frac{3Hc_3 p_n^2}{4W^2}\left(\frac{\phi_c}{W} - 1\right)\begin{bmatrix} \frac{\gamma_c}{2\sqrt{\Psi_c}} \bar{e}_N \\ 1 \end{bmatrix}, \quad \omega_n = \frac{n}{b} m_n^{-1}, \quad (16.4.48)$$

$$\nu_n = \frac{3Hc_3 p_n^2 j e^{j2\delta_n}}{4m_{2n} W^2 (\omega_{2n} - 2\omega_n)}\left(\frac{\phi_c}{W} - 1\right)\begin{bmatrix} \Theta_{2n} \\ 0 \end{bmatrix}, \quad (16.4.49)$$

where, if $2n > N$ in (16.4.49), $2n$ should be replaced by $\eta = (2N+1-2n)$ in expressions of m_{2n} and ω_{2n}, and

$$u_n^{\mathrm{T}} = \begin{bmatrix} u_{n_1} & u_{n_2} \end{bmatrix}, \quad \delta_n = \tan^{-1}\left(\frac{u_{n_2}}{u_{n_1}}\right).$$

THEOREM 16.6
For $1 \leq n \leq N$, the SCV for the nth pair of critical modes corresponding to rotating stall is given by

$$\tilde{\lambda}_2^{(n)} = \frac{3Hc_3 p_n^2}{4m_n W^3}\left[1 - \frac{6Hc_3 \gamma_c}{W\sqrt{\Psi_c}}\left(\frac{\phi_c}{W} - 1\right)^2\right] \quad (16.4.50)$$

where $\omega_n = \frac{n}{b} m_n^{-1}$. For $N < 2n \leq 2N$, $\tilde{\lambda}_2^{(n)}$ is the same as above except that $2n$ is replaced by $\eta = 2(N-n)+1$. Suppose that $c_1 + 3c_3 = 0$, $c_3 < 0$, and thus ϕ_c and Ψ_c are the same as in (16.3.40). Then $\tilde{\lambda}_2^{(n)} < 0$ for $1 \leq n \leq N$, if and only if

$$K > \frac{c_0 + 10c_3}{6c_3} W \quad (16.4.51)$$

where K is the feedback gain as in (16.4.41).

PROOF It is straightforward to obtain

$$\ell_n Q[x,y] = \frac{3Hc_3}{m_n W^2}\left(\frac{\phi_c}{W}-1\right)\ell_n(x\cdot y), \qquad (16.4.52)$$

$$\ell_n C[x,y,z] = \frac{Hc_3}{m_n W^3}\ell_n(x\cdot y\cdot z). \qquad (16.4.53)$$

With μ_n and ν_n obtained as in (16.4.48) and (16.4.48), there hold

$$r_n \cdot \mu_n = \frac{3Hc_3\gamma_c p_n^2 r_n}{8W^2\sqrt{\Psi_c}}\left(\frac{\phi_c}{W}-1\right),$$

$$\bar{r}_n \cdot \nu_n = \frac{3Hc_3 p_n^2 jr_n}{8W^2 m_{2n}(\omega_{2n}-2\omega_n)}\left(\frac{\phi_c}{W}-1\right).$$

It follows that

$$\mathrm{Re}\,[\ell_n(r_n\cdot\mu_n)] = \frac{3Hc_3\gamma_c p_n^2}{8W^2\sqrt{\Psi_c}}\left(\frac{\phi_c}{W}-1\right),$$

$$\mathrm{Re}\,[\ell_n(r_n\cdot\bar{r}_n\cdot r_n)] = \frac{p_n^2}{2},\quad \mathrm{Re}\,[\ell_n(\bar{r}_n\cdot\nu_n)] = 0. \qquad (16.4.54)$$

Hence the expression of $\tilde{\lambda}_2^{(n)}$ in (16.4.50) can be verified following (16.2.5). Since $c_3 < 0$, $\tilde{\lambda}_2^{(n)} < 0$ if and only if

$$\frac{6Hc_3\gamma_c}{W\sqrt{\Psi_c}}\left(\frac{\phi_c}{W}-1\right)^2 = \frac{6Hc_3\gamma_c}{W\sqrt{\Psi_c}} < 1.$$

Substituting γ_c as in (16.4.42) and letting $\Psi_c = H(c_0 + c_1 + c_3)$ and $\phi_c = 2W$ yields the inequality (16.4.51) that ensures stability for each projected dynamics with $n = 1, 2, \cdots, N$.

As mentioned earlier, $\tilde{\lambda}_2^{(n)} < 0$ for $n = 1, 2, \cdots, N$ may not imply stability of the rotating stall dynamics because of the coupling between each different pair of the critical modes. Hence Lyapunov method needs applied, which yields only a sufficient condition. It is interesting to note that the condition (16.2.8) in Theorem 16.3 is satisfied, and thus the condition in (16.4.51) is both necessary and sufficient to ensure local asymptotic stability of the rotating stall dynamics.

THEOREM 16.7
Under the condition $c_1 + 3c_3 = 0$ and $c_3 < 0$, the rotating stall dynamics is locally asymptotically stabilized by feedback control in (16.4.41), if and only if condition (16.4.51) is satisfied.

PROOF It is noted that μ_{n_l} and ν_{n_l} as defined in (16.2.6) for the local feedback compression model are given by

$$\mu_{n_l} = \frac{3Hc_3 p_n p_l j e^{j(\delta_n - \delta_l)}}{4m_{|n-l|}W^2(\omega_{|n-l|} - \omega_n + \omega_l)} \left(\frac{\phi_c}{W} - 1\right) \begin{bmatrix} \Theta_{n-l} \\ 0 \end{bmatrix}, \quad (16.4.55)$$

$$\nu_{n_l} = \frac{3Hc_3 p_n p_l j e^{j(\delta_n + \delta_l)}}{4m_{n+l}W^2(\omega_{n+l} - \omega_n - \omega_l)} \left(\frac{\phi_c}{W} - 1\right) \begin{bmatrix} \Theta_{n+l} \\ 0 \end{bmatrix}, \quad (16.4.56)$$

where $n, l = 1, 2, \cdots, N$, $n \neq l$, and if $n + l > N$, $\omega_{n+l} = \omega_\eta$, $m_{n+l} = m_\eta$, and $\Theta_{n+l} = \Theta_{-\eta}$ with $\eta = 2N + 1 - n - l$. It should be mentioned that the above is true even if $c_1 + 3c_3 \neq 0$. Substituting $\phi_c = 2W$ and using straightforward calculation give

$$\ell_n(\bar{r}_l \cdot \nu_{n_l}) = jr_\nu \frac{m_n p_n^2 p_l^2}{8}(2N + 1),$$

$$\ell_n(r_l \cdot \mu_{n_l}) = jr_\mu \frac{m_n p_n^2 p_l^2}{8}(2N + 1), \quad (16.4.57)$$

for $n \neq l$ where r_μ, r_ν are some real numbers. Thus

$$\text{Re}\left[\ell_n(r_l \cdot \mu_{n_l})\right] = \text{Re}\left[\ell_n(\bar{r}_l \cdot \nu_{n_l})\right] = 0$$

for $n, l = 1, 2, \cdots, N$. It follows that

$$\chi_{nl} = 16\text{Re}\left(\ell_n Q[r_n, \mu_l] + \frac{3}{4}\ell_n C[r_n, r_l, \bar{r}_l]\right)$$

at $\gamma_o = \gamma_c$. Using the expression of r_n, and $\phi_c = 2W$,

$$r_n \cdot \mu_l = -\frac{3Hc_3 \tau p_l^2}{4W^2 \kappa^2 l_c} r_n, \quad r_n \cdot r_l \cdot \bar{r}_l = \frac{p_l^2}{2} r_n.$$

By (16.4.52), (16.4.53), and relation $\ell_n r_n = 1$,

$$\ell_n Q[r_n, \mu_l] = -\frac{\tau}{m_n l_c}\left(\frac{3Hc_3 p_l}{2W^2 \kappa}\right)^2, \quad \ell_n C[r_n \cdot r_l \cdot \bar{r}_l] = \frac{Hc_3 p_l^2}{2m_n W^3}.$$

It can now be readily verified that the condition (16.2.8) in Theorem 16.3 is satisfied with $f_2(l) = p_l^2$, so (16.4.51) is both necessary and sufficient to ensure local stability of the rotating stall dynamics at the criticality.

16.5 Simulation Results and Discussions

To illustrate the effectiveness of the feedback control law as in (16.4.41), a Simulink program is compiled in MATLAB. Due to the physical nature of

γ_o, an additional condition is enforced on the feedback gain K to ensure positivity of γ_o. Hence we limit the feedback gain K as

$$\frac{7}{9}W = \frac{c_0 + 10c_3}{6c_3}W < K < \phi_c = 2W$$

with $c_0 = 3/8$, and $c_1 = -3c_3 = 1.5$ as in [22]. Other physical parameters are also taken from [22] which are listed below:

$$m = 1.75, \quad H = 0.18, \quad W = 0.25, \quad B = 0.1, \quad a = 1/3.5, \quad l_c = 8, \quad l_F = \infty.$$

By setting $K = 0$ as in (16.4.51), Theorem 16.7 also gives a necessary and sufficient condition for stability of the rotating stall dynamics at the criticality. For the parameters considered here, the critical operating point is clearly unstable as shown in the following picture:

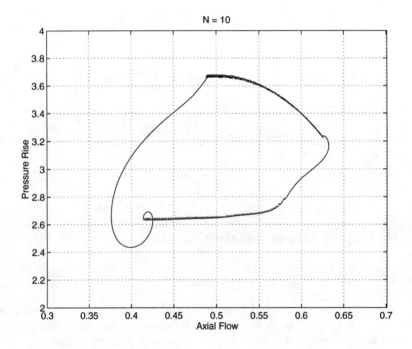

FIGURE 16.3
Hysteresis loop associated with rotating stall for $N = 10$.

Feedback control as in (16.4.41) is employed to improve the performance. The throttle value γ_o is reduced to $0.89\gamma_c$ with γ_c as in (16.4.42), before moving back to $1.2\gamma_c$. The time responses of the averaged mass flow rate Φ and pressure rise Ψ are plotted next. It is seen that both quantities come back to the performance characteristic curve that is in contrast to the uncontrolled case. However, it should be commented that our results show only that the critical operating point is locally stabilized. If γ_o is reduced below $0.85\gamma_c$, there appears to be a secondary Hopf bifurcation that induces hysteresis loop as well, which deserves further investigations.

Acknowledgment

This research was supported in part by AFOSR and ARO.

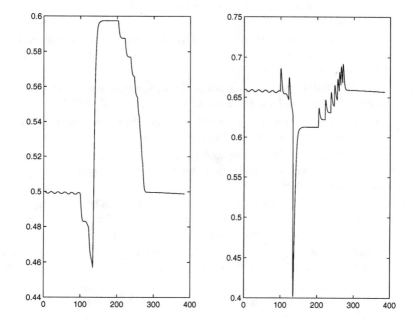

FIGURE 16.4
Time response of mass flow (left) and pressure rise (right).

References

[1] E. H. Abed and J.-H. Fu, "Local feedback stabilization and bifurcation control, I. Hopf bifurcation," *Systems and Control Letters*, vol. 7, pp. 11-17, 1986.

[2] E. H. Abed and J.-H. Fu, "Local feedback stabilization and bifurcation control, II. stationary bifurcation," *Systems and Control Letters*, vol. 8, pp. 467-473, 1986.

[3] E. H. Abed, P. K. Houpt, and W. M. Hosny, "Bifurcation analysis of surge and rotating stall in axial flow compressors," *J. of Turbomachinery*, vol. 115, pp. 817-824, 1993.

[4] R. A. Adomaitis and E. H. Abed, "Bifurcation analysis of nonuniform flow patterns in axial-flow gas compressors," *Proc. 1st World Congress of Nonlinear Analysis*, Aug. 1992.

[5] D. Aeyels, "Stabilization of a class of nonlinear systems by a smooth feedback control," *Systems and Control Letters*, vol. 5, pp. 467-473, 1985.

[6] O. O. Badmus, S. Chowdhury, and C. N. Nett, "Nonlinear control of surge in axial compression systems," *Automatica*, 1994.

[7] A. Banaszuk, H. A. Hauksson, and I. Mezic, "A Backstepping controller for Moore-Greitzer PDE model describing stall and surge in compressors," technical report, 1996.

[8] R. W. Brokett, "Asymptotic stability and feedback stabilization," in R. W. Brokett, R.S. Millman, and H.J. Sussmann (Eds.), *Differential Geometry Control Theory*, Birkhäuser, Boston, 1983, pp. 181-191.

[9] X. Chen, G. Gu, P. Martin, and K. Zhou, "Rotating stall control via bifurcation stabilization," vol 34, 437-443, April 1998.

[10] A. H. Epstein, J. E. F. Williams, and E. M. Greitzer, "Active suppression of aerodynamic instabilities in turbomachinery," *J. of Propulsion*, vol. 5, pp. 204-211, 1989.

[11] J.-H. Fu, "Lyapunov functions and stability criteria for nonlinear systems with multiple critical modes," *Mathematics of Contr., Sig., and Syst.*, vol. 7, pp. 255-278, 1994.

[12] G. Gu, S. Banda, and A. Sparks, "An overview of rotating stall and surge control for axial flow compressors," *Proc. of IEEE Conf. on Dec. and Contr.*, Kobe, Japan, Dec. 1996, pp. 2786-2791.

[13] G. Gu, A. Sparks, and S. Banda, "Bifurcation based nonlinear feedback control for rotating stall in axial flow compressors," *Int. J. of Contr.*, vol. 6, pp. 1241-1257, 1997.

[14] X. Chen, G. Gu, P. Martin, and K. Zhou, "Rotating stall control via bifurcation stabilization," *Automatica*, vol. 34, 437-443, April 1998.

[15] G. Gu, X. Chen, A. Sparks, and S. Banda, "Bifurcation stabilization with local output feedback," *SIAM J. of Optimiz. Contr.*, vol. 37, 934-956, 1999.

[16] G. Iooss and D. D. Joseph, *Elementary Stability and Bifurcation Theory*, Springer-Verlag, New York, 1980.

[17] W. Kang, "Bifurcation and normal form of nonlinear control systems – Parts I and II, *SIAM J. of Control and Optimization*, vol. 36, pp. 193-232, 1998.

[18] M. Krstic, J. M. Protz, J. D. Paduano, and P. V. Kokotovic, "Backstepping designs for jet engine stall and surge control," *Proc. of IEEE Conf. on Dec. and Contr.*, 1995, pp. 3049-3055.

[19] D.-C. Liaw and E. H. Abed, "Active control of compressor stall inception: A bifurcation-theoretical approach," *Automatica*, vol. 32, pp. 109-116, 1996.

[20] C. A. Mansoux, J. D. Setiawan, D. L. Gysling, and J. D. Paduano, "Distributed nonlinear modeling and stability analysis of axial compressor stall and surge," *Proc. of American Control Conf.*, 1994.

[21] F. E. McCaughan, "Bifurcation analysis of axial flow compressor stability," *SIAM J. of Applied Math.*, vol. 20, pp. 1232-1253, 1990.

[22] F. K. Moore and E. M. Greitzer, "A theory of post-stall transients in axial compressors: Part I – Development of the equations," *ASME J. of Engr. for Gas Turbines and Power*, vol. 108, pp. 68-76, 1986.

[23] J. D. Paduano, A. H. Epstein, L. Valavani, J. P. Longley, E. M. Greitzer, and G. R. Guenette, "Active control of rotating stall in a low-speed axial compressor," *J. of Turbomachinery*, vol. 115, 48-56, 1993.

[24] H. O. Wang, R. A. Adomatis, and E. H. Abed, "Nonlinear analysis and control of rotating stall in axial flow compressors," *Proc. of American Control Conf.*, 1994, pp. 2317-2321.

17

Bifurcations of Control Systems in Normal Form

Wei Kang

Department of Mathematics
Naval Postgraduate School
Monterey, CA 93943 USA
wkang@nps.navy.mil

Abstract

The state feedback control of bifurcations with quadratic or cubic degeneracy is addressed for systems with a single uncontrollable mode. Using invariants, stability characterizations are derived for transcritical bifurcation and pitchfork bifurcation. Results on the saddle-node and the cusp bifurcations are referred to [15, 17, 12]. References on related results such as bifurcations of systems under perturbed feedbacks are also given. An example of engine compressor control is introduced to illustrate the method of bifurcation control developed in this chapter.

17.1 Introduction

Nonlinear dynamical systems exhibit complicated performance around bifurcation points. As the parameter of a system is varied, changes may occur in the qualitative structure of its solutions around a point of bifurcation. Using a feedback to stabilize a system with bifurcations has been studied by many authors (see, for instance, [1, 4, 7, 9, 21, 24]). Many engineering applications of bifurcation control can be found in the literature (e.g., control of surge and rotating stall in engine compressors, flight control under high angle-of-attack). Quadratic and cubic feedbacks were introduced in [1] for the stabilization of bifurcated equilibria. It was proved in [1] that the periodic solution of a Hopf bifurcation can be stabilized by using state feedbacks. For the period doubling bifurcation, the method of harmonic balance was introduced in [7]. A feedback design method for delaying and stabilizing period doubling bifurcations was obtained. In [24], control laws were designed for the suppression of chaos in a thermal convection system model. A review of bifurcation and chaos in control systems can be found in [4]. More references on related topics can be found in [3], a bibliography of publications on bifurcation and chaos in control systems.

This chapter introduces a framework for the analysis and control of bifurcations. What makes this chapter unique is the approach based on the normal form and the invariants of nonlinear control systems. The two successful methods in the classical bifurcation theory are the normal form method and the projection method. In the normal form approach of bifurcation theory, dynamical systems are transformed into simple equivalent systems, which are called normal forms. The bifurcations of the systems in normal forms represent the qualitative behavior of general systems, thus classifies the bifurcations for general systems.

Different from the Poicaré's normal form of dynamical systems, the normal form of control systems is the canonical form under both the change of coordinates and the state feedback. The transformation group is larger than that used in dynamical systems, because the state feedback also serves as a transformation for systems with control inputs. An advantage of using the control system normal form is that the stability around bifurcations for a family of control systems is equivalent to the stability of their normal forms. This equivalence relation significantly simplifies the problem. It enables us to study a family of control systems with various bifurcations in a unified approach. A complete classification of bifurcations with quadratic or cubic degeneracy for systems having a single uncontrollable mode is summarized in [15, 17, 12].

Another advantage of the normal form approach is that the set of all equilibria of a control system (without feedback) in normal form can be found and it is approximately a quadratic surface. The geometry of equilibrium sets clearly shows the way in which feedbacks change the distribution of the equilibrium points in the closed-loop system. This is important because the graph of the equilibrium points determines the type of the bifurcation. For certain systems, the position of an equilibrium point determines its stability as well. Furthermore, understanding the geometry of equilibrium sets enables us to characterize bifurcations under feedbacks which are not zero at the critical point.

The Poincaré normal form of dynamical systems is the foundation of the normal form approach of bifurcation theory. The control system normal form adopted in the present chapter is different from those used in the literature of nonlinear dynamical systems without control inputs. Why is it necessary to introduce the control system normal form instead of adopting the Poincaré normal form of vector fields? In fact, even for a linear control system $\dot{x} = Ax + Bu$, the controller normal form is more useful than the diagonal form of A in the feedback design. The normal form of nonlinear control systems generalizes the linear controller form. An affine control system $\dot{x} = f(x) + g(x)u$ has two vector fields $f(x)$ and $g(x)$. Therefore, the normal form of a control system requires the simplification of both f and g simultaneously. The simplification of f does not necessarily result in a simple form for g. Furthermore, the transformation group of control systems consists of changes of coordinates and feedbacks. This is different from the normal form theory of dynamical systems where feedbacks are not considered. The resonant terms defined for the control system normal form characterize the nature of a control system because they are invariant under both changes of coordinates and state feedbacks. The results obtained in this chapter are intrinsic. They link the qualitative properties such as the bifurcation of control systems and its stability with their invariants.

17.2 Problem Formulation

Consider the following control system with a parameter

$$\dot{x} = f(x, \mu) + g(x, \mu)u, \quad f(0,0) = 0, \qquad (17.2.1)$$

where $x \in \mathbb{R}^n$ is the state variable, $u \in \mathbb{R}^m$ is the control input, and μ is the parameter. We assume that the rank of $g(x)$ is m at the point of interest. Unless it is otherwise specified, all vector fields and state feedbacks in this chapter are C^k for some sufficiently large $k > 0$. System (17.2.1) is said to be *linearly controllable* at $(x, \mu) = (0, 0)$ if its linearization (A, B),

$$A = \frac{\partial f}{\partial x}(0, 0), \quad B = g(0, 0)$$

is controllable. The origin $(x,\mu) = (0,0)$ is called an equilibrium or equilibrium point of (17.2.1) because $x(t) = 0$ is a constant solution if $\mu = 0$ and $u = 0$. Constant solutions may exist for other values of (x,μ,u). The *equilibrium set* is defined by

$$E = \{(x,\mu) | \text{there exists } u_0 \in \mathbb{R} \text{ such that} f(x,\mu) + g(x,\mu)u_0 = 0 \}.$$

A point (x,μ) is in E if and only if $f(x,\mu) + g(x,\mu)u_0 = 0$ for some constant number u_0. A point in E is called an *equilibrium* or *equilibrium point*. Feedbacks are not involved in this definition. In general, the value of u_0 corresponding to an equilibrium in E differs from point to point. If the control input u is substituted by a feedback $u = u(x)$, a *closed-loop equilibrium*, (x_0, μ_0), is defined by $f(x_0,\mu_0) + g(x_0,\mu_0)u(x_0) = 0$. The set of all closed-loop equilibria under a given feedback $u(x)$ is

$$E_c = \{(x,\mu) | f(x,\mu) + g(x,\mu)u(x) = 0\}.$$

The concept of an equilibrium set plays an important role in this chapter. It is known that the closed-loop equilibrium set E_c, in general, is changed if the feedback is varied. However, the set E_c of a given state feedback must be a subset of E. So, E consists of all possible closed-loop equilibria. The topology of E_c is induced from E.

The classical bifurcation theory studies the change of qualitative properties of dynamical systems as the parameters are varied. Qualitative properties include the topology of the equilibrium set, the stability, the existence of periodic solutions, etc. Control systems have two types of qualitative properties, which are those invariant under regular feedbacks (for example, the controllability, the stabilizability, and the topology of E) and those determined by the closed-loop system (for example, the closed-loop equilibria and the stability under a state feedback). Studying how these properties are changed with parameters leads to the following two bifurcation problems for control systems.

Problem 1. Bifurcation of control systems. It focuses on the change of qualitative properties of control systems (such as controllability, stabilizability, and the topology of E).

Problem 2. Bifurcation control using feedbacks. The problem focuses on the feedback design to achieve the stability around a critical point, or to achieve the desired performance by qualitatively changing a bifurcation.

Problem 1 was addressed in [13, 14] for systems with a single uncontrollable mode. In the present chapter, we focus on Problem 2. It is proved that the same control system may exhibit several different kinds of bifurcations under different feedbacks. Instead of focusing on a single bifurcation, we ask the following questions. What kinds of bifurcations can occur in

Problem Formulation

a given control system, and what is the relationship between bifurcations and control laws? It is a different view point from the existing bifurcation control approaches. It is known that local bifurcations at a linearly controllable point can be either removed or delayed by pole placement. In this chapter, we study systems which are not linearly controllable. The work is motivated by engineering problems such as engine compressors and submersible vehicles [22, 21, 18]. In addition to the engineering applications, our research on uncontrollable systems is also motivated by the fact that qualitative properties such as controllability and stabilizability of control systems are generic (they are not changed by a small variation of parameters) at a linearly controllable point. If a system is not linearly controllable at a point, nonlinear phenomena such as bifurcations are expected around the critical point.

It is assumed throughout this chapter that there exists a single uncontrollable mode (denoted by λ) in the linearization. The dimension of the state space is at least two ($n \geq 2$). If $\lambda \neq 0$, the sign of λ determines the stabilizability of the uncontrollable dynamics. Therefore, the variation of μ does not change the stability, i.e., there is no stationary bifurcation at $\mu = 0$. If $\lambda = 0$, the stability of the system depends on the value of the parameter. Different kinds of bifurcations occur in the performance. So, we focus on systems with $\lambda = 0$ in the following sections. Under a suitable linear change of coordinates and linear feedback, a system with a single uncontrollable mode $\lambda = 0$ can be transformed into one of the following forms (see [14]):

$$\dot{z} = f_1(z, x, \mu) + g_1(z, x, \mu)u$$
$$\dot{x} = A_2 x + B_2 u + f_2(z, x, \mu) + g_2(z, x, \mu)u \qquad (17.2.2)$$

or

$$\dot{z} = \mu + f_1(z, x, \mu) + g_1(z, x, \mu)u$$
$$\dot{x} = A_2 x + B_2 u + f_2(z, x, \mu) + g_2(z, x, \mu)u, \qquad (17.2.3)$$

where f_1, f_2 and their first derivatives equal zero at the origin $(z, x, \mu) = (0, 0, 0)$, g_1 and g_2 equal zero at the origin. The pair (A_2, B_2) is in the following Brunovsky form

$$A_2 = \begin{bmatrix} 0 & 1 & 0 & \cdots & 0 \\ 0 & 0 & 1 & \cdots & 0 \\ \cdot & \cdot & \cdot & & \cdot \\ 0 & 0 & 0 & \cdots & 1 \\ 0 & 0 & 0 & \cdots & 0 \end{bmatrix}_{(n-1)\times(n-1)} \quad B_2 = \begin{bmatrix} 0 \\ 0 \\ \vdots \\ 1 \end{bmatrix}. \qquad (17.2.4)$$

A feedback

$$u = \alpha(z, x, \mu) \qquad (17.2.5)$$

for bifurcation control is a smooth function of (z, x, μ) such that $\alpha(0,0,0) = 0$. The linearization of a feedback is in the following form:

$$\alpha(z, x, \mu) = a_z z + \sum_{i=1}^{n-1} a_i x_i + a_\mu \mu + O(z, x, \mu)^2. \qquad (17.2.6)$$

Notice that the value of μ is not always available. The function $\alpha(z, x, \mu)$ involves the parameter μ for two reasons. (1) Introducing μ in $\alpha(z, x, \mu)$ makes the theory more general. Feedbacks independent of μ form a subset of the feedbacks defined by (17.2.5). (2) By transforming the original system into its normal form, the linearization of the last equations in (17.2.2) and (17.2.3) has only one term which is the input u. However, the original system may have the terms with μ in the last equation. In this case, the term μ is absorbed in (17.2.5). So, a_μ in (17.2.6) comes from the original model. It is not necessarily zero.

In this chapter, a system is said to be *stable* if it is locally asymptotically stable. To achieve stability, feedbacks in this chapter are assumed to satisfy the following assumption.

Assumption A1: The state feedback (17.2.6) places the controllable poles in the left half plane, i.e., the eigenvalues of the matrix

$$A_2 + B_2 \begin{bmatrix} a_1 & a_2 & \cdots & a_{n-1} \end{bmatrix} \qquad (17.2.7)$$

are all in the left half plane.

It is known that $(-1)^n a_1$ equals the multiplication of all eigenvalues of the matrix (17.2.7). From A1, these eigenvalues are on the left half plane. So, we have the following lemma.

LEMMA 17.1
If a feedback (17.2.6) satisfies A1, then $a_1 < 0$.

17.3 Normal Forms and Invariants

In this section, nonlinear invariants are defined by the coefficients of resonant terms. Then, quadratic normal forms in [13, 14] are introduced without proof.

17.3.1 Resonant Terms and Invariants

In the classical theory of dynamical systems, a set of resonant terms was found for the homogeneous parts of nonlinear systems. The coefficients of resonant terms are invariant under homogeneous transformations. For systems with the Hopf bifurcation, the values of invariants determine the stability of the periodic solutions. For control systems, the invariants were introduced in [16] for linearly controllable systems. In [15, 17, 12], a set of invariants is found for systems which are not linearly controllable. They play a key role in the stability analysis for control systems with bifurcations. These invariants are introduced in this section.

The quadratic and cubic terms in the Taylor expansion of vector fields are used in the proofs of many results. The homogeneous parts of degree d for f_i and g_i in (17.2.2) and (17.2.3) are denoted by $f_i^{[d]}$ and $g_i^{[d]}$. For instance, the quadratic terms in the Taylor expansion of $f_1 + g_1 u$ have the form $f_1^{[2]}(z, x, \mu) + g_1^{[1]}(z, x, \mu)u$. The components of $f_i^{[d]}$ and $g_i^{[d-1]}$ are homogeneous polynomials of degree d and $d-1$, respectively. A homogeneous transformation of degree d for control systems consists of the change of coordinates and state feedbacks in the following form

$$z = \bar{z} + \phi_1^{[d]}(\bar{z}, \bar{x}, \mu), \quad x = \bar{x} + \phi_2^{[d]}(\bar{z}, \bar{x}, \mu),$$
$$u = \bar{u} + \alpha^{[d]}(\bar{z}, \bar{x}, \mu) + \beta^{[d-1]}(\bar{z}, \bar{x}, \mu)\bar{u},$$
(17.3.1)

where $\bar{z}$ and $\bar{x}$ are the new coordinates, and $\bar{u}$ is the new control input introduced by the regular feedback. A transformation of degree d does not change the terms of degree less than d in a control system. If $d = 2$, (17.3.1) is called a quadratic transformation. If $d = 3$, it is called a cubic transformation.

DEFINITION 17.1 *Consider (17.2.2) or (17.2.3). A homogeneous term in $f_i^{[d]}(z, x, \mu)$ or $g_i^{[d-1]}(z, x, \mu)u$, is called a resonant term if any transformation of the form (17.3.1) leaves the coefficient of the term invariant. The coefficient of a resonant term is called an invariant.*

For instance, if (17.3.1) is applied to (17.2.2), it can be proved that the term $\bar{z}^2$ in $\bar{f}_1^{[2]}$ of the resulting system has the same coefficient as the term z^2 in $f_1^{[2]}$ of (17.2.2). So, z^2 in $f_1^{[2]}$ is resonant. Notice that the resonant terms of control systems are different from the resonant terms in the classical theory of dynamical systems. A resonant term in Definition 17.1 is invariant under both changes of coordinates and state feedbacks. However, the classical dynamic systems theory does not deal with any feedback. In the next theorem, resonant terms are found for (17.2.2) and (17.2.3). Define

$$R^{[d]}(z, x_1, \mu) = f_1^{[d]}(z, x, \mu)|_{x_2 = x_3 = \cdots = x_{n-1} = 0},$$
$$R_1^{[d]}(z, x_1) = R(z, x_1, 0),$$
(17.3.2)

where $f_1^{[d]}(z, x, \mu)$ is the homogeneous vector field of degree d from the Taylor expansion of $f_1(z, x, \mu)$ in (17.2.2) and (17.2.3).

THEOREM 17.2
In (17.2.2), all terms of $R^{[d]}(z, x_1, \mu)$ are resonant. In (17.2.3), all terms of $R_1^{[d]}(z, x_1)$ are resonant.

PROOF Consider the system (17.2.2). Suppose that (17.2.2) is transformed into the following system by (17.3.1):
$$\dot{\bar{z}} = \bar{f}_1(\bar{z}, \bar{x}, \mu) + \bar{g}_1(\bar{z}, \bar{x}, \mu)\bar{u}$$
$$\dot{\bar{x}} = A_2\bar{x} + B_2\bar{u} + \bar{f}_2(\bar{z}, \bar{x}, \mu) + \bar{g}_2(\bar{z}, \bar{x}, \mu)\bar{u}.$$

It was proved in [11] that the homogeneous parts of f_1 and $\bar{f}_1$ satisfy the homological equation
$$\frac{\partial \phi_1^{[d]}(z, x, \mu)}{\partial x} A_2 x = f_1^{[d]}(z, x, \mu) - \bar{f}_1^{[d]}(z, x, \mu).$$

However,
$$\frac{\partial \phi_1^{[d]}}{\partial x} A_2 x = \frac{\partial \phi_1^{[d]}}{\partial x_1} x_2 + \cdots + \frac{\partial \phi_1^{[d]}}{\partial x_{n-2}} x_{n-1}. \qquad (17.3.3)$$

Therefore, every term of $f_1^{[d]} - \bar{f}_1^{[d]}$ has at least one of the variables $x_2, \ldots, x_{n-1}$. The terms in $R(z, x_1, \mu)$ do not appear in $f_1^{[d]} - \bar{f}_1^{[d]}$. This implies that the function $R^{[d]}(z, x_1, \mu)$ in $f_1^{[d]}$ is invariant under (17.3.1).

Now, consider the system (17.2.3). The homological equation for $f_1^{[d]}$ is
$$\frac{\partial \phi_1^{[d]}(z, x, \mu)}{\partial z} \mu + \frac{\partial \phi_1^{[d]}(z, x, \mu)}{\partial x} A_2 x = f_1^{[d]}(z, x, \mu) - \bar{f}_1^{[d]}(z, x, \mu).$$

From (17.3.3), every nonzero term in $f_1^{[d]} - \bar{f}_1^{[d]}$ has at least one of the variables $\mu, x_2, \ldots, x_{n-1}$. This implies that the coefficients in $R_1^{[2]}(z, x_1)$ are not changed by (17.3.1). □

In the following, the coefficients of resonant terms are denoted by γ with corresponding subindices. For example, the coefficient of z^2 in $R_1^{[2]}(z, x_1)$ or $R^{[2]}(z, x_1, \mu)$ is γ_{zz}, the coefficient of $zx_1\mu$ is $\gamma_{zx_1\mu}$, etc. Based on this theorem, the coefficients $\gamma_{zz}, \gamma_{zx_1}, \gamma_{z\mu}, \gamma_{x_1x_1}, \gamma_{x_1\mu}, \gamma_{\mu\mu}$ in (17.2.2) and the coefficients $\gamma_{zz}, \gamma_{zx_1}, \gamma_{x_1x_1}$ in (17.2.3) are called (quadratic) invariants. They are part of the quadratic invariants introduced in [16, 14] using Lie brackets. The quadratic functions of resonant terms $R^{[2]}(z, x_1, \mu)$ and $R_1^{[2]}(z, x_1)$ determine two symmetric matrices,

$$Q = \begin{bmatrix} \gamma_{zz} & \dfrac{\gamma_{zx_1}}{2} & \dfrac{\gamma_{z\mu}}{2} \\ \dfrac{\gamma_{zx_1}}{2} & \gamma_{x_1x_1} & \dfrac{\gamma_{x_1\mu}}{2} \\ \dfrac{\gamma_{z\mu}}{2} & \dfrac{\gamma_{x_1\mu}}{2} & \gamma_{\mu\mu} \end{bmatrix}, \quad Q_1 = \begin{bmatrix} \gamma_{zz} & \dfrac{\gamma_{zx_1}}{2} \\ \dfrac{\gamma_{zx_1}}{2} & \gamma_{x_1x_1} \end{bmatrix}. \qquad (17.3.4)$$

Normal Forms and Invariants

In this chapter, the quadratic function defined by Q (or Q_1) is also denoted by $Q(x,y,z)$ (or $Q_1(x,y)$), i.e.,

$$Q(x,y,z) = \begin{bmatrix} x & y & z \end{bmatrix} Q \begin{bmatrix} x & y & z \end{bmatrix}^T,$$
$$Q_1(x,y) = \begin{bmatrix} x & y \end{bmatrix} Q_1 \begin{bmatrix} x & y \end{bmatrix}^T. \tag{17.3.5}$$

Equivalently, $Q(z,x_1,\mu) = R^{[2]}(z,x_1,\mu)$, $Q_1(z,x_1) = R_1^{[2]}(z,x_1)$.

17.3.2 Quadratic Normal Forms

Since systems with the same normal form have equivalent bifurcations, most proofs in this chapter are given for quadratic normal forms. From [14] (17.2.2) and (17.2.3) can be transformed into a unique system in normal form by a suitable quadratic transformation of the form (17.3.1) with $d = 2$.

For (17.2.2), the normal form is

$$\dot{z} = \sum_{i=2}^{n-1} \gamma_{x_i x_i} x_i^2 + Q(z, x_1, \mu) + O(z, x, \mu, u)^3,$$
$$\dot{x} = A_2 x + B_2 u + \tilde{f}_2^{[2]}(x) + O(z, x, \mu, u)^3. \tag{17.3.6}$$

For (17.2.3), the normal form is

$$\dot{z} = \mu + \sum_{i=2}^{n-1} \gamma_{x_i x_i} x_i^2 + \gamma_{x_1 \mu} x_1 \mu + Q_1(z, x_1) + O(z, x, \mu, u)^3,$$
$$\dot{x} = A_2 x + B_2 u + \tilde{f}_2^{[2]}(x) + O(z, x, \mu, u)^3, \tag{17.3.7}$$

where $\tilde{f}_2^{[2]}(x)$ is in the extended controller form of [16]. Details are omitted since it is not used in this chapter. Before the end of this section, we introduce the following well-known result on stationary bifurcations (see [8, 10, 25]).

THEOREM 17.3
Consider the following one-dimensional dynamical system with parameter μ:

$$\dot{x} = f(x, \mu), \quad f(0,0) = 0, \quad x \in \mathbb{R}. \tag{17.3.8}$$

(i) *It has a saddle-node bifurcation at the origin if*

$$f_x(0,0) = 0, \quad f_\mu(0,0) \neq 0, \quad f_{xx}(0,0) \neq 0. \tag{17.3.9}$$

(ii) It has a transcritical bifurcation at the origin if

$$f_x(0,0) = 0, \quad f_\mu(0,0) = 0,$$

$$f_{xx}(0,0) \neq 0, \quad f_{x\mu}^2(0,0) - f_{xx}(0,0)f_{\mu\mu}(0,0) > 0. \tag{17.3.10}$$

(iii) It has a pitchfork bifurcation at the origin if

$$f_x(0,0) = 0, \tag{17.3.11}$$

$$f_\mu(0,0) = 0, \quad f_{xx}(0,0) = 0,$$

$$f_{x\mu}(0,0) \neq 0, \quad f_{xxx}(0,0) \neq 0. \tag{17.3.12}$$

If $f_{xxx}(0,0) < 0$, the pitchfork bifurcation is supercritical. If $f_{xxx}(0,0) > 0$, it is subcritical.

17.4 Bifurcations of System with Quadratic Degeneracy

As an illustrating example, the bifurcations of (17.2.2) are studied in this section and § 17.5 using the method of normal form and invariants. As we know that the locations of the closed-loop equilibrium points depend on the choice of the feedback function $u = \alpha(x, \mu)$, different feedback may result in different equilibrium set E_c. However, under any state feedback, a closed-loop equilibrium point must lie in the set E. In [14], it was proved that the projection of the (open-loop) equilibrium set E into $zx_1\mu$-space is a homeomorphism. For the normal form (17.3.6), the projection of E is approximately a cone defined by

$$Q(z, x_1, \mu) = 0,$$
$$x_i = 0, \text{ for } i = 2, 3, \cdots, n-1, \tag{17.4.1}$$

provided that Q is indefinite with full rank. The third and higher degree terms of (z, x_1, μ) are omitted in the approximation. Under the feedback $\alpha(z, x, \mu) = \sum a_i x_i + a_z z + a_\mu \mu + O(z, x, \mu)^2$, the closed-loop equilibrium set E_c is the intersection between E and the surface $\alpha(z, x, \mu) = 0$. Therefore, E_c is approximated by the intersection between the plane

$$a_z z + a_1 x_1 + a_\mu \mu = 0 \tag{17.4.2}$$

and the cone (17.4.1). Understanding the geometry of E is important for the study of the variation of the closed-loop equilibrium set E_c under state feedbacks. For example, the geometry of E_c for normal form (17.3.6) has two generic cases, which are (1) the intersection between the cone E and the plane (17.4.2) consists of two lines, which indicate a transcritical bifurcation; and (2) the only intersection point is the origin, no bifurcation occurs. The first case is characterized by (ii) of Theorem 17.4. The second case is proved in (i) of Theorem 17.4.

Define a matrix $\bar{Q}$ from the linearization of a feedback and the quadratic invariants of (17.2.2) by

$$\bar{Q} = \begin{bmatrix} a_1 & -a_z & 0 \\ 0 & -a_\mu & a_1 \end{bmatrix} Q \begin{bmatrix} a_1 & 0 \\ -a_z & -a_\mu \\ 0 & a_1 \end{bmatrix}, \qquad (17.4.3)$$

where Q is defined by (17.3.4). The matrix $\bar{Q}$ is used in the next theorem to characterize the bifurcation. Following the notation introduced in § 17.2, E_c represents the set of closed-loop equilibrium points, i.e.,

$$E_c = \{(x, \mu) | f(x, \mu) + g(x, \mu)\alpha(x, \mu) = 0\}. \qquad (17.4.4)$$

THEOREM 17.4
Consider a closed-loop system (17.2.2) – (17.2.6) satisfying A1. Suppose

$$Q_1(a_1, -a_z) \neq 0. \qquad (17.4.5)$$

(i) If $\bar{Q}$ is sign definite, then $(z, x, \mu) = (0, 0, 0)$ is an isolated equilibrium point of the closed-loop system. It is unstable.
(ii) If $\bar{Q}$ is indefinite with full rank, then the closed-loop system has a transcritical bifurcation around the origin.
(iii) Assume that the feedback satisfies the condition in (ii). Given any $(z, x, \mu) \in E_c$ in a neighborhood of the origin, it is locally asymptotically stable if

$$\begin{bmatrix} a_1 & -a_z & 0 \end{bmatrix} Q \begin{bmatrix} z & x_1 & \mu \end{bmatrix}^T > 0. \qquad (17.4.6)$$

The system is unstable if

$$\begin{bmatrix} a_1 & -a_z & 0 \end{bmatrix} Q \begin{bmatrix} z & x_1 & \mu \end{bmatrix}^T < 0. \qquad (17.4.7)$$

This result ruled out the saddle-node bifurcation for normal form (17.3.6). In fact, the saddle-node bifurcation and the cusp bifurcation exist only in the systems with normal form (17.3.7). Details can be found in [15, 17, 12].

Since (17.2.2) is equivalent to its normal form (17.3.6), it is important to find a center manifold and the reduced system on it for the normal form (17.3.6). This is obtained in the following lemma.

LEMMA 17.5
Consider the quadratic normal form (17.3.6). Under the state feedback (17.2.6), the center manifold of the closed-loop system satisfies

$$x_1 = -\frac{a_z}{a_1}z - \frac{a_\mu}{a_1}\mu + O(z, \mu)^2, \quad x_i = O(z, \mu)^2 \quad for \ i = 2, \cdots, n-1.$$

$$(17.4.8)$$

The reduced system on the center manifold satisfies

$$\dot{z} = \frac{1}{a_1^2} [\, z \;\; \mu \,] \bar{Q} [\, z \;\; \mu \,]^T + O(z, \mu)^2. \tag{17.4.9}$$

PROOF From [2], the center manifold is determined by a function $x = \pi(z, \mu)$, where $\pi(z, \mu)$ can be approximated by polynomials. The function $\pi(z, \mu)$ satisfies an equation of the following form

$$A\pi(z, \mu) + Bu(z, \pi, \mu) + O(z, \pi, \mu)^2 = \frac{\partial \pi}{\partial z} O(z, \pi, \mu)^2.$$

Denote the linear part of $\pi(z, \mu)$ by $\pi^{[1]}(z, \mu)$. The linearization of this equation is

$$\pi_2^{[1]} = 0, \quad \pi_3^{[2]} = 0, \quad \cdots, \quad a_z z + a_1 \pi_1^{[1]}(z, \mu) + a_\mu \mu = 0. \tag{17.4.10}$$

It is easy to check that the linear part of (17.4.8) satisfies (17.4.10). The functions in (17.4.8) are equivalent to

$$\begin{bmatrix} z \\ x_1 \\ \mu \end{bmatrix} = \frac{1}{a_1} \begin{bmatrix} a_1 & 0 \\ -a_z & -a_\mu \\ 0 & a_1 \end{bmatrix} \begin{bmatrix} z \\ \mu \end{bmatrix} + O(z, \mu)^2,$$

$$x_i = O(z, \mu)^2, \quad \text{for } 2 \le i \le n - 1.$$

Substituting this relation into the z dynamical equation in (17.3.6), we obtain (17.4.9) as the reduced system on the center manifold.

PROOF (The proof of Theorem 17.4). Since (17.2.2) can be transformed into its normal form (17.3.6) by a quadratic transformation, and since the conditions in the theorem are invariant under quadratic transformations, it is enough to prove the result for the quadratic normal form (17.3.6).

(i) The closed-loop system is equivalent to its reduced system (17.4.9) on the center manifold. Denote the right side of (17.4.9) by $f_c(z, \mu)$. If $\bar{Q}$ is sign definite, then $(z, \mu) = (0, 0)$ is the unique local solution of $f_c(z, \mu) = 0$. Therefore, the origin is an isolated equilibrium point. In this case, the reduced system (17.4.9) at $\mu = 0$ is

$$\dot{z} = \frac{1}{a_1^2} Q_1(a_1, -a_z) z^2 + O(z)^3.$$

Since $det(\bar{Q}) > 0$ and since $Q_1(a_1, -a_z)$ is the first diagonal entry in $\bar{Q}$, we know that $Q_1(a_1, -a_z) \ne 0$. The system is unstable. Part (i) is proved.

(ii) Now, assume that $det(\bar{Q}) < 0$. It is obvious that

$$\frac{\partial f_c}{\partial z}(0, 0) = 0, \quad \frac{\partial f_c}{\partial \mu}(0, 0) = 0. \tag{17.4.11}$$

It is easy to check that

$$\frac{\partial^2 f_c}{\partial z^2}(0,0) = \frac{1}{a_1^2} Q_1(a_1, -a_z) \neq 0. \tag{17.4.12}$$

$$\frac{\partial^2 f_c}{\partial z^2}(0,0)\frac{\partial^2 f_c}{\partial \mu^2}(0,0) - \left(\frac{\partial^2 f_c}{\partial z \mu}(0,0)\right)^2 = 4\det(\bar{Q}) < 0. \tag{17.4.13}$$

Therefore, the conditions in (ii) of Theorem 17.3 are satisfied. This implies that the closed-loop system has a transcritical bifurcation.

(iii) The stability of the closed-loop system agrees with the reduced system on the center manifold. It is easy to check that

$$\frac{\partial f_c}{\partial z} = \frac{2}{a_1^2} [\,1 \ \ 0\,] \bar{Q} [\,z \ \ \mu\,]^T + O(z,\mu)^2. \tag{17.4.14}$$

If (z, x, μ) is in E_c, then (z, μ) is an equilibrium point on the center manifold,

$$[\,z \ \ \mu\,]\bar{Q}[\,z \ \ \mu\,]^T + O(z,\mu)^3 = 0, \quad x_1 = -\frac{a_z}{a_1}z - \frac{a_\mu}{a_1}\mu + O(z,\mu)^2. \tag{17.4.15}$$

Therefore,

$$[\,z \ \ \mu\,] = [\,z_0 \ \ \mu_0\,]t + O(t)^2, \quad x_1 = (-\frac{a_z}{a_1}z_0 - \frac{a_\mu}{a_1}\mu_0)t + O(t)^2, \tag{17.4.16}$$

where $t \in \mathbb{R}$ and $(z_0, \mu_0) \neq (0,0)$ satisfies

$$[\,z_0 \ \ \mu_0\,]\bar{Q}[\,z_0 \ \ \mu_0\,]^T = 0. \tag{17.4.17}$$

From (17.4.3) and (17.4.16), we have

$$[\,a_1 \ \ -a_z \ \ 0\,]Q[\,z \ \ x_1 \ \ \mu\,]^T = \frac{1}{a_1}[\,1 \ \ 0\,]\bar{Q}[\,z_0 \ \ \mu_0\,]^T t + O(t)^2 \tag{17.4.18}$$

If we can prove that

$$[\,1 \ \ 0\,]\bar{Q}[\,z_0 \ \ \mu_0\,]^T \neq 0 \tag{17.4.19}$$

then the sign of $\frac{1}{a_1^2}[\,1 \ \ 0\,]\bar{Q}[\,z_0 \ \ \mu_0\,]^T t$, which agrees with that of $\frac{\partial f_c}{\partial z}$, is opposite to the sign of the number given by

$$[\,a_1 \ \ -a_z \ \ 0\,]Q[\,z \ \ x_1 \ \ \mu\,]^T$$

because $a_1 < 0$. Therefore, (17.4.6) implies that $\frac{\partial f_c}{\partial z} < 0$ at the point in E_c around zero. The closed-loop system is locally asymptotically stable. Similarly, (17.4.7) implies that the system is unstable.

Now, we prove (17.4.19) by contradiction. Suppose

$$\begin{bmatrix} 1 & 0 \end{bmatrix} \bar{Q} \begin{bmatrix} z_0 & \mu_0 \end{bmatrix}^T = 0. \qquad (17.4.20)$$

Because $\det(\bar{Q}) \neq 0$, and because $(z_0, \mu_0) \neq (0, 0)$, we have $\bar{Q} \begin{bmatrix} z_0 & \mu_0 \end{bmatrix}^T \neq 0$. Equation (17.4.17) implies that $\begin{bmatrix} z_0 & \mu_0 \end{bmatrix} \bar{Q} \begin{bmatrix} z_0 & \mu_0 \end{bmatrix}^T = 0$. Comparing this equation with (17.4.20), we have $\mu_0 = 0$ and $z_0 \neq 0$. Therefore, (17.4.20) implies $q_{1\,1} = 0$, where $q_{1\,1}$ is the entry in $\bar{Q}$ at the upper left corner. However, $q_{1\,1} = Q(a_1, -a_z)$. It contradicts (17.4.5). Therefore, (17.4.19) is true.

REMARK 17.6 From (17.4.14), it is obvious that the system is locally asymptotically stable at $(z, x, \mu) \in E_c$ if $\begin{bmatrix} 1 & 0 \end{bmatrix} \bar{Q} \begin{bmatrix} z & \mu \end{bmatrix} < 0$. This is another method of stability testing.

17.5 Bifurcations of System with Cubic Degeneracy

Theorem 17.4 deals with bifurcation control under the assumption $Q_1(a_1, -a_z) \neq 0$. In the following, we study the case in which $Q_1(a_1, -a_z) = 0$. Since $(z, x_1, \mu) = (a_1, -a_z, 0)$ is a point in the intersection of (17.4.2) and the zx_1-plane, the condition $Q_1(a_1, -a_z) = 0$ implies that the closed-loop equilibrium set E_c has a branch tangent to the zx_1-plane and orthogonal to the μ axis. The set E_c is not transversal to the zx_1-plane. In the following, it is proved that the quadratic part of the reduced system on the center manifold is degenerated. The bifurcation is pitchfork. Because of the cubic degeneracy nature of the bifurcation, we need cubic invariants. The function of cubic resonant terms, $R^{[3]}(z, x_1)$, of the normal form (17.3.6) is denoted by $C(z, x_1)$, which is

$$C(z, x_1) = R_1^{[3]}(z, x_1) = f_1^{[3]}(z, x, \mu)|_{x_2 = x_3 = \cdots = x_{n-1} = \mu = 0}, \qquad (17.5.1)$$

where $f_1^{[3]}$ represents the cubic part in the first equation of (17.3.6). The state feedback for bifurcation control is

$$\begin{aligned} u &= \alpha(z, x, \mu), \\ \alpha(z, x, \mu) &= a_z z + a_1 x_1 + \cdots + a_{n-1} x_{n-1} + a_\mu x_\mu \\ &\quad + \alpha^{[2]}(z, x, \mu) + O(z, x, \mu)^3, \end{aligned} \qquad (17.5.2)$$

where $\alpha^{[2]}$ is a quadratic homogeneous polynomial. The coefficients in $\alpha^{[2]}$ are denoted by a_{zz}, $a_{z\mu}$, $a_{x_1 x_1}$, $a_{x_1 \mu}$, etc. The following quadratic function from $\alpha^{[2]}$ is useful:

$$\alpha^{[2]}_{zx_1}(z, x_1) = a_{zz}z^2 + a_{zx_1}zx_1 + a_{x_1x_1}x_1^2, \qquad (17.5.3)$$

i.e., $\alpha^{[2]}_{zx_1}$ is the restriction of $\alpha^{[2]}$ to the zx_1-plane. To simplify the notation, we define

$$D = a_1 C(a_1, -a_z) + (2a_z\gamma_{x_1x_1} - a_1\gamma_{zx_1})\alpha^{[2]}_{zx_1}(a_1, -a_z), \qquad (17.5.4)$$

where $\gamma_{x_1x_1}$ and γ_{zx_1} are quadratic invariants in (17.3.4).

THEOREM 17.7
Consider a closed-loop system (17.3.6) − (17.5.2) *satisfying A1.*
(i) *Suppose*

$$Q_1(a_1, -a_z) = 0. \qquad (17.5.5)$$

Then the closed-loop system has a pitchfork bifurcation at the origin provided

$$\begin{aligned} &D \neq 0, \\ &\begin{bmatrix} a_1 & -a_z & 0 \end{bmatrix} Q \begin{bmatrix} 0 & -a_\mu & a_1 \end{bmatrix}^T \neq 0. \end{aligned} \qquad (17.5.6)$$

(ii) *The pitchfork bifurcation is supercritical if $D < 0$. It is subcritical if $D > 0$.*

The proof of the theorem can be found in [15, 17]. The basic idea is the same as the proof of 17.4, namely, finding the approximation of the center manifold and the reduced system in terms of invariants. Then, apply Theorem 17.3.

Since $\alpha^{[2]}_{zx_1}$ is in the feedback, its coefficients are adjustable. There always exist suitable quadratic functions $\alpha^{[2]}_{zx_1}$ which render the pitchfork bifurcation supercritical, provided

$$2a_z\gamma_{x_1x_1} - a_1\gamma_{zx_1} \neq 0. \qquad (17.5.7)$$

This condition is related to the rank of Q_1. From $Q_1(a_1, -a_z) = 0$, we have

$$\begin{bmatrix} a_1, -a_z \end{bmatrix} \begin{bmatrix} \gamma_{zz}a_1 - \dfrac{\gamma_{zx_1}}{2}a_z & \dfrac{\gamma_{zx_1}}{2}a_1 - \gamma_{x_1x_1}a_z \end{bmatrix}^T = 0. \qquad (17.5.8)$$

Therefore, if $2a_z\gamma_{x_1x_1} - a_1\gamma_{zx_1} = 0$, (17.5.8) implies $a_1(\gamma_{zz}a_1 - \dfrac{\gamma_{zx_1}}{2}a_z) = 0$. Since $a_1 \neq 0$, we have $Q_1 \begin{bmatrix} a_1 & -a_z \end{bmatrix}^T = \begin{bmatrix} 0 & 0 \end{bmatrix}^T$. So, $rank(Q_1) < 2$. This is equivalent to saying that $rank(Q_1) = 2$ implies (17.5.7). Therefore, we have

COROLLARY 17.8
Suppose that (17.3.6) − (17.5.2) *satisfies* (17.5.5) *and* (17.5.6). *If* Q_1 *has full rank, then there exists a quadratic function* $\alpha^{[2]}_{zx_1}(z, x_1)$ *for the nonlinear feedback such that the closed-loop system has a supercritical pitchfork bifurcation.*

REMARK 17.9 If $C(a_1, -a_z) > 0$, then $\alpha^{[2]}_{zx_1}(z, x_1) = 0$ implies $D < 0$, so that a linear feedback renders the pitchfork bifurcation supercritical.

17.6 Application Example of Bifurcation Control

Now, we introduce the following Moore-Greitzer model of engine compressors as an example of Theorem 17.4. The system exhibits various bifurcation phenomena (see, for instance, [22, 21, 19, 5]). The bifurcations effectively reduce the performance of aeroengines and limit further improvements on reliability and efficiency of future airplanes. Using feedback to control compressors in the presence of bifurcations has been studied by many authors [20, 21, 19]. In this section, we use the compressor model to illustrate some ideas of feedback design based on the results of § 17.4 and § 17.5. In the following, the results obtained in [19] are proved using Theorem 17.4. The simplest model that describes the system is a three-state O.D.E. in Moore and Greitzer [23],

$$\dot{R} = \sigma R(-2\phi - \phi^2 - R),$$
$$\dot{\phi} = -\psi - \frac{3}{2}\phi^2 - \frac{1}{2}\phi^3 - 3R\phi - 3R, \qquad (17.6.1)$$
$$\dot{\psi} = \phi - \sqrt{\psi + \psi_0}(\frac{2}{\sqrt{\psi_0}} + \mu + u) + 2,$$

where $R \geq 0$ is the normalized stall cell squared amplitude, ϕ is the mass flow, ψ is the pressure rise, and ψ_0 and σ are constant positive numbers. The control input u can be changed by varying the throttle opening. The system has an uncertain parameter μ. The values of ϕ, ψ, and u are shifted by a constant [20] so that the origin is the focal bifurcation point. A simple linear change of coordinates $z = R$, $x_1 = \phi$, and $x_2 = -\psi - 3R$ transforms the dynamics into the following system in the form of (17.2.3):

$$\dot{z} = -2\sigma z x_1 - \sigma z^2 - \sigma z x_1^2,$$
$$\dot{x}_1 = x_2 - \frac{3}{2}x_1^2 - \frac{1}{2}x_1^3 - 3zx_1, \qquad (17.6.2)$$
$$\dot{x}_2 = -\frac{3}{\psi_0}z - x_1 - \frac{1}{\psi_0}x_2 + \sqrt{\psi_0}(\mu + u) + O(z, x_1, x_2, \mu, u)^2.$$

Application Example of Bifurcation Control

Although the quadratic part of the system is not in normal form, the invariant matrices Q and Q_1 are found from the resonant terms

$$Q = \begin{bmatrix} Q_1 & 0 \\ 0 & 0 \end{bmatrix}, \quad Q_1 = \begin{bmatrix} -\sigma & -\sigma \\ -\sigma & 0 \end{bmatrix}.$$

So, Q_1 is indefinite. In zx_1-plane (or equivalently $R\phi$-plane), the graph of the equilibrium set E is shown in Figure 17.1a.

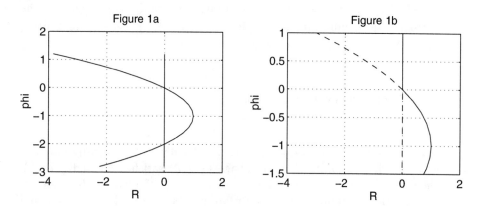

FIGURE 17.1
The equilibrium set of the engine compressor model.

The state of the real system always stays in the region $R \geq 0$. Therefore, it is desired to find state feedbacks which render the system asymptotically stable at equilibrium points with $R > 0$. The equilibrium points with $R < 0$ are meaningless. Consider a state feedback

$$u = k_1 R + k_2\phi + k_3\psi + O(R,\phi,\psi)^2. \quad (17.6.3)$$

In the coordinates (z, x_1, x_2), (17.6.3) is

$$u = (k_1 - 3k_3)z + k_2 x_1 - k_3 x_2 + O(z, x_1, x_2)^2. \quad (17.6.4)$$

Substituting (17.6.4) into (17.6.2), the closed-loop system satisfies

$$a_z = -\frac{3}{\psi_0} + \sqrt{\psi_0}(k_1 - 3k_3), \quad a_1 = -1 + \sqrt{\psi_0}k_2,$$
$$a_2 = -\frac{1}{\psi_0} - \sqrt{\psi_0}k_3, \quad a_\mu = \sqrt{\psi_0}. \quad (17.6.5)$$

It is easy to check that

$$Q_1(a_1, -a_z) = -\sigma a_1(a_1 - 2a_z), \quad \bar{Q} = \sigma \begin{bmatrix} a_1(a_1 - 2a_z) & -a_1 a_\mu \\ -a_1 a_\mu & 0 \end{bmatrix}. \quad (17.6.6)$$

Equations in (17.6.6) imply that the conditions in (ii) of Theorem 17.4 are satisfied if $a_1 \neq 2a_z$. Therefore, the bifurcation is transcritical. In the following, we use the result (iii) of Theorem 17.4 to find all feedbacks which stabilize the system at the equilibrium points with $R > 0$. From (17.6.2), the equilibrium points with $z > 0$ satisfy $z = -(2x_1 + x_1^2)$, and

$$\begin{bmatrix} a_1 & -a_z \end{bmatrix} Q_1 \begin{bmatrix} z & x_1 \end{bmatrix}^T = \sigma\left((a_1 - 2a_z)x_1 + (a_1 - a_z)x_1^2\right), \quad x_1 < 0. \tag{17.6.7}$$

So, (17.6.7) and Theorem 17.4 imply that, to guarantee the stability of the system at the equilibrium points with $z > 0$, we need $a_z > \dfrac{a_1}{2}$. Substituting (17.6.4) into this inequality, we have

$$2k_1 - k_2 - 6k_3 > (6 - \psi_0)/\psi_0^{\frac{3}{2}}. \tag{17.6.8}$$

Another branch of equilibrium points satisfies $z = 0$, so

$$\begin{bmatrix} a_1 & -a_z \end{bmatrix} Q_1 \begin{bmatrix} z & x_1 \end{bmatrix}^T = -\sigma a_1 x_1.$$

On this curve, the closed-loop system is locally asymptotically stable if $x_1 > 0$. In summary, if the feedback (17.6.3) satisfies A1 and (17.6.8), then the closed-loop system has a transcritical bifurcation. It is locally asymptotically stable at the equilibrium points where $R > 0$ or $R = 0$ and $\phi > 0$. The closed-loop equilibrium points are shown in Figure 17.1b. The system is locally stable on the solid curve and unstable on the dotted curve.

17.7 Conclusions

In this chapter, the method of bifurcation control using normal forms and invariants is illustrated for systems with a single uncontrollable mode. The normal form (17.3.6) is adopted as an example. A complete classification of bifurcations for normal forms (17.3.6) and (17.3.7) can be found in [15, 17, 12]. Feedbacks can be designed based on the results in these references to achieve the stability or to achieve the desired bifurcation pattern. Another result which is not included in the present chapter is the bifurcation analysis for feedbacks which are not zero at the bifurcation point. It is shown in [15] that a transcritical bifurcation of (17.3.6) is bifurcated into two saddle-node bifurcations if the feedback is perturbed from zero at the bifurcation point. For the system (17.3.7) with cubic degeneracy, a perturbation of a feedback causes either a hysteresis or no bifurcation, depending on the value of the feedback at the critical point.

The approach based on control system normal forms can certainly be used for the study of other bifurcations which are not addressed here. For

Conclusions 387

instance, the normal form of control systems with uncontrollable dynamics of dimension greater or equal to two is derived in [6]. It includes some interesting cases such as a pair of imaginary uncontrollable modes or a double zero uncontrollable mode. Applications of normal forms and invariants in the study of chaos and discrete time systems are also interesting topics for further research.

References

[1] E. H. Abed and J.-H. Fu, "Local feedback stabilization and bifurcation control, I-II: Stationary bifurcation, *Systems & Control Letters*, vol. 8, pp. 467-473, 1987.

[2] J. Carr, *Application of Center Manifold Theory*, Springer-Verlag, New York, 1981.

[3] G. Chen, "Control and synchronization of chaotic systems (bibliography)" ECE Dept., Univ. of Houston, TX - available from ftp: "ftp.egr.uh.edu/pub/TeX/chaos.tex".

[4] G. Chen and J. L. Moiola, "An overview of bifurcation, chaos and nonlinear dynamics in control systems," *J. of Franklin Instit.*, vol. 331B, pp. 819-858, 1994.

[5] K. M. Eveker and C. N. Nett, "Control of compression system surge and rotating stall: A laboratory-based 'hands-on' introduction," *Proc. of American Control Conference*, Baltimore, MD, 1994, pp. 1307-1311.

[6] O. E. Fitch, *The Control of Bifurcations with Engineering Applications*, Ph.D. Dissertation, Naval Postgraduate School, Monterey, CA 93943, 1997.

[7] R. Genesio, A. Tesi, H. O. Wang, and E. H. Abed, "Control of period doubling bifurcations using harmonic balance," *Proc. of IEEE Conf. Decision and Control*, San Antonio, Texas, 1993, pp. 492-497.

[8] P. Glendinning, *Stability, Instability and Chaos: An introduction to the Theory of Nonlinear Differential Equations*, Cambridge University Press, 1994.

[9] G. Gu, X. Chen, A. Sparks, and S. Banda, "Bifurcation stabilization with local output feedback," *SIAM J. of Control and Optimization*, 1999, in press.

[10] J. Hale and H. Koçak, *Dynamics and Bifurcations*, Springer-Verlag, New York, 1991.

[11] W. Kang, *Extended Controller Normal Form, Invariants and Dynamic Feedback Linearization of Nonlinear Control Systems*, Ph.D. Dissertation, University of California at Davis, 1991.

[12] W. Kang, "Invariants and stability of control systems with transcritical and saddle-node bifurcations," *Proc. of IEEE Conf. on Decision and Control*, San Diego, California, December 10-12, 1997.

[13] W. Kang, "Bifurcation and normal form of nonlinear control systems − part I," *SIAM J. of Control and Optimization*, vol. 36, pp. 193-212, 1998.

[14] W. Kang, "Bifurcation and normal form of nonlinear control systems − part II," *SIAM J. Control and Optimization*, vol. 36, pp. 213-232, 1998.

[15] W. Kang, "Bifurcation control via state feedback for systems with a single uncontrollable mode," preprint, 1999.

[16] W. Kang and A. J. Krener, "Extended quadratic controller normal form and dynamic feedback linearization of nonlinear systems," *SIAM J. of Control and Optimization*, vol. 30, pp 1319-1337, 1992.

[17] W. Kang and K. Liang, "The stability and invariants of control systems with pitchfork or cusp bifurcations," *Proc. of IEEE Conf. on Decision and Control*, San Diego, California, December 10-12, 1997.

[18] W. Kang and F. Papoulias, "Bifurcation and normal forms of dive plane reversal of submersible vehicles," *Proc. of International Offshore and Polar Engineering Conference*, Honolulu, Hawaii, 1997.

[19] A. J. Krener, "The feedbacks which soften the primary bifurcation of MG 3," preprint.

[20] M. Krstic, J. M. Protz, J. D. Paduano, and P. V. Kokotovic, "Backstepping designs for jet engine stall and surge control," *Proc. of IEEE Conf. Decision and Control*, New Orleans, LA, 1995, pp. 3049-3055.

[21] D.-C. Liaw and E. H. Abed, "Stability analysis and control of rotating stall," *Proc. of IFAC Nonlinear Control Systems Design Symposium*, Bordeaux, France, June, 1992.

[22] F. E. McCaughan, "Bifurcation analysis of axial flow compressor stability," *SIAM J. of Applied Math.*, vol. 20, pp. 1232-1253, 1990.

[23] F. K. Moore and E. M. Greitzer, "A theory of post-stall transients in axial compression systems − Part I: Development of equations," *ASME J. of Engineering for Gas Turbines and Power*, vol. 108, pp. 68-76, 1986.

[24] H. O. Wang and E. H. Abed, "Bifurcation control of a chaotic system," *Automatica*, vol. 31, pp. 1213-1226, 1995.

[25] S. Wiggins, *Introduction to Applied Nonlinear Dynamical Systems and Chaos*, Springer-Verlag, 1990.

18

Controlling Bifurcations in Nonsmooth Dynamical Systems

Mario di Bernardo[1] and Guanrong Chen[2]

[1]Department of Engineering Mathematics
University of Bristol, Bristol BS8 1TR, UK
M.diBernardo@bristol.ac.uk

[2]Department of Electrical and Computer Engineering
University of Houston, Houston, Texas, 77204 USA
gchen@uh.edu

Abstract

> In this chapter, some typical dynamical phenomena such as border-collision and grazing bifurcations in piecewise smooth (PWS) systems are first described. Some strategies for the control of dynamical PWS systems are then introduced, in which a unified framework for local state feedback control of PWS systems is proposed. Some general controllability conditions are stated. Two typical PWS systems, Chua's circuit and the DC/DC buck converter, are used as examples for the study. Finally, some other new control techniques, potentially useful for various PWS systems, are briefly discussed.

18.1 Introduction

A large number of physical systems in engineering and applied science are characterized by the occurrence of discrete events affecting their dynamical evolution. For instance, power electronic systems, vibro-impacting mechanical machines, and structures in earthquake engineering, to name just a few, are naturally characterized by switchings, impacts, and other nonsmooth events, which are fundamental in organizing their dynamical behaviors. From a mathematical viewpoint, these systems cannot be appropriately modeled by systems of smooth ordinary differential equations (ODEs). As a matter of fact, they are typically modeled by systems of piecewise smooth (PWS) ODEs [12, 14, 22, 34, 33, 25].

This class of dynamical systems, which include the familiar piecewise linear models (e.g., Chua's circuits), has recently been shown to exhibit, under certain condition, some rich and complex dynamical behaviors, including different types of bifurcations and chaos.

Generally the phase space of a PWS system can be divided into different regions separated by some boundaries; in each region the system is described by a smooth functional form. It has been shown that the system dynamical behavior undergoes dramatic changes when, as some parameters are varied, the trajectory (or part of it) becomes tangent to one of the phase-space boundaries [17, 20, 32, 31]. Among these complex dynamical phenomena are novel types of bifurcations, called *border-collisions* or *grazing bifurcations*, which have been observed in several systems of relevance in some applications [22, 12].

A classification of these phenomena has been reported in [17, 18, 19, 20, 13], for both continuous-time and discrete-time dynamical systems. One of the most striking possible consequences of a border-collision bifurcation is the occurrence of a sudden jump from a periodic to a chaotic evolution. This phenomenon has been detected both numerically and experimentally in many engineering systems, in such applications as power electronic converters [14], impact oscillators [33], and human heart models [36].

The wide-spread use of such systems motivates new investigations on possible strategies for solving the problem of controlling bifurcations and chaos within the context of PWS dynamical systems. For instance, in power electronics, pulse width modulated (PWM) feedback controllers are usually exploited for the control of power converters and other switching devices. The resulting systems are piecewise smooth and have been shown to undergo border-collisions including a sudden jump onto a chaotic attractor [14]. It should be noted that the aim of the conventional PWM control strategy is usually to stabilize the system dynamics onto a one-periodic solution in the presence of noise and parameter variations. Therefore, appropriate *chaos control* techniques could be used, for instance, to extend the range of operation of these devices; thus allowing the system parameters to be varied beyond the point at which a sudden jump to chaos (and the corresponding broadening of the output spectrum) occurs.

To our knowledge, there does not seem to be a general theory on the stabilization and control of nonsmooth systems. In fact, several open problems remain unsolved, which are the subject of much ongoing research [2]. Toward this goal, in this chapter, we propose and discuss some strategies for the control of bifurcations of dynamical PWS systems. Specifically, after describing the most common phenomena exhibited by this class of dynamical systems in Section 18.2, we present the issue of controlling the occurrence of border-collision bifurcations in Section 18.4 via an appropriate local action. Then, we present a unified framework, in Section 18.5, for a promising state feedback control approach to tackling the problem. We further describe this approach for the control of two typical PWS systems in Section 18.3: the Chua circuit and the DC/DC buck converter. Then, we state some general controllability conditions for feedback control of PWS systems. Finally, we discuss some other new control techniques in Section 18.6. The ideas and techniques developed in this chapter are useful for control and anticontrol of bifurcations and chaos – two active research topics in the current literature [4, 7].

18.2 Some Typical Dynamical Phenomena in PWS Systems

Consider a general PWS system of the form

$$\dot{x} = f(x, t, \mu), \qquad (18.2.1)$$

where $x \in R^n$ is the state vector, $\mu \in R^p$ is the parameter vector, and $f : R^{n \times 1 \times p} \mapsto R^n$ is a piecewise smooth function.

The phase space of this PWS system can be divided into countably many regions, R_i, $i = 1, 2, \cdots$, corresponding to the different smooth functional forms of the system. On the boundaries, Φ_i, of these regions, f can be either discontinuous or nonsmooth with a discontinuous first derivative. An example is the displacement $y(t)$ of a mechanical oscillator with an obstacle at $y = s$, which (an impact oscillator) can be described by

$$m\ddot{y} + c\dot{y} + ky = a\sin(\omega t), \quad \text{for } y < s \qquad (18.2.2)$$

with $\dot{y}^+ \to -r\dot{y}^-$, $r \in (0,1)$ if $y = s$. Thus, the phase space for an impact oscillator is characterized by the presence of a boundary corresponding to the position $y = s$ of the obstacle. On such a boundary, the functional form of the system is discontinuous since the velocity undergoes a jump, i.e., $\dot{y}^+ \to -r\dot{y}^-$.

18.2.1 Border-Collision and Grazing Bifurcations

As mentioned in the introduction, a dramatic change of the system dynamical behavior is often observed when, as some parameter values are varied, a system trajectory becomes tangent to one of the phase space boundaries. When this tangency occurs, a border-collision bifurcation is said to have taken place if the system flow is continuous but nonsmooth across the boundary curve. If, instead, the flow is discontinuous, then a grazing bifurcation is observed (see Figure 18.1).

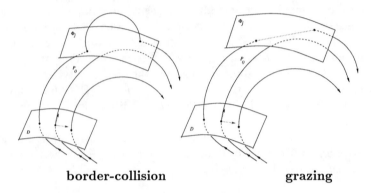

FIGURE 18.1
Schematic diagram of a C-bifurcation (left) and a grazing bifurcation (right).

In both cases, following a border-collision or grazing bifurcation, similar dynamical scenarios can be detected [13, 32]. These can be classified as the following three cases:

1. A continuous transition from the orbit, involved in the bifurcation, to an orbit of a similar or different periodicity which contains one or more trajectory sections in another region of the phase space (e.g., the transition from a periodic orbit to one of a similar type having an additional section lying on the other side of the boundary, or a period-doubling phenomenon).
2. The merging of two different solution orbits (existing on both sides of a phase space boundary), followed by their disappearance.
3. A sudden transition from a periodic orbit to a chaotic evolution.

Also, it has been shown that when a system orbit becomes tangent to one of the phase space boundaries, the Jacobian of the corresponding Poincaré map (from the switching plane to itself) has a singularity [14, 29]. In particular, one of the Jacobian eigenvalues diverges to infinity while the other approaches the origin. Hence, *infinite local stretching* is introduced along one direction in the phase space. This phenomenon has been shown to be the cause of many other interesting features induced by grazing and border-collision bifurcations, such as chaotic attractors characterized by fingered structures, period adding cascades, hysteretic phenomena (due to switching between coexisting attractors), and sudden jumps to chaos.

Despite the apparent similarity between grazing and border-collision bifurcations, they are usually studied separately in the literature. In particular, while a grazing bifurcation is associated with a normal form characterized by a square-root singularity [29], border-collisions are described by piecewise linear normal forms [32, 31, 13].

In what follows we discuss only the classification and control of border-collision bifurcations, since we assume that the systems of relevance in engineering are typically continuous (although not necessarily smooth) across the boundaries between different phase space regions.

18.2.2 Classification of Border-Collision Bifurcations

An effective method for classifying and predicting the occurrence of a border-collision bifurcation was introduced by M. I. Feigin in the late seventies and has recently been put into the context of the modern theory of nonlinear dynamics [13].[1] This method of classification is based on the definition of an appropriate local map describing the system dynamics in a neighborhood of a border-collision bifurcation. A characterization of the bifurcation is obtained by studying the eigenvalues of this local map.

More precisely, suppose that for a given parameter value, $\mu = \mu^*$, a periodic orbit, P_0, of the system becomes tangent to one of the phase space boundaries, say Φ. Label by M_0 the corresponding fixed point on an appropriately chosen Poincaré section, Π (see Figure 18.2). Assume, without loss of generality, that by perturbing the system parameter in a neighborhood of μ^*, the system orbit P_0 does not touch the boundary if $\mu < \mu^*$, while crossing it if $\mu > \mu^*$. As the parameter varies, the fixed point M_0, associated with P_0, will move accordingly on Π: from the point M^- (associated with the orbit that does not cross the boundary) to the point M^+ (associated with the orbit that crosses the boundary).

It should be noted that the Poincaré section Π cannot be chosen to be the switching surface itself, since before the occurrence of the tangency, $\mu < \mu^*$, the system orbit does not cross Π and, therefore, no fixed point M^- can be isolated.

Now, let J_α be the Jacobian of the fixed point M^- on Π for $\mu < \mu^*$, and let J_β be the Jacobian of the fixed point M^+ on Π which corresponds

---

[1] In his pioneering work, Feigin carried out detailed studies of the so-called "C-bifurcations" (from the Russian word *shivanije*, meaning "sewing"), which are indeed equivalent to the "border-collision" studied in the western literature.

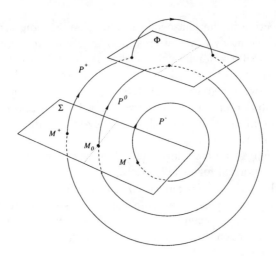

FIGURE 18.2
Local dynamics of a border-collision bifurcation.

to the system orbit crossing the boundary for $\mu > \mu^*$. Also, let σ_α^+ and σ_β^+ be the numbers of eigenvalues greater than 1, and let σ_α^- and σ_β^- be the numbers of eigenvalues less than -1, of J_α and J_β, respectively. Then, according to Feigin's method, after a border-collision the orbit involved in the bifurcation will behave as follows:

1. Smoothly change into one containing an additional section in the other region of the phase space, if $\sigma_\alpha^+ + \sigma_\beta^+$ is even.
2. Suddenly disappear after touching the boundary, if $\sigma_\alpha^+ + \sigma_\beta^+$ is odd.
3. Undergo a period-doubling, if $\sigma_\alpha^- + \sigma_\beta^-$ is odd.

Note also that, as shown in [20, 13], these three elementary conditions can be used to build bifurcation scenarios of increasing complexity, which can include a sudden jump to chaos as mentioned in the previous section.

In what follows, we assume that the system dynamics locally confined to a border-collision bifurcation can be described by a map of the form

$$x_{n+1} = \begin{cases} A_1 x_n + c\mu, & \text{if } L x_n \leq 0 \\ A_2 x_n + c\mu, & \text{if } L x_n > 0, \end{cases} \quad (18.2.3)$$

where $x_n \in R^n$ and $L \in R^{1 \times n}$. This is typically the case in most systems of interest in engineering applications [13, 32].

18.2.3 Sliding Mode Solutions

In addition to the novel class of bifurcations introduced in the previous section, PWS systems can, under certain conditions, exhibit a peculiar type of solution named *sliding mode*. These solutions are characterized by their remaining within the discontinuity set of the differential equations. An extensive study of basic properties of such solutions and their possible use in designing sliding mode controllers can be found in, e.g., [35, 21].

It is worth mentioning that recently the existence of periodic solutions characterized by sliding sections (*sliding orbits*) has been outlined in [12, 28]. These solutions have been shown to play an important role in organizing the dynamics of different PWS systems, such as friction oscillators and DC/DC converters. In the latter case, for instance, it has been shown that the formation of a sliding orbit is a codimension-two phenomenon that organizes some peculiar double-spiral bifurcation diagrams. These diagrams describe the accumulation of periodic orbits characterized by an increasing number of switchings per period onto a sliding solution (characterized by the limit of infinitely many switchings). An analytical explanation of such bifurcation structures can be found in [12], along with some detailed numerical evidence of their existence.

18.3 Two Examples: Chua's Circuit and the Buck Converter

We now summarize the main characteristics of two typical PWS systems which are considered particularly relevant in engineering applications of nonlinear dynamics: Chua's circuit and the buck converter.

18.3.1 Chua's Circuit

Chua's circuit is a simple nonlinear circuit consisting of a linear inductance, L, a linear resistor, R, two capacitors, C_1 and C_2, and a nonlinear resistor, N_R, as shown in Figure 18.3.

The circuit dynamics can be described by [6]

$$C_1 \frac{dv_{C_1}}{dt} = \frac{1}{R}(v_{C_2} - v_{C_1}) - g(v_{C_1})$$
$$C_2 \frac{dv_{C_2}}{dt} = \frac{1}{R}(v_{C_1} - v_{C_2}) + i_L \quad (18.3.4)$$
$$L \frac{di_L}{dt} = -v_{C_2},$$

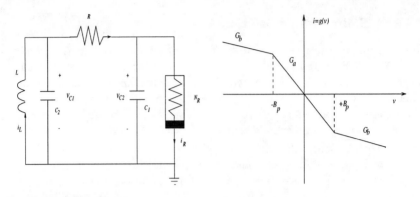

FIGURE 18.3
Chua's circuit (left) and the nonlinear resistor characteristic (right).

where $g(\cdot)$ is a piecewise linear function (a special resistor N_R) defined by

$$g(v_R) = G_b\, v_R + \frac{1}{2}(G_a - G_b)\big[|v_R + B_p| - |v_R - B_p|\big]$$

and is depicted in Figure 18.3. This function is continuous and piecewise smooth.

Without loss of generality, dimensionless version of (18.3.5) may be used:

$$\begin{aligned}
\dot{x} &= p\,(y - x - f(x)) \\
\dot{y} &= x - y + z \\
\dot{z} &= -q\,y\,.
\end{aligned} \qquad (18.3.5)$$

in which the dot indicates a derivative with the respect to τ, where

$$x = v_{C_1}/B_p, \quad y = v_{C_2}/B_p, \quad z = Ri_L/B_p, \quad \tau = t/(RC_2),$$
$$m_0 = RG_b, \quad m_1 = RG_a, \quad p = C_2/C_1, \quad q = C_2 R^2/L,$$

and

$$f(x) = m_0 x + \frac{1}{2}(m_1 - m_0)\big[|x + 1| - |x - 1|\big]\,.$$

It is known that with $p = 9$, $q = 14\frac{2}{7}$, $m_0 = -\frac{5}{7}$, and $m_1 = -\frac{8}{7}$, the circuit is chaotic and displays a double scroll (strange attractor) around two unstable equilibria, as well as an unstable saddle-type periodic orbit encompassing the attractor [6]. Figure 18.4 shows the chaotic attractor of the circuit.

18.3.2 DC/DC Buck Converter

Another typical PWS system, particularly relevant in engineering applications, is a simple power electronic circuit called the *DC/DC buck converter*.

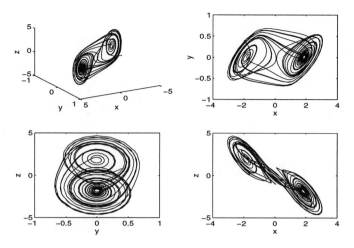

FIGURE 18.4
The chaotic attractor of Chua's circuit.

DC/DC converters are among the most widely used devices in power electronics. Power converters are used to convert electrical energy from one form to another. Since electrical sources can be either DC or AC, there are four basic types of converters: AC/DC, DC/AC, AC/AC, and DC/DC. In particular, DC/DC converters are used in all those situations where there is a need to stabilize a given DC voltage to a desired value. This is generally achieved by "chopping" and filtering the input voltage through an appropriate "switching" action, generally implemented via a Pulse Width Modulation (PWM). There are three basic second-order converter topologies which are classified according to whether they adjust the input voltage to a higher (*boost*), lower (*buck*), or generic (*buck-boost*) value.

An elementary buck converter can be implemented by circuitry as illustrated by Figure 18.5. This consists of a basic RLC circuit, a diode, and a switching element. The aim of the circuit is to maintain a desired voltage, lower than that provided by the input battery E, across the load resistance R. This can be achieved by appropriately turning on and off the switch S, so that the circuit is repeatedly forced by the external forcing voltage source E.

The switching action is usually implemented through a PWM feedback law. A linear combination of the two system states, $v_c(t) = g_1 i(t) + g_2 v(t)$ (in the present work $g_1 = 1, g_2 = 0$), is compared with a given asymmetric sawtooth (ramp) signal of assigned period T, as shown in Figure 18.6. The circuit switch, S, is then turned on whenever the ramp signal becomes greater than the combination of the two states, and turned off when the ramp signal falls below this combination. Hence, a switching occurs when-

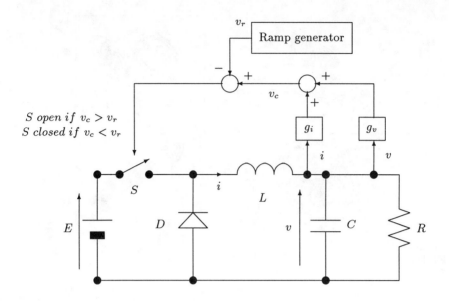

FIGURE 18.5
The DC/DC buck converter.

ever $v_c(t)$ crosses the ramp (either within the ramp period or at the times $t = nT$, $n = 1, 2, \cdots$).

Whether the switch is on or off, the buck converter can always be described as a second order linear system. Its states are the voltage v across the capacitor and the current i along the inductor. The equations take the form

$$\frac{dv}{dt} = -v/RC + i/C,$$

$$\frac{di}{dt} = -v/L + \begin{cases} 0 & \text{if } v > v_r(t) = \alpha + \beta(t \bmod T) \quad \text{(OFF)} \\ E/L & \text{if } v < v_r(t) = \alpha + \beta(t \bmod T) \quad \text{(ON)}. \end{cases}$$

(18.3.6)

Equation (18.3.6) is only one possibility among a class of buck converters modeled by ODEs of the form

$$\dot{y} = \begin{cases} A_1 y + B_1 u, & \text{if } v_c(t) := g_1 y_1 + g_2 y_2 < v_r(t), \\ A_2 y + B_2 u, & \text{if } v_c(t) := g_1 y_1 + g_2 y_2 > v_r(t), \end{cases}$$

(18.3.7)

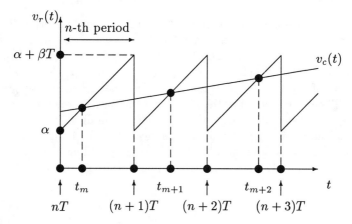

FIGURE 18.6
Standard PWM operating conditions of a DC/DC buck converter: one switching per ramp cycle.

where, in our case, $y = (y_1 \ y_2)^\top = (v, i)^\top$, and

$$A_1 = A_2 = \begin{pmatrix} 0 & -1/L \\ 1/C & -1/(RC) \end{pmatrix},$$

$$B_1 = \begin{pmatrix} 1/L \\ 0 \end{pmatrix}, \qquad B_2 = \begin{pmatrix} 0 \\ 0 \end{pmatrix},$$

$$g_1 = 1, \qquad g_2 = 0,$$

$$u(t) = E, \qquad v_r(t) = \alpha + \beta(t \bmod T).$$

18.4 Control of Border-Collision Bifurcations

As mentioned in the introduction, in some applications we may be interested in achieving control of a PWS system, in the sense of controlling the occurrence of its border-collision and grazing bifurcations. In this section, we show that this is possible by exploiting the nature of the controlled PWS system. Our main interest is to develop a local state feedback control strategy to control the complex dynamical behavior of a PWS system near a border-collision bifurcation. The main idea is to use a switching state feedback controller to effectively modify the normal form of the controlled system near a border-collision.

18.4.1 A Local State Feedback Control Approach

Suppose that as some parameter μ is increased, a system orbit, P_0, undergoes a border-collision bifurcation at $\mu = \mu^*$. Then, locally the system dynamical evolution is described by Equation (18.2.3) and, accordingly, the type of dynamical behavior of the system for $\mu > \mu^*$ depends on the eigenvalues of A_1 and A_2 in this equation, as described in the previous section. Therefore, changing the eigenvalues of these dynamical matrices will cause a consistent change of the system dynamical behavior at the border-collision. In particular, if the matrices A_1 and A_2 are stabilizable, we can control the system at a border-collision by adding a state feedback controller to system (18.2.3), through an appropriate matrix B, in the form

$$x_{n+1} = Bu_n + \begin{cases} A_1 x_n + c\mu, & \text{if } Lx_n \leq 0 \\ A_2 x_n + c\mu, & \text{if } Lx_n > 0, \end{cases} \qquad (18.4.8)$$

where we choose

$$u_n = \begin{cases} -K_1 x_n, & \text{if } x \leq 0 \\ -K_2 x_n, & \text{if } x > 0, \end{cases} \qquad (18.4.9)$$

so that, locally,

$$x_{n+1} = \begin{cases} (A_1 - BK_1)x_n + c\mu, & \text{if } x \leq 0 \\ (A_2 - BK_2)x_n + c\mu, & \text{if } x > 0. \end{cases} \qquad (18.4.10)$$

Using the pole placement technique, we can determine K_1 and K_2 such that the eigenvalues of the dynamical matrices in (18.4.8) satisfy one of the conditions described in Section 18.2. In so doing, we can select any of the possible dynamical evolutions following a border-collision bifurcation. In other words, the occurrence of a border-collision bifurcation can be used to control the system dynamical evolution.

Notice that the conditions on the eigenvalues of the map, listed in Section 18.2, do not involve their exact numerical values; only the sums of the eigenvalues that are greater than 1 or -1, respectively, are required to be either odd or even. Thus, the control objective can be achieved by a controller that is simpler than that in (18.4.10). This is because knowing the eigenvalues of the map on one side of the boundary is sufficient to control the positions of the eigenvalues on the other side. Therefore, assuming the knowledge of the eigenvalues of A_1, the control objective can be achieved, without loss of generality, by using the following controller:

$$u_n = \begin{cases} 0, & \text{if } x \leq 0 \\ -K_2 x_n, & \text{if } x > 0. \end{cases} \qquad (18.4.11)$$

Using (18.4.11), we can select the eigenvalues of the system when $x > 0$ and, thereafter, decide which of the conditions listed in Section 18.2 is satisfied.

It is worth mentioning that the conditions outlined in Section 18.2 identify only the simplest possible dynamical scenario following a border-collision. Hence, a *trial and error* procedure is usually advised for the synthesis of the control gains. Nevertheless, as will be seen in the following section, this method represents an effective way of steering the system dynamics to achieve the desired control objective.

18.4.2 An Example of a Two-Dimensional Map

Consider the map studied in [30], which is a suitable normal form for border-collisions of two-dimensional PWS maps. This map takes on the form

$$x_{n+1} = \begin{cases} \begin{pmatrix} a_1 & 1 \\ b_1 & 0 \end{pmatrix} x_n + \begin{pmatrix} 1 \\ 0 \end{pmatrix} \mu, & \text{if } \begin{pmatrix} 0 & 1 \end{pmatrix} x_n \leq 0 \\ \begin{pmatrix} a_2 & 1 \\ b_2 & 0 \end{pmatrix} x_n + \begin{pmatrix} 1 \\ 0 \end{pmatrix} \mu, & \text{if } \begin{pmatrix} 0 & 1 \end{pmatrix} x_n > 0. \end{cases} \quad (18.4.12)$$

It has been shown [30] that this map undergoes a border-collision bifurcation when $\mu = 0$ and that the dynamics following this bifurcation are specified by the values of the parameters a_1, b_1, a_2, b_2 in the map (18.4.12). These corresponds to the traces, a_1 and a_2, and the determinants $-b_1$ and $-b_2$, of the system matrices on both sides of the phase space boundary determined by $x_{n_1} = 0$.

For instance, by selecting $a_1 = 1.3$, $b_1 = 0.4$, $a_2 = 1.15$, and $b_2 = 0.3$ at the border-collision point (with $\mu = 0$), a stable equilibrium point existing on the left of the boundary will turn into a stable equilibrium existing on the right of the boundary. This is confirmed by Figure 18.7 (a), where the bifurcation diagram of the map (18.4.12) is shown for these values of a_1, b_1, a_2, b_2. We can see that, as μ crosses zero, a branch of stable equilibria turns into another branch of a similar type.

We now apply a controller of the form 18.4.11, with $K = (2\ 1)$, to the map. As shown in Figure 18.7 (b), when the control is activated, the bifurcation diagram changes abruptly and the system orbit suddenly jumps onto a chaotic attractor at the border-collision point. In fact, with this choice of the feedback gain, the controlled map satisfies the conditions formulated in [24] that predicts such a sudden jump to occur. Similarly, other types of dynamical behavior can also be obtained by simply tuning the feedback gain to different values.

Notice that the buck converter presented in Section 18.5.2 can be studied by using a suitable two-dimensional Poincaré map [14] and, as proposed in [37], border-collision bifurcations cause the system to exhibit a sudden transition to chaos. Hence, we believe that the technique presented in this chapter can be easily used to control border-collision bifurcations in the

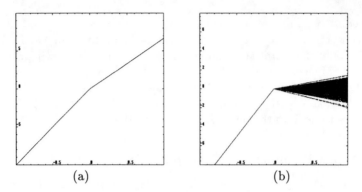

FIGURE 18.7
Bifurcation diagram: (a) before and (b) after feedback control is applied.

buck converter as well as in other electronic devices. This is the subject of work in progress and will be presented elsewhere.

Finally, it is relevant to point out that this type of controller can be particularly effective in solving the problem of anti-controlling chaos [4, 7]. In other words, given a system that is evolving along a regular orbit, we are able to push it to switch to a chaotic attractor by applying a simple state feedback control. For this purpose, we can literally use the theory of border-collision bifurcation classification [13, 30, 24] as a controller design criterion.

18.5 Feedback Control of PWS Chaotic Systems

Knowing that the control strategy outlined in the previous section is a trial-and-error approach in nature, and with the desire of controlling the global dynamics of a given PWS system (which can be chaotic as those presented above), in this section we present a possible alternative strategy. We try to employ an appropriate state feedback control methodology, which has been shown to be particularly effective in controlling chaotic dynamics of smooth dynamical systems and is convenient to implement in most engineering applications. In fact, recently a state feedback control method has been used successfully to achieve the control of some PWS systems such as the buck converter discussed above [16].

18.5.1 Problem Formulation

As mentioned above, many PWS engineering systems commute between two different configurations. For example, the dynamics of Chua's circuit switch among the three pieces of straight lines of the nonlinear resistor, and the power converter switches between the ON and OFF phases whenever a linear combination of the system states crosses a prescribed periodic signal.

Therefore, for our design, we consider a general nonlinear nonautonomous switching dynamical system of the form

$$\dot{\mathbf{x}} = \mathbf{f}(\mathbf{x},t) = \begin{cases} \mathbf{f}_1(\mathbf{x},t), & \sigma(\mathbf{x}(t),t) > 0 \\ \mathbf{f}_2(\mathbf{x},t), & \sigma(\mathbf{x}(t),t) < 0, \end{cases} \qquad (18.5.13)$$

where $\sigma(\mathbf{x},t)$ is a switching function, and the system is assumed to possess a periodic orbit $\tilde{\mathbf{x}}$ of period T: $\tilde{\mathbf{x}}(t+T) = \tilde{\mathbf{x}}(t)$ for all $0 \le t < \infty$. In a general situation, it is also assumed that the switching function $\sigma(\mathbf{x},t)$ is itself periodic of period τ, with $\tau = T/m$ for an integer $m \ge 0$. This means that within one period of T, the switching function may switch m times.

We want to design a feedback controller of the form

$$\mathbf{u}(t) = K(\mathbf{x} - \tilde{\mathbf{x}}) + \mathbf{g}(\mathbf{x} - \tilde{\mathbf{x}}, t), \qquad (18.5.14)$$

with a constant gain matrix K and possibly a (simple) nonlinear controller $\mathbf{g}$, so that the controlled system

$$\dot{\mathbf{x}} = \mathbf{f}(\mathbf{x},t) + \mathbf{u} = \mathbf{f}(\mathbf{x},t) + K(\mathbf{x} - \tilde{\mathbf{x}}) + \mathbf{g}(\mathbf{x} - \tilde{\mathbf{x}}, t) \qquad (18.5.15)$$

can track the target periodic orbit $\tilde{\mathbf{x}}$, in the sense that

$$\lim_{t \to \infty} ||\mathbf{x}(t) - \tilde{\mathbf{x}}(t)|| = 0, \qquad (18.5.16)$$

where, and throughout, $||\cdot||$ denotes the Euclidean norm.

Since the target periodic orbit $\tilde{\mathbf{x}}$ is itself a solution of the original system, it satisfies

$$\dot{\tilde{\mathbf{x}}} = \mathbf{f}(\tilde{\mathbf{x}},t) = \begin{cases} \mathbf{f}_1(\tilde{\mathbf{x}},t), & \sigma(\tilde{\mathbf{x}}(t),t) > 0 \\ \mathbf{f}_2(\tilde{\mathbf{x}},t), & \sigma(\tilde{\mathbf{x}}(t),t) < 0, \end{cases} \qquad (18.5.17)$$

a subtraction of (18.5.17) from (18.5.15) gives

$$\dot{\mathbf{x}}_\mathbf{e} = \mathbf{F}(\mathbf{x}_\mathbf{e},t) + K\mathbf{x}_\mathbf{e} + \mathbf{g}(\mathbf{x}_\mathbf{e},t), \qquad (18.5.18)$$

where $\mathbf{x_e} := \mathbf{x} - \tilde{\mathbf{x}}$ and

$$\mathbf{F}(\mathbf{x_e}, t) := \mathbf{f}(\mathbf{x}, t) - \mathbf{f}(\tilde{\mathbf{x}}, t) = \begin{cases} \mathbf{f}_1(\mathbf{x}, t) - \mathbf{f}_1(\tilde{\mathbf{x}}, t), \\ \qquad \sigma(\mathbf{x}(t), t) > 0, \sigma(\tilde{\mathbf{x}}(t), t) > 0 \\ \mathbf{f}_1(\mathbf{x}, t) - \mathbf{f}_2(\tilde{\mathbf{x}}, t), \\ \qquad \sigma(\mathbf{x}(t), t) > 0, \sigma(\tilde{\mathbf{x}}(t), t) < 0 \\ \mathbf{f}_2(\mathbf{x}, t) - \mathbf{f}_1(\tilde{\mathbf{x}}, t), \\ \qquad \sigma(\mathbf{x}(t), t) < 0, \sigma(\tilde{\mathbf{x}}(t), t) > 0 \\ \mathbf{f}_2(\mathbf{x}, t) - \mathbf{f}_2(\tilde{\mathbf{x}}, t), \\ \qquad \sigma(\mathbf{x}(t), t) < 0, \sigma(\tilde{\mathbf{x}}(t), t) < 0. \end{cases}$$

Here, it must be noted that the target periodic function $\tilde{\mathbf{x}}$ is a fixed function once it is given, and so is not a system variable. Since the only system variable is $\mathbf{x}$, the new function $\mathbf{F}(\cdot, t)$ defined above is a function of the new variable $\mathbf{x_e}$ (so, implicitly, a function of $\mathbf{x}$). It is also clear that $\mathbf{F}(0, t) = 0$ for all $t \in [0, \infty)$.

Next, to apply the linear stability theory for nonlinear systems (with or without a periodic linear part: see, e.g., [5], we piecewise Taylor-expand the function $\mathbf{F}$ of the system (18.5.18) at $\mathbf{x_e} = 0$ (i.e., at $\mathbf{x} = \tilde{\mathbf{x}}$). Note that this is only a reformulation but not a linearization of the original system function since, when we apply the linear stability theory to the system later, we will not simply truncate the residual of the expansion. Suppose the nonlinear controller to be designed satisfies $\mathbf{g}(0, t) = 0$. Then we obtain

$$\dot{\mathbf{x}}_\mathbf{e} = \left[\mathbf{A}(\tilde{\mathbf{x}}, t) + K \right] \mathbf{x_e} + \mathbf{h}(\mathbf{x_e}, t), \qquad (18.5.19)$$

where

$$\mathbf{A}(\tilde{\mathbf{x}}, t) := \left. \frac{\partial \mathbf{F}(\mathbf{x_e}, t)}{\partial \mathbf{x_e}} \right|_{\mathbf{x_e}=0} = \begin{cases} \left. \frac{\partial \left[\mathbf{f}_1(\mathbf{x},t) - \mathbf{f}_1(\tilde{\mathbf{x}},t) \right]}{\partial \mathbf{x}} \right|_{\mathbf{x}=\tilde{\mathbf{x}}} \\ \left. \frac{\partial \left[\mathbf{f}_1(\mathbf{x},t) - \mathbf{f}_2(\tilde{\mathbf{x}},t) \right]}{\partial \mathbf{x}} \right|_{\mathbf{x}=\tilde{\mathbf{x}}} \\ \left. \frac{\partial \left[\mathbf{f}_2(\mathbf{x},t) - \mathbf{f}_1(\tilde{\mathbf{x}},t) \right]}{\partial \mathbf{x}} \right|_{\mathbf{x}=\tilde{\mathbf{x}}} \\ \left. \frac{\partial \left[\mathbf{f}_2(\mathbf{x},t) - \mathbf{f}_2(\tilde{\mathbf{x}},t) \right]}{\partial \mathbf{x}} \right|_{\mathbf{x}=\tilde{\mathbf{x}}} \end{cases}$$

$$= \begin{cases} A_1(\tilde{\mathbf{x}}, t), & \sigma(\mathbf{x}(t), t) > 0, \sigma(\tilde{\mathbf{x}}(t), t) > 0 \\ A_2(\tilde{\mathbf{x}}, t), & \sigma(\mathbf{x}(t), t) > 0, \sigma(\tilde{\mathbf{x}}(t), t) < 0 \\ A_3(\tilde{\mathbf{x}}, t), & \sigma(\mathbf{x}(t), t) < 0, \sigma(\tilde{\mathbf{x}}(t), t) > 0 \\ A_4(\tilde{\mathbf{x}}, t), & \sigma(\mathbf{x}(t), t) < 0, \sigma(\tilde{\mathbf{x}}(t), t) < 0 \end{cases}$$

is T-periodic and τ-switching, and $\mathbf{h}(\mathbf{x_e}, t)$ is the rest of the Taylor expansion plus the nonlinear controller $\mathbf{g}$, which is a function of t and $O(\mathbf{x_e})$.

Feedback Control of PWS Chaotic Systems

To this end, our design is to determine both the constant control gain matrix K and, if needed, the (simple) nonlinear controller $\mathbf{g}(\mathbf{x_e}, t)$, based on the expanded model (18.5.19), such that $\mathbf{x_e} \to 0$ (i.e., $\mathbf{x} \to \tilde{\mathbf{x}}$) as $t \to \infty$. This means that when the controller is being applied to the original system (18.5.15), the goal (18.5.16) can be achieved. Upon this formulation, different approaches may be taken to accomplish the design of the controller [4].

18.5.2 Chua's Circuit and the Buck Converter

We verify here that the two examples studied in Section 18.3 are special cases of the formulation given in Section 18.5.1. Therefore, whatever control methodology derived for the above general formulation applies to these two examples.

Chua's circuit

In Chua's circuit (18.3.5), if a linear state feedback controller of the form

$$\mathbf{u}(t) = -K \begin{bmatrix} x - \bar{x} \\ y - \bar{y} \\ z - \bar{z} \end{bmatrix}, \quad K = \text{diag}\{K_{11}, K_{22}, K_{33}\}$$

is used, the controlled circuit is

$$\begin{cases} \dot{x} = p(-x + y - f(x)) - K_{11}(x - \bar{x}), \\ \dot{y} = x - y + z - K_{22}(y - \bar{y}), \\ \dot{z} = -qy - K_{33}(z - \bar{z}). \end{cases}$$

Furthermore, if an unstable periodic orbit $(\bar{x}, \bar{y}, \bar{z})$ is chosen as the target for tracking control, then, since it is itself a solution of the circuit, we have

$$\begin{cases} \dot{\bar{x}} = p(-\bar{x} + \bar{y} - f(\bar{x})), \\ \dot{\bar{y}} = \bar{x} - \bar{y} + \bar{z}, \\ \dot{\bar{z}} = -q\bar{y}, \end{cases}$$

so that a subtraction yields

$$\begin{cases} \dot{X} = p(-X + Y - \tilde{f}(x, \bar{x})) - K_{11} X, \\ \dot{Y} = X - Y + Z - K_{22} Y, \\ \dot{Z} = -qY - K_{33} Z, \end{cases}$$

where $X = x - \bar{x}$, $Y = y - \bar{y}$, $Z = z - \bar{z}$, and

$$\tilde{f}(x,\bar{x}) = \begin{cases} m_0(x - \bar{x}) & x \geq 1, \bar{x} \geq 1 \\ m_0 x - m_1 \bar{x} + m_1 - m_0 & x \geq 1, -1 \leq \bar{x} \leq 1 \\ m_0(x - \bar{x}) + 2(m_1 - m_0) & x \geq 1, \bar{x} \leq -1 \\ m_1 x - m_0 \bar{x} - m_1 + m_0 & -1 \leq x \leq 1, \bar{x} \geq 1 \\ m_1(x - \bar{x}) & -1 \leq x \leq 1, -1 \leq \bar{x} \leq 1 \\ m_1 x - m_0 \bar{x} + m_1 - m_0 & -1 \leq x \leq 1, \bar{x} \leq -1 \\ m_0(x - \bar{x}) - 2(m_1 - m_0) & x \leq -1, \bar{x} \geq 1 \\ m_0 x - m_1 \bar{x} - m_1 + m_0 & x \leq -1, -1 \leq \bar{x} \leq 1 \\ m_0(x - \bar{x}) & x \leq -1, \bar{x} \leq -1 \end{cases}$$

in which $m_1 < m_0 < 0$. Obviously, this is a special case of system (18.5.15) or (18.5.19).

The Buck Converter

The PWM controlled buck converter can be described by

$$\dot{\mathbf{x}}(t) = \begin{cases} A_1 \mathbf{x}(t) + B_1 \mathbf{u}(t), & \mathbf{c}^T \mathbf{x}(t) > \sigma(t) \\ A_2 \mathbf{x}(t) + B_2 \mathbf{u}(t), & \mathbf{c}^T \mathbf{x}(t) < \sigma(t), \end{cases} \quad (18.5.20)$$

where $\mathbf{x} \in R^2$, $\mathbf{c} = [c_1 \ c_2]^T$, $\sigma(t) = \sigma(t+T)$ is T-periodic, and $\mathbf{u}$ is a control input.

Define a switch function,

$$s(\mathbf{x},\mathbf{u},t) := \mathrm{sgn}\left\{\mathbf{c}^T \mathbf{x}(t) - \sigma(t)\right\} = \begin{cases} 1, & \mathbf{c}^T \mathbf{x}(t) > \sigma(t) \\ -1, & \mathbf{c}^T \mathbf{x}(t) < \sigma(t), \end{cases} \quad (18.5.21)$$

and rewrite the converter model (18.5.20) as

$$\dot{\mathbf{x}}(t) = \left[\frac{1}{2} A_1 + \frac{1}{2} A_2\right] \mathbf{x}(t) + \mathbf{g}(\mathbf{x},\mathbf{u},t), \quad (18.5.22)$$

where

$$\mathbf{g}(\mathbf{x},\mathbf{u},t) = \left[\frac{s(\mathbf{x},\mathbf{u},t)}{2} A_1 - \frac{s(\mathbf{x},\mathbf{u},t)}{2} A_2\right] \mathbf{x}(t).$$

Clearly, the constant coefficient matrix in model (18.5.22) is T-periodic, and so system (18.5.22) is a special case of system (18.5.19).

At this stage, a general design procedure for solving the control problem formulated in this section is not yet available. As is well known, effective methods for guaranteed stability of general nonsmooth systems are still a main subject of much ongoing research [23]. We limit our discussion by only stating, in the next subsection, some general controllability conditions whose implementation and application are subject to further investigations.

18.5.3 Some General Controllability Conditions

The results summarized here are classical in nature, which can be proven by employing Lyapunov function methods [4, 8].

THEOREM 18.1
Suppose that in system (18.5.19), $A_i(\tilde{\mathbf{x}}, t) = A_i$ (constant matrices), or consider system (18.5.22), and in both systems, all eigenvalues of A_i are assumed to have negative real parts, $i = 1, 2, 3, 4$. In both systems, suppose that $\mathbf{h}(0, t) = 0$, and let P be a common positive definite and symmetric solution of the Lyapunov equations

$$PA_i + A_i^\top P = -I, \qquad i = 1, 2, 3, 4, \qquad (18.5.23)$$

where I is the identity, with the maximum eigenvalue $\lambda_{\max}(P) > 0$. If

$$\|\mathbf{h}(\mathbf{x_e}, t)\| \leq c \|\mathbf{x_e}\| \qquad (18.5.24)$$

for a constant $c < \frac{1}{2}\lambda_{\max}(P)$ for all $0 \leq t < \infty$, then the controller $\mathbf{u}(t)$ defined in (18.5.14) will drive the trajectory $\mathbf{x}$ of the controlled system (18.5.15) to the target orbit $\tilde{\mathbf{x}}$ as $t \to \infty$.

Note that if there are four positive definite and symmetric solution matrices P_i ($i = 1, 2, 3, 4$) satisfying the Lyapunov equation (18.5.23) individually but there is no common solution matrix P that satisfies (18.5.23) for all $i = 1, 2, 3, 4$ simultaneously, then the above result may not hold.

To state the next theorem, we first recall some concepts from linear algebra. For system

$$\dot{\mathbf{x}} = A(t)\mathbf{x}, \qquad (18.5.25)$$

its fundamental matrix is defined to be

$$\Phi(t, t_0) = \begin{bmatrix} \mathbf{x}_1(t) & \cdots & \mathbf{x}_n(t) \end{bmatrix}$$

which consists of n linearly independent solution vectors, $\mathbf{x}_1, \cdots, \mathbf{x}_n$, of the system (18.5.25), with a given initial condition $\mathbf{x}_0 = \mathbf{x}(t_0)$. Using this fundamental matrix, any solution of system (18.5.25) can be expressed as

$$\mathbf{x}(t) = \Phi(t, t_0)\mathbf{x}_0 + \int_{t_0}^{t} \Phi(t, s) \mathbf{g}((\mathbf{x}, s), s) \, ds. \qquad (18.5.26)$$

Next, for system (18.5.19), there is always a T-periodic nonsingular matrix $M(\tilde{\mathbf{x}}, t)$ and a constant matrix Q such that the solution fundamental matrix of the system, associated with the matrix $\mathbf{A}(\tilde{\mathbf{x}}, t)$, is given by

$$\Phi(\tilde{\mathbf{x}}, t) = M(\tilde{\mathbf{x}}, t) e^{tQ}.$$

The eigenvalues of the constant matrix e^{TQ} are called the *Floquet multipliers* of the system.

THEOREM 18.2
In system (18.5.19), *suppose that* $\mathbf{h}(0,t) = 0$ *and that both* $\mathbf{h}(\mathbf{x_e},t)$ *and* $\partial \mathbf{h}(\mathbf{x_e},t)/\partial \mathbf{x_e}$ *are continuous in a bounded region* $||\mathbf{x_e}|| < \infty$. *Assume, moreover, that*

$$\lim_{||\mathbf{x_e}|| \to 0} \frac{||\mathbf{h}(\mathbf{x_e}, K, t)||}{||\mathbf{x_e}||} = 0, \qquad (18.5.27)$$

uniformly with respect to $t \in [0, \infty)$. *Then the nonlinear controller* (18.5.14) *designed under the condition that all the Floquet multipliers of the system* (18.5.19) *satisfy*

$$|\lambda_j| < 1, \qquad j = 1, \cdots, n, \qquad \forall \; t \in [0, \infty), \qquad (18.5.28)$$

will drive the orbit $\mathbf{x}$ *of the original controlled system* (18.5.15) *to the target orbit* $\widetilde{\mathbf{x}}$ *as* $t \to \infty$.

THEOREM 18.3
In system (18.5.22), *suppose that there are two positive constants, a and b, such that*

$$||\Phi(t,s)|| \leq a\, e^{-b(t-s)}, \qquad \forall \; t_0 \leq s \leq t < \infty. \qquad (18.5.29)$$

and the nonlinear controller (18.5.14) *is so designed that it satisfies*

$$\lim_{||\mathbf{x}|| \to 0} \frac{||\mathbf{g}(\mathbf{x}, \mathbf{u}, t)||}{||\mathbf{x}||} = 0, \qquad (18.5.30)$$

uniformly with respect to $t \in [0, \infty)$. *Then*

$$||\mathbf{x}(t)|| \leq a\, e^{-c(t-t_0)} \qquad (18.5.31)$$

for a constant $c > 0$, *implying that* $\mathbf{x} \to 0$ *exponentially fast as* $t \to \infty$.

18.6 Other Control Techniques

Alternative strategies to control the occurrence of various nonlinear phenomena, in particular for different classes of PWS systems, have recently appeared in the literature (e.g., the chapter by Suykens et al. in this book). In this section, we very briefly summarize some of these strategies and discuss some alternative techniques which deserve further investigation.

18.6.1 Parameter Variation Techniques

The well known OGY method for chaos control has been applied to controlling chaotic impacts in the so-called Nordmark map [3]. This map is piecewise continuously differentiable and exhibits several types of grazing bifurcations and chaos. It models the behavior of forced mechanical oscillators that undergo impacts with a hard obstacle. Using small parameter variation in accordance with the OGY control strategy, it is possible to steer the system dynamics away from chaos via the stabilization of some unstable periodic orbits embedded in the chaotic attractor.

An alternative parameter variation technique is presented in [26], where a switching feedback controller, as the one discussed in Section 18.4, is used to vary the accessible system parameter in order to stabilize the chaotic motion exhibited by a discontinuous mechanical oscillator onto a periodic orbit that grazes the switching plane. It is shown that such a grazing orbit can indeed be stabilized by an appropriate piecewise-control strategy.

Similarly, in [1], the stabilization of a chaotic DC/DC converters onto an unstable periodic orbit is achieved by altering the switching logic of the control device. This method consists in the synthesis of an open-loop strategy that, by passing the original switching logic, forces the system to switch among some target unstable periodic orbits. Note that this method can be simply implemented and used as an alternative to the OGY method when the controlled system is not evolving within a chaotic regime (e.g., for the control of some bifurcations).

18.6.2 Hybrid Systems Control

The parameter variation control methods discussed above, though highly affected by the presence of noise and external disturbances, give their open-loop nature. Moreover, the controllers are usually designed ad hoc for many specific systems. For industrial applications, such as the control of power electronic converters, more general and robust strategies need to be used. The study of this and other related issues is the subject of the emerging field of hybrid systems control.

In the past few years, the control of so-called hybrid automata has been the subject of much ongoing research [2, 27]. These systems are characterized by the occurrence of several discrete events and can be seen as a generalization of the PWS systems. New stability results have recently been suggested and hopefully can allow for a more general approach to the solution of the problems stated in this chapter. As of today, this remains a subject for further research.

18.6.3 Controlling a Smooth System by a Discontinuous Action

Up to now, we have considered piecewise smoothness as a property of the dynamical system whose behavior is to be controlled. It is worth mentioning that sometimes making a smooth chaotic dynamical system discontinuous, through the action of a switching controller, can be a useful strategy for taming bifurcations and chaos. This is, for instance, the aim of the discontinuous adaptive controllers [15, 10, 11], where it is shown that through an appropriate switching action (whose amplitude is adaptively estimated), both controlling and synchronizing chaos can be successfully achieved.

18.7 Conclusions

In the last decade, control of discontinuous and hybrid systems has been an attractive subject of much research effort in control systems engineering. Within this context, the control strategies studied in this chapter represent a first attempt at devising a general approach to the control of bifurcations and chaos in piecewise smooth dynamical systems. In particular, the control of novel bifurcations, such as border-collisions, has been achieved through a local state feedback control action. Moreover, a unified state feedback control framework has been outlined. These techniques and their possible application to physical devices, such as DC/DC converters, is still a subject of current investigations.

Acknowledgments

Mario di Bernardo is supported by the Nuffield Foundation Grant, UK under the scheme 'NUF-NAL'. G. Chen is supported by the U.S. Army Research Office under the grant DAAG55-98-1-0198.

References

[1] C. Battle-Arnau, E. Fossas, and G. Olivar, "Stabilization of periodic orbits in variable structure systems: application to dc-dc power converters," *Int. J. of Bifur. Chaos*, vol. 6, pp. 2635–2643, 1996.

[2] M. S. Branicky, "Multiple Lyapunov functions and other analysis tools for switched and hybrid systems," *IEEE Trans. on Auto. Contr.*, vol. 43, pp. 475–482, 1998.

[3] F. Casas and C. Grebogi, "Control of chaotic impacts," *Int. J. of Bifur. Chaos*, vol. 7, pp. 951–955, 1997.

[4] G. Chen, "On some controllability conditions for chaotic dynamics control," *Chaos, Solitons and Fractals*, vol. 8, pp. 1461-1470, 1997.

[5] G. Chen, "Stability of nonlinear systems," in *Wiley Encyclopedia of Electrical and Electronics Engineering*, Wiley, New York, 1999.

[6] G. Chen and X. Dong, *From Chaos to Order: Methodologies, Perspectives and Applications*, World Scientific Pub. Co., Singapore, 1998.

[7] G. Chen and D. Lai, "Feedback anticontrol of discrete chaos," *Int. J. of Bifur. Chaos*, vol. 8, pp. 1585-1590, 1998; see also G. Chen and D. Lai, "Anticontrol of chaos via feedback," *Proc. of 36th IEEE Conf. on Decision and Control*, San Diego, CA, USA, Dec. 1997, pp.367–472.

[8] G. Chen and M. di Bernardo, work in progress.

[9] L. O. Chua, M. Komuro, and T. Matsumoto, "The double scroll family: I and II," *IEEE Trans. on Circ. Sys.*, vol. 33, pp. 1072–1118, 1986.

[10] M. di Bernardo, "An adaptive approach to the control and synchronization of continuous-time chaotic systems," *Int. J. of Bifur. Chaos*, vol. 6, pp. 557–568, 1996.

[11] M. di Bernardo, "A purely adaptive controller to synchronize and control chaotic systems," *Phys. Lett. A*, vol. 214, pp. 139–144, 1996.

[12] M. di Bernardo, C. J. Budd, and A. R. Champneys, "Grazing, skipping and sliding: analysis of the nonsmooth dynamics of the DC/DC buck converter," *Nonlinearity*, vol. 11, pp. 858-890, 1998.

[13] M. di Bernardo, M. I. Feigin, S. J. Hogan, and M. E. Homer, "Local analysis of C-bifurcations in n-dimensional piecewise-smooth dynamical systems," *Chaos, Solitons and Fractals*, 1998, in press.

[14] M. di Bernardo, F. Garofalo, L. Glielmo, and F. Vasca, "Switchings, bifurcations and chaos in DC/DC converters," *IEEE Trans. on Circ. Sys.*, I, vol. 45, pp. 133-141, 1998.

[15] M. di Bernardo and D. P. Stoten, "A new extended minimal control synthesis algorithm with an application to the control of a chaotic system," *Proc. of 35th IEEE Conf. on Decision and Control*, San Diego, CA, USA, Dec. 1997.

[16] C. Fang and E. H. Abed, "Limit cycle stabilization in PWM DC-DC converters," *Proc. of 37th IEEE Conf. on Decision and Control*, Tampa, FL, USA, Dec. 1998.

[17] M. I. Feigin, "Doubling of the oscillation period with C-bifurcations in piecewise continuous systems," *Prikladnaya Matematika i Mekhanika* (J. of Applied Math. Mech.), vol. 34, pp. 861–869, 1970.

[18] M. I. Feigin, "On the generation of sets of subharmonic modes in a piecewise continuous system," *Prikladnaya Matematika i Mekhanika* (J. of Applied Math. Mech.), vol. 38, pp. 810–818, 1974.

[19] M. I. Feigin, "On the structure of C-bifurcation boundaries of piecewise continuous systems," *Prikladnaya Matematika i Mekhanika* (J. of Applied Math. Mech.), vol. 42, pp. 820–829, 1978.

[20] M. I. Feigin, "The increasingly complex structure of the bifurcation tree of a piecewise-smooth system," *Prikladnaya Matematika i Mekhanika* (J. of Applied Math. Mech.), vol. 59, pp. 853–863, 1995.

[21] A. F. Filippov, *Differential Equations with Discontinuous Righthand Sides*. Kluwer Academic Press, Boston, 1988.

[22] E. Fossas and G. Olivar, "Study of chaos in the buck converter," *IEEE Trans. on Circ. Sys.*, I, vol. 43, pp. 13–25, 1996.

[23] A. Kh. Gelig and A. N. Churilov, *Stability and Oscillations of Nonlinear Pulse-Modulated Systems*, Birkhäuser, Boston, 1998.

[24] C. Grebogi, E. Ott, and J. A. Yorke, "Fractal basin boundaries, long-lived chaotic transients and unstable-unstable pair bifurcations," *Phys. Rev. Lett.*, vol. 50, pp. 935–938, 1983.

[25] S. Hogan, "On the dynamics of rigid-block motion under harmonic forcing," *Proc. Roy. Soc. London A*, vol. 425, pp. 441–476, 1989.

[26] H. Hu, "Controlling chaos of a dynamical system with discontinuous vector field," *Physica D*, vol. 106, pp. 1–8, 1997.

[27] Y. Hui, A. N. Michel, and L. Hou, "Stability theory for hybrid dynamical systems," *IEEE Trans. on Auto. Contr.*, vol. 43, pp. 461–473, 1998.

[28] K. H. Johansson, *Relay Feedback and Multivariable Control*. Ph.D. thesis, Lund Institute of Technology, Sweden, 1997.

[29] A. B. Nordmark, "Non-periodic motion caused by grazing incidence in impact oscillators," *J. of Sound and Vibration*, vol. 2, pp. 279–297, 1991.

[30] H. E. Nusse, E. Ott, and J. A. Yorke, "Border collisions bifurcations–an explanation for observed bifurcation phenomena," *Phys. Rev. E*, vol. 49, pp. 1073–1076, 1994.

[31] L. E. Nusse and J. A. Yorke, "Border-collision bifurcations including 'period two to period three' for piecewise smooth systems," *Physica D*, vol. 57, pp. 39–57, 1992.

[32] L. E. Nusse and J. A. Yorke, "Border–collision bifurcations for piece-wise smooth one-dimensional maps," *Int. J. of Bifur. Chaos*, vol. 5, pp. 189–207, 1995.

[33] M. Oestreich, N. Hinrichs, K. Popp, and C. J. Budd, "Analytical and experimental investigation of an impact oscillator," *Proc. of ASME Biennal Conf. on Mech. Vibrations and Noise*, 1996.

[34] K. Popp, N. Hinrichs, and M. Oestreich, "Dynamical behaviour of friction oscillators with simultaneous self and external excitation," *Sadhana* (Indian Academy of Sciences), vol. 20, pp. 627–654, 1995.

[35] V. I. Utkin, *Sliding Modes and Their Application in Variable Structure Systems*. MIR Publishers, Moscow, 1978.

[36] H. O. Wang, D. Chen, and G. Chen, "Anticontrol of bifurcations," *Proc. of IEEE Conf. on Decision and Control*, Tampa, FL, USA, Dec. 1998.

[37] G. Yuan, S. Banerjee, E. Ott, and J. A. Yorke, "Border-collision bifurcations in the buck converter," *IEEE Trans. on Circ. Sys.*, I, 1999, in press.

19

Adaptive Observer–Based Synchronization

This Chapter is dedicated to our friend and coworker Professor Ilya I. Blekhman on the occasion of his 70th birthday

A. L. Fradkov,[1,2] **H. Nijmeijer,**[2,3] **A. Yu. Pogromsky**[1,3]

[1]Institute for Problems of Mechanical Engineering,
Russian Academy of Sciences
61, Bolshoy, V.O., 199178, St. Petersburg, Russia
alf@ccs.ipme.ru, sasha@ccs.ipme.ru

[2]Faculty of Mathematical Sciences,
University of Twente
P.O.Box 217, 7500 AE Enschede, The Netherlands
h.nijmeijer@math.utwente.nl

[3]Faculty of Mechanical Engineering,
Eindhoven University of Technology
P.O.Box 513, 5600 MB Eindhoven, The Netherlands

Abstract

An approach to the unidirectional synchronization based on the concept of adaptive observer is presented. A brief exposition of adaptive observer design for nonlinear systems by means of filtered transformations is given. For the important special case of synchronizing Lur'e systems, a simple adaptive observer structure is described and synchronization conditions including the persistency of excitation condition are formulated. The proposed design is illustrated by example of information transmission by a pair of adaptively synchronized Chua circuits.

19.1 Introduction

Synchronization is commonly understood as correlated or corresponding in time behavior of two or more processes. Synchronization has recently attracted much attention from different scientific and engineering communities like physics, biology, mechanical engineering, telecommunications, etc.

One reason for such interest is that, besides traditional applications [3, 4, 26], new application fields appeared, e.g., telecommunications where synchronization of chaotic systems has become popular [8, 19].

The term "synchronization" has a slightly different meaning in different communities. For instance, physicists study the synchronization phenomenon and its properties, while engineers try to build a system where this phenomenon can be observed and provide it with the desired properties.

In fact the engineering task is simultaneously more simple and more complex. It is more simple because the engineer has additional freedom: some of the systems to be synchronized are not given to him but rather are to be designed (e.g., the structure of the receiver should be determined given the structure of the transmitter). On the other hand it may be more difficult to choose the structure and system than to just determine appropriate values of the coupling parameters which can often be done by means of computer simulations.

In both cases mathematics plays an important role for obtaining a correct solution to the problem. An accurate problem statement and solvability conditions not only guarantee desired system behavior but also lead to a better understanding of the dynamics responsible for one or another type of behavior.

In this chapter we consider synchronization of systems on the basis of so-called observer-based synchronization. We understand observer-based synchronization systems as those where the model of one of the interacting subsystems (called transmitter, drive system, or master system) is fixed and generates a signal (output) available for measurements, while the other subsystem (called receiver, response system, or slave system) is to be designed in order to estimate (reconstruct) the states of the first subsystem based on the available measurements. Such a problem is well known in control theory where it is usually referred to as the *observer problem*, and the dynamical system generating the estimate of the transmitter state (i.e., the receiver) is called *observer*. Observer design methods for certain classes of nonlinear systems were proposed in [24, 17, 7]; see also [30]. A general approach to observer-based synchronization was introduced in [34]. The pioneering study of chaotic synchronization by Pecora-Carroll [35] also belongs to the class of observer-based schemes as well as some other proposals for concrete systems, e.g., [32, 18].

The problem of adaptive synchronization arises when some uncertainties are present in the interacting subsystems. The uncertainties may be caused by different reasons, e.g. lack of information about some transmitter parameters or their unpredictable change in time. The uncertainty may also be of artificial origin, if it is caused by the modulation of transmitter parameters with a message as a binary signal. Clearly, the message is known at the transmitter side, but is unknown at the receiver side, which makes the problem quite similar to the previous case.

An important feature of the receiver providing adaptive synchronization is the presence of an adaptation algorithm which generates estimates of the unknown parameters (or some functions of them). The convergence of the parameter estimates to the true parameter values is, in fact, *not necessary* for synchronization. However, if convergence is established, it provides the receiver with additional features, e.g., it allows an increase in speed and accuracy of the message reconstruction.

The problem of adaptive synchronization is close to that of model reference adaptive control for nonlinear systems which is well studied in control theory (see, e.g., [12, 25, 30, 15, 16]). In [13, 14] the general definition of adaptive synchronization was proposed as well as the design methods based on the previous results from nonlinear and adaptive control [12]. The extended versions of designs [13, 14] can be found in [31, 36, 15].

Finally, a general mathematical formulation of the synchronization problem was proposed in [5]. It encompasses as a special case different versions appearing in the literature and initiates a possibly unified framework for studying synchronization.

In this chapter adaptive observer–based synchronization will be presented based both on previous results of [14, 34] and on existing nonlinear adaptive observer techniques [29, 30]. It heavily relies on passivity and passification concepts. The proposed methodology may have applications in various fields; as an example, we consider the design of an adaptive receiver for messages transmitted by modulation of a chaotic carrier generated by Chua circuit.

19.2   General Definition of Synchronization

Below we present a definition of the synchronization which is a particular case of the general definition given in [5] and was inspired by the classical definition given in the seminal book [3].

Consider k dynamical systems described by the k *interconnected* systems of ordinary differential equations:

$$S_i: \quad \dot{x}_i = F_i(x_1, x_2, \ldots, x_k, t), \quad i = 1, \ldots, k \qquad (19.2.1)$$

where $F_i : \mathbf{R}^{n_1} \times \ldots \times \mathbf{R}^{n_k} \times \mathbf{R}^+ \to \mathbf{R}^{n_i}$. Associated with the set of systems (19.2.1) consider some time-dependent functional Q_t defined on the solutions $x_i(\cdot)$ to these systems: $Q_t : \mathcal{X}_1 \times \ldots \times \mathcal{X}_k \times \mathbf{R}^+ \to \mathbf{R}$, where $\mathcal{X}_i \subset \{x_i : \mathbf{R}^+ \to \mathbf{R}^{n_i}\}$. Loosely speaking, the solutions to the systems S_i are synchronized with respect to the functional Q_t if the value of Q_t is identically zero for these solutions for all $t \geq 0$.

DEFINITION 19.1 *Solutions $x_1(t), \ldots, x_k(t)$ of the systems $S_1, \ldots, S_k$ with initial conditions $x_1(0), \ldots, x_k(0)$ are called* synchronized *with respect to the functional Q_t if*

$$Q_t(x_1(\cdot), \ldots, x_k(\cdot)) \equiv 0, \qquad (19.2.2)$$

for all $t \in \mathbf{R}^+$.

The solutions $x_1(t), \ldots, x_k(t)$ of the systems $S_1, \ldots, S_k$ with initial conditions $x_1(0), \ldots, x_k(0)$ are asymptotically synchronized *with respect to the functional Q_t, if*

$$\lim_{t \to \infty} (Q_t(x_1(\cdot), \ldots, x_k(\cdot))) = 0 \qquad (19.2.3)$$

Remark A. The general definition of [5] also allows possible time (phase) shifts between state variables of the synchronized subsystems.

In the case of arbitrary initial conditions $x_i(0)$ this phenomenon will be referred to as *global* synchronization, while for certain admissible initial conditions we will speak about *conditional synchronization*.

If the equations of the interconnected systems (19.2.1) are given, then the problem of synchronization is one of analysis. A different situation appears when $S_1 \ldots, S_k$ are controlled systems, and the problem is to design the controller that yields synchronization. Let the interacting systems $S_1 \ldots, S_k$ be described by equations

$$S_i : \quad \dot{x}_i = F_i(x_1, x_2, \ldots, x_k, u, t), \ y_i = h_i(x_i, u, t) \ i = 1, \ldots, k \quad (19.2.4)$$

where $F_i : \mathbf{R}^{n_1} \times \ldots \times \mathbf{R}^{n_k} \times \mathbf{R}^m \times \mathbf{R}^+ \to \mathbf{R}^{n_i}$, the function $u(t) \in \mathbf{R}^m$ is the input and outputs $y_i(t) \in \mathbf{R}^{l_i}$ are available for measurement. The problem of output feedback synchronization can be posed as follows: find controller equations

$$u(t) = \mathcal{U}(y_1(s), \ldots, y_k(s), t, 0 \leq s \leq t) \qquad (19.2.5)$$

where $\mathcal{U}$ is some operator such that the relation (19.2.3) in the system (19.2.4), (19.2.5) is fulfilled. Thus the relation (19.2.3) is now considered as a goal. It will be referred to as *synchronization goal*. If the controller (19.2.5) ensures the goal (19.2.3) for arbitrary initial conditions, the feedback (19.2.5) will be referred to as a *globally synchronizing feedback*.

Note that in controlled synchronization problems, the structure of the system (19.2.4) is still fixed. More "flexible" situations are also of interest when some of the subsystems S_i are not fixed, i.e., they are to be determined in order to achieve synchronization. Such a situation arises in case of observer-based synchronization [34].

In what follows we will consider synchronization of two dynamical systems ($k = 2$). An important class of the synchronization problems arises when $n_1 = n_2 = n$ and the functional Q_t is given as a function:

$$Q_t(x_1(\cdot), x_2(\cdot)) = ||x_1(t) - x_2(t)||. \qquad (19.2.6)$$

It corresponds to what sometimes is called *identical* or *exact* synchronization. In this case, if all solutions $x_1(t), x_2(t)$ are bounded functions of time and $Q_t(x_1(\cdot), x_2(\cdot)) \to 0$ as $t \to \infty$, then the overall system often possesses an asymptotically stable set which is a compact subset of the "diagonal" set defined as $\{(x_1, x_2) \in \mathbf{R}^{2n} : x_1 = x_2\}$. However, we emphasise that the existence of asymptotically stable "diagonal" set does not follow from asymptotic stability of solution $e(t) = x_1(t) - x_2(t)$.

A more general case when $n_1 \neq n_2$ and the functional Q_t is defined as

$$Q_t(x_1(\cdot), x_2(\cdot)) = ||G(x_1(t)) - x_2(t)||. \qquad (19.2.7)$$

with a prespecified mapping $G : X_1 \to X_2$ is called *generalized synchronization*, see [1, 37, 23].

Finally, a special case of controlled synchronization with unidirectional coupling arises where (19.2.4) takes the form

$$\begin{cases} \dot{x}_1 = F_1(x_1), & y_1 = h_1(x_1) \\ \dot{x}_2 = F_2(x_2, u), & y_2 = h_2(x_2) \end{cases} \qquad (19.2.8)$$

with $h_2(x_2) = x_2 \in \mathbf{R}^n$, $F_2(x_2, u) = u \in \mathbf{R}^n$, and Q_t is defined as in (19.2.6). The problem of finding an output feedback $u = U(y_1, y_2)$ in that case is one of observer design for the first subsystem of (19.2.8), i.e., it corresponds to an observer-based synchronization.

19.3 Adaptive Observers

The problem of observer design or state reconstruction for linear dynamical systems was first considered by Kalman [21] (full-order observers) and Luenberger [28] (reduced-order observers).

The problem of adaptive observer design arises when the system model depends on unknown parameters and we need good estimation of the system state no matter what the true values of parameters are. An adaptive observer usually includes an estimation subsystem for the unknown parameters. Therefore, a secondary goal, convergence of the estimates to the true values, may be considered. However, for applications in telecommunications this secondary goal of parameter reconstruction is of principal interest.

Adaptive observers for linear systems were proposed in [27]; see [33]. For nonlinear systems the general problem is unsolved but a relatively complete solution was given in [29]. Some definitions and results are given in the following [29, 30]:

Consider a single output nonlinear system with unknown constant parameters $\theta_i, i = 1, \ldots, N$.

$$\dot{x} = f(x) + g(x) + \sum_{i=1}^{N} \theta_i q_i(x), \quad y = h(x) \tag{19.3.9}$$

where $x(t) \in \mathbf{R}^n$, $y(t) \in \mathbf{R}^1$, $f(0) = 0$, $h(0) = 0$, $\theta = (\theta_1, \ldots, \theta_N)^\top$ is the vector of all unknown parameters.

DEFINITION 19.2 [30] *A global adaptive observer for system* (19.3.9) *is a finite dimensional system*

$$\begin{cases} \dot{w} = F_1(w, \widehat{\theta}, y) \\ \dot{\widehat{\theta}} = F_2(w, \widehat{\theta}, y) \\ \widehat{x} = H(w, \widehat{\theta}, y) \end{cases} \tag{19.3.10}$$

where $w(t) \in \mathbf{R}^r, r \geq n$, $\widehat{\theta}(t) \in \mathbf{R}^N$, $\widehat{x}(t) \in \mathbf{R}^n$, *the vector–functions* F_1, F_2, H *have appropriate dimensions, if for every* $x(0) \in \mathbf{R}^n$, $w(0) \in \mathbf{R}^r$, $\widehat{\theta}(0) \in \mathbf{R}^N$ *and any value of unknown parameters* θ *providing boundedness of the state vector of* (19.3.9) $x(t)$, $0 \leq t < \infty$, *the state vector of* (19.3.10) $w(t), \widehat{\theta}(t)$ *is also bounded and*

$$\lim_{t \to \infty} \|x(t) - \widehat{x}(t)\| = 0. \tag{19.3.11}$$

The solution of the adaptive observer design problem will be given for systems (19.3.9) which can be transformed by a θ-independent diffeomorphism into the following form:

$$\dot{z} = A_0 z + \phi_0(y) + \sum_{i=1}^{N} \theta_i \phi_i(y) \tag{19.3.12}$$

where

$$A_0 = \begin{bmatrix} 0 & 0 & \cdots & 0 & 0 \\ 1 & 0 & \cdots & 0 & 0 \\ 0 & 1 & \cdots & 0 & 0 \\ \vdots & \vdots & \ddots & \vdots & \vdots \\ 0 & 0 & \cdots & 1 & 0 \end{bmatrix}, \quad C_0 = [0, \ldots, 0\ 1],$$

$\phi_i(y(t)) \in \mathbf{R}^n$, $i = 0, 1, \ldots, N$ are vectors of nonlinearities depending on the output $y = y(t)$.

The linear part of the transformed system is in the so-called Brunovsky observer form. Necessary and sufficient conditions allowing the transformation of (19.3.9) into (19.3.12) via θ-independent nonlinear coordinate change can be found in [29, 30]. In what follows, we assume that the system already is in the form (19.3.12).

We first consider a special case, the so-called *adaptive observer form*, which is given as

$$\dot{z} = A_0 z + \phi_0(y) + B \sum_{i=1}^{N} \theta_i \psi_i(y), \quad y = C_0 z \qquad (19.3.13)$$

where $\psi_i(y)$ are smooth scalar functions, and $B = [b_1, b_2, \ldots, b_n]^\mathsf{T} \in \mathbf{R}^n$ is a constant vector such that $b_n > 0$ and the polynomial $b_n \lambda^{n-1} + \ldots + b_1$ is Hurwitz, i.e., all its roots have negative real parts. In [30] the conditions of transformability of (19.3.9) into the adaptive observer form (19.3.13) can be found.

A solution to the adaptive observer design problem is provided by the following theorem[30].

THEOREM 19.1
The system

$$\begin{cases} \dot{\hat{z}} = A_0 \hat{z} + \phi_0(y) + B \sum_{i=1}^{N} \hat{\theta}_i \psi_i(y) + K(C_0 \hat{z} - y) \\ \dot{\hat{\theta}} = \Gamma \psi(y)(y - C_0 \hat{z}) \end{cases} \qquad (19.3.14)$$

is a global adaptive observer for system (19.3.13), *where* $\hat{z}(t) \in \mathbf{R}^n$, $\hat{\theta}(t) \in \mathbf{R}^N$, *if* $\Gamma = \Gamma^\mathsf{T} > 0$ *is any* $N \times N$ *symmetric positive definite matrix and the vector* $K \in \mathbf{R}^n$ *is defined as*

$$K = (A_0 B + \mu B)/b_n$$

with $\mu > 0$.

To design an adaptive observer for the more general class of systems

(19.3.12) the so called *filtered transformation*

$$\begin{cases} \eta_j = z_j - \sum_{i=1}^{N} \xi_j[i]\theta_i, & 1 \le j \le n-1 \\ \eta_n = z_n \end{cases} \qquad (19.3.15)$$

is introduced where the vectors $\xi[i] \in \mathbf{R}^{n-1}, i = 1, \ldots, N$ satisfy the equation

$$\dot{\xi}[i] = D\xi[i] + D'\phi_i(y), \quad \xi[i](0) = 0, \qquad (19.3.16)$$

with $(n-1) \times (n-1)$ matrix D and $(n-1) \times n$ matrix D' of the form

$$D = \begin{bmatrix} 0 & 0 & \cdots & 0 & -b_1 \\ 1 & 0 & \cdots & 0 & -b_2 \\ 0 & 1 & \cdots & 0 & -b_3 \\ \vdots & \vdots & \ddots & \vdots & \vdots \\ 0 & 0 & \cdots & 1 & -b_{n-1} \end{bmatrix} \quad D' = \begin{bmatrix} 1 \\ 0 \\ \vdots \\ 0 \\ 0 \end{bmatrix} D$$

It is easy to check (see [30]) that the θ-independent and time-independent transformation $z \mapsto \eta$ transforms the system (19.3.12) into the adaptive observer form

$$\begin{cases} \dot{\eta} = A_0\eta + \phi_0(y) + B \sum_{i=1}^{N}(\xi_{n-1}[i] + \phi_{in}(y))\theta_i \\ y = C_0\eta \end{cases} \qquad (19.3.17)$$

where $B = [b_1, \ldots, b_{n-1}, 1]^\top$, $\psi_i(y,t) = \xi_{n-1}[i] + \phi_{in}(y)$. If the polynomial $\lambda^{n-1} + b_{n-1}\lambda^{n-2} + \ldots + b_2\lambda + b_1$ is chosen to be Hurwitz, then for bounded state $z(t)$ of the system (19.3.12), the mapping (19.3.15) is smooth, globally defined, and globally invertible. Therefore, any adaptive observer for the system (19.3.17) (e.g., of form (19.3.14)) will provide an estimate $\hat{\eta}$ for the vector $\bar{z} = z + \sum_{i=1}^{N} \xi[i]\theta_i$.

However, this is not sufficient for reconstruction of the initial state $z(t)$. To solve the initial problem an additional assumption is needed which guarantees the convergence of the parameter estimate $\hat{\theta}$ to its true value θ. A standard assumption of such kind, commonly used in the control literature, is the so-called *persistency of excitation (PE)* condition.

DEFINITION 19.3 A vector-function $f : [0, \infty) \to \mathbf{R}^m$ satisfies the *persistency of excitation (PE) condition* on $[0, \infty)$, if it is piecewise continuous, bounded and if there exist positive constants $\alpha > 0, T > 0$ such that

$$\int_t^{t+T} f(s)f(s)^\top ds \ge \alpha I_m$$

for all $t \ge 0$.

It follows from the "convergence under PE" theorems which can be found in many texts, e.g., [39, 30, 12, 33, 38, 12] that if the vector function

$$f(t) = [\xi_{n-1}[1](t) + \phi_{1n}(y(t)), \ldots, \xi_{n-1}[N](t) + \phi_{Nn}(y(t))]^\top$$

satisfies the PE condition, then the estimates $\widehat{\theta}_i$ provided by the adaptive observer (19.3.14), where $\psi_i(y) = \xi_{n-1}[i] + \phi_{in}(y)$ will converge to the true parameter values θ_i. As a consequence we get the convergence

$$|\widehat{\eta}(t) + \sum_{i=1}^{N} \xi[i](t)\widehat{\theta}_i(t) - z(t)| \to 0, \qquad (19.3.18)$$

i.e., a solution to the observer design problem for the system (19.3.12).

19.4 Adaptive Synchronization of Lur'e Systems

In this section a special case of the adaptive synchronization problem will be considered admitting a simple solution. We will be dealing with the unidirectional observer-based scheme where the transmitter is described by so-called Lur'e system (linear system with output-dependent nonlinearities):

$$\begin{cases} \dot{x}_d = Ax_d + \varphi_0(y_d) + B \sum_{i=1}^{m} \theta_i \varphi_i(y_d), \\ y_d = Cx_d \end{cases} \qquad (19.4.19)$$

where $x_d \in \mathbf{R}^n$ is the transmitter state vector, $y_d \in \mathbf{R}^l$ is the vector of outputs (transmitted signals), $\theta = (\theta_1, \ldots, \theta_m)^\top$ is the vector of transmitter parameters (possibly representing a message). It is assumed that the nonlinearities $\varphi_i(\cdot)$, $i = 0, 1, \ldots, m$, matrices A, C, and vector B are known.

The receiver will be designed as another dynamical system which provides estimates $\hat{\theta}_i$, $i = 1, \ldots, m$ of the transmitter parameters based on the observations of the transmitted signal $y_d(t)$. The problem is to design receiver equations

$$\dot{z} = F(z, y_d), \qquad (19.4.20)$$

$$\hat{\theta} = h(z, y_d), \qquad (19.4.21)$$

ensuring convergence

$$\lim_{t \to \infty} \left[\hat{\theta}(t) - \theta\right] = 0 \qquad (19.4.22)$$

where $\hat{\theta}(t) = \left(\hat{\theta}_1(t), \ldots, \hat{\theta}_m(t)\right)^\top$ is the vector of parameter estimates.

The proposed receiver is a kind of adaptive observer. Its simplest version for the case when A, B, C are known is as follows:

$$\dot{x} = Ax + \varphi_0(y_d) + B\left[\sum_{i=1}^{m} \hat{\theta}_i \varphi_i(y_d) + \hat{\theta}_0 G(y_d - y)\right],$$
$$y = Cx, \qquad (19.4.23)$$

$$\dot{\hat{\theta}}_i = \psi_i(y_d, y), \quad i = 0, 1, \ldots, m, \qquad (19.4.24)$$

where $x \in \mathbf{R}^n, y_d \in \mathbf{R}^l, \theta_0 \in \mathbf{R}$, and $G \in \mathbf{R}^l$ is the vector of weights. The adaptation algorithm (19.4.24) will be determined later. Thus, the state of receiver is $z = \left[x, \hat{\theta}_0, \hat{\theta}_1, \ldots, \hat{\theta}_m\right]$; the right-hand side of (19.4.20) is determined from (19.4.23) and (19.4.24).

Since the structure of (19.4.23) is similar to (19.4.19), a natural secondary goal might be

$$\lim_{t\to\infty} e(t) = 0, \qquad (19.4.25)$$

where $e(t) = x(t) - x_d(t)$ is the observation error.

Although (19.4.25) is not necessary in order to provide (19.4.22), it may give a hint how to choose a Lyapunov function for a proper design of an adaptation algorithm (19.4.24).

To solve the problem we write down the error equation:

$$\begin{cases} \dot{e} = Ae + B\left[\sum_{i=1}^{m} \tilde{\theta}_i \varphi_i(y_d) + \hat{\theta}_0 G \tilde{y}\right], \\ \tilde{y} = Ce \end{cases} \qquad (19.4.26)$$

where $\tilde{\theta}_i = \hat{\theta}_i - \theta_i, i = 1, \ldots, m$ are the parameter errors. The adaptation algorithm is provided by standard gradient (or speed-gradient) techniques as follows:

$$\dot{\hat{\theta}}_i = -\gamma_i(y - y_d)\varphi_i(y_d), \quad i = 1, \ldots, m, \qquad (19.4.27)$$

$$\dot{\hat{\theta}}_0 = -\gamma_0(y - y_d)^2. \qquad (19.4.28)$$

In order to formulate the conditions required for a successful applicability of the proposed scheme, we need some definitions and auxiliary results.

DEFINITION 19.4 [12] *The system $\dot{x} = \bar{A}x + \bar{B}u$, $y = \bar{C}x$ with transfer matrix $W(\lambda) = \bar{C}(\lambda I - \bar{A})^{-1}\bar{B}$, where $u, y \in \mathbf{R}^l$ and $\lambda \in \mathbf{C}$ is called hyper-minimum-phase if it is minimum-phase (i.e., the polynomial $\varphi(\lambda) = \det(\lambda I - \bar{A}) \det W(\lambda)$ is Hurwitz), and the matrix $\bar{C}\bar{B} = \lim_{\lambda\to\infty} \lambda W(\lambda)$ is symmetric and positive definite.*

Note that for $l = 1$ the system of order n is hyper-minimum-phase if the numerator of its transfer function is a Hurwitz polynomial of degree $n - 1$ with positive coefficients.

LEMMA 19.2 *[11]*
Let matrices $\bar{A}, \bar{B}, \bar{C}, G$ of sizes $n \times n$, $n \times m$, $l \times n$, $m \times l$ be given. Assume $\text{rank}(\bar{B}) = m$. Then there exists a positive definite $n \times n$ matrix $P = P^T > 0$ and an $m \times l$ matrix θ_* such that

$$PA_* + A_*^T P < 0, \quad P\bar{B} = \bar{C}^T G^T, \quad A_* = \bar{A} + \bar{B}\theta_* \bar{C}$$

if and only if the system $\dot{x} = \bar{A}x + \bar{B}u$, $y = G\bar{C}x$ is hyper-minimum-phase.

Lemma 19.2 establishes conditions for the existence of a feedback $u = \theta_* y + v$ making the closed loop system with input v and output Gy strictly passive. It is closely related to the Kalman-Yakubovich lemma and can be called a "Feedback Kalman-Yakubovich lemma" (see [2]).

LEMMA 19.3 *[39]*
Consider vector-functions $f, \tilde{\theta} : [0, \infty) \to \mathbf{R}^m$. Assume that $\tilde{\theta}(t)$ is continuously differentiable, $\dot{\tilde{\theta}}(t) \to 0$ as $t \to \infty$ and f is PE. Then $\tilde{\theta}(t) \to 0$ as $t \to \infty$ provided that $\tilde{\theta}(t)^T f(t) \to 0$ as $t \to \infty$.

THEOREM 19.4
Assume that all the trajectories of the transmitter (19.4.19) are bounded and the linear systems with the transfer function $W(\lambda) = GC(\lambda I - A)^{-1}B$ be hyper-minimum-phase. Then all the trajectories of the receiver (19.4.23), (19.4.27), (19.4.28) are bounded and the relation (19.4.25) holds. If, in addition, the vector-function $[\varphi_1(y_d), \ldots, \varphi_m(y_d)]$ satisfies the PE condition, then (19.4.22) also holds.

PROOF To prove the theorem consider the Lyapunov function candidate

$$V(x, \hat{\theta}_0, \hat{\theta}, t) = \frac{1}{2} e^T P e + \frac{1}{2} \sum_{i=0}^{m} \|\hat{\theta}_i - \theta_i\|^2 / \gamma_i$$

$$+ \|\hat{\theta}_0 - \theta_{*0}\|^2 / \gamma_0 \qquad (19.4.29)$$

where a matrix $P = P^T > 0$ and a number θ_{*0} are to be determined. Calculation of $\dot{V}$ gives that $\dot{V} < 0$ for $e \neq 0$ if and only if the following conditions are valid:

$$\begin{cases} \dot{\hat{\theta}}_0 = -\gamma_0 e^T PBGCe, \\ \dot{\hat{\theta}}_i = -\gamma_i e^T PB\varphi_i(y_d), \\ e^T(PA_* + A_*^T P)e < 0. \end{cases} \quad (19.4.30)$$

Using Lemma 19.2 we obtain that $\dot{V} < 0$ for $e \neq 0$ if and only if the adaptation algorithm has the form (19.4.27), (19.4.28) and the system $\dot{x} = Ax + Bu$, $y = Cx$ is hyper-minimum-phase. Therefore, under the given conditions the function $V(t) = V(x(t), \hat{\theta}_0(t), \hat{\theta}(t), t)$ is bounded. Since $\varphi_i(y_d(t)), i = 1, \ldots, m$ are bounded, the functions $e(t)$, $\hat{\theta}_i(t)$ are bounded, too. Equations (19.4.30) imply that $\dot{V} = e^T(PA_* + A_*^T P)e \leq -\mu \|e(t)\|^2$ for some $\mu > 0$. Integration of the last inequality over the interval $[0, t]$ gives: $V(t) - V(0) \leq -\mu \int_0^t \|e(s)\|^2 ds$. Taking into consideration that $V \geq 0$ we obtain: $V(0) \geq \mu \int_0^t \|e(s)\|^2 ds$. This yields the inequality

$$\int_0^\infty \|e(t)\|^2 dt < \infty. \quad (19.4.31)$$

Since $\varphi_i(y_d), i = 1, \ldots, m$ are bounded, $\dot{e}(t)$ is also bounded in view of (19.4.26). From (19.4.31) and Barbalat's lemma we obtain that the goal (19.4.25) is achieved.

To prove (19.4.22) we first note that $\dot{\tilde{\theta}}(t) \to 0$ as $t \to \infty$ from (19.4.27) and (19.4.25). Differentiating (19.4.26), from boundedness of functions $e, \tilde{\theta}, \varphi_d, \tilde{y}, \hat{\theta}_0$ and their time-derivatives we conclude that $\ddot{e}(t)$ is bounded. Barbalat's lemma then implies that $\dot{e}(t) \to 0$ as $t \to \infty$. This and (19.4.27) yield $\tilde{\theta}(t)^T \varphi_d(t) \to 0$ as $t \to \infty$. Hence (19.4.22) follows from the PE condition and Lemma 19.3.

Remark B. Theorem 19.4 in fact gives necessary and sufficient conditions for the existence of a Lyapunov function of the form (19.4.29) with the properties

$$\begin{cases} V(x, \hat{\theta}_0, \hat{\theta}, t) > 0 \text{ for } e \neq 0, \\ \dot{V}(x, \hat{\theta}_0, \hat{\theta}, t) < 0 \text{ for } e \neq 0. \end{cases} \quad (19.4.32)$$

It means that it is not possible to find another adaptation algorithm based on the Lyapunov function (19.4.29) with the properties (19.4.32).

The proposed design can be extended to the case when the matching conditions with respect to unknown parameters are violated. Namely, using the results of the previous section, the system can be transformed into the adaptive observer form by means of filtered transformation (19.3.15), (19.3.16). Combining this result with the Theorem 19.4 we obtain that under PE condition the convergence of parameters (19.4.22) will be ensured and the system (19.3.14),(19.3.15), (19.3.16) will serve as adaptive observer for system (19.4.19), providing synchronization with identification of parameters.

As an example we consider the problem of synchronizing two Chua circuits with unknown parameters and incomplete measurements.

19.5 Signal Transmission and Reconstruction

In recent years much attention has been devoted to methods for secure communications utilizing chaos [35, 7, 26]. Various methods for transmitting signals via chaotic synchronization were proposed like chaotic signal masking [35, 7], chaotic binary communications [7, 26], etc.

A possible application of the synchronization scheme proposed in Section 19.4 to chaotic binary communications algorithms goes as follows and is based on the dependence of the synchronization effect on the matching of the corresponding parameters of the systems. The transmitter and receiver have identical structure as in the previous section. The basic idea is to modulate this coefficient with an information-bearing binary waveform and transmit the chaotic signal. At the receiver side the coefficient modulation will produce a synchronization error between the received signal and the corresponding transmitter reconstructed signal: if the coefficients of transmitter and receiver are identical, the signals will synchronize; otherwise, synchronization fails. Using the synchronization error the modulation can be detected. Security of communications is possibly enhanced by a set of other transmitter parameters.

Consider as an example of information transmission where both transmitter and receiver system are implemented as a Chua circuit, similar to [26]. The transmitter model in dimensionless form is given as

$$\dot{x}_{d_1} = p[x_{d_2} - x_{d_1} + f(x_{d_1}) + s f_1(x_{d_1})]$$
$$\dot{x}_{d_2} = x_{d_1} - x_{d_2} + x_{d3} \qquad (19.5.33)$$
$$\dot{x}_{d_3} = -q x_{d_2}$$

where $f(z) = M_0(z) + 0.5(M_1 - M_0)f_1(z)$, $f_1(z) = |z+1| - |z-1|$, M_0, M_1, p, q are the transmitter parameters, $s = s(t)$ is the signal to be reconstructed in the receiver. Assume that the transmitted signal is $y_d(t) = x_{d_1}(t)$, and the values of the parameters p, q are known.

The parameters M_0, M_1 are assumed to be *a priori* unknown which motivates the use of an adaptation for the receiver design. The receiver designed according to the results of Section 19.4 is modeled as

$$\dot{x}_1 = p[x_2 - x_1 + f(y_d) + c_1 f_1(y_d) + c_0(x_1 - y_d)],$$

$$\dot{x}_2 = x_1 - x_2 + x_3, \qquad (19.5.34)$$

$$\dot{x}_3 = -qx_2,$$

where c_0, c_1 are the adjustable parameters. The adaptation algorithm (19.4.27), (19.4.28), takes the form

$$\begin{aligned} \dot{c}_0 &= -\gamma_0 (y_d - x_1)^2, \\ \dot{c}_1 &= -\gamma_1 (x_1 - y_d) f_1(y_d), \end{aligned} \qquad (19.5.35)$$

where γ_0, γ_1 are the adaptation gains.

First we examine the ability of the system (19.5.34), (19.5.35) to receive and decode messages. To this end we verify the conditions of Theorem 19.4 assuming that $s(t) = \text{const}$. Clearly, if $s(t)$ is a time-varying binary signal, we can only expect that the results of Theorem 19.4 can be used if the parameter estimation is fast enough, at least much faster than the actual parameter modulation. Writing the error equations yields

$$\begin{cases} \dot{e}_1 &= p[e_2 - e_1 + (c_1 - s)f_1(y_d) + c_0 e_1] \\ \dot{e}_2 &= e_1 - e_2 + e_3 \\ \dot{e}_3 &= -qe_2, \end{cases} \qquad (19.5.36)$$

where $e_i = x_i - x_{d_i}$, $i = 1, 2, 3$. The system (19.5.36) is obviously in Lur'e form (19.4.26), where

$$A = \begin{bmatrix} -p & p & 0 \\ 1 & -1 & 1 \\ 0 & -q & 0 \end{bmatrix}, \quad B = \begin{bmatrix} 1 \\ 0 \\ 0 \end{bmatrix}, \quad C = [1\ 0\ 0],$$

$\hat{\theta}_1 = c_1$, $\theta_1 = s$, $\theta_0 = c_0$.

The transfer function of the linear part is

$$W(\lambda) = \frac{\lambda^2 + \lambda + q}{\lambda^3 + (p+1)\lambda^2 + q\lambda + pq} \qquad (19.5.37)$$

We see that the order of the system is $n = 3$, while the numerator polynomial is Hurwitz and has degree 2 for all $q > 0$ and all real p. Therefore, the hyper-minimum-phase condition holds for $q > 0$ and any p, M_0, M_1. Thus, Theorem 19.4 yields the boundedness of all receiver trajectories $x(t)$ and convergence of the observation error: $e(t) \to 0$. In particular, $y_d(t) - x_1(t) \to 0$. Furthermore, to be able to reconstruct the signal $s(t)$, the receiver should provide convergence $c_1(t) - s \to 0$ for constant s. According to Theorem 19.4, this will be the case if the PE condition (see Definition 19.3) holds, which reads as

$$\int_{t_0}^{t_0+T} f_1^2(y_d(t))\, dt \geq \alpha \qquad (19.5.38)$$

for some $T > 0$, $\alpha > 0$ and all $t_0 \geq 0$. To verify (19.5.38), we note that condition (19.5.38) basically means that the trajectory of the transmitter $x_d(t)$ does not converge to the plane $x_{d_1} = 0$ when $t \to \infty$. This is not the case, at least when the system (19.5.33) exhibits chaotic behavior. Indeed, in this case the value $x_{d_1}(t)$ leaves the interval $(-1, 1)$ (where $f_1(z)$ is linear) infinitely many times, say at t_k, $k = 1, 2, \ldots$. The time intervals $\Delta t_k = t_{k+1} - t_k$ between t_k can be overbounded by constant, if the trajectory does not converge to the set $x_{d_1} = 0$.

We may also evaluate a lower bound for α in (19.5.38):

$$\alpha_0 = \liminf{}_{T \to \infty} \frac{1}{T} \int_0^T f_1^2(x_{d_1}(t)) \, dt. \qquad (19.5.39)$$

The value of α_0 characterizes the parameter convergence rate. It follows from the standard convergence rate results (see, e.g., [38]) that if $\alpha_0 > 0$, then the convergence $c_1(t) - s \to 0$ is exponential, with rate $\gamma_1 \alpha_0$, at least for sufficiently small $\gamma_1 > 0$. Ergodicity arguments suggest that

$$\alpha_0 \geq \frac{\overline{x}_{d_1}^2}{\mu}, \qquad (19.5.40)$$

where $\overline{x}_{d_1}^2$ is the average value of $x_{d_1}^2(t)$ over the attractor Ω, and $\mu = \sup_{x \in \Omega} |x_{d_1}(t)|$.

We carried out simulations for the above scheme. Parameter values in dimensionless model (19.5.33), (19.5.34) were selected as $p = 10; q = 15.6; M_0 = 0.33; M_1 = 0.22$. These values are close to those used in [26] after standard rescaling [6] (below the time unit is $1ms$). For these parameter values, the system (19.5.33) possesses a chaotic attractor.

The initial conditions for the transmitter were taken as $[0.3, 0.3, 0.3]^T$. For the receiver, zero initial conditions were chosen both for the state x_0 as well as for the adjustable parameters $c_0(0)$, $c_1(0)$. In order to eliminate the influence of initial conditions, no message was transmitted during the first 20 ms ("tuning" or "calibration" of the receiver), i.e., $s(t) \equiv 1$ for $0 \leq t \leq 20$ ms. The time history of observation errors (Figure 19.1) and parameter estimates (Figure 19.2) during tuning show that all observation errors and parameter estimation error $c_1(t) - s$ tend to zero rapidly. The value $c_0(t)$ tends to some constant value.

After the tuning period, the square wave message

$$s(t) = s_0 + s_1 \operatorname{sign} \sin\left(\frac{2\pi t}{T_0}\right), \qquad (19.5.41)$$

where $s_0 = 1.005$, $s_2 = 0.005$ was sent. Simulation results for $T_0 = 5$ s, $\gamma_1 = 1$ are shown in Figure 19.3 and Figure 19.4. It is seen that the reconstructed signal $y(t)$ coincides with the transmitted signal $y_d(t)$ with very good accuracy. However, both observation errors (Figure 19.3) and estimation errors (Figure 19.4) do not decay completely during the interval when $s(t)$ is constant. Nevertheless, a reliable reconstruction of the signal $s(t)$ is very well possible. The accuracy of estimation can be improved by increasing the adaptation gain γ_1, which is confirmed by simulation results for $\gamma_1 = 5$ (Figures 19.5 and 19.6). The reconstruction time delay for $\gamma_1 = 5$ is about 0.3–0.5 ms, which is about ten times less than that obtained in [26].

19.6 Conclusions

The proposed adaptive observer-based synchronization scheme demonstrates good signal reconstruction abilities. It allows achieving a high information transmission rate (close to the upper bound of the carrier spectrum).

Acknowledgments

This work was supported in part by the Dutch Organization for Pure Research (NWO), the RFBR Grant 96-01-01151, and the Russian Federal Programme "Integration" (project 2.1-589).

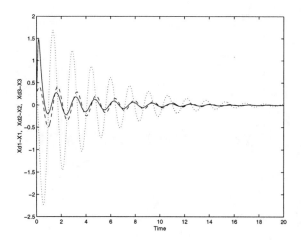

FIGURE 19.1
Time history of observation errors during tuning.

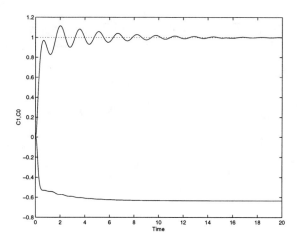

FIGURE 19.2
Time history of parameter estimates during tuning.

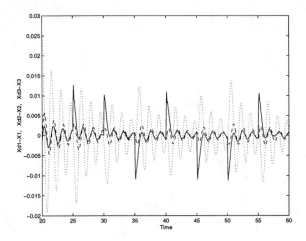

FIGURE 19.3
Time history of observation errors, $\gamma = 1$.

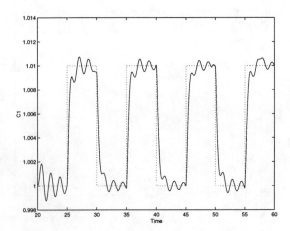

FIGURE 19.4
Time history of parameter estimates, $\gamma = 1$.

Conclusions

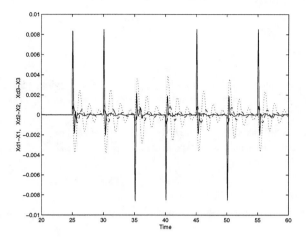

FIGURE 19.5
Time history of observation errors, $\gamma = 5$.

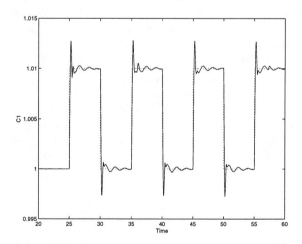

FIGURE 19.6
Time history of parameter estimates, $\gamma = 5$.

References

[1] V. S. Afraimovich, N. N. Verichev, and M. I. Rabinovich, "Synchronization of oscillations in dissipative systems," *Radiophysics and Quantum Electronics*, pp. 795-803, 1987.

[2] B. R. Andrievsky, A. N. Churilov, and A. L. Fradkov, "Feedback Kalman-Yakubovich lemma and its applications," *Proc. IEEE Conf. on Dec. Contr.*, Kobe, 1996, pp. 4537–4542.

[3] I. I. Blekhman, *Synchronization of Dynamical Systems*, (in Russian) Nauka, Moscow, 1971.

[4] I. I Blekhman, *Synchronization in Science and Technology*, ASME Press, New York, 1988.

[5] I. I. Blekhman, A. L. Fradkov, H. Nijmeijer, and A. Yu. Pogromsky, "On self-synchronization and controlled synchronization," *Systems and Control Letters*, vol. 31, pp. 299–306, 1997.

[6] L. O. Chua, M. Komuro, and T. Matsumoto, "The double scroll family. Parts I and II," *IEEE Trans. Circ. Sys.*, vol. 33, pp. 1072–1118, 1986.

[7] G. Cicarella, M. D. Mora, and A. Germani, "A Luenberger-like observer for nonlinear systems," *Int. J. of Control*, vol. 57, pp. 537-556, 1993.

[8] G. Columban, M. P. Kennedy, and L. O. Chua, "The role of synchronization in digital communications using chaos," *IEEE Trans. Circ. Sys.*, vol. 44, pp. 927-936, 1997.

[9] K. Cuomo, A. V. Oppenheim, and S. H. Strogatz, "Synchronization of Lorenz-based chaotic circuits with application to communications," *IEEE Trans. Circ. Sys.*, II, vol. 40, pp. 626-633, 1993.

[10] H. Dedieu, M. Kennedy, and M. Hasler, "Chaos shift keying: Modulation and demodulation of chaotic carrier using self-synchronized Chua's circuits," *IEEE Trans. Circ. Sys.*, II, vol. 40, pp. 634–642, 1993.

[11] A. L. Fradkov, "Quadratic Lyapunov functions in the adaptive stabilization problem of a linear dynamic plant," *Siberian Math. J.*, pp. 341-348, 1976.

[12] A. L. Fradkov, *Adaptive Control of Complex Systems*, (in Russian), Nauka, Moscow, 1990.

[13] A. L. Fradkov, "Nonlinear adaptive control: Regulation - tracking - oscillations," *Proc. of IFAC Workshop on "New Trends in Design of Control Systems,"* Smolenice, pp. 426-431, 1994.

[14] A. L. Fradkov, "Adaptive synchronization of hyper-minimum-phase systems with nonlinearities," *Proc. of 3rd IEEE Mediterranean Symp. on New Directions in Control*, Limassol, 1995, vol.1, pp. 272-277.

[15] A. L. Fradkov and A. Yu. Pogromsky, *Introduction to Control of Oscillations and Chaos*, World Scientific Pub. Co., Singapore, 1998.

[16] A. L. Fradkov, I. V. Miroshnik, and V. O. Nikiforov, *Nonlinear and Adaptive Control of Complex Systems*, Kluwer, Dordrecht, 1999.

[17] J. P. Gauthier, H. Hammouri, and S. Othman, "A simple observer for nonlinear systems with applications to bioreactors," *IEEE Trans. Aut. Contr.*, vol. 37, pp. 75–880, 1992.

[18] G. Grassi and S. Mascolo, "Synchronization of hyperchaotic oscillators using a scalar signal," *IEEE Trans. Circ. Sys.*, I, vol. 44, pp. 1011–1014, 1997.

[19] M. Hasler, "Synchronization of chaotic systems and transmission of information," *Int. J. Bifurcation and Chaos*, vol. 8, pp. 647–659, 1998.

[20] A. Isidori, *Nonlinear Control Systems*, 2nd ed., Springer-Verlag, 1989.

[21] R. Kalman, "A new approach to linear filtering and prediction problems," *Trans. ASME, Part D.: J. of Basic Engineering*, vol. 82, pp. 35-45, 1960.

[22] L. Kocarev, K. S. Halle, K. Eckert, and L. O. Chua. "Experimental demonstration of secure communication via chaotic synchronization," *Int. J. Bif. Chaos*, vol. 2, pp. 709–713, 1992.

[23] L. Kocarev and U. Parlitz, "Generalized synchronization, predictability and equivalence of unidirectionally coupled dynamical systems," *Phys. Rev. Letts.*, vol. 76, pp. 816–1819, 1996.

[24] A. Krener and A. Isidori, "Linearization by output injection and nonlinear observers," *Systems and Control Letters*, vol. 3, pp. 47-52, 1983.

[25] M. Krstic, I. Kanellakopoulos, and P. V. Kokotovic, *Nonlinear and Adaptive Control Design*, Wiley, New York, 1995.

[26] W. Lindsey, *Synchronization Systems in Communications and Control*, Prentice-Hall, NJ, 1972.

[27] G. Luders and K. S. Narendra, "An adaptive observer and identifier for a linear system." *IEEE Trans. Auto. Contr.*, vol. 18, pp. 496–499, 1973.

[28] D. Luenberger, "Observing the state of a nonlinear system," *IEEE Trans. on Military Electronics*, vol. 8, pp. 74–80, 1961.

[29] R. Marino, "Adaptive observers for single output nonlinear systems," *IEEE Trans. Auto. Contr.*, vol. 35, pp. 1054–1058, 1990.

[30] R. Marino and P. Tomei, *Nonlinear Control Systems Design*, Prentice-Hall, NJ, 1995.

[31] A. Yu. Markov and A. L. Fradkov, "Adaptive synchronization of chaotic systems based on speed-gradient and passification," *IEEE Trans. Circ. Sys.*, I, vol. 44, pp. 905–912, 1997.

[32] Ö. Morgül and E. Solak, "Observer based synchronization of chaotic systems," *Phys. Rev. E*, vol. 54, pp. 4803–4811, 1996.

[33] K. Narendra and A. Annaswamy, *Stable Adaptive Systems*, Prentice-Hall, NJ, 1989.

[34] H. Nijmeijer and I. M. Y. Mareels, "An observer looks at synchronization," *IEEE Trans. Circ. Sys.*, I, vol. 44, pp. 882–890, 1997.

[35] L. Pecora and T. Carroll, "Synchronization in chaotic systems," *Physics Rev. Lett.*, vol. 64, pp. 821-824, 1990.

[36] A. Yu. Pogromsky, "Passivity based design of synchronizing systems," *Int. J. Bifurcation Chaos*, vol. 8, pp. 295–319, 1998.

[37] N. F. Rulkov, M. M.Suschik, L. S. Tsirling, and H. D. I. Abarbanel, "Generalized synchronization of chaos in directionally coupled systems," *Phys. Rev. E*, vol. 51, pp. 980–994, 1995.

[38] S. S. Sastry and M. Bodson, *Adaptive Control: Stability, Convergence and Robustness*, Prentice-Hall, NJ, 1989.

[39] J. Yuan and W. Wonham, "Probing signals for model reference identification," *IEEE Trans. Auto. Contr.*, vol. 22, pp. 530–538, 1977.

20

Discrete-Time Observers and Synchronization

H.J.C. Huijberts,[1] **H. Nijmeijer,**[2,3] **and A. Yu. Pogromsky**[3,4]

[1]Faculty of Mathematics and Computing Science
Eindhoven University of Technology
P.O. Box 513, 5600 MB Eindhoven, The Netherlands
hjch@win.tue.nl

[2]Faculty of Mathematical Sciences
University of Twente
P.O. Box 217, 7500 AE Enschede, The Netherlands
h.nijmeijer@math.utwente.nl

[3]Faculty of Mechanical Engineering
Eindhoven University of Technology
P.O. Box 513, 5600 MB Eindhoven, The Netherlands

[4]Institute for Problems of Mechanical Engineering
61, Bolshoy, V.O., 199178, St. Petersburg, Russia
sasha@ccs.ipme.ru

Abstract

The synchronization problem for complex discrete-time systems is revisited from a control perspective and it is argued that the problem may be viewed as an observer problem. It is shown that a solution for the synchronization (observer) problem exists for several classes of systems. Also, by allowing past measurements, a dynamic mechanism for state reconstruction is provided.

20.1 Introduction

Since the work of Pecora and Carroll [18], a huge interest in (chaos) synchronization has arisen. Among others, this is illustrated by the appearance of a number of special issues of journals devoted to the subject, cf. [29, 28, 30]. One clear motivation for this widespread interest lies in the fact that Pecora and Carroll indicated that chaos synchronization might be useful in communications. Although this claim is not yet fully justified, several interesting applications of (chaos) synchronization are envisioned.

Synchronization as it was introduced by Pecora and Carroll has been studied from various viewpoints. Following [18], often a receiver-transmitter (or master-slave) formalism is taken, where typically the receiver system is an exact copy of the transmitter system and the aim is to synchronize the receiver response with that of the transmitter, provided the receiver dynamics are driven by a scalar signal from the transmitter; see [18, 4, 26].

More recently, the above method was recast in an *active-passive decomposition*, see [17], where the decomposition idea has to be understood in the way that part of the transmitter state needs to be transmitted, while the "passive" part then will be derived asymptotically.

Another idea to achieve synchronization between (identical) transmitter and receiver dynamics is to include (linear) feedback of the drive signal in the receiver system; see [16] and [11] where a number of sucessfull experimental settings of this type are discussed.

A third way to achieve synchronization between transmitter and receiver was recently put forward in [6] and essentially advertises the idea of system inversion for (state) synchronization.

Notwithstanding the widespread interest in the synchronization problem, the problem leaves some ambiguity in how to make an active-passive decomposition or how to successfully build an (stable) inverse system. Indeed, this ambiguity disappears when the synchronization problem is viewed as the question of how to reconstruct the full state trajectory of the transmitter system, given some (scalar) drive signal from the transmitter. This is essentially the observer problem from control theory, and has, following the earlier attempts [5, 19, 13], by now obtained a prominent place within recent synchronization literature; see, for instance, [14] and various other observer-based synchronization papers.

The purpose of the present chapter is to revisit the synchronization problem for discrete-time systems using (discrete-time) observers. Synchronization of complex/chaotic discrete-time systems has been the subject of various publications; see, e.g., [2, 1, 7, 21, 27], but only little attention for an observer-based viewpoint exists (see, however, [24, 25] where this viewpoint *is* taken, and [26], which may be interpreted as a particular application of the observer-based viewpoint (although this is not mentioned explicitly in

[26]). One could argue, however, that the synchronization problem for discrete-time systems is as important as the continuous-time counterpart. First, for communications of *binary* signals one can very well base oneself on discrete-time transmitter systems instead of continuous-time transmitters. A second motivation to look at discrete-time synchronization is that many continuous-time models are in the end – for instance, for simulation and implementation – discretized or sampled. A third motivation is that discrete-time dynamics are obtained when one considers the Poincaré map at a suitably defined Poincaré section of a chaotic transmitter system.

As stated, we pursue an observer-based view on (discrete-time) synchronization. Although some clear analogies exist between discrete-time and continuous-time observers, there are various results available in either context which do not admit a proper analogon in the other domain.

This chapter is organized as follows. In Section 20.2, we treat some preliminaries and give our problem statement. Section 20.3 is devoted to nonlinear discrete-time transmitters of a special form, the so-called Lur'e form. It is shown that for this kind of system, the construction of an observer is relatively easy. In Section 20.4, we study the question when a given nonlinear discrete-time transmitter is equivalent to a system in Lur'e form by means of a coordinate transformation. In Section 20.5, we introduce a so-called extended Lur'e form, indicate how observers for transmitters in this extended Lur'e form may be constructed, and give conditions under which a nonlinear discrete-time transmitter may be transformed into an extended Lur'e form. Section 20.6 treats the observer design for perturbed linear transmitters. Section 20.7, finally, contains some conclusions.

20.2 Preliminaries and Problem Statement

Throughout this chapter, we consider discrete-time nonlinear (transmitter) dynamics of the form

$$x(k+1) = f(x(k)), \quad x(0) = x_0 \in \mathbf{R}^n \qquad (20.2.1)$$

where the state transition map f is a smooth mapping from $\mathbf{R}^n$ into itself. Note that direct extensions of (20.2.1) are possible by allowing the state to belong to an open subset of $\mathbf{R}^n$ or to a differentiable manifold. The solution $x(k, x_0)$ of (20.2.1) is not directly available, but only an *output* is measured, say

$$y(k) = h(x(k)) \qquad (20.2.2)$$

where $y \in \mathbf{R}^p$ and $h : \mathbf{R}^n \to \mathbf{R}^p$ is the smooth output map. Though in the sequel there is no restriction in assuming the transmitted signal $y(k)$

to be p-dimensional, we will for simplicity – and following most work on synchronization – take $p = 1$.

The observer problem for (20.2.1,20.2.2) now deals with the question how to reconstruct the state trajectory $x(k, x_0)$ on the basis of the measurements $y(k)$. A *full observer* (or briefly *observer*) for the system (20.2.1, 20.2.2) is a dynamical system of the form

$$\hat{x}(k+1) = \hat{f}(\hat{x}(k), y(k)), \quad \hat{x}(0) = \hat{x}_0 \in \mathbf{R}^n \qquad (20.2.3)$$

where $\hat{x} \in \mathbf{R}^n$, and $\hat{f}$ is a smooth mapping on $\mathbf{R}^n$ parametrized by y, such that the error $e(k) := x(k) - \hat{x}(k)$ asymptotically converges to zero as $k \to \infty$ for all initial conditions x_0 and $\hat{x}_0$. Moreover, we require that if $e(k_0) = 0$ for some k_0, then $e(k) = 0$ for all $k \geq k_0$.

20.3 Systems in Lur'e Form

The problem of observer design in its full generality is a problem that is difficult to solve. Basically, only the observer design problem for linear systems has been solved in its full generality; see [20]. Therefore, we start our survey of possible approaches to observer-based synchronization by considering a class of nonlinear systems that is slightly more general than linear systems, namely, systems in so-called Lur'e form.

Assume that the master dynamics are governed by the following system of difference equations

$$x(k+1) = Ax(k) + \varphi(y(k)), \quad y(k) = Cx(k), \qquad (20.3.4)$$

where $x(k) \in \mathbf{R}^n$ is the state, $y(k) \in \mathbf{R}^1$ is the scalar output, $\varphi : \mathbf{R}^1 \to \mathbf{R}^n$ is a smooth function, and A, C are constant matrices of appropriate dimensions. Dynamics of the form (20.3.4) are referred to as dynamics in *Lur'e form*. The question we now pose is, under what conditions is it possible to design an observer for (20.3.4)? As a possible observer candidate one can build a copy of (20.3.4) augmented with so-called *output injection*:

$$\hat{x}(k+1) = A\hat{x}(k) + \varphi(y(k)) + L(y(k) - \hat{y}(k)), \quad \hat{y}(k) = C\hat{x}(k), \qquad (20.3.5)$$

where $\hat{x}(k) \in \mathbf{R}^n$ is the estimate of $x(k)$ and L is a $n \times 1$ matrix; see [10].

The solutions of systems (20.3.4) and (20.3.5) will synchronize if for all initial conditions the error $e(k) := x(k) - \hat{x}(k)$ tends to zero when k tends to infinity. Substracting (20.3.5) from (20.3.4), one can easily see that the error vector $e(k)$ obeys the following linear difference equation:

$$e(k+1) = (A - LC)e(k). \qquad (20.3.6)$$

Therefore, if all eigenvalues of $A - LC$ lie in the open unit disc (i.e., the set $\{z \in \mathbf{C} \mid |z| < 1\}$), then (20.3.5) is an observer for (20.3.4). In other words, for the system (20.3.4) the synchronization problem can be reduced to the following question: given A, C, under what conditions does there exist a matrix L such that $A - LC$ has all eigenvalues in the open unit disc? This linear algebraic problem has a simple solution. Namely, a sufficient condition for existence of L is the invertibility of the following linear mapping

$$\mathcal{O}(x) := \begin{bmatrix} C \\ CA \\ CA^2 \\ \vdots \\ CA^{n-1} \end{bmatrix} x. \tag{20.3.7}$$

In linear control theory, a pair of matrices (C, A) such that $\mathcal{O}(x)$ in (20.3.7) is invertible, is said to be an *observable pair*. Using this terminology we can formulate the following result.

THEOREM 20.1
Assume that the pair (C, A) is observable. Then the system (20.3.4) admits an observer (20.3.5) with the exponentially stable linear error dynamics (20.3.6).

The proof of this result can be found in any textbook on linear control theory (see, e.g., [20]). It is worth mentioning that the proof is constructive. Namely, the linear mapping $\mathcal{O}$ defines a similarity transformation such that the matrix $A - LC$ is similar to the following matrix in Frobenius form:

$$\begin{bmatrix} 0 & \cdots & 0 & a_1 - l_1 \\ 1 & \cdots & 0 & a_2 - l_2 \\ \vdots & \ddots & \vdots & \vdots \\ 0 & \cdots & 1 & a_n - l_n \end{bmatrix}$$

where $\mathrm{col}(l_1, l_2, \ldots, l_n) = \mathcal{O}(L)$, and the a_i are the coefficients of the characteristic polynomial of A. Since $a_i - l_i$ are the coefficients of the characteristic polynomial of $A - LC$, it is always possible to locate the eigenvalues of $A - LC$ in the open unit disc by means of an appropriate choice of the matrix L.

It is worth mentioning that the condition of observability is, in fact, a sufficient but not necessary condition to allow design of an observer. Namely, the system may have $\mathcal{O}(x)$ of rank lower than n, but at the same time, it may admit an observer. This situation occurs when the so-called unobservable dynamics are exponentially stable. In the control literature, linear systems with exponentially stable unobservable dynamics are referred to as *detectable* (see [20]). In practice it often means that such systems can be transformed to an observable system via model reduction.

Example 20.1
Consider the following discrete-time dynamics in Lur'e form:

$$\begin{bmatrix} z_1(k+1) \\ z_2(k+1) \end{bmatrix} = \underbrace{\begin{bmatrix} 0 & -\alpha \\ 1 & 1+\alpha \end{bmatrix}}_{A} \begin{bmatrix} z_1(k) \\ z_2(k) \end{bmatrix} + \underbrace{\begin{bmatrix} 0 \\ -\beta \cos y(k) \end{bmatrix}}_{\varphi(y(k))}$$

$$y(k) = \underbrace{\begin{bmatrix} 0 & 1 \end{bmatrix}}_{C} z(k)$$

(20.3.8)

where $\alpha, \beta > 0$. In this case, we obtain

$$\begin{bmatrix} C \\ CA \end{bmatrix} = \begin{bmatrix} 0 & 1 \\ 1 & 1+\alpha \end{bmatrix}$$

(20.3.9)

which clearly is an invertible matrix. Thus, one may construct an observer for (20.3.8) of the following form:

$$\begin{bmatrix} \widehat{z}_1(k+1) \\ \widehat{z}_2(k+1) \end{bmatrix} = \underbrace{\begin{bmatrix} 0 & -\alpha \\ 1 & 1+\alpha \end{bmatrix}}_{A} \begin{bmatrix} \widehat{z}_1(k) \\ \widehat{z}_2(k) \end{bmatrix} + \underbrace{\begin{bmatrix} 0 \\ -\beta \cos y(k) \end{bmatrix}}_{\varphi(y(k))} +$$

$$+ L(y(k) - \widehat{y}(k))$$

$$\widehat{y}(k) = \underbrace{\begin{bmatrix} 0 & 1 \end{bmatrix}}_{C} \widehat{z}(k)$$

(20.3.10)

where $L = \text{col}(l_1, l_2)$, and l_1 and l_2 are chosen such that all eigenvalues of the matrix

$$A - LC = \begin{bmatrix} 0 & -\alpha - l_1 \\ 1 & 1+\alpha - l_2 \end{bmatrix}$$

(20.3.11)

lie in the open unit disc.

20.4 Transformation into Lur'e Form

In the previous section we learned that if the transmitter dynamics are in Lur'e form (20.3.4) and the pair (C, A) is observable, then it is always possible to design a receiver system which synchronizes with (20.3.4).

The result presented in the previous section is very simple. However, the following question remains open: what can we do if the transmitter dynamics are not in the form (20.3.4)? In this section we will present a partial answer to this question.

First of all, notice that the representation (20.3.4) is coordinate dependent. This means that if one rewrites system (20.3.4) in a new coordinate system via a (nonlinear) coordinate change $z = T(x)$, then a new representation of the *same* dynamical system is not necessarily in the form (20.3.4).

By the same token, however, this may also mean that it is possible to transform a system into Lur'e form by means of a nonlinear coordinate change $z = T(x)$. Hence, we arrive at the following problem.

Let a discrete-time system (20.2.1, 20.2.2) with scalar output be given, and assume that $f(0) = 0$, $h(0) = 0$. The problem is to find conditions ensuring existence of an invertible coordinate change $z = T(x)$ such that the system (20.2.1) is locally (or globally) equivalent to the following Lur'e system

$$z(k+1) = Az(k) + \varphi(y(k)), \quad y(k) = Cz(k) \qquad (20.4.12)$$

where the pair (C, A) is observable.

As one can see from the problem statement, the coordinate change $z = T(x)$ can be either locally or globally defined (i.e., the inverse mapping T^{-1} can exist on a neighborhood of the origin or everywhere). In the first case the systems (20.2.1, 20.2.2) and (20.4.12) are equivalent if for all k one has that $||x(k)||$ is sufficiently small. In the second case there are no such restrictions.

The following result from [12] gives a (local) solution to the problem.

THEOREM 20.2
A discrete-time system (20.2.1, 20.2.2) with single output is locally equivalent to a system in Lur'e form (20.4.12) with observable pair (C, A) via a coordinate change $z = T(x)$ if and only if

(i) *the pair $(\partial h(0)/\partial x, \partial f(0)/\partial x)$ is observable,*
(ii) *the Hessian matrix of the function $h \circ f^n \circ \mathcal{O}^{-1}(s)$ is diagonal, where*

$x = \mathcal{O}^{-1}(s)$ is the inverse map of

$$\mathcal{O}(x) = \begin{bmatrix} h(x) \\ h \circ f(x) \\ \vdots \\ h \circ f^{n-1}(x) \end{bmatrix}, \quad (20.4.13)$$

with $h \circ f(x) := h(f(x))$, $f^1 := f$, $f^j := f \circ f^{j-1}$.

It is important to notice that condition (i) means that the Jacobian $\partial \mathcal{O}(0)/\partial x$ is invertible. In an equivalent form it can be rewritten in the form

$$\dim \left(\mathrm{span} \left\{ \frac{\partial h}{\partial x}(0), \frac{\partial h \circ f}{\partial x}(0), \ldots, \frac{\partial h \circ f^{n-1}}{\partial x}(0) \right\} \right) = n.$$

The condition (ii) may be interpreted in the following way. As indicated above, if condition (i) holds, the transformation $s = \mathcal{O}(x)$ is a local diffeomorphism. Thus, s forms a new set of local coordinates for the dynamics (20.2.1) around the origin. It is straightforwardly checked that, in these new coordinates, the system (20.2.1, 20.2.2) takes the form

$$\begin{cases} s_1(k+1) &= s_2(k) \\ &\vdots \\ s_{n-1}(k+1) &= s_n(k) \\ s_n(k+1) &= f_s(s(k)) \\ y(k) &= s_1(k) \end{cases} \quad (20.4.14)$$

where $f_s(s) := h \circ f^n \circ \mathcal{O}^{-1}(s)$. In the literature (see [15]), the form (20.4.14) is referred to as the *observable form* of the system (20.2.1, 20.2.2). Condition (ii) then is equivalent to the local existence of functions $\varphi_1, \cdots, \varphi_n : \mathbf{R} \to \mathbf{R}$ such that

$$f_s(s) = \varphi_1(s_1) + \varphi_2(s_2) + \cdots + \varphi_n(s_n). \quad (20.4.15)$$

With the functions $\varphi_1, \cdots, \varphi_n$ at hand, the transformation

$$z_i := s_{n+1-i} - \sum_{k=i+1}^{n} \varphi_k(s_{k-i}) \quad (i = 1, \cdots, n) \quad (20.4.16)$$

then transforms the observable form (20.4.14) into the following Lur'e form:

$$\begin{cases} z_1(k+1) &= \varphi_1(y(k)) \\ z_2(k+1) &= z_1(k) + \varphi_2(y(k)) \\ &\vdots \\ z_n(k+1) &= z_{n-1}(k) + \varphi_n(y(k)) \\ y(k) &= z_n(k) \end{cases} \quad (20.4.17)$$

The mapping $\mathcal{O}$ in (20.4.13) and the observable form play an important role in the observer design for nonlinear discrete-time systems. As one can easily see that, in the linear case, the mapping $\mathcal{O}$ is exactly the linear operator (20.3.7) introduced in the previous section. Since the Jacobian of $\mathcal{O}$ is invertible around $x = 0$ the mapping $\mathcal{O}$ is a local diffeomorphism. If one is interested in finding a coordinate change $z = T(x)$ which is globally defined, it is sufficient to check that $\mathcal{O}$ is a global diffeomorphism from $\mathbf{R}^n$ to $\mathbf{R}^n$ and the functions $\varphi_1, \cdots, \varphi_n$ satisfying (20.4.15) exist globally.

Example 20.2
[(Bouncing ball)]

Consider the following discrete-time model which describes the bouncing ball system [23, 3]:

$$\begin{cases} x_1(k+1) = x_1(k) + x_2(k) \\ x_2(k+1) = \alpha x_2(k) - \beta \cos(x_1(k) + x_2(k)) \end{cases} \quad (20.4.18)$$

where $x_1(k)$ is the phase of the table at the k-th impact, $x_2(k)$ is proportional to the velocity of the ball at the k-th impact, the parameter α is the coefficient of restitution, and $\beta = 2\omega^2(1+\alpha)A/g$. Here ω is the angular frequency of the table oscillation, A is the corresponding amplitude, and g is the gravitational acceleration. For some values of the parameters the system can exhibit very complex behavior. However, we will show that this is not an obstacle for the design of an observer.

Suppose only the first variable x_1 (the phase) is available for measurement. The question is: can we reconstruct the second variable? Clearly the system (20.4.18) is not in Lur'e form. However, using the theory presented in this section, we will show that there exists a coordinate change that transform (20.4.18) into Lur'e form.

So, we assumed that

$$y(k) = h(x(k)) = x_1(k).$$

Let us check the conditions of Theorem 20.2. A simple calculation gives

$$\frac{\partial h(0)}{\partial x} = \begin{bmatrix} 1 & 0 \end{bmatrix}, \quad \frac{\partial f(0)}{\partial x} = \begin{bmatrix} 1 & 1 \\ 0 & \alpha \end{bmatrix}$$

and this pair is clearly observable. Hence, condition (i) is satisfied.

To check condition (ii), let us find the mapping $\mathcal{O}$. Obviously,

$$\mathcal{O}(x) = \begin{bmatrix} 1 & 0 \\ 1 & 1 \end{bmatrix} x \quad (20.4.19)$$

with $x = \text{col}(x_1, x_2)$. This mapping is linear, it is invertible, and, therefore, it is a global diffeomorphism. Introducing $s = \text{col}(s_1, s_2) := \mathcal{O}(x)$ we see

that
$$f_s(s) := h \circ f^2 \circ \mathcal{O}^{-1}(s) = -\alpha s_1 + (1+\alpha)s_2 - \beta \cos s_2$$
and it is clear that the Hessian of this function is diagonal. Thus, condition (ii) is satisfied as well. Note that, in view of (20.4.15), we obtain $f_s(s) = \varphi_1(s_1) + \varphi_2(s_2)$, with $\varphi_1(s_1) := -\alpha s_1$, $\varphi_2(s_2) := (1+\alpha)s_2 - \beta \cos s_2$. Therefore there exists a coordinate change which locally transforms the system (20.4.18) into Lur'e form. Moreover, the mapping $\mathcal{O}$ is a global diffeomorphism *and* the functions φ_1, φ_2 are globally defined, which implies that this coordinate change is, in fact, global.

From (20.4.19) and (20.4.16) we obtain the following coordinate change:
$$\begin{cases} z_1 = -\alpha x_1 + x_2 + \beta \cos x_1 \\ z_2 = x_1 \end{cases} \qquad (20.4.20)$$

with the output $y = z_2 = x_1$. In the new coordinate system the original system (20.4.18) has the following form
$$\begin{cases} z_1(k+1) = -\alpha z_2(k) \\ z_2(k+1) = z_1(k) + (1+\alpha)z_2(k) - \beta \cos z_2(k). \end{cases} \qquad (20.4.21)$$

Note that the dynamics (20.4.21) are identical to the dynamics (20.3.8). Therefore, an observer for (20.4.21) is given by (20.3.10).

The estimates $\widehat{x}_1, \widehat{x}_2$ for x_1, x_2 are given by the following relations, which immediately follow from (20.4.20)
$$\begin{cases} \widehat{x}_1 = \widehat{z}_2 \\ \widehat{x}_2 = \widehat{z}_1 + \alpha \widehat{z}_2 - \beta \cos \widehat{z}_2 \end{cases} \qquad (20.4.22)$$

with $\widehat{z}_1, \widehat{z}_2$ the observer state for (20.4.21). Moreover, by means of an appropriate choice of l_1, l_2 one can achieve arbitrarily fast convergence of $\widehat{x}(k)$ to $x(k)$.

20.5 Transformation into Extended Lur'e Form

In the previous section we found that if the observability mapping $\mathcal{O}$ is a diffeomorphism and condition (ii) of Theorem 20.2 holds, then there exists a coordinate change transforming the system (20.2.1, 20.2.2) into Lur'e form, which makes the observer design a simple linear algebraic problem. Condition (ii) of Theorem 20.2 is especially restrictive. Therefore, the question arises whether, and in what way, this condition may be relaxed.

To answer this question, we will assume, in this section, that at time k not only $y(k)$ but also the past output measurements $y(k-1), \cdots, y(k-N)$

for some $N > 0$ are available. We first consider nonlinear dynamics of the following form:
$$\begin{cases} x(k+1) &= Ax(k) + \varphi(y(k), y(k-1), \cdots, y(k-N)) \\ y(k) &= Cx(k) \end{cases} \quad (20.5.23)$$
where $x(k) \in \mathbf{R}^n$, $y(k) \in \mathbf{R}^1$, $\varphi : \mathbf{R}^{N+1} \to \mathbf{R}^n$ is a smooth mapping, and A, C are matrices of appropriate dimensions. Note that the dynamics (20.5.23) for $N = 0$ are just the dynamics (20.3.4). Therefore, we refer to dynamics of the form (20.5.23) as dynamics in *extended Lur'e form with buffer* N. Assume that the pair (C, A) is observable. As we have seen in Section 20.3 there then exists a matrix L such that all eigenvalues of $A - LC$ lie in the open unit disc. Along the same lines as in Section 20.3, it may then be shown that the following dynamics are an observer for (20.5.23):
$$\begin{cases} \widehat{x}(k+1) &= A\widehat{x}(k) + \varphi(y(k), \cdots, y(k-N)) + L(y(k) - \widehat{y}(k)) \\ \widehat{y}(k) &= C\widehat{x}(k) \end{cases}$$
$$(20.5.24)$$

As in Section 20.4, we now ask ourselves the question under what conditions the discrete-time system (20.2.1, 20.2.2) may be transformed into an extended Lur'e form for some $N \geq 0$. The transformations we are going to use here are more general than the transformation in Section 20.4, in the sense that we also allow them to depend on the past output measurements $y(k-1), \cdots, y(k-N)$. More specifically, we will be looking at parametrized transformations $z = T(x, \xi_1, \cdots, \xi_N)$, where $z \in \mathbf{R}^n$, with the property that (locally or globally) there exists a mapping $T^{-1}(\cdot, \xi_1, \cdots, \xi_N) : \mathbf{R}^n \to \mathbf{R}^n$ parametrized by $(\xi_1, \cdots, \xi_N)$, such that for all $(\xi_1, \cdots, \xi_N)$ we have
$$T(T^{-1}(z, \xi_1, \cdots, \xi_N), \xi_1, \cdots, \xi_N) = z$$
A mapping having this property will be referred to as an *extended coordinate change*. We will then say that the system (20.2.1, 20.2.2) may be transformed into an extended Lur'e form with buffer N if there exists an extended coordinate change $T(\cdot, \xi_1, \cdots, \xi_N) : \mathbf{R}^n \to \mathbf{R}^n$ parametrized by $(\xi_1, \cdots, \xi_N)$ such that the variable
$$z(k) := T(x(k), y(k-1), \cdots, y(k-N)) \quad (20.5.25)$$
satisfies (20.5.23), where the pair (C, A) is observable. As pointed out above, one may then build an observer (20.5.24) for $z(k)$ in (20.5.25). From this observer, one then obtains estimates $\widehat{x}(k)$ for $x(k)$ by inverting the extended coordinate change T:
$$\widehat{x}(k) := T^{-1}(\widehat{z}(k), y(k-1), \cdots, y(k-N)) \quad (20.5.26)$$

The following result from [9] (see also [8]) gives conditions under which a system (20.2.1, 20.2.2) may be transformed into an extended Lur'e form with buffer N.

THEOREM 20.3

Consider a discrete-time system (20.2.1, 20.2.2), and assume that the mapping $\mathcal{O}$ in (20.4.13) is a local diffeomorphism. Let $N \in \{0, \cdots, n-1\}$ be given. Then (20.2.1, 20.2.2) may be locally transformed into an extended Lur'e form with buffer N if and only if there locally exist functions $\varphi_{N+1}, \cdots, \varphi_n : \mathbf{R}^{N+1} \to \mathbf{R}$ such that the function f_s in the observable form (20.4.14) satisfies

$$f_s(s_1, \cdots, s_n) = \sum_{i=N+1}^{n} \varphi_i(s_i, \cdots, s_{i-N}) \qquad (20.5.27)$$

The proof of the above theorem is constructive. Namely, assume that functions $\varphi_{N+1}, \cdots, \varphi_n$ satisfying (20.5.27) exist, and define an extended coordinate change by

$$z_i := \begin{cases} s_{n-i+1} - \sum_{j=N+1}^{n} \varphi_j(s_{j-i}, \cdots, s_{j-i-N}) & (i = 1, \cdots, N-1) \\ \\ s_{n-i+1} - \sum_{j=i+1}^{n} \varphi_j(s_{j-i}, \cdots, s_{j-i-N}) & (i = N, \cdots, n) \end{cases}$$
$$(20.5.28)$$

It is then straightforwardly checked that in these new extended coordinates the observable form (20.4.14) takes the following extended Lur'e form:

$$\begin{cases} z_1(k+1) &= 0 \\ z_2(k+1) &= z_1(k) \\ &\vdots \\ z_N(k+1) &= z_{N-1}(k) \\ z_{N+1}(k+1) &= z_N(k) + \varphi_{N+1}(y(k), \cdots, y(k-N)) \\ &\vdots \\ z_n(k+1) &= z_{n-1}(k) + \varphi_n(y(k), \cdots, y(k-N)) \\ y(k) &= z_n(k) \end{cases} \qquad (20.5.29)$$

Theorem 20.3 gives necessary and sufficient conditions for the local existence of an extended Lur'e form with buffer N for (20.2.1, 20.2.2). For global existence of an extended Lur'e form with buffer N, the mapping $\mathcal{O}$ in (20.4.13) needs to be a global diffeorphism, and the mappings $\varphi_{N+1}, \cdots, \varphi_n$ satisfying (20.5.27) need to exist globally.

It is easily checked that for $N = n-1$, condition (20.5.27) is always satisfied globally. Thus, we have a system (20.2.1, 20.2.2) for which the mapping $\mathcal{O}$ in (20.4.13) is a local (global) diffeomorphism that may always be locally (globally) transformed into an extended Lur'e form with buffer $n-1$.

20.6 Observers for Perturbed Linear Systems

So far the design procedure for observers has been based on the assumption that for the discrete-time system under consideration the mapping $\mathcal{O}$ in (20.4.13) is a (local or global) diffeomorphism. In the sequel, we consider a particular class of systems for which this might not be the case. Namely, we consider systems of the form

$$\begin{cases} x(k+1) &= Ax(k) + Bf(x(k)) \\ y(k) &= Cx(k) \end{cases} \quad (20.6.30)$$

where $x(k) \in \mathbf{R}^n$ is the state, $y(k) \in \mathbf{R}^1$ is the scalar output, the function $f : \mathbf{R}^n \to \mathbf{R}^1$ is smooth, A, B, C are matrices of appropriate dimensions, and the pair (C, A) is observable. Clearly, depending on the specific structure of f and B, the system (20.6.30) may have a mapping $\mathcal{O}$ that is not a diffeomorphism. Nevertheless, we may derive conditions on (20.6.30) that guarantee the existence of an observer.

Define the rational function $G(s)$ by

$$G(s) := C(sI - A)^{-1}B. \quad (20.6.31)$$

Then $G(s)$ has the form $G(s) = \frac{q(s)}{p(s)}$, where q and p are polynomials in s, with $\deg(p) > \deg(q)$. We now assume that $\deg(p) - \deg(q) = 1$. It may be shown that this is equivalent to the fact that $CB \neq 0$. To obtain an observer for (20.6.30), we first define new coordinates in the following way. Since $CB \neq 0$, there exists an $(n-1) \times n$ matrix N such that $NB = 0$ and the matrix $S := \begin{bmatrix} C^T & N^T \end{bmatrix}^T$ is invertible. Thus, $(\xi, z) := (Cx, Nx)$ forms a new set of coordinates for (20.6.30). It is straightforwardly checked that in these new coordinates the system (20.6.30) takes the form

$$\begin{cases} \xi(k+1) &= \bar{f}(\xi(k), z(k)) \\ z(k+1) &= A_1\xi(k) + A_2 z(k) \\ y(k) &= \xi(k) \end{cases} \quad (20.6.32)$$

where

$$\bar{f}(\xi, z) = C \left[AS^{-1} \begin{bmatrix} \xi \\ z \end{bmatrix} + CBf\left(S^{-1}\begin{bmatrix} \xi \\ z \end{bmatrix}\right) \right]$$

and

$$\begin{bmatrix} A_1 & A_2 \end{bmatrix} = NAS^{-1}$$

We now assume the following:

A1 The mapping $\bar{f}$ in (20.6.32) is globally Lipschitz with respect to z, i.e., there exists an $L > 0$ such that

$$(\forall \xi \in \mathbf{R})(\forall z, \bar{z} \in \mathbf{R}^{n-1})(|\bar{f}(\xi, \bar{z}) - \bar{f}(\xi, z)| < L\|\bar{z} - z\|)$$

A2 All zeros of the polynomial $q(s)$ are located in the open unit disc.

As an observer candidate we take the following system:
$$\begin{cases} \widehat{\xi}(k+1) &= \bar{f}(y(k), \widehat{z}(k)) \\ \widehat{z}(k+1) &= A_1 y(k) + A_2 \widehat{z}(k) \end{cases} \quad (20.6.33)$$

We then have the following result.

THEOREM 20.4
*Assume that for (20.6.30) we have that the pair (C, A) is observable, that $CB \neq 0$, and that assumptions **A1** and **A2** hold. Then (20.6.33) is an observer for (20.6.32).*

PROOF Defining the error signals
$$e_\xi(k) := \xi(k) - \widehat{\xi}(k), \quad e_z(k) := z(k) - \widehat{z}(k),$$
we obtain the following error equations:
$$\begin{cases} e_\xi(k+1) &= \bar{f}(\xi(k), e_z(k) + \widehat{z}(k)) - \bar{f}(\xi(k), \widehat{z}(k)) \\ e_z(k+1) &= A_2 e_z(k) \end{cases} \quad (20.6.34)$$

It is easily checked that assumption **A2** implies that all eigenvalues of A_2 are in the open unit disc. This implies, on its turn, that there exist $\gamma > 0, 0 < \lambda < 1$ such that $e_z(k)$ satisfies
$$\|e_z(k)\| \leq \gamma \lambda^k \|e_z(0)\| \quad (20.6.35)$$

Using assumption **A1** and (20.6.35), we then obtain
$$|e_\xi(k)| = |\bar{f}(\xi(k-1), e_z(k-1) + \widehat{z}(k-1)) - \bar{f}(\xi(k-1), \widehat{z}(k-1))|$$
$$< L\|e_z(k-1)\| \leq L\gamma \lambda^{k-1} \|e_z(0)\| \quad (20.6.36)$$

Since $0 < \lambda < 1$, it follows from (20.6.35) and (20.6.36) that $e_\xi(k), e_z(k) \to 0$ for $k \to +\infty$, and thus (20.6.33) is an observer for (20.6.32).

Remark A.
The result in this section may be generalized to systems (20.6.30) for which we have $\deg(p) - \deg(q) > 1$. This generalization will be given in a forthcoming paper.

20.7 Conclusions

Following a similar line of research as in [14] we develop an observer perspective on the synchronization problem for nonlinear (complex) discrete-time systems. For several classes of discrete-time systems it is shown that a suitable observer can be found. In case such an observer does not exist, or cannot be found analytically, we propose to use an extended observer. The latter method follows [8] (see also [9]), and presents an observer that also uses past measurements and can be applied under fairly general conditions. Like the continuous-time paper [14], it seems that control theory might be a very valuable tool in the study of synchronization.

Acknowledgments

This work was supported in part by the RFBR Grant 96-01-01151, the Russian Federal Programme "Integration" (project 2.1–589).

References

[1] P. J. Aston and C. M. Bird, "Synchronization of coupled systems via parameter perturbations," *Physical Review E*, vol. 57, pp. 2787-2794, 1998.

[2] P. Badola, S. S. Tambe, and B. D. Kulkarni, "Driving systems with chaotic signals," *Phys. Rev. A*, vol. 46, pp. 6735-6738, 1992.

[3] L. Cao, K. Judd, and A. Mees, "Targeting using global models built from nonstationary data," *Phys. Lett. A*, vol. 231, pp. 367-372, 1997.

[4] K. Cuomo, A. Oppenheim, and S. Strogatz, "Robustness and signal recovery in a synchronized chaotic system," *Int. J. Bifur. Chaos*, vol. 3, pp. 1629-1638, 1993.

[5] M. Ding and E. Ott, "Enhancing synchronization of chaotic systems," *Phys. Rev. E*, vol. 49, pp. 945-948, 1994.

[6] U. Feldmann, M. Hasler, and W. Schwarz, "Communication by chaotic signals: the inverse system approach," *Int. J. Circ. Theory Appl.*, vol. 24, pp. 551-579, 1996.

[7] T. Hogg and B. A. Huberman, "Generic behavior of coupled oscillators," *Phys. Rev. A*, vol. 29, pp. 275-281, 1984.

[8] H. J. C. Huijberts, T. Lilge, and H. Nijmeijer, "A control perspective on synchronization and the Takens-Aeyels-Sauer Reconstruction Theorem," *Phys. Rev. E*, vol. 59, pp. 4691-4694, 1999.

[9] H. J. C. Huijberts, T. Lilge, and H. Nijmeijer, "Synchronization and observers for nonlinear discrete time systems," *Proc. Eur. Control Conf.*, Karlsruhe, Germany, 1999, in press.

[10] A. De Angeli, R. Genesio, and A. Tesi, "Dead-beat chaos synchronization in discrete-time systems," *IEEE Trans. Circ. Sys., I*, vol. 42, pp. 54-56, 1995.

[11] T. Kapitaniak, *Controlling Chaos*, Academic Press, New York, 1996.

[12] W. Lin and C. I. Byrnes, "Remarks on linearization of discrete-time autonomous systems and nonlinear observer design," *Systems & Control Letters*, vol. 25, pp. 31-40, 1995.

[13] O. Morgül and E. Solak, "Observer based synchronization of chaotic signals," *Phys. Rev. E*, vol. 54, pp. 4803-4811, 1996.

[14] H. Nijmeijer and I.M.Y. Mareels, "An observer looks at synchronization," *IEEE Trans. Circ. Systems I*, vol. 44, pp. 882-890, 1997.

[15] H. Nijmeijer and A. J. van der Schaft, *Nonlinear Dynamical Control Systems*, Springer, New York, 1990.

[16] M. J. Ogarzalek, "Taming chaos - Part I: Synchronization," *IEEE Trans. Circ. Sys., I*, vol. 40, pp. 693-699, 1993.

[17] U. Parlitz and L. Kocarev, "Synchronization of chaotic systems," in H. G. Schuster (Ed.), *Handbook of Chaos Control*, Springer-Verlag, Berlin, 1998.

[18] L. M. Pecora and T. L. Carroll, "Synchronization in chaotic systems," *Phys. Rev. Lett.*, vol. 64, pp. 821-824, 1990.

[19] J. H. Peng, E. J. Ding, M. Ding, and W. Yang, "Synchronizing hyperchaos with a scalar transmitted signal," *Phys. Rev. Lett.*, vol. 76, pp. 904-907, 1996.

[20] J. W. Polderman and J. C. Willems, *Introduction to Mathematical Systems Theory - A Behavioral Approach*, Springer, New York, 1998.

[21] M. de Sousa Vieira, A. J. Lichtenberg, and M. A. Lieberman, "Synchronization of regular and chaotic systems," *Phys. Rev. A*, vol. 46, pp. 7359-7362, 1992.

[22] T. Stojanovski, U. Parlitz, L. Kocarev, and R. Harris, "Exploiting delay reconstruction for chaos synchronization", *Phys. Lett. A*, vol. 233, pp. 355-360, 1997.

[23] N. B Tufillaro, T. Abbott, and J. Reilly, *An Experimental Approach to Nonlinear Dynamics and Chaos*, Addison–Wesley, Reading, MA, 1992.

[24] T.Ushio, "Synthesis of chaotically synchronized systems based on observers," *Proc. Int. Conf. Nonl. Bifurc. Chaos*, Lodz, Poland, pp. 251-254, 1996.

[25] T. Ushio, "Design of chaotically synchronized systems based on observers," *Electr. Comm. in Japan*, vol. 81, pp. 51-57, 1996.

[26] C. W. Wu and L. O. Chua, "A simple way to synchronize chaotic systems with applications to secure communication systems," *Int. J. Bifur. Chaos*, vol. 3, pp. 1619-1627, 1993.

[27] Y. Zhang, M. Dai, Y. Hua, W. Ni, and G. Du, "Digital communication by active-passive-decomposition synchronization in hyperchaotic systems," *Phys. Rev. E*, vol. 58, pp. 3022–3027, 1998.

[28] Special Issue, "Chaos in nonlinear electrical circuits, Part A: Tutorials and reviews", *IEEE Trans. Circ. Sys.*, I, vol. 40, pp. 637-786, 1993.

[29] Special Issue, "Chaos synchronization and control: theory and applications," *IEEE Trans. Circ. Sys.*, I, vol. 44, pp. 853-1039, 1997.

[30] Special Issue, "Control of chaos and synchronization," *Sys. Control Lett.*, vol. 31, pp. 259-322, 1997.

21

Separating a Chaotic Signal from Noise and Applications

Hervé Dedieu, Thomas Schimming, and Martin Hasler
Department of Electrical Engineering
Circuits and Systems Group
Swiss Federal Institute of Technology
CH 1015 Lausanne, Switzerland
dedieu@epfl.ch

Abstract

We describe various methods that aim at separating a chaotic signal from observation noise. We assume that the chaotic dynamical system is known in advance. After reviewing different noise decontamination tools, optimal and suboptimal, we explain how to use them in the context of modulation-demodulation of chaotic signals.

21.1 Introduction

Separating a signal from noise is a central problem in signal processing and there exists a very large literature on this subject. The separation has to rely on some features that distinguish the signal from the noise. In most cases, the separation cannot be performed entirely and, therefore, it would be better to use the term noise reduction instead of separation.

Conventional methods such as linear filtering use differences between the spectra of the signal and noise to separate them. Most often the noise and the signal do not occupy distinct frequency bands, but the noise energy is distributed over a large frequency interval, whereas the signal energy is concentrated in a small frequency band. Therefore, applying a filter whose output retains only the signal frequency band reduces the noise considerably.

We consider the problem of separation of a chaotic signal from noise. In this case spectrum-based methods are not applicable, because both the noise and the signal occupy a wide frequency band. Therefore, we have to rely on other properties that distinguish a chaotic signal from noise. What will be mainly exploited in the sequel is the assumption that the chaotic dynamical system that has produced the signal is known, whereas the system that has produced the noise is unknown and, therefore, modeled by a stochastic process.

In the literature, different problems depending on the degree of knowledge about the signal producing system have been studied. The extreme cases are

> the chaotic dynamical system is completely known;
>
> the chaotic dynamical system is of low order, but otherwise unknown.

We shall essentially concentrate on the first case, because it is pertinent for the transmission of information using chaotic signals, since the receiver knows the transmitter system. In the second case, the difference between the signal and the noise is that the signal is produced by a low order and the noise by a high order dynamical system.

For the sake of simplicity, we use as chaotic system iterations of one-dimensional map $f : R \longrightarrow R$ which leaves a bounded interval I of the real line invariant. It is not difficult to generalize to higher dimensional discrete time chaotic systems. Consequently, the possible trajectories are

$$x_1, \ x_2 = f(x_1), \ \ldots, \ x_n = f(x_{n-1}), \ \ldots \qquad (21.1.1)$$

To each initial condition

$$x_1 \in I \qquad (21.1.2)$$

there corresponds exactly one trajectory. We shall limit our attention to trajectories of finite length and consider them as column vectors.

$$\boldsymbol{x} = [x_1,\ x_2,\ \ldots,\ x_{N-1},\ x_N]^\top \qquad (21.1.3)$$

21.2 Definition of the Problem

Suppose a chaotic system given by the map

$$f: I \longrightarrow I,\ I \subset R \qquad (21.2.4)$$

produces a trajectory $\boldsymbol{x}$, but we actually observe the perturbed vector

$$\boldsymbol{y} = [y_1,\ y_2,\ \ldots,\ y_N]^\top \qquad (21.2.5)$$

The problem is to determine $\boldsymbol{x}$ "as well as possible", having observed $\boldsymbol{y}$.

In order to be able to formulate the problem more precisely, we model the noise as a random vector $\boldsymbol{W}$ composed of N independent Gaussian random variables w_n, $n = 1,\ \ldots,\ N$ with zero mean and variance σ_w^2. Consequently, the observed signal also is a random vector $\boldsymbol{Y} = \boldsymbol{x} + \boldsymbol{W}$. Depending on the application, other noise models may be more appropriate.

The aim is to find an algorithm which, for each observation $\boldsymbol{y}$, produces an estimate

$$\widehat{\boldsymbol{x}} = \phi(\boldsymbol{y}),\ \phi: R^N \longrightarrow R^N \qquad (21.2.6)$$

which is on the average as close as possible to the "true" trajectory $\boldsymbol{x}$. As a measure of "closeness," we use the euclidean distance

$$e = \|\widehat{\boldsymbol{x}} - \boldsymbol{x}\| = \sqrt{\sum_{n=1}^{N} (\widehat{x}_n - x_n)^2} \qquad (21.2.7)$$

Depending on the application, the constraint may be imposed that $\widehat{\boldsymbol{x}}$ is itself a true trajectory.

The average of this estimation error is taken over repeated observations. Usually, each observation is the noise-contaminated version of a different trajectory. Consequently, we also assume that the true trajectories are produced by some stochastic law and thus they are modeled by a random vector $\boldsymbol{X} = [X_1, ..., X_N]$.

Note that the random variables X_n are by no means independent, since $X_n = f(X_{n-1})$. In fact, the N-dimensional probability measure of X is concentrated on the corresponding 1-dimensional manifold and generated by the probability measure of X_1.

The choice of the probability measure for X_1 depends on the application. Often, and in particular for the information transmission application, the different trajectories x are finite segments of the same very long chaotic trajectory. In this case, the probability measure of X_1 should be chosen as the asymptotic distribution of the points x_n of the trajectory. For ergodic dynamical systems, this is the natural invariant measure. In the case of iterations of one-dimensional maps, this measure is usually given by a density function ρ on I.

This completes the probabilistic framework. The underlying probability space is given by the $N+1$ independent random variables $X_1, W_1, \ldots, W_N$ with the joint probability density

$$p(x_1, w_1, \ldots, w_N) = \rho(x_1) \frac{1}{(2\pi)^{\frac{N}{2}} \sigma_w^N} \exp^{-\frac{w_1^2 + \ldots + w_N^2}{2\sigma_w^2}} \quad (21.2.8)$$

The residual noise variance, after applying the estimation

$$\widehat{x} = \phi(y) \quad (21.2.9)$$

is

$$E\left(||\widehat{X} - X||^2\right) = E\left(||\phi(Y) - X||^2\right) \quad (21.2.10)$$

and the processing gain

$$P_G = \frac{E\left(||Y - X||^2\right)}{E\left(||\phi(Y) - X||^2\right)} = \frac{N\sigma_w^2}{E\left(||\phi(Y) - X||^2\right)} \quad (21.2.11)$$

which is a quantitative measure of the noise reduction.

21.3 Optimal Solution without Dynamic Constraint on the Estimator

The problem consists in finding the optimal estimate

$$\widehat{x} = \phi(y), \quad \phi : R^N \longrightarrow R^N \quad (21.3.12)$$

such that

$$E\left(\sum_{n=1}^{N}(\widehat{X}_n - X_n)^2\right) = E\left(\sum_{n=1}^{N}(\phi(Y))_n - X_n)^2\right) \quad (21.3.13)$$

is minimal. This is a well-known problem, and its solution is the Bayesian estimator

$$\widehat{x} = \phi(y) = E(X|Y = y) \quad (21.3.14)$$

More explicitly,
$$\widehat{x}_n = \int_I f^{n-1}(x_1)p(x_1|Y=y)dx_1 \qquad (21.3.15)$$

Applying Bayes' rules, it is straightforward to show that the estimation can be rewritten as
$$\widehat{x}_n = (\phi(y))_n = \frac{\int_I f^{n-1}(x_1)\rho(x_1)p(y|x_1)dx_1}{\int_I \rho(x_1)p(y|x_1)dx_1} \qquad (21.3.16)$$

where
$$p(y|x_1) = \frac{1}{(2\pi)^{\frac{N}{2}}\sigma_w^N} \exp^{-\frac{\sum_{n=1}^N (y_n - f^{n-1}(x_1))^2}{2\sigma_w^2}} \qquad (21.3.17)$$

The maximal processing gain is consequently
$$P_G = \frac{E(\|Y-X\|^2)}{E(\|E(X|Y)-X\|^2)} = 1 + \frac{E(\|E(X|Y)-Y\|^2)}{E(\|E(X|Y)-X\|^2)} \qquad (21.3.18)$$

This is a measure of the maximal noise reduction that is achievable due to the knowledge of the dynamics. In (21.3.18), the term $E\left[\|E(X|Y)-X\|^2\right]$ can be expressed

$$E\left[\|E(X|Y)-X\|^2\right]$$
$$= \int_I \rho(x_1) \int_{-\infty^N}^{+\infty^N} p(y|x_1) \sum_{n=1}^N ((\phi(y))_n - f^{n-1}(x_1))^2 \, dy\, dx_1 \quad (21.3.19)$$

Example 21.1
Let us take as a dynamical system the iterations of the skew tent map (see Figure 21.1)
$$f(x) = \begin{cases} \frac{x}{a} & \text{for } 0 \le x < a \\ \frac{1-x}{1-a} & \text{for } a \le x < 1 \end{cases} \qquad (21.3.20)$$

Let us find the optimal estimator ϕ depending on the noise level σ_w for values of $N = 2, 4, 8$ as given by Equation (21.3.16). Due to the piecewise linearity of the map f, the integrals over I in (21.3.16) and (21.3.19) can be computed over linear regions, yielding terms involving $erfs$. In this way, Equation (21.3.16) can be solved analytically, but its solution is omitted here due to its complexity. Unfortunately, the variance of the estimator (Equation (21.3.19)) cannot be analytically expressed even for the simple case $N = 2$.

Figure 21.2 shows the computation of $E(\|\phi(Y)-X\|^2)$ with respect to σ_w for $N = 2, 4, 8$, $a = 0.3$. We observe two limit cases: $\sigma_w \longrightarrow 0$

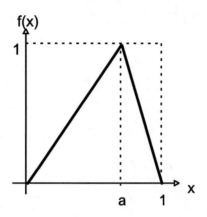

FIGURE 21.1
The skew tent map.

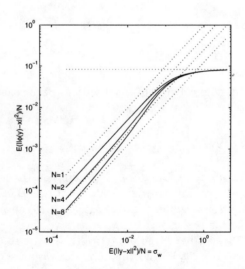

FIGURE 21.2
Variance of the optimal estimator $\phi(y)$ w.r.t. σ_w for $N = 2, 4, 8$.

and $\sigma_w \longrightarrow \infty$, and a transition region. If σ_w is small, the dependence of $E\left(||\phi(Y) - X||^2\right)$ on σ_w^2 is linear and the processing gain approaches N [1]. In Figure 21.2, this linear dependence appears as dotted lines with slopes $\frac{1}{N}$ ($P_G = N$). Observe that the lower the N is, the better is the matching $E\left(||\phi(Y) - X||^2\right) \approx \frac{1}{N} E\left(||Y - X||^2\right)$ for a given σ_w. If σ_w is large it can be shown that $E\left(||\phi(Y) - X||^2\right)$ saturates, i.e., $E\left(||\phi(Y) - X||^2\right) \longrightarrow \frac{N}{12}$, due to the strictly bounded distribution $\rho(x_1)$ as opposed to the noise energy which grows without bound. As we have normalized the plot by dividing $E\left(||\phi(Y) - X||^2\right)$ by N, the saturation appears as a single dotted line of ordinate $\frac{1}{12}$.

21.4 Optimal Solution with Dynamic Constraint on the Estimator

The problem consists in finding the optimal estimate

$$\widehat{x} = \phi_C(y), \ \phi_C : R^N \longrightarrow R^N \tag{21.4.21}$$

such that

$$E\left(\sum_{n=1}^{N}(\widehat{X}_n - X_n)^2\right) = E\left(\sum_{n=1}^{N}(\phi_C(Y))_n - X_n)^2\right) \tag{21.4.22}$$

is minimal under the constraint that the estimate $\phi_C(y)$ must be a trajectory of the dynamical system f. Since for any estimate we can decompose

$$E\left(||\phi_C(Y) - X||^2\right) = E\left(||\phi_C(Y) - E(X|Y)||^2\right) \\ + E\left(||(E(X|Y) - X||^2\right) \tag{21.4.23}$$

we have to minimize

$$E\left(||\phi_C(Y) - E(X|Y)||^2\right) \tag{21.4.24}$$

This is achieved by minimizing for each vector y the distance

$$||\widehat{x} - E(X|Y = y)||^2 \tag{21.4.25}$$

under the constraint that $\widehat{x}$ is a trajectory. Since the trajectories constitute a one-dimensional manifold in R^N, the solution is simply the closest point of this manifold to the optimal estimate without constraint:

$$\widehat{x} = \phi_C(y) = \operatorname*{arg\,min}_{x \text{ trajectory}} ||x - E(X|Y = y)|| \tag{21.4.26}$$

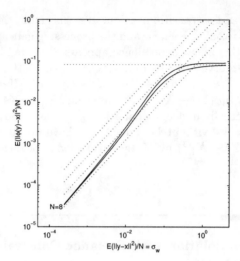

FIGURE 21.3
Variance of the constrained estimator $\phi_C(y)$ (upper curve) and the unconstrained estimator w.r.t. σ_w for $N = 8$, $a = 0.3$.

The residual noise variance for this estimate is

$$E\left(\underset{x\ trajectory}{\arg\min} \ \|x - E(X|Y)\|^2\right) + E\left(\|E(X|Y) - X\|^2\right)$$
(21.4.27)

This last expression shows that the variance of the constrained estimator is larger than that of the unconstrained one as given by 21.3.18.

Example 21.2
We consider the same example as before. The variance of the constrained estimator ϕ_C as given by (21.4.27) has been computed as a function of σ_w for $N = 8$. The variance behavior of ϕ_C is shown on Figure 21.3 and compared to the variance of the unconstrained estimator ϕ for the same N (already shown on Figure 21.2). Observe as expected by the theory that the variance of ϕ_C is larger than the variance of ϕ. In the range of the chosen σ_w, the ratio between the two variances is between 1 and 1.3 for $N = 8$. Similar qualitative behavior can be observed for other values of N.

21.5 Practical Algorithms for Noise Cleaning: Deterministic Approach

Until now, we have described optimal methods for the derivation of constrained and unconstrained trajectory estimators. From a practical point of view, these methods are not well suited for "fast" implementation since their computational cost is proportional to ($O(2^N)$). In the following we propose to review some of the methods which have been developed so far and which permit obtaining a solution at a reasonable computational cost.

21.5.1 Deterministic Optimization for Approximately Constrained Estimators

The problem can be written as a standard optimization problem in which our goal is to find a pseudo-trajectory $\widehat{x}$

$$\widehat{x} = [\widehat{x}_1,\ \widehat{x}_2,\ \ldots\ \widehat{x}_N]^\top \qquad (21.5.28)$$

which satisfies two conditions:

Condition A1: $\widehat{x}$ is in the neighborhood of y according to some measure of closeness.

Condition A2: $\widehat{x}$ is as close as possible to a system true trajectory according to another measure of closeness.

The idea consists in building a cost function in which the two above mentioned design objectives **A1** and **A2** are combined. First, a measure of closeness between $\widehat{x}$ and y, i.e., $C_1(\widehat{x},\ y)$ is defined. This measure can be any suitable distance between $\widehat{x}$ and y, for instance,

$$C_1(\widehat{x},\ y) = ||\widehat{x} - y||^2 \qquad (21.5.29)$$

or

$$C_1(\widehat{x},\ y) = 1 - \frac{\widehat{x}^\top y}{\sqrt{\widehat{x}^\top \widehat{x}\ y^\top y}} \qquad (21.5.30)$$

Second, a measure $C_2(\widehat{x})$ gives the distance between $\widehat{x}$ and a true trajectory of the system. $C_2(\widehat{x})$ can take different forms; a convenient one can be

$$C_2(\widehat{x}) = \sum_{n=1}^{N-1} ||f(\widehat{x}_n) - \widehat{x}_{n+1}||^2 \qquad (21.5.31)$$

The overall cost function $C(\widehat{x},\ y)$ appears as a linear combination of $C_1(\widehat{x},\ y)$ and $C_2(\widehat{x})$, i.e.,

$$C(\widehat{x},\ y) = C_1(\widehat{x},\ y) + \Gamma\ C_2(\widehat{x}) \qquad (21.5.32)$$

where Γ is some real positive number *a priori* defined. Having defined a general cost function, it remains to derive a recursive method which finds a local minimum of $\mathcal{C}(\widehat{x},\ y)$. First order methods (gradient methods), or second order methods like Newton or Quasi-Newton methods can solve this problem. Let us define as $\widehat{x}(k)$ the recursive estimate of $\widehat{x}$. A very common method is to use a gradient descent method, i.e.,

$$\widehat{x}(0) = y$$
$$\widehat{x}(k) = \widehat{x}(k-1) - \mu \frac{\partial \mathcal{C}(\widehat{x},\ y)}{\partial \widehat{x}}\Big[\widehat{x}=\widehat{x}(k-1)\Big] \qquad (21.5.33)$$

where μ is a positive real number. Provided that μ is chosen small enough (in general, μ is chosen as a decreasing function of k), $\widehat{x}(k)$ converges toward a local minimum of $\mathcal{C}(\widehat{x},\ y)$. Such methods have been proposed by Kostelich and Yorke [17] and Lee and Williams [19] among others. The approach is very simple and does not require any matrix inversion. Improvement in the speed of convergence can be obtained by using second order methods which require an inversion of the Hessian matrix of the cost function.

21.5.2 Deterministic Optimization for Exactly Constrained Estimators

In order to take into account the trajectory constraints reflecting condition **A2**, the method consists in searching for the minimization of $C_1(\widehat{x},\ y)$ with respect to $\widehat{x}$ subject to $N - 1$ equality constraints

$$f(\widehat{x}_n) = \widehat{x}_{n+1} \qquad n = 1 \ldots N - 1 \qquad (21.5.34)$$

The problem becomes a standard optimization problem with equality constraints which force the solution to be a trajectory of the system. Its solution $\widehat{x}_*$ is therefore a stationary point of the Lagrange function $L(\widehat{x},\ \Lambda)$, i.e.,

$$L(\widehat{x},\ \Lambda) = C_1(\widehat{x},\ y) + \sum_{n=1}^{N-1} \lambda_n^\top [f(\widehat{x}_n) - \widehat{x}_{n+1}] \qquad (21.5.35)$$

where λ_n is the Lagrange multiplier associated with the $n-th$ constraint and

$$\Lambda = [\lambda_1,\ \lambda_2,\ \ldots,\ \lambda_{N-1}]^\top \qquad (21.5.36)$$

The problem consists in finding the $2N - 1$ unknowns, i.e., N pseudo-trajectory points and $N - 1$ Lagrange multipliers. These unknowns can be determined by setting to zero the derivatives of $L(\widehat{x},\ \Lambda)$ with respect to $\widehat{x}$ and Λ which leads to solving a set of $2N - 1$ nonlinear equations. Details of the algorithm can be found in [8].

Although this method seems to be very attractive, it suffers from drawbacks which prevent its use for many signal processing real time applications :

1. In order to find the stationary point of the Lagrange function, an iterative solution is derived. This iterative solution is based on a Newton-Raphson technique [8] which linearizes the set of $2N-1$ equations around the previous solution at each iteration. Therefore finding the next solution requires a $2N-1 \times 2N-1$ matrix inversion. For large N this matrix inversion prevents use of the method for real time decontamination.
2. As most optimization methods, the method furnishes a local minimum. This local optimization is due to the iterative Newton-Raphson procedure which linearizes the Jacobian around the current operating point.

21.6 Practical Algorithms for Noise Cleaning: Probabilistic Approach

To our knowledge, noise reduction methods based on probabilistic approaches have been underemployed. Nevertheless, as we shall see, they are powerful methods for which standard tools such as the Viterbi algorithm furnish elegant and low-cost computational solutions which are familiar to communication engineers.

The main idea is to take into account the probabilistic properties of the attractor generated by the dynamical signal we observe. These probabilistic properties are expressed in terms of the invariant distribution of data points on the attractor [7]. Although the term probability is not fully adequate, it is used in the sense that these invariant distributions act like probability distributions [20]. The distributions can be evaluated using two techniques.

1. For some simple piecewise linear systems, they can be computed analytically.
2. For most systems, they can be computed by producing a very long reference clean orbit.

21.6.1 State Space Quantization and Noise Cleaning

Without loss of generality we assume that the state space of the signal x_k can be normalized on an interval of length 1, i.e., $[\frac{-1}{2}, \frac{+1}{2}]$. Suppose that we describe the state space $[\frac{-1}{2}, \frac{+1}{2}]$ with N_B nonoverlapping "balls" such as the ball number k, $\boldsymbol{B}(k)$ is the interval

$$\boldsymbol{B}(k) = \left](k-1)\Delta_B - \frac{1}{2},\ k\Delta_B - \frac{1}{2}\right] \qquad \Delta_B = \frac{1}{N_B} \qquad (21.6.37)$$

the center of which is defined as $B_c(k)$, i.e.,

$$B_c(k) = \left(\frac{2k-1}{2}\right)\Delta_B - \frac{1}{2} \qquad (21.6.38)$$

We will say that at time t the system is in state $s_t = k$ if $x_t \in \boldsymbol{B}(k)$.

If N_B is chosen large enough we could see our noise cleaning problem as: given N consecutive observations $y_1 \ldots y_n$ find the most likely state sequence $\widehat{\boldsymbol{s}} = [\widehat{s}_1,\ \widehat{s}_2, \ldots \widehat{s}_N]$. (Ideally x_i should belong to $\boldsymbol{B}(\widehat{s}(i))$ for $i = 1 \ldots N$.) If a method exists to find the location of these most likely balls we will define the clean estimated signal $\widehat{\boldsymbol{x}}$ as the sequence of the most likely ball centers, i.e.,

$$\widehat{\boldsymbol{x}} = [B_c(\widehat{s}_1),\ B_c(\widehat{s}_2), \ldots, B_c(\widehat{s}_N)]^\top \qquad (21.6.39)$$

(If desirable this obtained cleaned quantized signal could be even refined using the above described deterministic techniques.) Let $\boldsymbol{s}$ be a path of N consecutive states, i.e.,

$$\boldsymbol{s} = [s_1,\ s_2, \ldots, s_N]^\top \qquad (21.6.40)$$

We now have to define what kind of criterion should be used in order to find the best path among the N consecutive states (balls). A natural criterion is to find $\widehat{\boldsymbol{s}}$ which satisfies

$$p(\widehat{\boldsymbol{s}}|\boldsymbol{y}) = \max_{\boldsymbol{s}} p(\boldsymbol{s}|\boldsymbol{y}) \qquad (21.6.41)$$

Using Bayes' rules we have

$$p(\boldsymbol{s}|\boldsymbol{y}) = \frac{p(\boldsymbol{y}|\boldsymbol{s})\ p(\boldsymbol{s})}{p(\boldsymbol{y})} = \frac{p(\boldsymbol{s},\ \boldsymbol{y})}{p(\boldsymbol{y})} \qquad (21.6.42)$$

As the observation $\boldsymbol{y}$ is given, we have to find $\widehat{\boldsymbol{s}}$ which satisfies

$$p(\widehat{\boldsymbol{s}},\ \boldsymbol{y}) = \max_{\boldsymbol{s}} p(\boldsymbol{s},\ \boldsymbol{y}) = \max_{\boldsymbol{s}} p(\boldsymbol{y}|\boldsymbol{s})\ p(\boldsymbol{s}) \qquad (21.6.43)$$

21.6.2 Finding a Recurrent Expression for $p(\boldsymbol{s},\ \boldsymbol{y})$

Let us define as $\boldsymbol{s}_1^m$ the first m quantized states of $\boldsymbol{s}$, i.e.,

$$\boldsymbol{s}_1^m = [s_1,\ s_2, \ldots, s_m]^\top \qquad (21.6.44)$$

For sufficiently fine quantization, the "process" $\boldsymbol{s}_1^m$ can be approximated as a first order Markov process, i.e.,

$$p(\boldsymbol{s}_1^m) = p(s_m|s_{m-1})p(\boldsymbol{s}_1^{m-1}) \qquad (21.6.45)$$

where $p(s_m|s_{m-1})$ is the state transition probability from state s_{m-1} to state s_m.

Second, due to the *iid* nature of the noise, we have

$$p(y_1^m|s_1^m) = \prod_{k=1}^{m} p(y_k|s_k) = \prod_{k=1}^{m} p_\eta(y_k - B_c(s_k)) \qquad (21.6.46)$$

where $p(y_k|s_k)$ is the conditional observation probability and P_η is the probability distribution of the noise. (The last equation in (21.6.46) is true only if the number of balls is large enough.)

The probability $p(s, y)$ can be therefore expressed as a standard product of state transition probabilities and conditional observation probabilities.

$$p(s, y) = \prod_{k=1}^{N} p(s_k|s_{k-1}) \, p(y_k|s_k) \qquad (21.6.47)$$

21.6.3 Decoding the State Sequence with the Viterbi Algorithm

Finding s which maximizes (21.6.47) is a standard problem, the efficient solution of which is given by the Viterbi algorithm first introduced for enhancement of chaotic signals by Marteau and Abarbanel [20]. The Viterbi algorithm avoids an exhaustive search among the lattice of N_B^N paths. Let δ_t^i be the following probability quantity

$$\delta_t^i = \max_{s_1^{t-1}} p(s_1^{t-1}, \, s_t = i, \, y_1^T) \qquad (21.6.48)$$

The Viterbi algorithm is based on the fact that

$$\delta_t^j = \max_i [\delta_{t-1}^i \, a_{ij}] \, b_j(y_t) \qquad (21.6.49)$$

where

$$a_{ij} = p(s_t = j \mid s_{t-1} = i) \qquad (21.6.50)$$

and

$$b_j(y_t) = p(y_t|s_t) \qquad (21.6.51)$$

The Viterbi Algorithm works in two main passes, a forward one and a backward one (backtracking). During the forward pass Equation (21.6.49) is used to compute at time t the δ_t^j at the N_B lattice nodes $(t, \, j = 1 \ldots N_B)$. Among the N_B paths which can link nodes $(t-1, \, k = 1 \ldots N_B)$ to node $(t, \, j)$, only the best one is kept. After the termination $t = N$, we select the node $(N, \, k^*)$ which gives the highest probability. To actually retrieve the state sequence, we need to keep track of the argument that maximized (21.6.49), for each t and j. This can be done via the array ϕ_t^j in which optimal trajectory in node j at time t is stored. Let π_i be the *a priori* probability of $y(1)$ to be in state i ($\pi_i = \frac{1}{N_B}$); the complete Viterbi procedure can be stated as follows:

1. Initialization

$$\delta_1^i = \pi_i\, b_i(y_1) \quad 1 \leq i \leq N_B$$

$$\phi_1^i = 0$$

2. Forward pass

$$\delta_t^j = \max_{1 \leq i \leq N_B} [\delta_{t-1}^i\, a_{ij}]\, b_j(y_t)$$

$$2 \leq t \leq N \quad 1 \leq j \leq N_B$$

$$\phi_t^j = \arg \max_{1 \leq i \leq N_B} [\delta_{t-1}^i\, a_{ij}]\, b_j(y_t)$$

$$2 \leq t \leq N \quad 1 \leq j \leq N_B$$

3. Termination

$$\widehat{s}_N = \arg \max_{1 \leq i \leq N_B} [\delta_N^i]$$

4. Backward pass

$$\widehat{s}_t = \phi_{t+1}^{\widehat{s}_{t+1}} \quad t = N-1, \ldots 1$$

21.6.4 Applying the Viterbi Algorithm for Gaussian Noise

Applying the Viterbi algorithm requires the knowledge of the probability distribution of the observation noise p_η. If the noise of observation is Gaussian, the only parameter to be estimated is the standard deviation. The standard deviation can be, in general, estimated off line. However, we propose a recursive estimation of the noise variance. Starting for some *a priori* estimation of the noise variance, one can launch the Viterbi algorithm and find the most probable sequence. Once this optimal sequence is found, the re-estimation of the noise variance can be done. We launch again the Viterbi algorithm with this new value of the variance and so forth. The algorithm is halted when the noise variance converges. The recurrent application of the algorithm obeys the following procedure.

1. Find a crude approximation of σ_0 by some off line procedure. For instance choose

$$\sigma_0 = \sqrt{\sigma_y^2 - \sigma_x^2}$$

Communication Applications

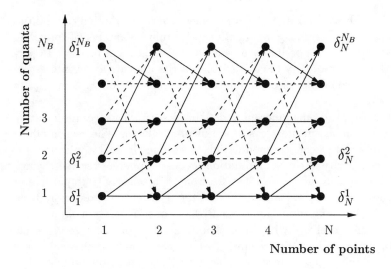

FIGURE 21.4
Lattice for the decoding of N points with N_B balls.

2. Initialize $i = 0$

3. Repeat the Viterbi Algorithm and find the optimal sequence $\hat{s}$. The observation probabilities are computed using

$$P_\eta(y) = \frac{1}{\sigma_i \sqrt{2\pi}} \exp^{-\frac{1}{2} \frac{y^2}{\sigma_i^2}}$$

4. Re-estimate the noise variance

$$\sigma_{i+1} = \sqrt{\frac{1}{N} \sum_{k=1}^{N} (B_c(s_k) - y_k)^2}$$

5. Go to Step 3 until $|\sigma_{i+1} - \sigma_i| \leq \epsilon$, where ϵ is some tolerance factor fixed in advance.

Experiments have shown that the convergence of the variance estimate is very fast even with bad estimations of σ_0; three to five iterations of this procedure are sufficient.

21.7 Communication Applications

Many applications of noise cleaning have been suggested recently. In particular, communication applications have received a great deal of attention.

Indeed the rich dynamics of chaos provides several attractive capabilities such as spread spectrum communications with infinite number of spreading sequences [2] and the possibility of achieving some level of privacy. In particular, the nonperiodicity of chaotic times series appears as an interesting feature which could be used for information transmission in a multipath environment.

However, despite some inherent and claimed advantages, communicating with chaos remains, in most cases, a difficult problem. The reason which still prevents use of chaotic modulation techniques in real applications is the great sensitivity of the demodulation process. Coherent receivers which are based on synchronization [15] suffer from high sensitivity to parameter mismatches between the transmitter and the receiver and even more from signal distortion/contamination caused by the communication channel.

Noncoherent receivers which use chaotic signals for their good decorrelation properties have been shown to be more robust to channel noise; however, as they rely mainly on (long term) decorrelation properties of chaotic carriers, their performances are very close to those of standard noncoherent receivers using pseudo-random or random signals. In other terms, the present noncoherent receivers do not rely on any dynamical constraint which could be known *a priori* at the receiver. For instance, if the receiver knows in advance the particular type of chaotic generator at the transmitter, it could take this information into account to restore the carrier before decision. In both cases, coherent or noncoherent, it has been shown that noise cleaning can drastically improve the performance (bit error rate) of communication systems. In principle, when the receiver knows the system that operates at the emitter, we could use the noise cleaning framework described until now to perform noise cleaning. There is however one difficulty we should solve before; when transmitting digital information one modulates in some way a chaotic system, e.g., by switching a system parameter according to the bit of information one wants to transmit. Therefore, modulation is equivalent to use, according to the information $+1$ or 0 we want to transmit, two different systems. These two different systems can be supposed known in advance at the receiver. Our previous framework cannot be used directly however since the receiver does not know in advance what system is used at the transmitter at a given time. We show below one example in which the demodulation and noise cleaning can be combined in an efficient way. For a detailed treatment of communication applications using chaotic carriers, the reader is referred to the Ph.D. thesis of Joerg Schweizer [24].

Example 21.3

We borrow the present example from [14]. The goal in this section is to give a flavor of the general concept which can be used to clean a modulated chaotic signal. We refer the reader to [14] for a more detailed version of this example.

Let us first present the general scheme of modulation which is based on the classical chaotic shift keying technique (CSK) [4]. At the emitter, two chaotic generators, M_1 and M_2 are running in parallel. According to the current bit we want to transmit we send to the channel either the M_1 or the M_2 output. The modulation process is represented at the left-hand side of Figure 21.5. Let us suppose for instance that each bit to be sent is mapped into N consecutive outputs of M_1 or M_2. The channel introduces noise and distortion in such a way that it would be useful to clean the data before taking any decision.

To simplify the presentation, suppose that we can achieve bit synchronization, i.e., we are sure of the exact time at which a piece of bit waveform begins at the receiver. If one can achieve bit synchronization, one simple idea we could adopt is to launch two cleaning processes in parallel on the received waveform.

One cleaning process would be carried out assuming that M_1 was used while the second one would be done by assuming that M_2 was used. By comparing the distance between cleaned and original corespondent trajectories one could base his decision by choosing the trajectory which is closest to its cleaned version assuming M_1 or M_2. This scheme would assume that we rebuild two cleaned trajectories and would be one possible scheme we could implement with the deterministic approaches we have presented in Section 21.5.

However, one could remark that it is not necessary to rebuild the cleaned trajectories if we could get a measure of likelihood of system M_1 and M_2 given the observations at the receiver. This measure is given directly by the probabilistic approach described in Section 21.6 after the forward pass of the Viterbi algorithm; the backward pass which rebuilds the trajectory is not needed. Therefore, in the probabilistic approach, one does not have to care about noise cleaning (which could be a byproduct of the algorithm); one has only to compare the two following likelihood measures:

$$\gamma_1 = p(s, Y|M_1) = \prod_{k=1}^{N} p(s_k|(s_{k-1}, M_1)) \ p(y_k|s_k) \qquad (21.7.52)$$

$$\gamma_2 = p(s, Y|M_2) = \prod_{k=1}^{N} p(s_k|(s_{k-1}, M_2)) \ p(y_k|s_k) \qquad (21.7.53)$$

and decide that the most likely bit which has been transmitted corresponds to the system which conveys the greater likelihood. The demodulation process is presented in the right-hand side of Figure 21.5.

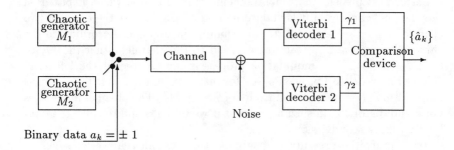

FIGURE 21.5
Modulation-demodulation of chaotic signals with implicit noise cleaning.

21.8 Conclusions

Separating a chaotic signal from noise when the chaotic dynamical system is known in advance has been investigated in this chapter. We first presented optimal solution for two fundamental classes of noise separation algorithms, 1) when the recovered signal is not forced to belong exactly to a true trajectory of the system; 2) when the solution is forced to belong to a true trajectory. The optimal tools which provide a global minimum of an optimization problem are generally too demanding in terms of computational cost. Therefore, we presented two general classes of practical methods for noise cleaning. The first general class is based on deterministic optimization and leads to two kinds of estimators, approximately constrained and exactly constrained. The second general class is based on a probabilistic approach which offers denoising capabilities at a very low computational cost. We have indicated how the denoising framework presented in this chapter could be used in communications applications in which demodulation has to be performed in parallel with noise decontamination. In this kind of application, probabilistic denoising is shown to offer attractive capabilities since very elegant and low cost algorithms can be implemented.

References

[1] H. D. I. Abarbanel, *Analysis of Observed Chaotic Data*, Springer Verlag, 1996.

[2] G. H. Bateni and C. D. McGillem, "A chaotic direct-sequence spread-spectrum communication system," *IEEE Trans. on Communications*, vol. 42, pp. 1524-1527, 1994.

[3] D. S. Broomhead, J. P.Huke, and R. Jones, "Signals in chaos: A method for the cancellation of deterministic noise from discrete signals," *Physica D*, vol. 80, pp. 413-432, 1995.

[4] H. Dedieu, M. P. Kennedy, and M. Hasler, "Chaos shift keying: Modulation and demodulation of a chaotic carrier using self-synchronizing Chua's circuits," *IEEE Trans. on Circ. Sys.*, II, vol. 40, pp. 634-642, 1993.

[5] H. Dedieu and M. Ogorzałek, "Overview of noise reduction algorithms for systems with known dynamics," *Procd. of Int. Symp. on Nonlinear Theory Appl.*, NOLTA'98, Crans-Montana, Switzerland, 1998, pp. 1297-1300.

[6] H. Dedieu and J. Schweizer, "Noise reduction in chaotic time series – An overview," *Proc. of Int. Workshop on Nonlinear Dynamics of Electronic Systems*, NDES'98, Budapest, 1998, pp. 53-62.

[7] J. P. Eckmann and D. Ruelle, "Ergodic theory of chaos and strange attractors," *Rev. Mod. Phys.*, vol. 57, pp. 617-656, 1985.

[8] J. D. Farmer and J. Sidorowich, "Optimal shadowing and noise reduction," *Physica D*, vol. 47, pp. 373-392, 1991.

[9] P. Grassberger, R. Hegger, H. Kantz, C. Schaffrath, and T. Schreiber, "On noise reduction methods for chaotic data," *Chaos*, vol. 3, pp.127-141, 1993.

[10] S. M. Hammel, "Noise reduction for chaotic systems," Tech Report, Naval Surface Warfare Center, Silver Spring, MD, 1989.

[11] M. Hasler, "Chaos shift keying in the presence of noise: A simple time example," *Proc. of IEEE Int. Symp. on Circ. Sys.*, Monterey, 1998.

[12] S. Haykin, *Communication Systems*, Wiley, New York, 1994.

[13] H. G. Kantz and T. Schreiber, *Nonlinear Time Series Analysis*, Cambridge Univesity Press, 1997.

[14] A. Kisel, H. Dedieu, and T. Schimming, "Maximum likelihood approaches for noncoherent communications with chaotic carriers," *IEEE Trans. on Circ. Sys.*, I, 1999, in press.

[15] G. Kolumban, M. P. Kennedy, and L. O. Chua, "The role of synchronization in digital communications using chaos – Part I: Fundamentals of digital communications," *IEEE Trans. on Circ. Sys.*, I, vol. 44, pp. 927-936, 1997.

[16] E. J. Kostelich and T. Schreiber, "Noise reduction in chaotic time-series data: A survey of common methods," *Phys. Rev. E*, vol. 48, pp. 1752-1763, 1993.

[17] E. J. Kostelich and J. A. Yorke, "Noise reduction in dynamical systems," *Physical Rev. A*, vol. 38, pp. 1649-1652, 1988.

[18] E. J. Kostelich and J. A. Yorke, "Noise reduction: Finding simplest dynamical system consistent with the data," *Physica D*, vol. 41, pp. 183-196, 1990.

[19] C. Lee and D. B. Williams, "Generalized iterative methods for enhancing contaminated chaotic systems," *IEEE Trans. on Circ. Sys.*, I, vol. 44, 1997.

[20] P. F. Marteau and H. D. I. Abarbanel, "Noise reduction in chaotic time series using scaled probabilistic method," *J. of Nonlinear Science*, vol. 1, pp. 313-343, 1991.

[21] C. Myers, S. Kay, and M. D. Richard, "Signal separation for nonlinear dynamical systems," *Proc. of ICASSP*, 1992, pp. IV-129, IV-132.

[22] E. Ott, T. Sauer, and J.A. Yorke, *Coping with Chaos: Analysis of Chaotic Data and the Exploitation of Chaotic Systems*, Wiley, New York, 1994.

[23] J. Schweizer, "The performance of chaos shift keying: Synchronization versus symbolic backtracking," *Proc. of IEEE Int. Symp. on Circ. Sys.*, Monterey, 1998.

[24] J. Schweizer, *Application of Chaos to Communications*, Doctoral thesis of the Swiss Federal Institute of Technology, March 1999.

22
Digital Communications Using Chaos

Michael Peter Kennedy[1] and Géza Kolumbán[2]

[1] Department of Electronic and Electrical Engineering
University College Dublin, Dublin 4, Ireland
peter.kennedy@ucd.ie

[2] Department of Measurement and Information Systems
Technical University of Budapest, H-1521 Budapest, Hungary
kolumban@mit.bme.hu

Abstract

During the past five years, there has been tremendous worldwide interest in the possibility of exploiting chaos in wideband communication systems. This chapter discusses the implications of using chaotic basis functions in digital communications. Preliminary performance results are given, potential benefits are discussed, and possible application domains are identified.

22.1 Motivation

In recent years, there has been explosive growth in personal communications, the aim of which is to guarantee the availability of voice and/or data services between mobile communication terminals. In order to provide these services, radio links are required for a large number of compact terminals in densely populated areas. As a result, there is a need to provide high-frequency, low-power, low-voltage circuitry. The huge demand for telecommunication results in a large number of users; therefore, today's telecommunication systems are limited primarily by interference from other users. In some applications, the efficient use of available bandwidth is extremely important, but in other applications, where the exploitation of communication channels is relatively low, a wideband communication technique having limited bandwidth efficiency can also be used.

Often, many users must be provided with simultaneous access to the same or neighboring frequency bands. The optimum strategy in this situation, where every user appears as interference to every other user, is for each communicator's signal to look like white noise which is as wideband as possible [27].

There are two ways in which a communicator's signal can be made to appear like wideband noise: by spreading each symbol using a pseudorandom sequence to increase the bandwidth of the transmitted signal, or by representing each symbol by a piece of "noiselike" waveform.

The conventional solution to this problem is the first approach: to use a synchronizable pseudorandom sequence to spread the transmitted signal, and to use a conventional modulation scheme based on Phase Shift Keying (PSK) or Frequency Shift Keying (FSK). Such Direct Sequence (DS) and Frequency Hopping (FH) Spread Spectrum (SS) schemes have processing gain associated with despreading at the receiver, and the possibility to provide multiple access by assigning mutually orthogonal sequences to different users. This is the basis of Code Division Multiple Access (CDMA) communication systems. Limitations are imposed by the need to achieve and maintain carrier and symbol synchronization, the periodic nature of the spreading sequences, the limited number of available orthogonal sequences, and the periodic nature of the carrier. One further problem is that the orthogonality of the spreading sequences requires the synchronization of all spreading sequences used in the same frequency band, i.e., the whole system must be synchronized. Due to different propagation times for different users, perfect synchronization can never be achieved in real systems.

An alternative approach to making a transmission "noiselike" is to represent the transmitted symbols not as weighted sums of periodic basis functions but as inherently nonperiodic *chaotic* basis functions [12]. This chapter focuses on digital modulation using chaotic carriers.

22.2 What is Chaos?

Deterministic dynamical systems are those whose states change with time in a deterministic way. They may be described mathematically by differential or difference equations, depending on whether they evolve in continuous or discrete time. Deterministic dynamical systems can produce a number of different steady-state behaviors including DC, periodic, and chaotic solutions [10].

DC is a nonoscillatory state. Periodic behavior is the simplest type of steady-state oscillatory motion. Sinusoidal signals, which are universally used as carriers in analog and digital communication systems, are periodic solutions of continuous-time deterministic dynamical systems.

Deterministic dynamical systems also admit a class of nonperiodic signals which are characterized by a continuous "noiselike" broad power spectrum; this is chaos [11]. In the time domain, chaotic signals appear "random. " Chaotic systems are characterized by "sensitive dependence on initial conditions;" a small perturbation eventually causes a large change in the state of the system. Equivalently, chaotic signals rapidly decorrelate with themselves. The autocorrelation function of a chaotic signal has a large peak at zero and decays rapidly.

Thus, while chaotic systems share many of the properties of stochastic processes, they also possess a deterministic structure which makes it possible to generate "noiselike" chaotic signals in a theoretically reproducible manner. In particular, continuous-time chaotic systems can be used to generate wideband carriers, i.e., basis functions for chaotic digital communication systems.

22.3 Potential Benefits of Chaotic Basis Functions in Digital Communications

The aim of this work is to describe the state of the art in digital modulation schemes which use chaotic rather than periodic basis functions. Problems and opportunities arising from the nonperiodicity of chaotic signals are highlighted, solutions are proposed, potential application domains are identified, and current research directions are discussed.

Since only analog telecommunication channels are available, digital information to be transmitted has to be mapped into analog sample functions in a communication system. The sample functions pass through the analog channel and are identified at the receiver in order to recover the transmitted digital information.

In a conventional digital communication system, several sample functions are used, i.e., the modulator represents each symbol to be transmitted as a weighted sum of a number of periodic basis functions. For example, two orthogonal signals, such as a sine and a cosine, may be used. The objective of the receiver is to recover the weights associated with the received signal and thereby to decide which symbol was transmitted [5].

The modulated signal consists of segments of periodic waveforms corresponding to the individual symbols. The segment of analog waveform corresponding to each symbol is unique. When sinusoidal basis functions are used without spread spectrum techniques, the transmitted signal is a narrowband signal. Consequently, multipath propagation (due to the reception of multiple copies of the transmitted signal traveling along different paths) can cause high attenuation or even dropout (in the case of catastrophic destructive interference) of the received narrowband signal.

A chaotic signal generator automatically produces a wideband noiselike signal with robust and reproducible statistical properties [8, 9]. Due to its wideband nature, a signal comprising chaotic basis functions is potentially more resistant to multipath propagation than one constructed of sinusoids.

One factor which limits the performance of all telecommunication systems is interference. In conventional systems based on periodic carrier signals, where the SS-CDMA technique is not used, signals can be made orthogonal by putting them in different frequency bands (FDMA technique), by ensuring that the basis functions are orthogonal to each other (using sine and cosine basis functions, for example), or by using orthogonal electromagnetic polarization, for example. If these requirements are not met, interference occurs.

In contrast with periodic signals, chaotic signals decorrelate rapidly with themselves and chaotic signals generated by different chaotic circuits are almost orthogonal. This means that the correlation, equivalently the interference, between two chaotic signals generated by unsynchronized chaotic circuits started from different initial conditions and/or having different circuit parameters is low.

22.4 Digital Communications Using Chaos

In a digital communication system, the information symbol to be transmitted is mapped by the modulator to an analog sample function and passed through an analog channel. The analog signal in the channel is subject to a number of disturbing influences including attenuation, bandpass filtering, and additive noise. The role of the demodulator is to decide, on the basis of the received corrupted sample function, which symbol was transmitted.

In a conventional communication system, the analog sample function of duration T which represents a symbol is a linear combination of sinusoidal basis functions and the symbol duration T is an integer multiple of the period of the basis function. In a chaos-based digital communication system, shown schematically in Figure 22.1, the analog sample function of duration T which represents a symbol is a *chaotic* basis function.

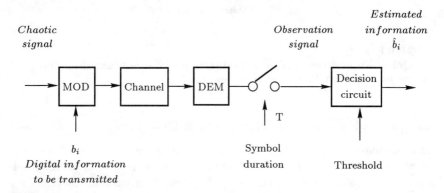

FIGURE 22.1
Block diagram of a chaotic communication scheme.

The decision as to which symbol was transmitted is made by estimating some property of the received sample function [26, 6]. That property might be the energy of the chaotic signal or the correlation measured between different parts of the transmitted signal, for example.

Since chaotic waveforms are not periodic, each sample function of duration T is different. This has the advantage that each transmitted symbol is represented by a unique analog sample function, and the correlation between chaotic sample functions is extremely low. However, it also produces a problem associated with estimating long-term statistics of a chaotic process from sample functions of finite duration. We will discuss this so-called *estimation problem* [17] in more detail in Section 22.5.1.

22.4.1 Channel Effects

In any practical communication system, the signal $r_i(t)$ which is present at the input to the demodulator differs from that which was transmitted, due to the effects of the channel.

The simplest realistic model of the channel is a linear bandpass channel with additive white Gaussian noise (AWGN). A block diagram of the bandpass AWGN channel model which we consider throughout this work is shown in Figure 22.2.

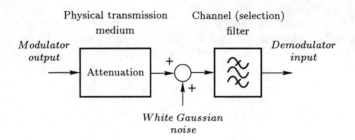

FIGURE 22.2
Model of an additive white Gaussian noise channel including the frequency selectivity of the receiver.

22.4.2 Coherent versus Noncoherent Detection

In chaos-based digital communication systems, as in conventional communication schemes, the transmitted symbols can be recovered using either coherent [13, 2, 3] or noncoherent [15, 25] demodulation techniques.

Coherent Detection (with Synchronization)

Coherent detection is accomplished by reproducing copies of the basis functions in the receiver, typically by means of a synchronization scheme [18]. When synchronization is exploited, the synchronization scheme must be able to recover the basis function(s) from the corrupted received signal.

If a sinusoidal basis function is used, then a *narrow-band* phase-locked loop (PLL) may be used to recover it [5]. Noise corrupting the transmitted signal is suppressed because of the low-pass property of the PLL. When an inherently wideband chaotic basis function is used, the synchronization circuit must be *wideband* in nature. Typically, both the "amplitude" and "phase" of the chaotic basis function must be recovered from the received signal. Because of the wideband property of the chaotic basis function, narrowband linear filtering cannot be used to suppress the noise.

Consider the binary Chaos Shift Keying (CSK) transmitter shown in Figure 22.3 [3]. The transmitted sample functions $s_1(t) = g_1(t)$ and $s_2(t) = g_2(t)$ (representing symbols "1" and "0," respectively) are the outputs of two free-running chaotic signal generators which produce basis functions $g_1(t)$ and $g_2(t)$.

Figure 22.4 shows a coherent (synchronization-based) receiver using binary Chaos Shift Keying (CSK) modulation. Synchronization circuits in the receiver attempt to reproduce the basis functions, given the received noisy sample function $r_i(t) = s_i(t) + n(t)$.

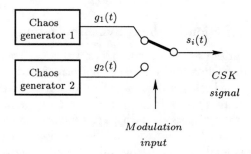

FIGURE 22.3
Block diagram of a CSK modulator.

An acquisition time T_S is allowed for the synchronization circuits to lock to the incoming signal. The recovered basis functions $\hat{g}_1(t)$ and $\hat{g}_2(t)$ are then correlated with $r_i(t)$ for the remainder of the bit duration T. A decision is made on the basis of the relative closeness of $r_i(t)$ to $\hat{g}_1(t)$ and $\hat{g}_2(t)$, as quantified by the observation variables z_{i1} and z_{i2}, respectively.

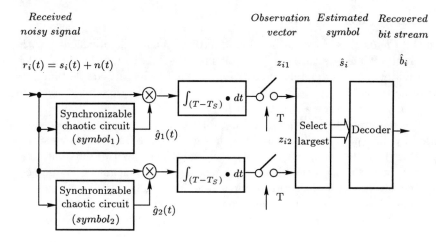

FIGURE 22.4
Block diagram of a coherent CSK receiver.

Recent studies of chaotic synchronization, where significant noise and filtering have been introduced in the channel, suggest that the performance of chaotic synchronization schemes is worse, at low signal-to-noise ratio (SNR), than that of the best synchronization schemes for sinusoids [18, 16, 19].

Noncoherent Detection (without Synchronization)

Synchronization (in the sense of carrier recovery) is not *required* for digital communication; demodulation can also be performed without synchronization. This is true for both periodic and chaotic sample functions.

In this case, the receiver estimates, from the noisy received signal, the property (such as the bit energy) which carries the transmitted information, without recovering the basis functions. The decision as to which symbol was transmitted is made by comparing this estimate against a threshold.

In the following section, we discuss four noncoherent chaotic communication receiver configurations: Chaos Shift Keying (CSK) [3], Chaotic On-Off Keying (COOK) [15], Differential Chaos Shift Keying (DCSK) [25], and FM-DCSK [22]. Of these schemes, DCSK and FM-DCSK have the best noise performance.

22.5 Survey of Noncoherent Chaotic Communication Schemes

22.5.1 Chaos Shift Keying

In the *Chaos Shift Keying* (CSK) modulation scheme, chaotic signals with different bit energies are used to transmit the binary information. The modulator is a very simple circuit: for bit "1," a chaotic sample function with mean bit energy E_{b1} is radiated; for bit "0," a chaotic sample function with mean bit energy E_{b2} is radiated.

The required chaotic signals having different bit energies can be generated by different chaotic circuits (as shown in Figure 22.3) or they can be produced by the same chaotic circuit and multiplied by two different constants. In both cases, the binary information to be transmitted is mapped to the bit energies of chaotic sample functions. The bit energy can be estimated by a correlator at the receiver, as shown in Figure 22.5.

The observation signal z_i which is used by the decision circuit is defined by

$$z_i = \int_T r_i^2(t)dt$$

$$= \int_T [s_i(t) + n(t)]^2 dt$$

$$= \int_T s_i^2(t)dt + 2\int_T s_i(t)n(t)dt + \int_T n^2(t)dt, \qquad (22.5.1)$$

where $\int_T$ denotes integration over one sample function.

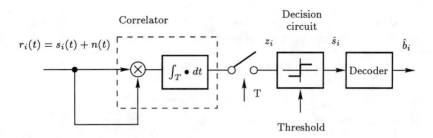

FIGURE 22.5
Block diagram of a noncoherent CSK receiver.

In the noise-free case, the second and third terms in (22.5.1) are zero. Consequently, z_i assumes the values $\int_T s_1^2(t)dt$ and $\int_T s_2^2(t)dt$ (equivalently the bit energies $E_b(s_1)$ and $E_b(s_2)$) when bits "1" and "0," respectively, are received.

Figures 22.6(a) and (b) show histograms of samples of the observation signal z_i for noise-free and noisy channels. Note that the samples of z_i associated with the transmitted symbols are not constant, even in the noise-free case. Rather, they are clustered about E_{b1} and E_{b2} with variances σ_1^2 and σ_2^2, respectively.

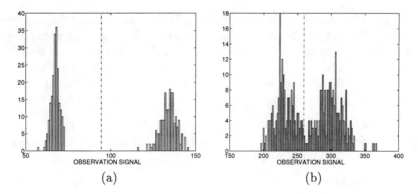

FIGURE 22.6
Histograms of samples of the observation signal in a noncoherent CSK receiver with (a) noise-free and (b) noisy channels.

In conventional modulation schemes using periodic basis functions, $s_i(t)$ is periodic and the bit duration T is an integer multiple of the period of the basis function; hence, $\int_T s_i^2(t)dt$ is constant. By contrast, chaotic signals are inherently nonperiodic, so $\int_T s_i^2(t)dt$ varies from one sample function of length T to another. This effect produces the nonzero variance in the samples of z_i in Figure 22.6(a).

The standard deviation of samples of $\int_T s_1^2(t)dt$ scales as $1/T$, as shown in Figure 22.7. Thus, the variance of estimation can be reduced by increasing the bit duration T.

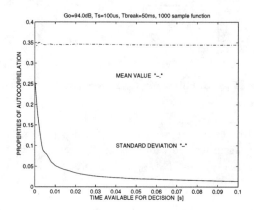

FIGURE 22.7
Variance of estimation versus bit duration.

Figure 22.8 shows the noise performance of a noncoherent CSK receiver for four different bit durations. For $T = 5$ ms and 10 ms, the variance of estimation is so large that the histograms associated with received bits "1" and "0" overlap, even for large ratios of energy per bit to noise spectral density (E_b/N_0). Consequently, the bit error rate (BER) is bounded from below.

Note, however, that the variance of estimation (Figure 22.7) and consequently the noise performance (Figure 22.8) are significantly improved for bit durations greater than 20 ms in this case.

A significant drawback of noncoherent CSK is that the threshold level required by the decision circuit depends on the signal-to-noise ratio (SNR). The optimum threshold of the level comparator for decision-making is denoted by the dash-dot line in Figure 22.6. Note that it is different in the noise-free and noisy cases, due to the third term in (22.5.1).

The distance between the peaks of the histogram is determined by the difference between the mean bit energies E_{b1} and E_{b2} associated with the two chaotic sample functions. For a given noise level and chaotic signal, the best noise performance can be achieved if the distance between the two mean values is maximized. This requirement can be satisfied by using *Chaotic On-Off-Keying*, described next [14].

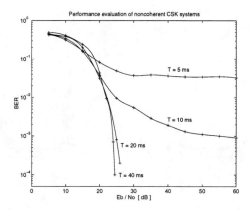

FIGURE 22.8
Noise performance of a noncoherent CSK system for four different bit durations T.

22.5.2 Chaotic On-Off-Keying

For a given E_b, the maximum distance between the elements of a binary signal set can be achieved if On-Off-Keying is used. In the Chaotic On-Off-Keying (COOK) scheme, the chaotic signal is switched on and off to transmit symbols "1" and "0," respectively, as shown in Figure 22.9.

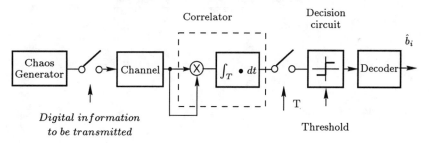

FIGURE 22.9
Block diagram of noncoherent COOK modulation scheme.

If the average bit energy is E_b and both symbols are equiprobable, then one symbol carries zero energy and the other has energy $2E_b$. Consequently, the distance between the elements of the signal set is $2E_b$. Histograms of samples of the observation signal z_i for noise-free and noisy channels are shown in Figures 22.10(a) and (b), respectively. The optimum decision threshold is denoted by a dash-dot line.

The noise performance of the noncoherent COOK scheme is shown in

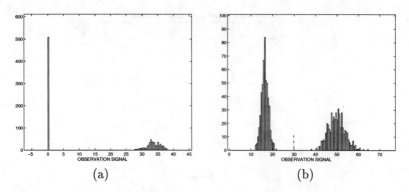

FIGURE 22.10
Histograms of samples of the observation signal in a noncoherent COOK system with (a) noise-free and (b) noisy channel.

Figure 22.11, where the parameter is the symbol duration. Note that for BER=10^{-3}, COOK requires an E_b/N_0 of 15.5 dB; this is 8 dB less than the corresponding value for noncoherent CSK [16].

The superior noise performance of the COOK scheme results from the fact that the distance between the elements of the signal set is increased compared to the CSK method.

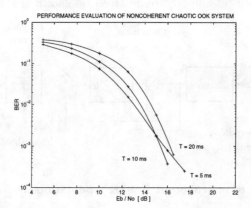

FIGURE 22.11
Noise performance of noncoherent COOK modulation scheme.

The major disadvantage of the CSK system, namely, that the threshold value of the decision circuit depends on the noise level, also appears in COOK. This means that by using COOK we can maximize the distance between the elements of the signal set, but the threshold level required by the decision circuit depends on the SNR. The threshold can be kept constant by applying the *Differential Chaos Shift Keying* method.

22.5.3 Differential Chaos Shift Keying

In Differential Chaos Shift Keying (DCSK), every bit to be transmitted is represented by two chaotic sample functions. The first sample function serves as a *reference* while the second one carries the information. Bit "1" is sent by transmitting a reference signal provided by a chaos generator twice in succession, while for bit "0", the reference chaotic signal is transmitted, followed by an inverted copy of the same signal. Thus,

$$s_i(t) = \begin{cases} x(t), & t_i \leq t < t_i + T/2 \\ +x(t - T/2), & t_i + T/2 \leq t < t_i + T \end{cases}$$

if symbol "1" is transmitted in $[t_i, t_i + T)$ and

$$s_i(t) = \begin{cases} x(t), & t_i \leq t < t_i + T/2 \\ -x(t - T/2), & t_i + T/2 \leq t < t_i + T. \end{cases}$$

if symbol "0" is transmitted in $[t_i, t_i + T)$.

Figures 22.12 and 22.13 show a block diagram of a DCSK modulator and a typical DCSK signal corresponding to the binary sequence 1100. In this example, the chaotic signal is produced by an analog phase-locked loop (APLL) [23] and the bit duration is 40 ms.

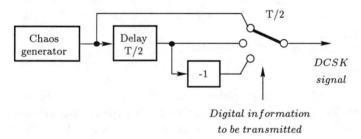

FIGURE 22.12
Block diagram of a DCSK modulator.

Since each bit is mapped to the correlation between successive segments of the transmitted signal of length $T/2$, the information signal can be recovered by a correlator. A block diagram of a DCSK demodulator is shown in Figure 22.14.

The received noisy signal is delayed by half of the bit duration $(T/2)$ and the correlation between the received signal and the delayed copy of itself is determined. The decision is made by a level comparator [25].

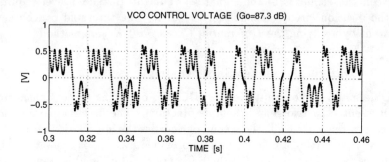

FIGURE 22.13
DCSK signal corresponding to binary sequence 1100.

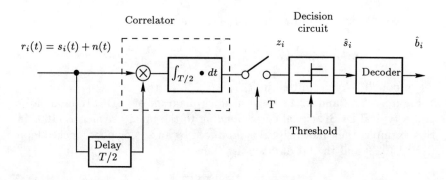

FIGURE 22.14
Block diagram of a DCSK receiver.

Figure 22.15 shows the correlator output for the transmitted signal shown in Figure 22.13. The observation variable z_i is positive at the decision time instants 0.32 s and 0.36 s, and is negative at 0.4 s and 0.44 s, allowing the transmitted bit sequence (1100) to be recovered.

In DCSK, the observation signal has the form

$$z_i = \int_{T/2} r_i(t) r_i(t - T/2) dt$$

$$= \int_{T/2} [s_i(t) + n(t)][s_i(t - T/2) + n(t - T/2)] dt.$$

When symbol "1" is transmitted, it becomes

$$z_i = \int_{T/2} s_i^2(t) dt + \int_{T/2} s_i(t)[n(t) + n(t - T/2)] dt + \int_{T/2} n(t) n(t - T/2) dt,$$

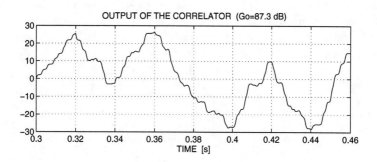

FIGURE 22.15
Correlator output corresponding to sequence 1100.

and for symbol "0"

$$z_i = -\int_{T/2} s_i^2(t)dt - \int_{T/2} s_i(t)[n(t) - n(t-T/2)]dt + \int_{T/2} n(t)n(t-T/2)dt.$$

The histogram of samples of the observation signal in the absence of channel noise is shown in Figure 22.16(a).

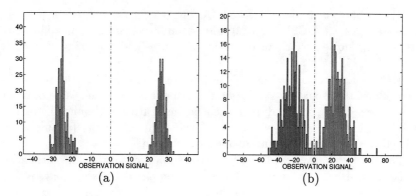

FIGURE 22.16
Histograms of samples of the observation signal in (a) noise-free and (b) noisy DCSK systems.

Note that this scheme also suffers from the estimation problem. The nonzero variance of z_i results from the nonperiodic nature of the chaotic signal: $\int_{T/2} s_i^2(t)dt$ is not constant, even if the same symbol is transmitted repeatedly.

The estimation problem also manifests itself in the noise performance characteristics shown in Figure 22.17. Note that if the bit duration T is decreased, the sensitivity of the decision process to noise increases, and the BER can be reduced only by increasing E_b/N_0.

For sufficiently large T, the noise performance of DCSK is comparable to that of a conventional sinusoid-based modulation scheme. In particular, $E_b/N_0 = 13.5$ dB is required for BER=10^{-3}.

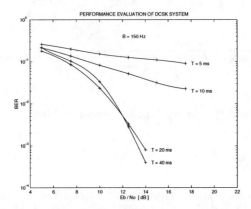

FIGURE 22.17
Noise performance of a DCSK system.

By contrast with the CSK and COOK schemes discussed in Sections 22.5.1 and 22.5.2, DCSK is an antipodal modulation scheme. In addition to its superior noise performance, the decision threshold is zero independently of E_b/N_0 [25].

A further advantage of DCSK results from the fact that the reference- and information-bearing sample functions pass through the same channel, thereby rendering the modulation scheme insensitive to channel distortion. DCSK can also operate over a time-varying channel if the parameters of the channel remain constant for the bit duration T.

The principal drawback of DCSK arises from the fact that the correlation is performed over half the bit duration. Compared to those conventional techniques where the elements of the signal set are available at the receiver, DCSK has half the data rate and only half the bit energy contributed to its noise performance [18, 19].

In the CSK, COOK, and DCSK modulation schemes, the information signal to be transmitted is mapped to chaotic sample functions of finite length. We have seen that the property required by the decision circuit at the receiver can only be *estimated* because of the nonperiodic nature of chaotic signals. The estimation has a nonzero variance even in the noise-free case; this puts a lower bound on the bit duration and thereby limits the data rate.

One way to improve the data rate is to use a multilevel modulation scheme such as those described in [20]. Alternatively, one may solve the estimation problem directly by modifying the modulation scheme such that the transmitted energy for each symbol is kept constant. FM-DCSK [21] is an example of the latter approach.

22.5.4 FM-DCSK

The objective of FM-DCSK is to generate a wideband DCSK signal with constant E_b. The instantaneous power of an FM signal does not depend on the modulation. Let the chaotic signal be the input of an FM modulator. If the wideband output of the FM modulator is varied using the DCSK technique, then the correlator output in the receiver has zero variance in the noise-free case and the estimation problem is solved.

Note that, as in the DCSK technique, every information bit is transmitted in two pieces; the first sample function serves as a reference, while the second carries the information. The operation of the modulator is the same as in DCSK, the only difference being that not the chaotic signal itself, but the FM modulated signal is the input to the DCSK modulator (see Figure 22.18).

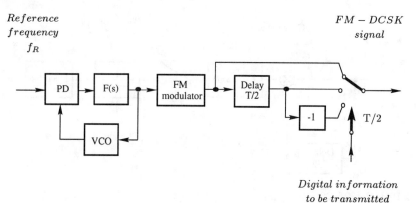

FIGURE 22.18
Block diagram of an FM-DCSK modulator. The low-frequency chaotic signal is generated by an APLL.

The input of the FM modulator is a chaotic signal, which can be generated by a chaotic analog phase-locked-loop (APLL), for example. If the chaotic signal is generated by an appropriately designed APLL then the output of the FM modulator, evaluated over a long time period, has a bandlimited spectrum with uniform power spectral density, as shown in Figure 22.19 [22].

The demodulator of an FM-DCSK system is a DCSK receiver. The only

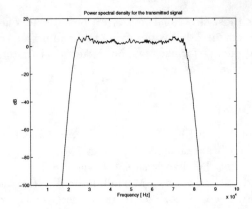

FIGURE 22.19
Power spectral density of the transmitted signal in an FM-DCSK scheme.

difference is that, instead of low-frequency chaotic signals, the FM signals are correlated directly, as shown in Figure 22.20.

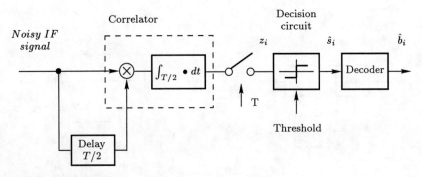

FIGURE 22.20
Block diagram of the FM-DCSK demodulator.

Figures 22.21(a) and (b) show histograms of the observation signal at the receiver for noise-free and noisy channels, respectively. Note that the variance of estimation is zero in the noise-free case, as expected. The threshold level required by the decision circuit is always zero, regardless of the noise level and the probabilities of the symbols.

The noise performance of the FM-DCSK system is shown in Figure 22.22.

The main advantage of FM-DCSK modulation over CSK, COOK, and DCSK is that the data rate is not limited by the properties of the chaotic signal.

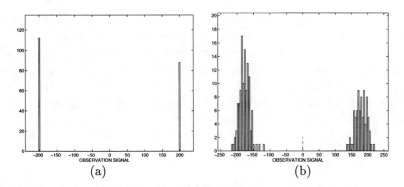

FIGURE 22.21
Histograms of samples of the observation signal z_i in an FM-DCSK receiver for (a) noise-free and (b) noisy channels.

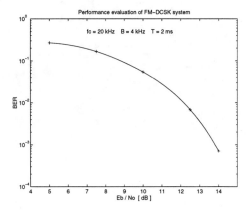

FIGURE 22.22
Noise performance of the FM-DCSK system.

22.6 Summary

Over the past few years, several methods have been proposed for using chaotic signals in digital communication systems. The oldest scheme is CSK, where different symbols are represented by different chaotic signals.

Coherent detection offers an advantage in terms of noise performance if synchronization can be achieved and maintained. Unfortunately, those chaos synchronization techniques which are currently available are not sufficiently robust to maintain synchronization in a noisy channel. By contrast, noncoherent detection based on estimating the energy per bit or statistical properties of chaotic basis functions represents a feasible solution to digital communication using CSK and COOK.

We have highlighted the drawbacks of coherent CSK and the fundamental problem associated with estimating long-term statistical properties of chaotic signals from sample functions of finite length. The estimation problem which we have described limits the data rate of chaotic communications.

We have shown how the noise performance of CSK can be improved by using the COOK technique. In DCSK, the threshold level of the decision device is kept constant and the distance between the elements of the signal set is maximized.

The noise performance of noncoherent CSK, COOK, and DCSK is summarized graphically in Figure 22.23.

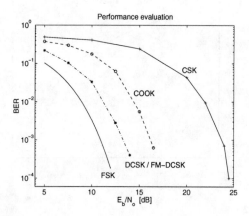

FIGURE 22.23
Noise performance of the CSK, COOK, and DCSK/FM-DCSK techniques. Noncoherent FSK is shown for comparison.

The upper bound on the data rate of DCSK can be increased by using multilevel modulation schemes or by keeping the transmitted energy constant for each symbol.

The FM-DCSK technique, which is an antipodal modulation scheme with constant bit energy, represents an optimal solution in the sense that the distance between the elements of the signal set is E_b. Its noise performance is equal to that of DCSK but the data rate is not limited by the properties of the underlying chaotic signal.

22.7 Open Problems and Expected Developments

Although FM-DCSK offers the best performance of the chaotic modulation techniques described here, it should be emphasized that the field of chaos communications is very young and that several significant improvements in performance are possible [12].

First, we have mentioned that existing chaos synchronization techniques are not sufficiently robust to permit coherent detection of chaotic transmissions. It should be emphasized that coherent antipodal chaotic modulation schemes have not yet been developed, nor have techniques been proposed for maintaining synchronization throughout the transmission. However, the field of chaos synchronization is undergoing rapid development and more robust chaos synchronization techniques and modulation schemes may emerge.

To our knowledge, FM-DCSK offers the best performance of the noncoherent modulation schemes which are available in the literature. The performance of this scheme should be theoretically quantified. Some results in this direction for simplified systems are beginning to appear [4, 1]. Proposed improvements to the DCSK/FM-DCSK modulation scheme suggest that its performance can approach that of coherent FSK [7].

It is not yet clear how well chaotic modulation schemes can perform in multipath and fading channels. Further research is required to verify the performance of chaos communications under poor propagation conditions.

Finally, on the implementation front, additional research and development work is required in the design, characterization, and implementation of chaotic signal sources. Low-pass chaotic signals can be generated by very simple circuits [9]. The chaotic analog phase-locked loop (APLL) can generate bandpass basis functions in any frequency band and at any power level [23, 24].

Potential application domains for low-cost robust wideband digital modulation schemes using chaos include wireless local area networks, indoor radio, factory automation, and powerline communications.

Acknowledgments

This work has been sponsored in part by the European Commission under the Open LTR initiative (Esprit Project 31103 - INSPECT) and by the National Scientific Research Foundation of Hungary under Grant number T-020522. Special thanks are due to Z. Jákó, G. Kis, and T. Kolumbán for producing the figures. The authors acknowledge extensive discussions with their colleagues in the Chaos Communications Collective.

References

[1] A. Abel, M. Götz, and W. Schwarz, "Statistical analysis of chaotic communication schemes," *Proc. of ISCAS'98*, Monterey, CA, 1998, pp. IV465–468.

[2] K. M. Cuomo, A. V. Oppenheim, and S. H. Strogatz. "Synchronization of Lorenz-based chaotic circuits with applications to communications," *IEEE Trans. on Circ. Sys.*, vol. 40, 1993.

[3] H. Dedieu, M. P. Kennedy, and M. Hasler, "Chaos shift keying: Modulation and demodulation of a chaotic carrier using self-synchronizing Chua's circuits," *IEEE Trans. on Circ. Sys.*, II, vol. 40, pp. 634–642, 1993.

[4] M. Götz, A. Abel, and W. Schwarz, "What is the use of Frobenius-Perron operator for chaotic signal processing?" *Proc. of NDES'97*, Moscow, June 1997, pp. 8–13.

[5] S. S. Haykin, *Communication Systems*, 3rd ed., Wiley, New York, 1994.

[6] S. H. Isabelle and G. W. Wornell, "Statistical analysis and spectral estimation techniques for one-dimensional chaotic signals," *IEEE Trans. on Signal Proc.*, vol. 45, 1997.

[7] Z. Jákó, "Performance improvement of DCSK modulation," *Proc. of NDES'98*, Budapest, 16–18 July 1998, pp. 119–122.

[8] M. P. Kennedy, "Three steps to chaos part I: Evolution," *IEEE Trans. on Circ. Sys.*, I, vol. 40, pp. 640–656, 1993.

[9] M. P. Kennedy, "Three steps to chaos part II: A Chua's circuit primer," *IEEE Trans. on Circ. Sys.*, I, vol. 40, pp. 657–674, 1993.

[10] M. P. Kennedy, "Basic concepts of nonlinear dynamics and chaos," in C. Toumazou (Ed.), *Circuits and Systems Tutorials*, IEEE Press, London, 1994, pp. 289–313.

[11] M. P. Kennedy, "Bifurcation and chaos," in W. K. Chen (Ed.), *The Circuits and Filters Handbook*, CRC Press, 1995, pp. 1089–1164.

[12] M. P. Kennedy, "Applications of chaos in communications," in D. Docampo, A. R. Figueiras-Vidal, and F. Perez-Gonzalez (Eds.), *Intelligent Methods in Signal Processing and Communications*, Birkhauser, Boston, 1997.

[13] M. P. Kennedy and H. Dedieu, "Experimental demonstration of binary chaos shift keying using self-synchronizing Chua's circuits," in A. C. Davies and W. Schwarz (Eds.), *Proc. of IEEE Int. Specialist Workshop on Nonlinear Dynamics of Electronic Systems*, Dresden, 23–24 July 1993, pp. 67–72.

[14] M. P. Kennedy, G. Kis, Z. Jákó, and G. Kolumbán, "Chaotic communications systems for unlicensed radio," *Proc. of NOLTA'97*, Hawaii, 29 November–3 December 1997, pp. 121–124.

[15] G. Kis, Z. Jákó, M. P. Kennedy, and G. Kolumbán, "Chaotic communications without synchronization," *Proc. of IEE Conference on Telecommunications*, Edinburgh, 29 March-1 April 1998, pp. 49-53.

[16] G. Kolumbán, H. Dedieu, J. Schweizer, J. Ennitis, and B. Vizvári, "Performance evaluation and comparison of chaos communication systems," *Proc. of Int. Workshop on Nonlinear Dynamics of Electronic Systems*, Sevilla, 27-28 June 1996, pp. 105-110.

[17] G. Kolumbán, M. P. Kennedy, and G. Kis, "Determination of symbol duration in chaos-based communications," *Proc. of NDES'97*, Moscow, Russia, 26-27 June 1997, pp. 217-222.

[18] G. Kolumbán, M. P. Kennedy, and L. O. Chua, "The role of synchronization in digital communications using chaos—Part I: Fundamentals of digital communications," *IEEE Trans. on Circ. Sys.*, I, vol. 44, pp. 927-936, 1997.

[19] G. Kolumbán, M. P. Kennedy, and L. O. Chua, "The role of synchronization in digital communications using chaos—Part II: Chaotic modulation and chaotic synchronization," *IEEE Trans. on Circ. Sys.*, I, vol. 45, pp. 1129-1140, 1998.

[20] G. Kolumbán, M. P. Kennedy, and G. Kis, "Multilevel differential chaos shift keying," *Proc. of NDES'97*, Moscow, Russia, 26-27 June 1997, pp. 191-196.

[21] G. Kolumbán, G. Kis, Z. Jákó, and M. P. Kennedy, "FM-DCSK: a robust modulation scheme for chaotic communications," *IEICE Transactions*, vol. E81-A, pp. 1798-1802, 1998.

[22] G. Kolumbán, G. Kis, M. P. Kennedy, and Z. Jákó, "FM-DCSK: a new and robust solution to chaos communications," *Proc. of NOLTA'97*, Hawaii, 29 November-3 December 1997, pp. 117-120.

[23] G. Kolumbán and B. Vizvári. "Nonlinear dynamics and chaotic behaviour of the analog phase-locked loop," *Proc. of Int. Workshop on Nonlinear Dynamics of Electronic Systems*, Dublin, 28-29 July 1995, pp. 99-102.

[24] G. Kolumbán and B. Vizvári, "Direct symbol generation by PLL for the chaos shift keying modulation," *Proc. of ECCTD'95*, Istanbul, 27-31 August 1995, pp. 483-486.

[25] G. Kolumbán, B. Vizvári, W. Schwarz, and A. Abel, "Differential chaos shift keying: A robust coding for chaotic communication," *Proc. of Int. Workshop on Nonlinear Dynamics of Electronic Systems*, Sevilla, 27-28 June 1996, pp. 87-92.

[26] H. Papadopoulos, G. W. Wornell, and A. V. Oppenheim, "Maximum likelihood estimation of a class of chaotic signals," *IEEE Trans. on Information Theory*, vol. 41, 1995.

[27] A. J. Viterbi, "Wireless digital communication: A view based on three lessons learned," *IEEE Communications Magazine*, pp. 33-36, Sept. 1991.

23

Synchronization in Arrays of Coupled Chaotic Circuits and Systems: Theory and Applications

Chai Wah Wu

IBM Thomas J. Watson Research Center
P. O. Box 218
Yorktown Heights, NY 10598 USA
chaiwah@watson.ibm.com

Abstract

This chapter studies synchronization phenomena in arrays of coupled chaotic circuits and oscillators. We show how chaotic synchronization is related to asymptotical stability of circuits and systems. We study synchronization by looking at the underlying coupling topology from a graph-theoretical viewpoint. Both static and dynamic coupling topologies are considered. Extensions to discrete-time systems are discussed. Two main approaches to studying synchronization in chaos are considered: the Lyapunov function approach and the Lyapunov exponents approach. Finally, an application of synchronized circuits to graph coloring is presented.

23.1 Introduction

In this chapter, we study synchronization in arrays of coupled circuits. We assume that the state equations of each circuit are of the form

$$\dot{x}_i = \hat{f}_i(x_i, t) \qquad (23.1.1)$$

where x_i is the state vector of the i-th circuit and assume that the coupled system can be written in the following form:

$$\begin{aligned}
\dot{x}_1 &= f_1(x_1, x_1, x_2, \ldots, x_n, t) \\
\dot{x}_2 &= f_2(x_2, x_1, x_2, \ldots, x_n, t) \\
&\vdots \\
\dot{x}_n &= f_n(x_n, x_1, x_2, \ldots, x_n, t)
\end{aligned} \qquad (23.1.2)$$

where f_i is obtained from $\hat{f}_i$ by adding the coupling between the circuits.

This is however too general a view and practically any array of coupled circuits can be written in this way. We are therefore interested in special cases such as when all the circuits are identical and the coupling is uniform. The underlying connection topology of such coupled circuits can often be expressed as a graph or a hypergraph, and properties of these graphs will be used in studying synchronization. A summary of basic terminology of graph theory is presented in Section 23.2.

We illustrate the relationship between stability and synchronization and consider two methods of studying stability which have been applied to studying synchronization: the use of Lyapunov energy functions and the calculation of Lyapunov exponents. We study both static and dynamic coupling topologies. Finally, we illustrate how synchronized circuits are used to color graphs.

23.2 Notation and Terminology

To make this chapter self-sufficient, we summarize some of the terminology used. A *hypergraph* $\mathcal{G}$ is a pair $\mathcal{G} = (V, E)$ where V is a set of *vertices*, and E is a set of edges where each edge $e \in E$ is a nonempty subset of V. We can think of the vertices in an edge e to be the vertices to which e is connected. The vertex degree of $v \in V$ is the number of edges connected to v. The edge degree of $e \in E$ is the number of elements in e. The (vertex-edge) incidence matrix E of a hypergraph is defined as $E_{ij} = 1$ if $v_i \in e_j$ and 0; otherwise, where v_i and e_j are the i-th vertex and the j-th edge, respectively. If a hypergraph $\mathcal{G}$ has incidence matrix

E, then the *dual* hypergraph of $\mathcal{G}$ (when it exists) has incidence matrix E^T. If all vertices have the same degree, then the hypergraph is *regular*. If all edges have the same degree, then the hypergraph is *uniform*. A *graph* is a uniform hypergraph where all edges have edge degree 2. The adjacency matrix of a graph is defined as $A_{ij} = 1$ if $i \neq j$ and vertex i is adjacent to vertex j and $A_{ij} = 0$ otherwise. The *Laplacian matrix* and the *algebraic connectivity* of a hypergraph with incidence matrix E are defined as $L = D_v - E D_e^{-1} E^T$ and the smallest nonzero eigenvalue of L, respectively, where D_v and D_e are diagonal matrices with the degrees of the vertices and the edges, respectively, on the diagonal [22]. When restricted to graphs, these definitions of Laplacian matrix and algebraic connectivity differ from definitions generally used in the literature [6] by a constant factor of $\frac{1}{2}$. A (vertex) coloring of a hypergraph $\mathcal{G}$ is an assignment of colors to the vertices of $\mathcal{G}$ such that no two adjacent vertices are given the same color. An edge coloring of $\mathcal{G}$ is a coloring of the dual of $\mathcal{G}$. A coloring of $\mathcal{G}$ with k colors is a k-coloring. The chromatic number of a graph is the least number of colors with which it can be colored.

We will mainly work with real matrices in this chapter. A square matrix A is positive definite (semidefinite)[1] if $x^T A x > 0$ ($x^T A x \geq 0$) for all nonzero x. We denote this by $A > 0$ ($A \geq 0$). Similar definitions exist for negative definite (semidefinite). The matrix I denotes the identity matrix whose size is dependent on context.

23.3 Synchronization of Chaotic Circuits and Systems

The topic of this chapter is synchronization of *chaotic* circuits. We assume that we can write state equations for these circuits and that the state equations have a unique solution for each initial condition. Since in general the trajectories are not periodic, we cannot use the classical definitions of synchronization used for coupled periodic oscillators. For chaotic attractors which exhibit a strong periodic component such as in the Rössler system, a generalization of the classical definition is possible [1]. In this chapter, we will use the following definition of synchronization which is commonly used in the literature:

DEFINITION 23.1 *A set of coupled chaotic circuits with state equations Equation (23.1.2) is synchronizing if the states of any two circuits, x_i and x_j, satisfy $\|x_i - x_j\| \to 0$ as $t \to \infty$.*

---

[1]Many authors include the requirement that A is symmetric into this definition. We will not require that A is necessarily symmetric.

For this definition to make sense, the states of the circuits must be of the same dimension. All the results in this chapter are presented as global results. Local versions can be obtained by restricting the states to some neighborhood.[2] There are other definitions of synchronization in the literature [2, 18, 14] which we will not discuss in this chapter.

Recall the definition of sensitive dependence on initial conditions (SDIC), a characteristic feature of chaotic systems: nearby trajectories diverge exponentially as time goes on. In a sense, SDIC can be thought of as the "opposite" of synchronization. Since SDIC is a measure of instability, it is reasonable that synchronization should be related to stability. Indeed, several authors have studied the relationship between Definition 23.1 and asymptotical stability [10, 24, 13] and several methods from control theory to make an unstable system stable have been applied to make chaotic systems synchronizing [19, 9, 11].

DEFINITION 23.2   *A system $\dot{x} = f(x,t)$ is (globally) asymptotically stable if for any two initial states x and y, their trajectories converge to each other as $t \to \infty$, i.e., $\|x(t) - y(t)\| \to 0$ as $t \to \infty$.*

The following theorem then follows trivially from Definition 23.2:

THEOREM 23.1
The coupled system (23.1.2) is synchronizing if $f_1 = f_2 = \ldots = f_n$ and $\dot{x} = f_1(x, \sigma_1(t), \ldots, \sigma_n(t), t)$ is asymptotically stable for all σ_i.

Several special cases of Theorem 23.1 are discussed in [24]. Of particular interest is the case when there are two coupled circuits ($n = 2$) because of its potential applications in communication systems, where one circuit acts as a transmitter and the other acts as a receiver.

Theorem 23.1 suggests the following way to design synchronizing arrays of chaotic circuits. Given a chaotic system $\dot{x} = \hat{f}(x,t)$, if $\hat{f}$ can be rewritten as

$$\hat{f}(x,t) = f(x, x, \ldots, x, t) \tag{23.3.3}$$

such that

$$\dot{x} = f(x, \sigma_1(t), \ldots, \sigma_n(t), t) \text{ is asymptotically stable for all } \sigma_i \tag{23.3.4}$$

---

[2]Note that local definitions of asymptotical stability require the additional condition of being Lyapunov stable. See [24] for more details.

then the coupled array of n chaotic circuits of the form:

$$\begin{aligned} \dot{x}_1 &= f(x_1, x_1, x_2, \ldots, x_n, t) \\ &\vdots \\ \dot{x}_n &= f(x_n, x_1, x_2, \ldots, x_n, t) \end{aligned} \qquad (23.3.5)$$

is synchronizing. Furthermore, at the synchronized state when $x_1(t) = x_2(t) = \ldots = x_n(t)$, the state vector $x_i(t)$ of each circuit follows the trajectory of the uncoupled system $\dot{x} = \hat{f}(x, t)$.

We give here a simple example of how $\hat{f}$ can be rewritten as f. Suppose $g(x, t)$ is a stabilizing feedback term such that $\dot{x} = \hat{f}(x, t) + g(x, t) + u(t)$ is asymptotically stable for all $u(t)$. Then f defined as

$$f(x, x_1, \ldots, x_n, t) = \hat{f}(x, t) + g(x, t) - \sum_{i=1}^{i=n} a_i g(x_i, t)$$

satisfies the conditions in Equation (23.3.3) and Equation (23.3.4) when $\sum_{i=1}^{i=n} a_i = 1$. Such feedback term $g(x, t)$ exists for systems with bounded Jacobian matrices [24]. Note that the case $n = 2$, $a_1 = 1$, $a_2 = 0$ corresponds to unidirectional coupling used in chaotic communication and control systems.

Consider the fully coupled system of Chua's oscillators [4] in Figure 23.1. This can be written in the form of Equation (23.1.2) if we choose

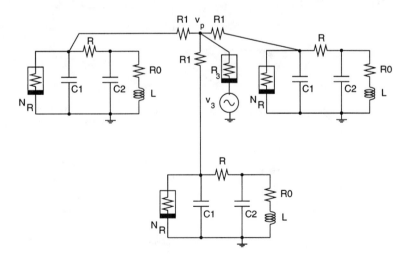

FIGURE 23.1
An array of fully coupled Chua's oscillators where each Chua's oscillator is coupled to every other Chua's oscillator.

$$\hat{f}_i(x) = \begin{cases} \frac{1}{RC_1}((x^2 - x^1) - Rf(x^1)) \\ \frac{1}{RC_2}(x^1 - x^2 + Rx^3) \\ \frac{-1}{L}(x^2 + R_0 x^3) \end{cases} \quad (23.3.6)$$

where $x = (x^1, x^2, x^3)^T$ and

$$f_i(x, v_1, v_2, v_3, t) = \hat{f}_i(x) - \frac{1}{RC_1}(x_1, 0, 0)^T + \frac{1}{RC_1}(v_p(t), 0, 0)^T$$

For circuit parameters such that $R, R_0, R_1, R_3, C_1, C_2, L$ are positive and the nonlinear resistor N_R has a globally Lipschitz characteristic, it can be shown [24] that a small enough R_1 satisfies the conditions of Theorem 23.1 and thus the array is synchronizing.

23.4 Static Coupling

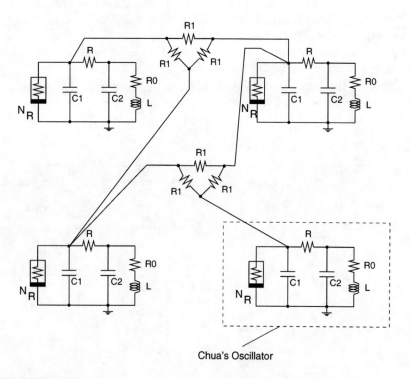

FIGURE 23.2
Array of Chua's oscillators with uniform linear static coupling.

Static Coupling

We distinguish two important classes of coupling: static coupling and dynamic coupling. In static coupling, the coupling terms do not introduce their own dynamics; the coupling elements consist of memoryless devices. An example of statically coupled array of Chua's oscillators is shown in Figure 23.2. The coupling elements consist of two-terminal linear resistors R_1.

23.4.1 Uniform Linear Static Coupling

In this section we will study the class of statically coupled circuits where all the circuits are identical and the coupling is linear and uniform.

A general form for the state equations of uniform linear statically coupled arrays is:

$$\dot{x} = \begin{pmatrix} f(x_1, t) \\ \vdots \\ f(x_m, t) \end{pmatrix} + (G \otimes D)x \qquad (23.4.7)$$

where $x = (x_1, \ldots, x_m)^T$.

The matrix G can be thought of as defining the coupling topology of the array while the matrix D defines the coupling between two circuits in the array. We call this type of coupling *uniform* since the matrix D is the same between any two coupled circuits. The state equations of Figure 23.2 can be written in the form of Equation (23.4.7) if f is given by $\tilde{f}$ in Equation (23.3.6), $D = (\frac{1}{R_1 C_1}, 0, 0)^T$ and

$$G = \begin{pmatrix} -2 & 1 & 1 & 0 \\ 1 & -4 & 2 & 1 \\ 1 & 2 & -4 & 1 \\ 0 & 1 & 1 & -2 \end{pmatrix}$$

DEFINITION 23.3 *Given a matrix V, a function $f(x, t)$ is V-uniformly increasing if V is symmetric positive definite and there exists $c > 0$ such that for all x, y and t:*

$$(x - y)^T V(f(x, t) - f(y, t)) \geq c\|x - y\|^2$$

f is V-uniformly decreasing if $-f$ is V-uniformly increasing. f is uniformly increasing (decreasing) if f is V-uniformly increasing (decreasing) and V is the identity matrix.

By Lyapunov's direct method, $\dot{x} = f(x, t)$ is asymptotically stable if f is V-uniformly decreasing. For the coupled system (23.4.7), the following theorem gives sufficient conditions for synchronization [25]:

THEOREM 23.2
Let T be a matrix such that $f(x,t) - Tx$ is V-uniformly decreasing for some V. Let U be an irreducible symmetric matrix with zero row sums and only nonpositive off-diagonal elements. System (23.4.7) is synchronizing in the sense that $x_i \to x_j$ as $t \to \infty$ if the matrix

$$(U \otimes V)(G \otimes D + I \otimes T) \quad \text{is negative semidefinite.} \quad (23.4.8)$$

The matrix T can be thought of as a stabilizing linear state feedback matrix such that $\dot{x} = f(x,t) - Tx$ is asymptotically stable.

PROOF Construct the Lyapunov function $g(x) = \frac{1}{2}x^T(U \otimes V)x$. It can be shown [26, 25] that U is positive semidefinite and has a zero eigenvalue of multiplicity 1 with eigenvector $(1,\ldots,1)^T$. Therefore, g is zero if $x_i = x_j$ for all i,j and positive elsewhere. The derivative of g along trajectories is

$$\dot{g} = x^T(U \otimes V)\dot{x} = x^T(U \otimes V)\begin{pmatrix} f(x_1,t) - Tx_1 \\ \vdots \\ f(x_m,t) - Tx_m \end{pmatrix}$$

$$+ x^T(U \otimes V)(G \otimes D + I \otimes T)x \quad (23.4.9)$$

It can be shown [26] that U can be decomposed as $M^T M$ where M is a matrix such that each row of M contains zeros and one entry α and one entry $-\alpha$ for some $\alpha \neq 0$. In other words, the first term of Equation (23.4.9) is of the form $\sum \alpha_{ij}^2 (x_i - y_j)^T V(f(x_i,t) - Tx_i - f(x_j,t) - Tx_j)$ which is nonpositive by the V-uniformly decreasing property of $f - T$. In fact, by the irreducibility of U, this term is zero exactly when $x_i = x_j$ for all i,j. Since the second term of Equation (23.4.9) is nonpositive by hypothesis, the theorem is proved by Lyapunov's direct method.

The above theorem is quite general and the condition in 23.4.8 depends on properties of a composite matrix depending on G and D. By making additional assumptions, we can separate the dependence on G from the dependence on D and obtain a condition which depends on the matrix properties of G and of D separately. Since G describes the coupling topology, this in effects decomposes the synchronization condition into a component which depends on G (the coupling topology) and a component which depends on D (the coupling matrix between two circuits). This allows us to study how the synchronization properties change by changing solely the coupling topology or solely the coupling between two circuits.

We consider two cases. In the first case, G is a symmetric irreducible matrix with zero row sums and nonnegative off-diagonal elements. By choosing $U = -G$, Theorem 23.2 is reduced to the following:

COROLLARY 23.3
Let T be a matrix such that $f(x,t) - Tx$ is V-uniformly decreasing for some V. Let G be an irreducible symmetric matrix with zero row sums and only nonnegative off-diagonal elements. System (23.4.7) is synchronizing if for all nonzero eigenvalues λ_i of G, the eigenvalues of $\lambda_i(VD + D^TV) + (VT + T^TV)$ are nonpositive.

By defining S as the set of complex numbers λ such that all eigenvalues of $\lambda(VD + D^TV) + (VT + T^TV)$ are nonpositive, the array is synchronizing if the spectrum of G lies in $S \cup \{0\}$.

In the second case, G is a normal matrix (i.e., $GG^T = G^TG$) such that $G + G^T$ has zero row sums, has nonnegative off-diagonal elements, and is irreducible. We also assume that D is a symmetric matrix which commutes with V. By choosing $U = -(G + G^T)$, Theorem 23.2 is reduced to

COROLLARY 23.4
Let T be a matrix such that $f(x,t) - Tx$ is V-uniformly decreasing for some V. Suppose G is a normal matrix such that $G + G^T$ has zero row sums, has nonnegative off-diagonal elements, and is irreducible. Suppose also that D is a symmetric matrix which commutes with V. System (23.4.7) is synchronizing if for all eigenvalues λ_i of G not on the imaginary axis, the eigenvalues of $2Re(\lambda_i)VD + (VT + T^TV)$ are nonpositive.

In this case, the array is synchronizing if the real parts of the spectrum of G are in $S \cup \{0\}$.

For the special case where $D = \beta T$ for some $\beta > 0$, the coupling is a proportion of the linear stabilizing state feedback T. If $VT + T^TV$ is positive semidefinite, then it is not hard to see that S will contain all sufficiently negative numbers. In other words, the array synchronizes if all the nonzero eigenvalues of G are negative enough. In particular,

COROLLARY 23.5
Let D be a matrix such that $f(x,t) - Dx$ is V-uniformly decreasing for some V. Let G be an irreducible symmetric matrix with zero row sums and only nonnegative off-diagonal elements. If VD is positive semidefinite, then system (23.4.7) is synchronizing if all nonzero eigenvalues of G are less than or equal to -1.

In Figure 23.2, the matrix G is proportional to the Laplacian matrix of the underlying coupling graph. The smallest nonzero eigenvalue of the Laplacian matrix is the algebraic connectivity of the graph and gives a measure of how tightly coupled the graph is. Thus Corollary 23.5 implies that given a fixed coupling matrix D, the synchronization condition is proportional to the algebraic connectivity of the underlying coupling graph; the array is synchronizing if the algebraic connectivity is large enough.

23.4.2 Uniform Nonlinear Static Coupling

The coupling in the array described by (23.4.7) can be made nonlinear by replacing the operator $G \otimes D$ by a nonlinear operator N. In this case Theorem 23.2 is still valid by a corresponding change to Equation (23.4.8): the array is synchronizing if the nonlinear operator $(U \otimes V)(N + I \otimes T)$ is negative semidefinite.[3]

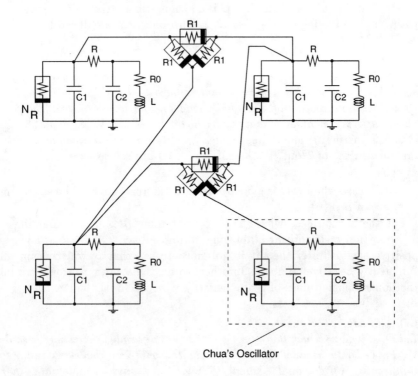

FIGURE 23.3
Array of Chua's oscillators coupled with nonlinear resistors R_1.

---

[3] A nonlinear operator $L : R^n \to R^n$ is negative semidefinite if $x^T L(x) \leq 0$ for all x.

For example, consider the array of Chua's oscillator coupled by nonlinear resistors shown in Figure 23.3. The state equations can be written in the form

$$\dot{x} = \begin{pmatrix} f(x_1, t) \\ \vdots \\ f(x_m, t) \end{pmatrix} + (I \otimes D)(M \otimes I)(I \otimes g)(M^T \otimes I)x \qquad (23.4.10)$$

where $(g(v_1), \cdots, g(v_n))^T$ is denoted as $(I \otimes g)v$, f is given by $\hat{f}$ in Equation (23.3.6),

$$M = \begin{pmatrix} 1 & 1 & 0 & 0 & 0 & 0 \\ -1 & 0 & 1 & -1 & 0 & 1 \\ 0 & -1 & -1 & 1 & 1 & 0 \\ 0 & 0 & 0 & 0 & -1 & -1 \end{pmatrix}$$

$g(v) = (-g_v(v_1), 0, 0))^T$ and g_v is the v-i characteristic of the nonlinear resistor R_1.

It can be shown that for circuit parameters where all the linear components are passive and the nonlinear resistors N_R have global Lipschitz characteristics, if R_1 is sufficiently passive in the sense that $g_v(v)v \geq Gv^2$ for a large enough G, then the array is synchronizing.

More general types of coupling and extensions to hypergraphs can be found in [26] and [22], respectively.

23.5 Dynamic Coupling

In dynamic coupling, the coupling elements can contain dynamic circuit elements and can have dynamics of their own. In this case, we distinguish two types of systems, the *cell* system and the *coupling* system.

23.5.1 Cell Systems

We consider the special case where the cell systems are not coupled to each other directly and similarly for the coupling systems. In other words, the coupling systems are used to couple several cell systems. For example, consider the array of Chua's oscillators coupled via relaxation oscillators shown in Figure 23.4. Such an array of cell systems coupled via the coupling systems is said to be *synchronizing* if the states of the cell systems converge to each other.

In some cases, Theorem 23.1 can still be applied. For example, consider the array of Chua's oscillators coupled via dynamic coupling elements as shown in Figure 23.5.

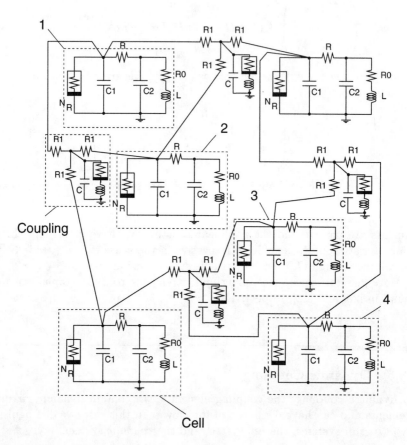

FIGURE 23.4
Array of Chua's oscillators (cell) coupled linearly via nonlinear oscillators (coupling).

By replacing the one-ports indicated by P_1 and P_2 with a corresponding time-varying voltage source and applying the substitution theorem [3], the state equations can be written as Equation (23.3.5) and for linear passive elements and globally Lipschitz N_R, the array is synchronizing when the resistance of R_1 is small enough.

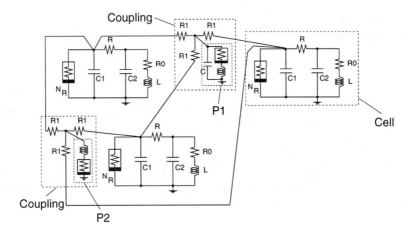

FIGURE 23.5
Array of Chua's oscillators coupled linearly via nonlinear oscillators for which Theorem 23.1 can be applied.

In more general cases, assume that the state equations can be written in the form:

$$\begin{aligned}
\dot{x}_1 &= f_1(x_1, y_1, y_2, \ldots, y_m, t) \\
&\vdots \\
\dot{x}_n &= f_n(x_n, y_1, y_2, \ldots, y_m, t) \\
\dot{y}_1 &= g_1(y_1, x_1, x_2, \ldots, x_n, t) \\
&\vdots \\
\dot{y}_m &= g_m(y_m, x_1, x_2, \ldots, x_n, t)
\end{aligned} \quad (23.5.11)$$

where x_i are the states of the cell systems and y_i are the states of the coupling systems.

DEFINITION 23.4 *System $\dot{x} = f(x, u(t), t)$ is said to be (globally) u- asymptotically stable if for $x_1(t)$ and $x_2(t)$ trajectories of $\dot{x} = f(x, u_1(t), t)$ and $\dot{x} = f(x, u_2(t), t)$, respectively, and $u_1 \to u$, $u_2 \to u$ as $t \to \infty$, we have $x_1 \to x_2$ as $t \to \infty$.*

Note that u-asymptotically stability implies asymptotical stability. Systems admitting certain quadratic Lyapunov functions are u-asymptotically stable [26].

An analog of Theorem 23.1 is as follows:

THEOREM 23.6
The coupled system

$$\begin{aligned}
\dot{x}_1 &= f_1(x_1, y_{s_{11}}, y_{s_{12}}, \ldots, y_{s_{1k}}, t) \\
&\vdots \\
\dot{x}_n &= f_n(x_n, y_{s_{n1}}, y_{s_{n2}}, \ldots, y_{s_{nk}}, t) \\
\dot{y}_1 &= g_1(y_1, x_{t_1}, \ldots, x_{t_l}, t) \\
&\vdots \\
\dot{y}_m &= g_m(y_m, x_{t_1}, \ldots, x_{t_l}, t)
\end{aligned} \qquad (23.5.12)$$

where $s_{ij} \in \{1, \ldots, m\}$ and $t_i \in \{1, \ldots, n\}$ is synchronizing in the sense that $|x_i(t) - x_j(t)| \to 0$ as $t \to \infty$ if

1. $f_1 = f_2 = \ldots = f_n$,
2. $\dot{x} = f_1(x, \sigma_1(t), \ldots, \sigma_k(t), t)$ is u-asymptotically stable for all σ_i's,
3. $g_1 = g_2 = \ldots = g_m$,
4. $\dot{y} = g_1(y, \eta_1(t), \ldots, \eta_l(t), t)$ is asymptotically stable for all η_i's.

If the conditions of Theorem 23.6 are satisfied, not only are the cell systems' trajectories converging to each other, but the coupling systems' trajectories are also converging to each other, i.e., $|y_i(t) - y_j(t)| \to 0$ as $t \to \infty$.

For example, consider the coupled array of Chua's oscillators coupled with relaxation oscillators in Figure 23.6. It is not hard to show that the state equations can be written in the form (23.5.12). For passive linear circuit elements and (active) nonlinear resistors which are globally Lipschitz, it can be shown that for a small enough R_1, the conditions of Theorem 23.6 are satisfied and thus the array is synchronizing.

Notice that the roles of cell systems and coupling systems can be swapped (i.e., changing x_i into y_i in Equation (23.5.11) and vice versa) resulting in a *dual* coupled system.

By considering the cell systems as vertices and coupling systems as edges, the underlying connection topology can be represented by a hypergraph. The dual coupled system would correspond to a dual hypergraph. For example the hypergraph corresponding to Figure 23.4 is shown in Figure 23.7 with incidence matrix

$$\begin{pmatrix} 1 & 1 & 0 & 0 \\ 1 & 0 & 1 & 0 \\ 1 & 1 & 0 & 0 \\ 0 & 0 & 1 & 1 \\ 0 & 1 & 0 & 1 \\ 0 & 0 & 1 & 1 \end{pmatrix} \qquad (23.5.13)$$

Dynamic Coupling

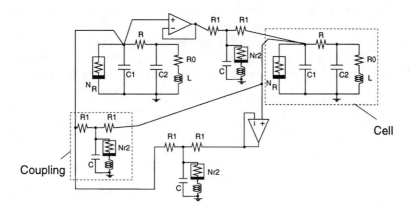

FIGURE 23.6
An array of Chua's oscillators coupled via relaxation oscillators to illustrate the application of Theorem 23.6.

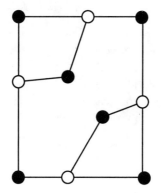

FIGURE 23.7
Hypergraph corresponding to the coupled array in Figure 23.4. The closed circles represent vertices while the open circles represent edges.

23.5.2 Uniform Linear Dynamic Coupling on Regular and Uniform Hypergraphs

Consider an array of uniform linear dynamically coupled circuits where the underlying hypergraph is regular and uniform. We can then write the Laplacian matrix of the hypergraph with incidence matrix E as $\alpha I - \frac{1}{\beta}EE^T$ with α and β being the vertex and edge degrees, respectively. Similar to Equation (23.4.7), a general form for the state equations is given by

$$\dot{x} = \begin{pmatrix} g(x_1) \\ \vdots \\ g(x_n) \end{pmatrix} + (E \otimes \gamma_1 T)y - (\alpha I \otimes \gamma_1 \phi(T))x$$

$$\dot{y} = \begin{pmatrix} h(y_1) \\ \vdots \\ h(y_m) \end{pmatrix} + (E^T \otimes \gamma_2 T^T)x - (\beta I \otimes \gamma_2 \phi(T^T))y \qquad (23.5.14)$$

where T is a matrix with nonnegative entries on the main diagonal and zeros everywhere else and $\phi(T)$ is a diagonal square matrix with the same number of rows as T and the row sums of T on the diagonal. The constants $\gamma_1 > 0$ and $\gamma_2 > 0$ indicate coupling strengths.

THEOREM 23.7 [22]
The coupled system (23.5.14) is synchronizing if there exists constants $\delta_1 > 0$, $\delta_2 > 0$, such that $\delta_1 \phi(T) - g$ and $\delta_2 \phi(T^T) - h$ are uniformly increasing and there exists $a_1 > 0$ and $a_2 > 0$ such that $\sigma_1 \geq \frac{\left(a_1 + \frac{\delta_1}{\gamma_1} - \alpha\right)\alpha a_2}{\beta^2} + \alpha$ and $\sigma_2 \geq \frac{\left(a_2 + \frac{\delta_2}{\gamma_2} - \beta\right)\beta a_1}{\alpha^2} + \beta$ where σ_1 and σ_2 are the algebraic connectivities of the hypergraph and its dual respectively.

Again, the conditions of this theorem also imply that the states of the coupling systems converge to each other.

For example, the state equations of the coupled array in Figure 23.4 can be written in the form of Equation (23.5.14) with $\alpha = 2$, $\beta = 3$, $\gamma_1 = \frac{1}{R_1 C_1}$, $\gamma_2 = \frac{1}{R_1 C}$, $T = (1, 0, 0)^T$, and E as in Equation (23.5.13).

23.6 Discrete-Time Systems

When the chaotic systems are modeled as discrete-time systems of the form[4] $x(k+1) = f(x(k), k)$, we obtain the following state equations:

---
[4]An alternative notation is $x(k+1) = f^{(k)}(x(k))$.

$$\begin{aligned} x_1(k+1) &= f_1(x_1, x_1, x_2, \ldots, x_n, k) \\ x_2(k+1) &= f_2(x_2, x_1, x_2, \ldots, x_n, k) \\ &\vdots \\ x_n(k+1) &= f_n(x_n, x_1, x_2, \ldots, x_n, k) \end{aligned} \qquad (23.6.15)$$

There are corresponding statements to Definition 23.2 and Theorem 23.1:

DEFINITION 23.5 *A system $x(k+1) = f(x(k), k)$ is (globally) asymptotically stable if for any two initial states x and y, their trajectories converge to each other as $k \to \infty$, i.e., $\|x(k) - y(k)\| \to 0$ as $k \to \infty$.*

THEOREM 23.8
The coupled system (23.6.15) is synchronizing if $f_1 = f_2 = \ldots = f_n$ and $x(k+1) = f_1(x(k), \sigma_1(k), \ldots, \sigma_n(k), k)$ is asymptotically stable for all σ_i.

DEFINITION 23.6 *For a zero row sum matrix G, let $L(G)$ denote the set of eigenvalues of G which do not correspond to the eigenvector $(1, \ldots, 1)^T$.*

In particular, if G has a zero eigenvalue of multiplicity 1, then $0 \notin L(G)$.

Similar to Theorem 23.2, synchronization in an array of coupled discrete-time systems can be deduced from the eigenvalues of the coupling matrix

THEOREM 23.9
Consider the following array of coupled discrete-time systems:

$$x(k+1) = \begin{pmatrix} x_1(k+1) \\ \vdots \\ x_n(k+1) \end{pmatrix} = (I - G \otimes I) \begin{pmatrix} f(x_1(k), k) \\ \vdots \\ f(x_n(k), k) \end{pmatrix}$$
$$= (I - G \otimes I) F(x(k), k) \qquad (23.6.16)$$

where G is a normal $n \times n$ matrix with zero row sums.

Let $c = \sup_{x,k} \|Df(\cdot, k)\|$. If $|1 - \lambda| < \frac{1}{c}$ for every eigenvalue λ of G in $L(G)$, then the coupled system (23.6.16) is synchronizing, i.e., $x_i(k) \to x_j(k)$ for all i, j as $k \to \infty$.

A graphical interpretation is that the coupled system is synchronizing if $L(G)$ lies in the interior of a circle of radius $\frac{1}{c}$ centered at 1 in the complex plane. For functions f which are not differentiable but Lipschitz continuous, c is defined as $\sup_{x \neq y, k} \frac{\|f(x,k) - f(y,k)\|}{\|x - y\|}$.

PROOF By normality of G, $G \otimes I$ has an orthonormal set of eigenvectors. In particular, they are of the form $g_i \otimes e_j$, where g_i and e_j are eigenvectors of G and I, respectively. Denote A as the subspace of vectors of the form $(1,\ldots,1)^T \otimes v$. It is easy to see that A is spanned by eigenvectors of $G \otimes I$ (since $(1,\ldots,1)^T$ is an eigenvector of G) and that A is in the kernel of $G \otimes I$. Let use denote the subspace orthogonal to A by B. Decompose $x(k)$ as $x(k) = y(k) + z(k)$ where $y(k) \in A$ and $z(k) \in B$. By the Mean Value Theorem $\|F(x(k),k) - F(y(k),k)\| \leq c\|z(k)\|$. We can decompose $F(x(k),k) - F(y(k),k)$ as $F(x(k),k) - F(y(k),k) = a(k) + b(k)$ where $a(k) \in A$ and $b(k) \in B$. Note that $F(y(k),k)$ and $a(k)$ are in A and therefore $(G \otimes I)F(y(k),k) = (G \otimes I)a(k) = 0$.

$$x(k+1) = (I - G \otimes I)F(x(k),k) = (I - G \otimes I)(a(k) + b(k)) + F(y(k),k)$$
$$= a(k) + F(y(k),k) + (I - G \otimes I)b(k)$$

Because of the orthogonality of $a(k)$ and $b(k)$, $\|b(k)\| \leq \|F(x(k),k) - F(y(k),k)\| \leq c\|z(k)\|$. Since $b(t) \in B$, $\|(I - G \otimes I)b(k)\| \leq |1 - \lambda|\|b(k)\| < \frac{\|b(k)\|}{c} \leq \|z(k)\|$ for the smallest nonzero eigenvalue λ of G. Since $a(k) + F(y(k),k)$ is in A, $z(k+1)$ is the orthogonal projection of $(I - G \otimes I)b(k)$ onto B and thus $\|z(k+1)\| \leq \|(I - G \otimes I)b(k)\| < \|z(k)\|$. In fact, $\|z(k+1)\|/\|z(k)\|$ is bounded away from 1 and thus $z(k) \to 0$ as $k \to \infty$. This means that $x(k)$ approaches A, the synchronization manifold.

When G is symmetric, all the eigenvalues of G are real, and we have the following corollary:

COROLLARY 23.10

If G is a symmetric matrix with zero row sums and a zero eigenvalue of multiplicity 1 and the nonzero eigenvalues of G are in the interval $(1 - \frac{1}{c}, 1 + \frac{1}{c})$ then the coupled system (23.6.16) is synchronizing.

When the individual systems are autonomous (i.e., f does not depend on k) and can be represented as maps, they are generally known as Coupled Map Lattices [12].

Note the difference between Corollary 23.5 and Corollary 23.10. In the continuous-time case, the synchronization condition is a condition on the smallest (in magnitude) nonzero eigenvalue of G, while in the discrete-time case, the synchronization condition is a condition on both the smallest and the largest nonzero eigenvalues of G.

Consider the special case where the individual systems lie on the nodes of a graph $\mathcal{G}$ and a uniform coupling exists between two systems if and only if there is an edge between the two corresponding nodes. The equation for the i-th system is given by:

$$x_i(k+1) = \left(1 - \epsilon \sum_j a_{ij}\right) f(x_i(k), k) + \epsilon \sum_j (a_{ij} f(x_j(k), k)) \quad (23.6.17)$$

The incidence matrix and the Laplacian matrix of the graph $\mathcal{G}$ are given by $A = \{a_{ij}\}$ and $\frac{1}{2}(V - A)$, respectively, where V is the diagonal matrix with the vertex degrees on the diagonal.

Corollary 23.10 can be used to prove the following corollary [21]:

COROLLARY 23.11
Let the graph $\mathcal{G}$ be connected and let α and β be the largest and the smallest nonzero eigenvalues of the Laplacian matrix of the graph, respectively. If $\frac{\alpha}{\beta} < \frac{2}{1-\frac{1}{c}} - 1$, then the system in Equation (23.6.17) is synchronizing with $\epsilon = \frac{1}{\alpha+\beta}$.

For a graph with n vertices constructed by connecting each vertex randomly to k vertices, with high probability the ratio $\frac{\alpha}{\beta}$ is approximately $1 + \frac{4}{\sqrt{k}}$ for even k and large n [7]. By Corollary 23.11, for finite c such graphs can be made synchronizing for large enough k by choosing an appropriate ϵ. This is true even when the ratio $\frac{k}{n}$ is arbitrarily small. Since a small $\frac{k}{n}$ indicates a sparsely coupled array, this shows that arbitrarily sparsely coupled array of discrete-time systems can be made synchronizing when the cells are coupled randomly and the array is large enough. This is in contrast to the continuous-time case (Equation (23.4.7)) which by Corollary 23.5 can always be made synchronizing by choosing the appropriate D if f has bounded Jacobian matrices and the underlying graph is connected.

23.7 Lyapunov Exponents Approach

While the Lyapunov function approach used so far can give mathematically rigorous results, it suffers from one drawback: there are no systematic ways for finding Lyapunov functions for all classes of circuits. On the other hand, the Lyapunov exponents approach [8, 15, 16], even though less rigorous and require numerical computations and approximations can be applied to a much larger class of systems. The main difference between the two approaches is that instead of using a Lyapunov function (Lyapunov's second method or direct method) to prove asymptotical stability, the Lyapunov exponents approach calculates Lyapunov exponents (when they exist) of trajectories to test for asymptotical stability. The basic idea is as follows:

if the Lyapunov exponents of some trajectory of $\dot{x} = f(x,t)$ are all negative, then $\dot{x} = f(x,t)$ is asymptotically stable (locally) near this trajectory.

Most of the results discussed so far have a counterpart using the Lyapunov exponents approach.

23.7.1 Continuous-Time Systems

Consider the state equations of an array of coupled circuits given by

$$\dot{x} = (I + p_1(G) \otimes D_1)F((I + p_2(G) \otimes D_2)x) + (p_3(G) \otimes D_3)x \quad (23.7.18)$$

where $F(x) = (f(x_1), \ldots, f(x_n))^T$ and p_i are polynomials such that $p_i(0) = 0$ and G is a zero row sum diagonalizable matrix.

In the Lyapunov exponents approach, synchronization is deduced from the Lyapunov exponents transverse to the synchronization subspace (we will simply call these the transverse Lyapunov exponents) which are calculated for a trajectory on the synchronization subspace. In particular, we conclude that the system is synchronizing if the transverse Lyapunov exponents are negative for a trajectory on the synchronization subspace.

For such a trajectory $x(t)$, the Jacobian matrix is given by

$$J(t) = (I + p_1(G) \otimes D_1)(I \otimes Df(t))(I + p_2(G) \otimes D_2) + p_3(G) \otimes D_3$$

If G is diagonalizable to Λ, then the variational equation $\dot{x} = J(t)x$ is linearly conjugate to

$$\dot{x} = [(I + p_1(\Lambda) \otimes D_1)(I \otimes Df(t))(I + p_2(\Lambda) \otimes D_2) + p_3(\Lambda) \otimes D_3]x$$

The matrix

$$(I + p_1(\Lambda) \otimes D_1)(I \otimes Df(t))(I + p_2(\Lambda) \otimes D_2) + p_3(\Lambda) \otimes D_3$$

is block-diagonal with the blocks equal to

$$H(\mu,t) = (I + p_1(\mu)D_1)Df(t)(I + p_2(\mu)D_2) + p_3(\mu)D_3$$

where μ ranges over the eigenvalues of G. The eigenvector $(1,\ldots,1)^T$ of G corresponds to the synchronization subspace and thus the transverse Lyapunov exponents can be found from the Lyapunov exponents of $\dot{x} = H(\lambda,t)x$ for each eigenvalue λ of G in $L(G)$ (Definition 23.6). Analogous to the set S in Section 23.4.1, we can define $\tilde{S}$ as the set of λ such that all Lyapunov exponents of $\dot{x} = H(\lambda,t)x$ are negative. Then the coupled system is synchronizing if $L(G) \subset \tilde{S}$.

If G has a zero eigenvalue of multiplicity larger than 1, the Lyapunov exponents of $\dot{x} = f(x)$ form a subset of the transverse Lyapunov exponents. This follows from the facts that $0 \in L(G)$ and $H(0,t) = Df(t)$. Thus, if in addition f has positive Lyapunov exponents then the array will not synchronize.

Special cases of interest are when two of the polynomials p_i are zero. For example, [17] considers the case when $p_1 = p_2 = 0$ and p_3 is the identity. In this case $H(\lambda, t)$ is simplified to $Df(t) + \lambda D_3$. Generally the relationship between the Lyapunov exponents of $\dot{x} = Df(t)x$ and the Lyapunov exponents of $\dot{x} = (Df(t) + \lambda D_3)x$ is nontrivial. However, when D_3 is the identity matrix, it can be shown that the Lyapunov exponents of $\dot{x} = Df(t)x$ and the Lyapunov exponents of $\dot{x} = (Df(t) + \lambda I)x$ differ by the constant λ.

23.7.2 Discrete-Time Systems

Similarly, the Lyapunov exponents approach can be used to infer synchronization in discrete-time systems.

Consider the following coupled discrete time system:

$$x(k+1) = (I + p_1(G) \otimes D_1)F_k((I + p_2(G) \otimes D_2)x(k)) + (p_3(G) \otimes D_3)x(k) \tag{23.7.19}$$

where $F_k(x) = (f_k(x_1), \ldots, f_k(x_n))^T$, p_i are polynomials such that $p_i(0) = 0$ and G is a diagonalizable zero row sum matrix.

For a trajectory $x(k) = (x_1(k), x_1(k), \ldots x_1(k))^T$ of the coupled system on the synchronization manifold, the Lyapunov exponents are given by

$$\lim_{n \to \infty} \ln |\lambda_i(n)|^{\frac{1}{n}} \tag{23.7.20}$$

where $\lambda_i(n)$ are the eigenvalues of $\Pi_{k=0}^{k=n-1} J(k)$ and $J(k) = (I + p_1(G) \otimes D_1)(I \otimes Df_k(x_1(k)))(I + p_2(G) \otimes D_2) + p_3(G) \otimes D_3$. If C is a matrix which diagonalizes G, then $(C \otimes I)J(k)(C^{-1} \otimes I)$ is block-diagonal and hence $(C \otimes I)\Pi_{k=0}^{k=n-1} J(k)(C^{-1} \otimes I)$ is block-diagonal and the eigenvalues of $\Pi_{k=0}^{k=n-1} J(k)$ are given by the eigenvalues of $H(\mu, n)$, where μ varies over the eigenvalues of G and $H(\mu, n)$ is defined as

$$H(\mu, n) = \Pi_{k=0}^{k=n-1} \left((I + p_1(\mu)D_1)Df(x_1(k))(I + p_2(\mu)D_2) + p_3(\mu)D_3 \right) \tag{23.7.21}$$

The synchronization manifold A (Definition 23.7) is invariant under Equation (23.7.19) and are eigenvectors of $\Pi_{k=0}^{k=n-1} J(k)$. Therefore, the Lyapunov exponents of the trajectory $x(k)$ transverse to A are given by Equation (23.7.20) where λ_i are the eigenvalues of $H(\mu, n)$ for each $\mu \in L(G)$ (Definition 23.6).

Similar to the continuous-time case, if G has a zero eigenvalue of multiplicity larger than 1, the Lyapunov exponents of f_k will be a subset of the transverse Lyapunov exponents, and thus if f_k has a positive Lyapunov exponent, then the coupled array will not synchronize. This follows from the facts that $0 \in L(G)$ and $H(0, n) = \Pi_{k=0}^{k=n-1} Df_k(x_1(k))$.

The expression for $H(\mu, n)$ can be simplified significantly if $p_3 = 0$ and the discrete-time systems are scalar systems, i.e., D_i are scalars and f_k

are real-valued functions of a single variable. In this case, $H(\mu, n) = (1 + p_1(\mu)D_1 + p_2(\mu)D_2 + p_1(\mu)p_2(\mu)D_1D_2)^n \Pi_{k=0}^{k=n-1} f'_k(x_1(k))$. Let $q(\mu) = \ln|1 + p_1(\mu)D_1 + p_2(\mu)D_2 + p_1(\mu)p_2(\mu)D_1D_2|$. In particular, the transverse Lyapunov exponents are $q(\mu) + \lambda_i$ where $\mu \in L(G)$ and λ_i are the Lyapunov exponents of f_k. Therefore, the coupled array is synchronizing if $-q(\mu)$ is bigger than the largest Lyapunov exponent of f_k for all $\mu \in L(G)$.

In [5] this approach was used to study synchronization in the case of arrays of coupled scalar maps where $p_1 = p_3 = 0$, p_2 is the identity, and D_2 is a scalar, in which case $q(\mu) = \ln|1 + \mu D_2|$.

23.8 Dynamics at Synchronization

We define the synchronization subspace as follows:

DEFINITION 23.7 *The* synchronization subspace *corresponding to* (23.1.2) *is the linear subspace of vectors of the form* $(x, ..., x)$.

From Definition 23.1, as $t \to \infty$, the trajectory of a synchronizing coupled system (23.1.2) converges toward a (not necessarily unique) solution in the synchronization subspace of the form $x_s(t) = (x(t), \ldots, x(t))$. Some of these $x_s(t)$ are *not* necessarily a solution of (23.1.2). However, in the static linear coupling case (23.4.7), a $x_s(t)$ exists which is also a solution of Equation (23.4.7). Furthermore, $x_s(t) = (x(t), \ldots, x(t))$ is given by $x(t)$ being a trajectory of the uncoupled individual system $\dot{x}_i = f(x_i, t)$. This is true since G has zero row sums and thus the synchronization subspace is in the nullspace of the linear operator $G \otimes D$, i.e., the coupling term $(G \otimes D)x$ is zero for any solution in the synchronization subspace. This is also true for Equation (23.7.18) and Equation (23.7.19).

When G has row sums μ for all rows, a simple substitution $f(x, t) \to f(x, t) - \mu D x$ will result in a matrix G with zero row sums.

In the case of dynamic coupling, because of the dynamics of the coupling systems, the cell systems' trajectories do not necessarily converge toward trajectories of uncoupled cell systems.

23.9 Synchronization of Clusters

So far we have discussed synchronization where *all* the circuits are identical and they are all synchronizing to each other. In some numerical studies of coupled systems, the object of interest is when they are not all synchronized, but the circuits form groups where circuits within a group are synchronizing to each other, but synchronization does not necessarily occur between circuits from different groups. This partitioning into groups can change with time and the coupled systems can form spatio-temporal patterns and wave phenomena.

We call the synchronization where the circuits are partitioned into groups and circuits within groups are synchronized *clustered synchronization*. Note that the dynamic coupling case studied in Section 23.5 is a special case of clustered synchronization with 2 clusters: a cluster consisting of circuits and a cluster consisting of the coupling elements.

Theorem 23.1 can be generalized to include clustered synchronization.

THEOREM 23.12
Consider the coupled array of circuits written in the form

$$
\begin{aligned}
\dot{x}_{11} &= f_1(x_{11}, x_{11}, \ldots, x_{kl}, \ldots, t) \\
\dot{x}_{12} &= f_1(x_{12}, x_{11}, \ldots, x_{kl}, \ldots, t) \\
&\vdots \\
\dot{x}_{1,n_1} &= f_1(x_{1,n_1}, x_{11}, \ldots, x_{kl}, \ldots, t) \\
\dot{x}_{21} &= f_2(x_{21}, x_{11}, \ldots, x_{kl}, \ldots, t) \\
\dot{x}_{22} &= f_2(x_{22}, x_{11}, \ldots, x_{kl}, \ldots, t) \\
&\vdots \\
\dot{x}_{2,n_2} &= f_2(x_{2,n_2}, x_{11}, \ldots, x_{kl}, \ldots, t) \\
&\vdots \\
\dot{x}_{i1} &= f_i(x_{i1}, x_{11}, \ldots, x_{kl}, \ldots, t) \\
&\vdots \\
\dot{x}_{i,n_i} &= f_i(x_{i,n_i}, x_{11}, \ldots, x_{kl}, \ldots, t) \\
&\vdots
\end{aligned}
\tag{23.9.22}
$$

If $\dot{x} = f_i(x, \sigma_1, \sigma_2, \ldots, t)$ is asymptotically stable for all σ_j, then $x_{ij} \to x_{ik}$ as $t \to \infty$.

What the above theorem says is that the circuits are partitioned into clusters $S_i = \{x_{i1}, \ldots, x_{i,n_i}\}$ and circuits within each cluster are synchronizing. It is left as an exercise for the reader to show that Chua's oscillator 1 synchronizes with Chua's oscillator 2 and Chua's oscillator 3 synchronizes with Chua's oscillator 4 in Figure 23.4 for small enough R_1 when N_R has a globally Lipschitz characteristic and all the linear elements are passive.

A generalization of Theorem 23.6 to this case is also possible.

THEOREM 23.13
Consider the coupled array of circuits written in the form

$$
\begin{aligned}
\dot{x}_1 &= f_1(x_1, x_1, \ldots, x_{n_m}, t) \\
\dot{x}_2 &= f_1(x_2, x_1, \ldots, x_{n_m}, t) \\
&\vdots \\
\dot{x}_{n_1} &= f_1(x_{n_1}, x_1, \ldots, x_{n_m}, t) \\
\dot{x}_{n_1+1} &= f_2(x_{n_1+1}, x_{s_1^{n_1+1}}, \ldots, x_{s_{k_2}^{n_1+1}}, t) \\
&\vdots \\
\dot{x}_{n_2} &= f_2(x_{n_2}, x_{s_1^{n_2}}, \ldots, x_{s_{k_2}^{n_2}}, t) \quad\quad (23.9.23)\\
&\vdots \\
\dot{x}_{n_{m-1}} &= f_{m-1}(x_{n_{m-1}}, x_{s_1^{n_{m-1}}}, \ldots, x_{s_{k_{m-1}}^{n_{m-1}}}, t) \\
\dot{x}_{n_{m-1}+1} &= f_m(x_{n_{m-1}+1}, x_{s_1^{n_{m-1}+1}}, \ldots, x_{s_{k_m}^{n_{m-1}+1}}, t) \\
&\vdots \\
\dot{x}_{n_m} &= f_m(x_{n_m}, x_{s_1^{n_m}}, \ldots, x_{s_{k_m}^{n_m}}, t)
\end{aligned}
$$

where $0 = n_0 < n_1 < n_2 < \ldots < n_m$ *and* $S_l = \{n_l + 1, \ldots n_{l+1}\}$. *The indices* s_j^i *are chosen such that* $s_j^i \in \{1, \ldots, n_l\}$ *if* $i \in S_l$.

If $\dot{x} = f_1(x, \sigma_1, \sigma_2, \ldots, t)$ *is asymptotically stable for all* σ_i *and* $\dot{x} = f_k(x, \eta_1, \eta_2, \ldots, t)$ *is u-asymptotically stable for all* η_i *and all* $k > 1$, *then* $x_a \to x_b$ *as* $t \to \infty$ *for a and b in the same set* S_l.

The synchronization in the above theorem occurs in m clusters with indices of the states in each cluster belonging to some S_l.

23.10 Applications to Graph Coloring

In this section, we show how synchronized array of circuits can be used to find a good coloring of graphs. As discussed earlier, the coupling topology of an array of circuits can be thought of as graphs with the circuits representing vertices. In Equation (23.4.7), an edge exists between vertex i and vertex j if $G_{ij} \neq 0$. In particular, the adjacency matrix A of the underlying graph is defined as

$$A_{ij} = \begin{cases} 1 & i \neq j, G_{ij} \neq 0 \\ 0 & \text{otherwise} \end{cases}$$

The following theorem gives conditions when the coupled array synchronizes into two clusters with opposite phases. By assigning two colors to the two phases we obtain a 2-coloring of the underlying graph.

THEOREM 23.14
Suppose that the following conditions are satisfied in Equation (23.4.7):

1. $G = -\alpha(A + \text{diag}(v(i)))$, *where $\alpha > 0$ and A is the adjacency matrix of the underlying graph and $\text{diag}(v(i))$ is the diagonal matrix with the degrees of the vertices on the diagonal,*
2. $f(x, t) = -f(-x, t)$ *for all x, t,*
3. *there exists matrices D and V such that $VD \geq 0$ and $f(x, t) - Dx$ is V-uniformly decreasing.*
4. *The underlying graph is connected and 2-colorable.*
5. *The algebraic connectivity of the underlying graph is larger than or equal to $\frac{1}{2\alpha}$.*

Then if x_i and x_j are adjacent in the underlying graph, they are synchronizing out of phase, i.e., $x_i \to -x_j$ as $t \to \infty$. In other words, the phases of the oscillators generate a 2-coloring of the graph.

PROOF For a 2-coloring of the graph, let B be the diagonal matrix such that B_{ii} is equal to 1 or -1 depending on the color of the i-th vertex. Since $A_{ij} \neq 0$ implies $B_{ii} = -B_{jj}$, we have $(BAB)_{ij} = B_{ii}A_{ij}B_{jj} = -A_{ij}$. Therefore, $BAB + B\text{diag}(v(i))B = -A + \text{diag}(v(i))$. Using the state transformation $x \to (B \otimes I)x$ and the fact that $f(-x, t) = -f(x, t)$, the state equation (23.4.7) can be rewritten as

$$\dot{x} = \begin{pmatrix} f(x_1, t) \\ \vdots \\ f(x_m, t) \end{pmatrix} + (B \otimes I)(G \otimes D)(B \otimes I)x \qquad (23.10.24)$$

The matrix $(B \otimes I)(G \otimes D)(B \otimes I)$ is equal to $BGB \otimes D = (-\alpha(\text{diag}(v(i)) - A)) \otimes D$. The algebraic connectivity condition implies that all the nonzero eigenvalues of $-\alpha(\text{diag}(v(i)) - A)$ are less than or equal to -1 and thus the conclusion follows from Corollary 23.5.

For an example of a coupled array of relaxation oscillators satisfying these conditions, see [23]. For graphs with chromatic numbers greater than

2, simulation results with small graphs indicate that the array in Equation (23.4.7) with

$$f\left(\begin{pmatrix} y \\ z \end{pmatrix}, t\right) = \begin{pmatrix} -z + 5\left(y - \frac{y^3}{3}\right) \\ y \end{pmatrix}, \quad D = \begin{pmatrix} 0 & 0 \\ 0 & 1 \end{pmatrix}$$

$$G = -0.21 \text{diag}(1/v(i))(A + \text{diag}(v(i)))$$

colors the graph with the minimal number of colors most of the time [23] and that the phases of the oscillators in the above array can be used to approximate the star chromatic number [20].

References

[1] V. S. Anishchenko, M. A. Safonova, and L. O. Chua, "Stochastic resonance in Chua's circuit," *Int. J. of Bifur. Chaos*, vol. 2, pp. 397–401, 1992.

[2] V. S. Afraimovich, N. N. Verichev, and M. I. Rabinovich, "Stochastically synchronized oscillations in dissipative systems," *Radiophysics*, vol. 29, pp. 1050–1060, 1986.

[3] L. O. Chua, C. A. Desoer, and E. S. Kuh, *Linear and Nonlinear Circuits*, McGraw-Hill, New York, 1987.

[4] L. O. Chua, C. W. Wu, A. Huang, and G. Q. Zhong, "A universal circuit for studying and generating chaos, part I: Routes to chaos," *IEEE Trans. on Circ. Sys.*, I, vol. 40, pp. 732–744, 1993.

[5] A. S. Dmitriev, M. Shirokov, and S. O. Starkov, "Chaotic synchronization in ensembles of coupled maps," *IEEE Trans. on Circ. Sys.*, I, vol. 44, pp. 918–926, 1997.

[6] M. Fiedler, "Algebraic connectivity of graphs," *Czechoslovak Math. J.*, vol. 23, pp. 298–305, 1973.

[7] J. Friedman, J. Kahn, and E. Szemerédi, "On the second eigenvalue in random regular graphs," *Proc. of Annual ACM Symp. on Theory of Comp.*, 1989, pp. 587–598.

[8] H. Fujisaka and T. Yamada, "Stability theory of synchronized motion in coupled-oscillator systems," *Progress of Theoretical Physics*, vol. 69, pp. 32–47, 1983.

[9] G. Grassi and S. Mascolo, "Nonlinear observer design to synchronize hyperchaotic systems via a scalar signal," *IEEE Trans. on Circ. Sys.*, I, vol. 44, pp. 1011–1014, 1997.

[10] R. He and P. G. Vaidya, "Analysis and synthesis of synchronous periodic and chaotic systems," *Phys. Rev. A*, vol. 46, pp. 7387–7392, 1992.

[11] H. Nijmeijer and I. M. Y. Mareels, "An observer looks at synchronization," *IEEE Trans. on Circ. Sys.*, I, vol. 44, pp. 882–890, 1997.

[12] K. Kaneko, "Overview of coupled map lattices," *Chaos*, vol. 2, pp. 279–282, 1992.

[13] L. Kocarev and U. Parlitz, "General approach for chaotic synchronization with applications to communications," *Phys. Rev. Lett.*, vol. 74, pp. 5028–5031, 1995.

[14] L. Kocarev and U. Parlitz, "Generalized synchronization, predictability, and equivalence of unidirectionally coupled dynamical systems," *Phys. Rev. Lett.*, vol. 76, pp. 1816–1819, 1996.

[15] L. M. Pecora and T. L. Carroll, "Synchronization in chaotic systems," *Phys. Rev. Lett.*, vol. 64, pp. 821–824, 1990.

[16] L. M. Pecora and T. L. Carroll, "Driving systems with chaotic signals," *Phys. Rev. A*, vol. 44, pp. 2374–2383, 1991.

[17] L. M. Pecora and T. L. Carroll, "Master stability functions for synchronized chaos in arrays of oscillators," *Proc. of IEEE Int. Symp. Circ. Sys.*, Monterey, CA, 1998, pp. IV562–567.

[18] N. F. Rulkov, M. M. Sushchik, L. S. Tsimring, and H. D. I. Abarbanel, "Generalized synchronization of chaos in directionally coupled chaotic system," *Physical Review E*, vol. 51, pp. 980–994, 1995.

[19] J. A. K. Suykens, P. F. Curran, J. Vandewalle, and L. O. Chua, "Robust nonlinear H_∞ synchronization of chaotic Lur'e systems," *IEEE Trans. on Circ. and Sys., I*, vol. 44, pp. 891–904, 1997.

[20] A. Vince, "Star chromatic number," *J. of Graph Theory*, vol. 12, pp. 551–559, 1988.

[21] C. W. Wu, "Global synchronization in coupled map lattices," *Proc. of IEEE Int. Symp. Circ. Sys.*, vol. 3, Monterey, CA, 1998, pp. III302–305.

[22] C. W. Wu, "Synchronization in arrays of chaotic circuits coupled via hypergraphs: Static and dynamic coupling," *Proc. of IEEE Int. Symp. Circ. Syst.*, Monterey, CA, 1998, pp. III287–290.

[23] C. W. Wu, "Graph coloring via synchronization of coupled oscillators," *IEEE Trans. on Circ. Sys., I*, vol. 45, pp. 974–978, 1998.

[24] C. W. Wu and L. O. Chua, "A unified framework for synchronization and control of dynamical systems," *Int. J. of Bifur. Chaos*, vol. 4, pp. 979–998, 1994.

[25] C. W. Wu and L. O. Chua, "Application of Kronecker products to the analysis of systems with uniform linear coupling," *IEEE Trans. on Circ. Sys., I*, vol. 42, pp. 775–778, 1995.

[26] C. W. Wu and L. O. Chua, "Synchronization in an array of linearly coupled dynamical systems," *IEEE Trans. on Circ. Sys., I*, vol. 42, pp. 430–447, 1995.

24

Chaos in Phase Systems: Generation and Synchronization

V. D. Shalfeev,[1] V. V. Matrosov,[1] and
M. V. Korzinova[2]

[1]Faculty of Radiophysics
Nizhni Novgorod State University
Nizhni Novgorod, 603600, Russia
shalfeev@hale.appl.sci-nnov.ru

[2]Institute for Nonlinear Science
University of California at San Diego
La Jolla, CA 92093 USA
korz@routh.ucsd.edu

Abstract

Some approaches to producing dynamical chaos in ensembles of coupled controlled phase systems (generation, synchronization, communication) are reviewed and discussed in this chapter.

24.1 Introduction

The discovery of dynamical (deterministic) chaos [24, 13] was undoubtedly the brightest event in nonlinear science in recent years. Comprehension of the fact that complex irregular behavior – dynamical chaos – may arise in nonlinear systems (even very simple ones) in the absence of any random actions dramatically changed traditional understanding of oscillatory-wave phenomena. It became apparent that chaotic oscillations rather than simple periodic motions are typical in most physical, chemical, biological, and other systems existing in nature.

Investigation of basic properties of dynamical chaos piqued an interest in its application to engineering, primarily in the fields of radio engineering and radio electronics. In this context, worthy of mentioning are the works on creating generators of chaotic oscillations, in particular, tunnel diode generators [17, 35], ring generators [9], inertial nonlinearity generators [4], Chua's circuits [7, 26], and some others. Of great interest are applications of dynamical chaos to problems involved with processing (neural networks, chaotic processors [1, 3]) and transmission of information [14, 11, 41].

The idea of using chaotic oscillations for data communication is very attractive. On one hand, it allows for application of traditional methods of information processing (e.g., correlation methods). On the other hand, one can employ nontraditional "dynamical" methods that are based on the properties of nonlinear dynamics.

Traditional devices for information transmission by means of regular signals employ synchronization phase systems – phase-locked loops (PLLs). High accuracy, reliability, noise resistance, controllability, capability to provide high power and frequency made PLLs an inherent part of nearly all communication systems [25]. Naturally, these properties of PLLs make them highly promising for data communication by means of chaotic signals too. However, the methods of chaos transmission using chaotic signals described in the literature [34, 18, 15, 43, 23, 10, 8, 16] almost ignore, with rare exceptions [22, 36, 12, 42], the use of PLLs for these purposes. This is evidently explained by insufficient knowledge of dynamical properties of PLLs in the regime of chaos generation.

The goal of this chapter is discussion of the issues concerned with generation of dynamical chaos in PLLs and synchronization of chaotic oscillations, as well as their application to transmission of information. It will be shown that these problems may be solved by using dynamical properties of ensembles of coupled PLLs [2, 20, 39].

24.2 A PLL System as a Generator of Chaotic Oscillations

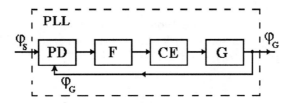

FIGURE 24.1
Block diagram of a phase-locked loop.

The PLL system is a standard automatic loop controlling generator frequency shown schematically in Figure 24.1. A reference signal having current phase φ_S is compared on phase discriminator PD with a signal from the controlled generator G having current phase φ_G. The output voltage of the PD depends on phase difference $\varphi = \varphi_S - \varphi_G$. (Usually it is a 2π-periodic function that will be supposed to be sinusoidal for definiteness.) The signal passes through filter F (transfer function $K(p)$, $p \equiv d/dt$) and arrives at the input of control element CE that directly changes the generator frequency. The equation for such a system is generally written in the form [25]

$$\frac{p\varphi}{\Omega} + K(p)\sin\varphi = \gamma. \qquad (24.2.1)$$

Here, Ω is the maximal deviation of generator frequency that may balance the control loop, $\gamma = (\Delta\omega)/\Omega$, and $\Delta\omega$ is the deviation of the frequency of free generator G from the reference frequency ω_S. The dynamics of such a PLL system depend significantly on parameters of the system, in particular, on parameters of the filter $K(p)$.

For the simplest first-order filters, the system (24.2.1) is considered on the cylindrical phase surface $\left(\varphi, y \equiv \frac{d\varphi}{d\tau}\right)$, where the following stable regimes may be realized [5]:

- a regime of synchronization of generator G by the reference signal (a stable equilibrium state with the $(\arcsin\gamma, 0)$ coordinates corresponds to this regime on the phase plane (φ, y));
- a regime of quasisynchronization, when the generator frequency is periodically modulated near an average frequency stabilized by the reference signal (a limit cycle of oscillatory type, i.e., without phase rotations, near equilibrium state corresponds to this regime on the phase plane (φ, y));

– a nonsynchronous regime of beats (a stable limit cycle with phase rotation on the phase plane (φ, y)).

The PLL dynamics is more complicated for the second-order filters [12, 6, 37, 19, 27, 29]. In this case, chaotic regimes may be realized in addition to the regular synchronous and nonsynchronous regimes mentioned above. Consider the simplest second-order filter of the 0/2 type with transfer function $K(p) = (a_2 p^2 + a_1 p + 1)^{-1}$. Equation (24.2.1) may be written in this case as

$$\frac{d\varphi}{d\tau} = y, \quad \frac{dy}{d\tau} = z, \quad \mu \frac{dz}{d\tau} = \gamma - \sin\varphi - y - \varepsilon z. \quad (24.2.2)$$

Here, $\mu = a_2 \Omega^2$, $\varepsilon = a_1 \Omega$, $\tau = \Omega t$, and $\varphi = \varphi_S - \varphi_G$. Projections on the (φ, y)-plane of some attractors typical of the system (24.2.2) [27] are given in Figure 24.2. Figures 24.3 and 24.4 demonstrate partitioning of

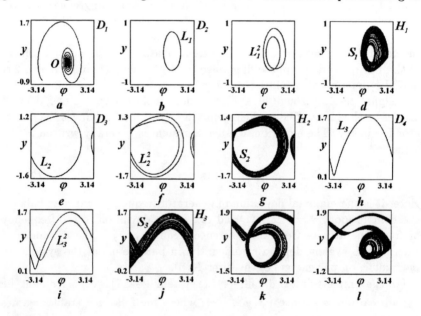

FIGURE 24.2
φ, y-projections of some attractors typical of a second-order filter PLL system.

the plane of the parameters μ, γ for $\varepsilon = $ const and of the plane of the parameters ε, γ for $\mu = $ const into the domains to which the attractors in Figure 24.2 correspond. The boundary curves separating different domains in Figures 24.3 and 24.4 correspond to the following bifurcations:

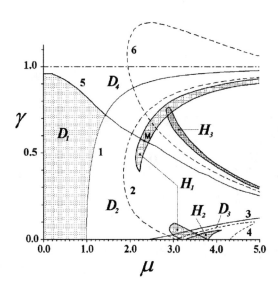

FIGURE 24.3
The plane of parameters $\varepsilon = 1$ for system (24.2.2).

curve 1 – the bifurcation of the loss of stability of equilibrium state O (Figure 24.2a) and of a soft onset of oscillatory cycle L_1 (Figure 24.2b);
curve 2 – the bifurcation of the first period doubling of cycle L_1 (Figure 24.2b) and of the transition to cycle L_1^2 (Figure 24.2c);
curve 3 – the bifurcation of the birth of cycle L_2 from crowding trajectories (saddle-node bifurcation of the birth of oscillatory cycle L_2) (Figure 24.2e);
curve 4 – the bifurcation of the first period doubling of cycle L_2 (Figure 24.2e) and of the transition to cycle L_2^2 (Figure 24.2f);
curve 5 – the bifurcation of the birth of rotatory cycle L_3 (Figure 24.2h) either from a separatrix loop (to the left of point M in Figure 24.3) or as a result of saddle-node bifurcation (to the right of point M in Figure 24.3);
curve 6 – the bifurcation of the first period doubling of cycle L_3 (Figure 24.2h) and of the transition to cycle L_3^2 (Figure 24.2i).

Note that only the bifurcation curves of the first period doublings of the cycles are shown in Figures 24.3 and 24.4; all the subsequent doublings are not depicted here. The cascade of period doublings of cycles L_1, L_2, L_3 in the system (24.2.2) gives rise to the following chaotic attractors: S_1 (Figure 24.2d) in domain H_1 (Figures 24.3 and 24.4); S_2 (Figure 24.2g) in domain H_2; and S_3 (Figure 24.2j) in domain H_3. Thus, in addition to the synchronous regime (Figure 24.2a) and the periodic quasisynchronous regimes (Figures 24.2b,c,e,f), chaotic quasisynchronous regimes (Figures 24.2d,g) may also exist in the system (24.2.2). In the latter case,

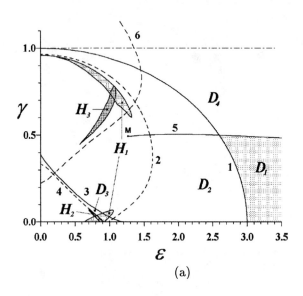

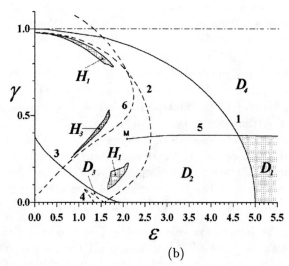

FIGURE 24.4
(a) The plane of parameters $\mu = 3$ for system (24.2.2). (b) The plane of parameters $\mu = 5$ for system (24.2.2).

the oscillations of generator G have angular chaotic modulation, with the frequency varying chaotically near an average value stabilized by the reference signal. As to nonsynchronous regimes (beats), chaotic beats with phase rotation (Figure 24.2j) as well as mixed oscillatory-rotatory chaotic beats (Figures. 24.2k,l) are possible besides regular beats (Figures 24.2h,i) and mixed oscillatory–rotatory periodic beats.

Note that, in addition to the motions mentioned above, the system (24.2.2) may also include more complicated, multiloop periodic and chaotic oscillatory (quasisynchronous) regimes, nonsynchronous multiloop periodic and chaotic rotatory regimes (beats), and mixed oscillatory-rotatory periodic and chaotic regimes. These motions have, as a rule, small existence domains in the space of the parameters and are omitted in Figures. 24.2–24.4 for simplicity.

The discussed features of the nonlinear dynamics of PLL systems with the simplest second-order filter draw one to an important conclusion. If we choose for the second-order filter PLL the parameters corresponding to chaotic attractors, then such a PLL may be regarded as a generator of chaotic oscillations. The regime of generation of quasisynchronous chaotic oscillations (domains H_1, H_2) seems to be the most interesting one for appications. In this case, quasisynchronous oscillations with chaotic angular modulation and the average frequency stabilized by the reference signal are generated at the PLL output. We will focus on this type of chaotic oscillations in PLLs. For brevity they will further be referred to as chaotically modulated oscillations (CMO).

Note that the existence domain of CMO (domains H_1, H_2 in Figures 24.3 and 24.4) is relatively narrow and small for the system (24.2.2). This can also be seen in Figure 24.5 that depicts the cross-section of domain H_1 by the (ε, γ)-plane for several values of $\mu = $ const. It is challenging to find effective ways for broadening the existence domains H_1, H_2 of CMO. For a single-loop PLL (Figure 24.1), we suggest the following alternatives: (i) make the filter $K(p)$ more complicated and (ii) change nonlinear characteristics of PD. The first one will inevitably lead to appearance in the system of new parameters and will thus give rise to additional difficulties in choosing the parameters needed for generation of chaotic signals. A possible expansion of the parameter domain corresponding to chaotic generation should be considered separately for each type of filter, although the data [21] obtained for the PLL with a delay element, that may be regarded as a high-order filter, indicate that the regions of the parameters providing CMO generation are small.

The second way – changing nonlinear characteristics of PD – may provide a certain gain in the size of CMO domains. The chaotic CMO domains H_1, H_2 for the system (24.2.2) possessing nonlinearity of the type $\sin^3 \varphi$ are demonstrated in Figure 24.6, (Figure 24.2b), and curves 8 and 9 are the bifurcations of the birth and of the first period doubling of oscillatory cycle,

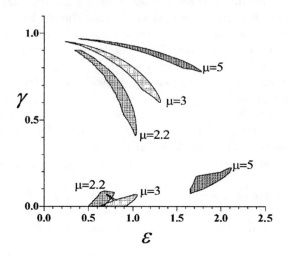

FIGURE 24.5
Comparison of existence domains of chaotically modulated oscillations in system (24.2.2) for different values of parameter μ.

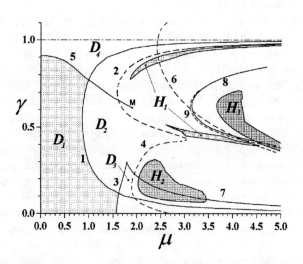

FIGURE 24.6
The plane of parameters $\varepsilon = 1$ for system (24.2.2) with PD nonlinearity of the type $\sin^3 \varphi$.

respectively. Comparison of Figures 24.3 and 6 shows that by changing characteristics of the PD one can substantially broaden domain H_2, but domain H_1 still remains small and narrow.

Thus, we can conclude that, with appropriate choice of parameters, a single-loop PLL may be regarded as a generator of CMO. However, because of relative narrowness of the existence domain of CMO in the space of parameters, it will be difficult to implement CMO generation, especially to control this regime in the case of uncontrolled variation of parameters.

We propose still another variant of PLL that would provide both reliable CMO generation in a broad range of parameter variation and sufficiently good controllability of such oscillations for synchronization. We propose to seak a solution in the class of two- or multiloop coupled PLLs [2] in which complex dynamical behavior and the transition to chaos may be attained by combining two or several simple (which is of primary importance) PLLs into a coupled ensemble possessing complex (chaotic) collective dynamics.

24.3 Chaotic Regimes in an Ensemble of Two Coupled PLLs

One can suggest quite a few variants to realize an ensemble of coupled PLLs. For example, by cascade coupling of PLLs, by parallel coupling, or by combined parallel-cascade coupling, with different types of couplings between neighboring PLLs. Apparently, even computer simulation will not provide a complete picture of all possible dynamical regimes for different versions of collective PLLs. Therefore, we will restrict ourselves to a simpler problem and will show on a few particular examples that chaotic oscillations may be effectively generated using the features of collective dynamics.

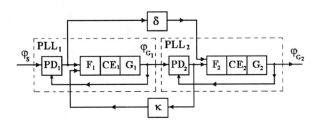

FIGURE 24.7
Cascade coupling of PLL_1 and PLL_2.

Consider a cascade coupling of two systems: PLL_1 and PLL_2 (Figure 24.7). Such a coupling is unidirectional because the input of PLL_2

is the output of PLL$_1$. In order to improve dynamical properties and enhance control of the dynamics of this cascade, it is reasonable to introduce additional couplings "forward" and "backward" [2, 37]. The "forward" coupling is implemented by transmitting a signal from the output of phase discriminator PD$_1$ in the PLL$_1$ system through the control loop to the control element of PLL$_2$ with coefficient δ. The "backward" coupling is realized by transmitting a signal from the output of phase discriminator PD$_2$ in PLL$_2$ through the control loop to the control element of PLL$_1$ with coefficient k. Under some simplifications, equations for such a system may be written in operator form as

$$\frac{p\varphi_1}{\Omega_1} + K_1(p)[\sin\varphi_1 + \kappa\sin(\varphi_2 - \varphi_1)] = \gamma_1, \qquad (24.3.3)$$

$$\frac{p\varphi_2}{\Omega_2} + K_2(p)[\sin(\varphi_2 - \varphi_1) + \delta\sin\varphi_1] = \gamma_2.$$

Stationary regimes of synchronization of the system (24.3.3) for $K_1(p) = K_2(p) = (1+mTp)/(1+Tp)$ were considered in [30, 31, 32, 33, 28] under different simplifications. Here, we address nonsynchronous regimes [20, 39, 38, 40].

We set $\Omega_1 = \Omega_2 = \Omega$, $K_1(p) = 1$, and $K_2(p) = 1/(1+Tp)$. Then, Equations (24.3.3) take the form

$$\frac{d\varphi_1}{d\tau} = \gamma_1 - \sin\varphi_1 - \kappa\sin(\varphi_2 - \varphi_1),$$

$$\frac{d\varphi_2}{d\tau} = y_2, \qquad (24.3.4)$$

$$\varepsilon\frac{dy_2}{d\tau} = \gamma_2 - y_2 - \sin(\varphi_2 - \varphi_1) - \delta\sin\varphi_1.$$

Here, $\tau = \Omega t$ and $\varepsilon = \Omega T$. Computer modeling of the system (24.3.4) showed that various regimes of chaotic oscillations may exist in this system in a rather broad region of the space of parameters. We will take interest in the values of the parameters at which the oscillations with chaotic angular modulation and the carrier frequency stabilized by the reference frequency (i.e., CMO) exist at the output of generator G$_2$.

A chaotic attractor of the system (24.3.4) in the $(\varphi_1, \varphi_2, y_2)$-space is shown in Figure 24.8a for $\gamma_1 = 1.7$, $\gamma_2 = 0.685$, $\varepsilon = 60$, $\delta = 0.1$, and $\kappa = 1.9$ under the initial conditions $y_2^0 = \gamma_2$ and arbitrary φ_1^0 and φ_2^0. Projections of this attractor on the $(\varphi_1, d\varphi_1/d\tau)$- and (φ_2, y_2)-planes are given in Figures 24.8b,c. Thus, the collective dynamics of the considered cascade coupling of two PLLs is chaotic. The chaotic beats of frequency of generator G$_1$ relative to the reference frequency occur at the output of generator G$_1$. At the output of G$_2$ there occur chaotic oscillations (without phase advance by 2π) at which the phase difference φ_2 fluctuates around a certain stationary value φ_2^* and the frequency difference y_2 is approximately

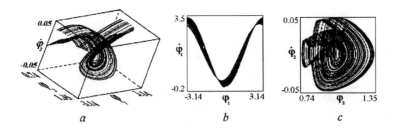

FIGURE 24.8
Oscillatory-rotatory chaotic attractor of system (24.3.4)

equal to zero. Consequently, CMO – the oscillations with chaotic angular modulation and average frequency stabilized near reference frequency – are observed at the output of generator G_2.

Consider now a more general case of the system (24.3.3) when $\Omega_1 = \Omega_2 = \Omega$, $K_1(p) = 1/(1+T_1 p)$, and $K_2(p) = 1/(1+T_2 p)$. In this case, the system of equations is written as

$$\frac{d\varphi_1}{d\tau} = y_1,$$

$$\varepsilon_1 \frac{dy_1}{d\tau} = \gamma_1 - y_1 - \sin\varphi_1 - \kappa \sin(\varphi_2 - \varphi_1), \qquad (24.3.5)$$

$$\frac{d\varphi_2}{d\tau} = y_2,$$

$$\varepsilon_2 \frac{dy_2}{d\tau} = \gamma_2 - y_2 - \sin(\varphi_2 - \varphi_1) - \delta \sin\varphi_1.$$

Modeling of the system (24.3.5) revealed that CMO may be generated both at the output of generator G_2 (Figure 24.9b) or at the output of generator G_1 (Figure 24.10a) and at the outputs of both generators G_1 and G_2 (Figures 24.11a,b). The power spectra of output chaotic oscillations are depicted in Figures 24.9c, 24.10c and 24.11c and Figures 24.9d, 24.10d and 11d display the existence domains of CMO on the corresponding planes of the parameters. The results are given for the coupling $\delta = 0$ because, as was confirmed by numerical modeling, introduction of $\delta > 0$ almost does not affect the dynamics. Thus, we can conclude that, under definite conditions, two cascade-coupled, rather simple PLLs may be regarded as a generator of chaotically modulated oscillations in a rather broad region of the parameters that provides generation of oscillations with chaotic angular modulation and the average frequency stabilized by the reference frequency.

The examples given above enable us to make an important conclusion. Chaotically modulated oscillations may be generated at generator output

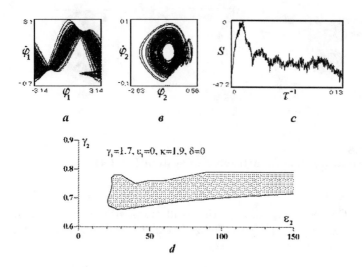

FIGURE 24.9
Chaotically modulated oscillations at the output of generator G_2 for system (24.3.5) and existence domain of these oscillations on the plane of parameters.

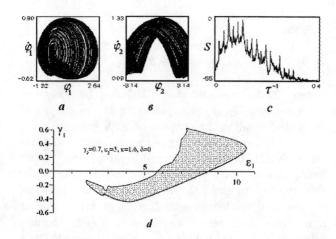

FIGURE 24.10
Chaotically modulated oscillations at the output of generator G_1 for system (24.3.5) and existence domain of these oscillations on the plane of parameters.

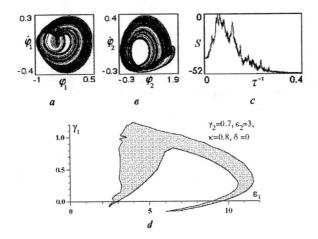

FIGURE 24.11
Chaotically modulated oscillations at the outputs of generators G_1 and G_2 for system (24.3.5) and existence domain of these oscillations on the plane of parameters.

in a single loop of a PLL with a high-order filter. However, because of a relatively narrow existence domain of these oscillations, fine adjustment of parameters of the system is needed to provide such a generation in practice. These difficulties may be eliminated by taking, instead of a single-loop PLL, an ensemble of coupled PLLs (in particular, cascade coupled PLLs). By using the features of collective dynamics one can attain generation of chaotically modulated oscillations in a rather broad parameter region at the output of different oscillators of a PLL ensemble.

It is also worthy to note that the number of the parameters related to filter inertia may be diminished in a collective PLL system. Consider as an example chaos generation by a system of three cascade-coupled PLLs having the simplest, idealized filters (Figure 24.12). Here, $K_1(p) = K_2(p) = K_3(p) \equiv 1$, $\Omega_1 = \Omega_2 = \Omega_3 \equiv \Omega$. Examples of CMO generation at the outputs of PLL$_2$ and PLL$_3$ are given in Figure 24.13 and at the outputs of all the three systems: PLL$_1$, PLL$_2$, and PLL$_3$ in Figure 24.14.

It is apparent that, under appropriate choice of parameters, an ensemble of coupled PLLs may be regarded as an effective generator of chaotic oscillations and, by choosing the number and parameters of the PLLs in the ensemble, it is possible to provide reliable generation of chaotic oscillations possessing the desired statistical characteristics.

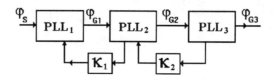

FIGURE 24.12
Cascade coupling of PLL_1, PLL_2 and PLL_3.

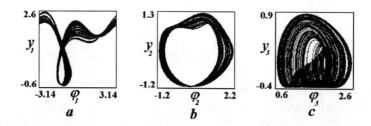

FIGURE 24.13
Oscillatory-rotatory chaotic regime in a system of three coupled PLLs.

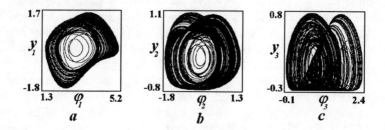

FIGURE 24.14
Oscillatory chaotic regime in a system of three coupled PLLs.

24.4 Synchronization of Chaotic Oscillations

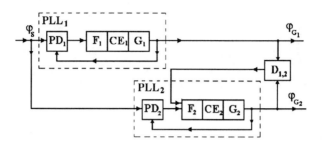

FIGURE 24.15
Variant of unidirectional coupling of PLL_1 and PLL_2 through additional discriminator $D_{1,2}$ for synchronization of chaotically modulated oscillations.

We now address the problem of synchronization of the chaotic oscillations generated by two generators of chaotic oscillations coupled unidirectionally: a master and a synchronized (response) generator.

We first consider a single loop PLL_1 as a master generator of CMO and a single loop PLL_2 as a response generator of CMO, both having close but not equal parameters. Under synchronization of two CMO we will understand "complete synchronization" [41] at which the output signals of the two chaotic generators fully coincide being chaotic.

For attaining synchronization we can provide unidirectional, difference couplings between the master PLL_1 and response PLL_2 generators along the (φ, y, z)-coordinates. A variant of unidirectional coupling of PLL_1 and PLL_2 through additional discriminator $D_{1,2}$ is presented in Figure 24.15. If the discriminator is phase discriminator $PD_{1,2}$, then we obtain, at the output, the signal $\delta_\varphi \sin(\varphi_2 - \varphi_1)$ that is used for controlling the frequency of generator G_2. If PLL_1 and PLL_2 are coupled through additional frequency discriminator $FD_{1,2}$, then we obtain, at the output, the signal $\delta_y \Phi(y_2 - y_1)$ that is transmitted to the control loop of generator G_2. The characteristics of the $FD_{1,2}$, may be considered to be linear to the first approximation and we can assume $\delta_y \Phi(y_2 - y_1) = \delta_y(y_2 - y_1)$. Finally, unidirectional coupling may be achieved along the z-coordinate, too, so the signal $\delta_z(z_2 - z_1)$ should be transmitted to the control loop of G_2. The use of such couplings is quite justified because the goal of controlling generator G_2 by couplings [41] is matching the oscillations of generator G_2 to the oscillations of generator G_1.

A mathematical model of PLL_1 and PLL_2 with such a coupling is a set of equations

$$\frac{d\varphi_1}{d\tau} = y_1, \quad \frac{dy_1}{d\tau} = z_1,$$

$$\mu_1 \frac{dz_1}{d\tau} = \gamma_1 - \sin\varphi_1 - y_1 - \varepsilon_1 z_1,$$

$$\frac{d\varphi_2}{d\tau} = y_2, \quad \frac{dy_2}{d\tau} = z_2, \qquad (24.4.6)$$

$$\mu_2 \frac{dz_2}{d\tau} = \gamma_2 - \sin\varphi_2 - y_2 - \varepsilon_2 z_2 -$$

$$-\delta_\varphi \sin(\varphi_2 - \varphi_1) - \delta_y \Phi(y_2 - y_1) - \delta_z(z_2 - z_1),$$

where $\Phi(y) = 2ay/(1 + a^2 y^2)$.

Results of modeling the dynamics (24.4.6) are given in Figure 24.16 for

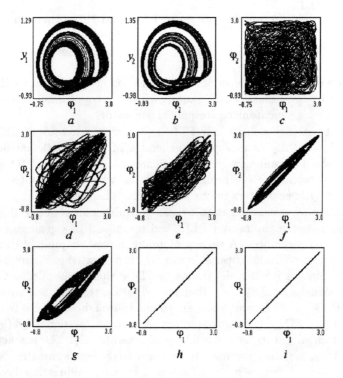

FIGURE 24.16
Results of synchronization of chaotic oscillations according to the scheme given in Figure 24.15 for different coupling coefficients δ_φ, δ_y, δ_z.

the following initial conditions: $y_1^0 = \gamma_1$, $y_2^0 = \gamma_2$, $z_1^0 = 0$, $z_2^0 = 0$; φ_1^0 and

φ_2^0 are arbitrary values. The parameters γ, ε, and μ in domain H_1 of PLL_1 and PLL_2 (Figures 24.3 and 24.4) were taken to be not equal but rather close values (differing in the limit of 1–5%).

The oscillations in PLL_1 and PLL_2 are not synchronized without coupling, or if a coupling is introduced only along the φ-coordinate (Figures 24.16a,b,c). With the use of y and z couplings, the oscillations in PLL_2 are matched to those of PLL_1, with the accuracy of matching (synchronization) increasing in proportion to coupling coefficients (Figure 24.16d – only linear coupling along z; Figure 24.16e – only linear coupling along y; Figures 24.16f,g,h – mixed linear control along z and y; Figure 24.16i – nonlinar coupling only along y). However, when the values of coupling coefficients are chosen to be sufficiently high, the synchronization may break down and the PLL_2 system will pass over to another attractor. This is due to the fact that synchronization occurs in a definite bounded region of parameter variation (synchronization region).

For estimation of the synchronization region, we set the values of the PLL_1 parameters in the domain of CMO generation, i.e., in H_1 (Figures 24.3 and 24.4), to be the following: $\varepsilon_1 = 1$, $\mu_1 = 2.2$, $\gamma_1 = 0.5$; the initial conditions φ_1, y_1, z_1 are on the chaotic attractor, the projection of which is shown in Figure 24.2d. Further, we will seek such values of parameters $\gamma_2, \varepsilon_2, \mu_2$ of the system PLL_2 and couplings δ_φ, δ_y, a under which the CMO generated in PLL_1 and PLL_2 differ by a small value of ε_c to an accuracy of a constant. We set the initial conditions for PLL_2 to be $\varphi_2 \in [-\pi, \pi], y_2 = \gamma_2, z_2 = 0$. Specific values of φ_2 are calculated either as random values obtained by means of a random-number generator or as coordinates of equidistant points in the interval $-\pi < \varphi < \pi$. After starting and some onset time τ_y, the deviations of the coordinates $\Delta\varphi = \varphi_2 - \varphi_1, \Delta y = y_2 - y_1, \Delta z = z_2 - z_1$ during the observation time τ_c are calculated. Then, the deviations are analyzed. If, for a certain point in the space of the parameters, the deviations $\Delta\varphi, \Delta y, \Delta z$ are within the set accuracy limits ε_c during time τ_c for all initial conditions z_2, y_2, φ_2 (N_φ points on the section $[-\pi, \pi]$), then we can say that this point belongs to the synchronization region. Bearing in mind that the calculated trajectories are chaotic, the observed coordinates are permitted to fall outside the limits of accuracy ε_c but the total time outside these limits should be less than 10% of τ_c. The sections of the synchronization region (δ_y, γ_2) obtained using this algorithm are given in Figure 24.17a for $\varepsilon_2 = 1, \mu_2 = 2.2, \delta_\varphi = 0, 0.1, 0.5$, the sections (μ_2, γ_2) are given in Figure 24.17b for $\varepsilon_2 = 1, \delta_\varphi = 0, \delta_y = 0.5, 1$, the sections $(\varepsilon_2, \gamma_2)$ are given in Figure 24.17c for $\mu_2 = 2.2, \delta_\varphi = 0, \delta_y = 0.5, 1$, respectively, at $N_\varphi = 30, \varepsilon_c = 0.025, \tau_y = 1000, \tau_c = 1000$. It is clear from Figure 24.17 that the synchronization, indeed, occurs in a rather broad region of the parameters, with the values of the parameters of PLL_2 not necessarily belonging to domain H_1. Note that the synchronization region depicted in

Figure 24.17a must be bounded at sufficiently large δ_y, but this portion of the boundary is not calculated in the figure.

It is known that an intrinsic feature of synchronization of regular oscillations is hysteresis, i.e., pull-in and pull-out occur at different values of initial mismatch of generator frequency relative to a reference generator. It would be a challenge to consider such a hysteresis for the synchronization of chaotic oscillations.

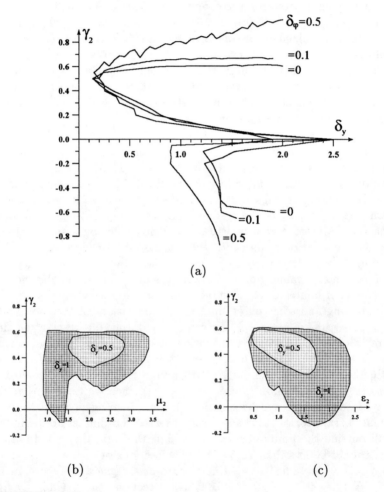

FIGURE 24.17
Cross-section in the space of parameters of the synchronization region of chaotically modulated signals of two coupled PLLs.

We performed experiments on calculation of the boundaries of synchronization region at a slow increase of parameter γ_2 (pull-out) and a slow decrease of parameter γ_2 (pull-in) in the synchronization regions of interest (Figure 24.17). It was revealed that the synchronization pull-in and pull-out occur at the same value of γ_2, i.e., no hysteresis is observed ($\gamma_2^{pull-in} = \gamma_2^{pull-out}$). The point is that a chaotic attractor is the only attractor of the system under the chosen values of the parameters in domain H$_1$ (Figures 24.3, 24.4). Hysteresis may take place ($\gamma_2^{pull-in} \neq \gamma_2^{pull-out}$) under other values of the parameters from the same domain. An example is given in Figure 24.18 for the PLL$_1$ parameters: $\gamma_1 = 0.7$, $\varepsilon_1 = 1$, $\mu_1 = 2.6$,

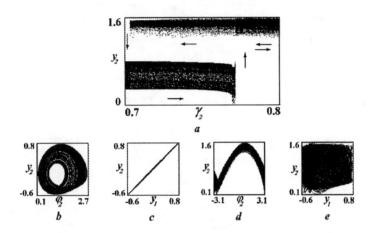

FIGURE 24.18
Demonstration of the existence of pull-in and pull-out regions during synchronization of chaotically modulated oscillations of two coupled PLLs.

the PLL$_2$ parameters: $\varepsilon_2 = 1.4$, $\mu_2 = 2.5$, and the control loop parameters: $\delta_y = 0.5$, $\delta_\varphi = 0$, $a = 20$. Parameter γ_2 first increases from 0.7 to 0.8 and then decreases from 0.8 to 0.7. The one-parametric bifurcation diagram $\{\gamma_1, y_2\}$ of Poincare map in Figure 24.18a demonstrates that, as γ_2 changes from 0.7 to 0.8, there first occur at the output of PLL$_2$ chaotic oscillations (Figure 24.18b) synchronized with the oscillations of PLL$_1$ (Figure 24.18c). Then, at $\gamma_2^{pull-out} = 0.77$ (the boundary of pull-out region), the synchronization breaks down and with a further decrease of γ_2 (the upper portion of Figure 24.18a) the chaotic oscillations of PLL$_2$ (Figure 24.18d) are no longer synchronized with the PLL$_1$ oscillations (Figure 24.18e). The subsequent pull-in of chaotic oscillations occurs at $\gamma_2^{pull-in} = 0.703$. Thus, the processes of CMO synchronization possess hysteresis, i.e., their space of the parameters has regions of pull-in and pull-out the boundaries of which do not coincide. Upon the whole, in spite of the fact that computer modeling verifies that it is possible to achieve complete synchronization of CMOs in two nonidentical PLLs having close parameters, it should be hypothesized that it will not be easy in practice to synchronize two chaotic PLLs with the second-order filters because of relative smallness of the existence domains of chaotic oscillatory attractors in the space of the parameters of a single PLL.

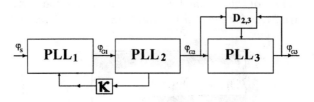

FIGURE 24.19
Synchronization of chaotically modulated oscillations of generators G_2 and G_3 (cascade coupling).

We now consider synchronization of the chaotic oscillations generated by cascade coupling of PLL_1 and PLL_2 (further referred to as PLL_1+PLL_2 for shortness). We take PLL_1+PLL_2 as a master chaotic generator and the cascade PLL_3+PLL_4 with close but not equal parameters as a response chaotic generator. For matching oscillations at the output of PLL_4 to oscillations at the output of PLL_2, i.e., for attaining synchronization, we can provide, similar to the case of a single-loop PLL, unidirectional difference couplings between the master cascade PLL_1+PLL_2 and the response cascade PLL_3+PLL_4. However, taking into account the specific features of cascade coupling we can propose a simpler solution to the synchronization problem.

In Figure 24.19 we show a variant of cascade coupling of three PLLs that allows us to obtain at the output of G_3, chaotic oscillations which are synchronous to the oscillations at the output of G_2. The synchronization is realized by the principle of "synchronous response" [43]. PLL_3 is cascade-coupled to the master cascade PLL_1+PLL_2. If we take in PLL_3 a low-inertia filter $K(p) = 1/(1 + T_3 p)$, then the chaotic oscillations stimulated by the chaotic oscillations at the input (i.e., at the output of G_2) appear at the output of PLL_3. For matching the oscillations of G_3 to the oscillations of G_2 we introduce control coupling from PPL_2 to PPL_3 through frequency discriminator $FD_{2,3}$. Equations for such a system are written in the form

$$\frac{d\varphi_1}{d\tau} = \gamma_1 - \sin\varphi_1 - \kappa \sin(\varphi_2 - \varphi_1),$$
$$\frac{d\varphi_2}{d\tau} = y_2,$$

$$\varepsilon_2 \frac{dy_2}{d\tau} = \gamma_2 - y_2 - \sin(\varphi_2 - \varphi_1), \tag{24.4.7}$$

$$\frac{d\varphi_3}{d\tau} = y_3,$$

$$\varepsilon_3 \frac{dy_3}{d\tau} = \gamma_3 - y_3 - \sin(\varphi_3 - \varphi_2) - \delta_y \Phi(y_3 - y_2),$$

where $\Phi(y_3 - y_2)$ is the nonlinear characteristic of frequency discriminator $FD_{2,3}$. Modeling of the system (24.4.7) showed that there is no synchronization between chaotic oscillations of PLL_2 and PLL_3 under the initial conditions $y_2^0 = \gamma_2, y_3^0 = \gamma_3$, arbitrary φ_2^0, φ_3^0 and in the absence of control coupling $\delta_y = 0$ (Figures 24.20a–d). However, by introducing the control

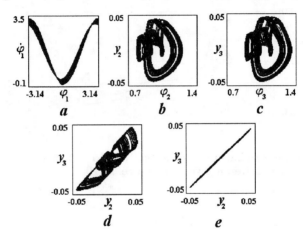

FIGURE 24.20
Results of synchronization according to the scheme given in Figure 24.19 in the absence (a,b,c,d) and in the presence (e) of control coupling δ_y.

coupling $\delta_y \neq 0$ (nonlinearity $\Phi(y_3 - y_2) = 2a(y_3 - y_2)/[1 + a^2(y_3 - y_2)^2]$) one can match the oscillations of G_3 to the oscillations of G_2 and attain a rather good synchronization (Figure 24.20e).

Results of calculations of the synchronization region for system (24.4.7) in the plane of the parameters (δ_y, γ_3) using the algorithm described above for synchronization of two chaotic PLLs are presented in Figure 24.21. One can clearly see from this figure that the synchronization region is limited on δ_y (the limiting values of δ_y were calculated for $a = 30$).

Still, another more efficient variant of solution to the synchronization problem using parallel coupling of PLL_3 to the $PLL_1 + PLL_2$ cascade is proposed in Figure 24.22. Here, PLL_3 is a complete analog of PLL_2 and the input signal from PLL_1 is transmitted both to the PLL_2 and PLL_3

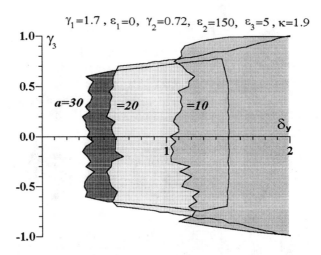

FIGURE 24.21
Region of synchronization of chaotically modulated oscillations according to the scheme given in Figure 24.19.

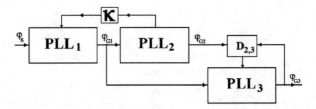

FIGURE 24.22
Synchronization of chaotically modulated oscillations of generators G_2 and G_3 (parallel coupling).

inputs. Consequently, the chaotic oscillations formed at the PLL_3 output should be the same as at the PLL_2 output because of the identity of PLL_2 and PLL_3. If there are deviations in the parameters of PLL_3 and PLL_2, the oscillations in G_3 should be matched to the oscillations in G_2 by means of additional coupling from G_2 to G_3 through frequency discriminator $FD_{2,3}$. Equations for such a system may be written in the form

$$\frac{d\varphi_1}{d\tau} = \gamma_1 - \sin\varphi_1 - \kappa\sin(\varphi_2 - \varphi_1),$$

$$\frac{d\varphi_2}{d\tau} = y_2,$$

$$\varepsilon_2 \frac{dy_2}{d\tau} = \gamma_2 - y_2 - \sin(\varphi_2 - \varphi_1), \qquad (24.4.8)$$

$$\frac{d\varphi_3}{d\tau} = y_3,$$

$$\varepsilon_3 \frac{dy_3}{d\tau} = \gamma_3 - y_3 - \sin(\varphi_3 - \varphi_1) - \delta_y \Phi(y_3 - y_2).$$

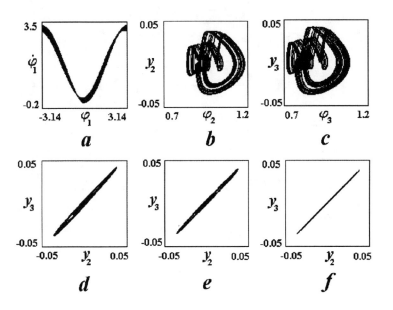

FIGURE 24.23
Results of synchronization of chaotically modulated oscillations according to the scheme given in Figure 24.22 for linear (d,c) and nonlinear (f) control.

In modeling the system (24.4.8), we chose the following initial conditions: $y_2^0 = \gamma_2$, $y_3^0 = \gamma_3$; φ_1, φ_2, φ_3 were arbitrary. The deviations of the parameters $\gamma_2 - \gamma_3 = \Delta\gamma \neq 0$, $\varepsilon_2 - \varepsilon_3 = \Delta\varepsilon \neq 0$ were taken within 5%. Results of modeling show that by using the coupling δ_y one can match the chaotic oscillations in G_3 to the oscillations in G_2 in a rather broad region of parameter variation (Figure 24.23). The regime of synchronization is demonstrated in Figure 24.23d (linear control coupling) and in Figure 24.23f (nonlinear control coupling) for different values of the coefficient δ_y.

Thus, the cascade-coupled PLLs may be a useful tool for solution of both the problem of generation of chaotically modulated oscillations and the synchronization problem. Since the traditional PLL scheme (for regular oscillations) gives a successful solution of the problems of modulation and demodulation of oscillations, there is a hope that the cascade-coupled PLLs will enable researchers to solve the entire complex of the problems concerned with transmission of information by means of chaotic signals.

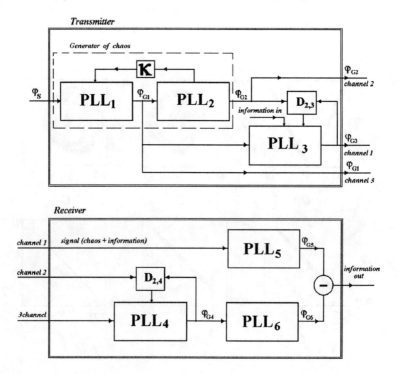

FIGURE 24.24
Transmission of information using chaotically modulated oscillations generated by cascade-coupled PLLs.

24.5 Application of Chaotic PLLs to Transmission of Information

A variant of the scheme for transmission of information using the chaotic signals generated by cascade-coupled PLLs is presented in Figure 24.24 [20, 39]. The transmitter consists of three PLLs. The PLL_1+PLL_2 cascade generates a carrier – chaotically modulated oscillations with the average frequency stabilized by the reference signal. PLL_3 plays the part of modulator.

Chaotic modulation of the carrier is supplemented at its output by information signal modulation. The parameters of the PLL_3 system should be close to those of PLL_2. There is a control circuit for matching the oscillations in G_3 to the oscillations in G_2 through frequency discriminator $FD_{2,3}$. The signals from the outputs of G_1, G_2, and G_3 are transmitted through data communication channels to the receiver. In the receiver, the signal from the output of G_3 is transmitted to the demodulator PLL_5 from which the voltage, proportional to the mixture of the chaotically modulated carrier signal and information signal, is read. The signals from the G_1 and G_2 outputs are transmitted to PLL_4 for reconstruction of a copy of the chaotically modulated carrier signal which is then demodulated in PLL_6. The voltage proportional to the chaotically modulated carrier is read from the output of demodulator PLL_6. The useful information signal is obtained by subtracting the signals from the demodulators PLL_5 and PLL_6.

Results of numerical modeling of the operation of the system for information transmission are shown in Figure 24.25. The parameters γ and ε in PLL_3 and PLL_4 were specified with admissible deviations from the parameters of PLL_2 within 5%. One can see in this figure the signals at the outputs of the phase discriminators of PLL_1–PLL_4 (Figures 24.25a-d), at the outputs of the demodulators PLL_5, PLL_6 (Figures 24.25e,f), transmitted (Figure 24.25g) and received (Figure 24.25h) signals.

Thus, if there is a pronounced difference between the loops of a PLL, for attaining synchronization, i.e., for matching chaotic signals during information transmission, additional control couplings are needed between the loops. This, in turn, demands the presence of three coupling channels between the transmitter and the receiver. However, more stringent requirements imposed on parameter deviations in the receiver and transmitter will diminish the needed number of channels that can be restricted to only two. In the latter case, PLL_4 may be eliminated and the signal from the second channel will be transmitted directly to demodulator PLL_6. The simplest scheme may be realized for transmission of a binary signal. One communication channel is sufficient in this case.

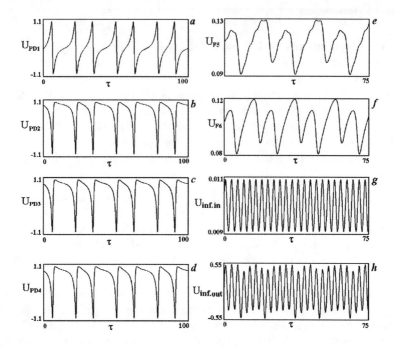

FIGURE 24.25
Results of modeling information transmission using chaotic signals according to the scheme shown in Figure 24.24.

24.6 Conclusion

Results of numerical modeling verify that a second-order filter PLL system may be regarded as a generator of chaotically modulated oscillations with the average frequency stabilized by the reference signal. However, such a generator possesses relatively narrow regions of the parameters corresponding to generation of chaotically modulated oscillations of the same type. The generator of CMO based on a small ensemble of cascade-coupled PLLs is free of this drawback. In addition, first-order filters are quite sufficient for control of such PLLs, which makes requirements to tuning the parameters much less stringent.

It has been established that chaotically modulated oscillations from two generators of chaos may be synchronized with the aid of a control loop, with hysteresis being intrinsic in such synchronization processes (the presence of pull-in and pull-out regions).

Results of modeling indicate that the dynamical features of the considered generators of CMOs (the possibility of synchronization) enable one to construct a system for transmission of information using dynamical chaos. The collective dynamics of small ensembles of PLLs provides rich opportunities for steady generation of chaotic oscillations and their control aimed at synchronization, which is highly significant for creating systems for information transmission.

Of course, it was not our goal to give specific recommendations on designing systems for coupling. We intended to draw the reader's attention to the exceptionally rich and promising potentialities of the collective dynamics of coupled PLL systems for generation and synchronization of chaotic oscillations and their possible applications to information transmission. Further research will depend significantly on results of experimental testing of the proposed schemes.

Acknowledgments

The research was supported in part by the Russian Foundation for Basic Research (grant 96–02–16559) and under the Program of Support of Leading Scientific Schools of the Russian Federation (grant 96–15–96593).

References

[1] H. D. I. Abarbanel, M. I. Rabinovich, A. I. Selverston, M. V. Bazhenov, R. Huerta, M. M. Sushchik, and L. L. Rubchinsky, "Synchronization of neural networks," *Phisics–Uspekhi*, vol. 39, pp. 337–362, 1996.

[2] V. S. Afraimovich, V. I. Nekorkin, G. V. Osipov, and V. D. Shalfeev, *Stability, Sructures and Chaos in Nonlinear Synchronization Networks*, World Scientific Pub. Co., Singapore, 1994.

[3] Yu. V. Andreev, A. S Dmitriev, and A. A. Kuminov, "Chaotic processors," (in Russian) *Zarubezhnaja Radio Electronika, Uspekhi Sovremennoi Radioelektroniki*, vol. 10, pp. 50–79, 1997.

[4] V. S. Anischenko, *Dynamical Chaos – Models and Experiments*, World Scientific Pub. Co., Singapore, 1995.

[5] L. N. Belyustina and V. N. Belykh, "Dynamic system on cylinder," (in Russian) *Diff. uravneniya.*, vol. 9, pp. 403–415, 1973.

[6] V. N. Belykh and V. I. Nekorkin, "Bifurcations in the third-order nonlinear equation of phase synchronization," (in Russian) *Prikladnaya Matematika*, vol. 42, pp. 808–815, 1978.

[7] L. O. Chua, "Global unfolding of Chua's circuit," *IEICE Trans. Fundamentals*, vol. E76A, pp. 704-734, 1993.

[8] H. Dedieu, M. P. Kennedy, and M. Hasler, "Chaos shift keying: Modulation and demodulation of a chaotic carrier using self-synchronizing Chua's circuits," *IEEE Trans. on Circ. Sys.*, I, vol. 40, pp. 634-642, 1993.

[9] A. S. Dmitriev and V. Y. Kislov, *Stochastic Oscillations in Radiophysic and Electronic Circuits*, Nauka, Moscow, 1989.

[10] A. S. Dmitriev, A. I. Panas, and S. O. Starkov, "Experiments on speach and music signals transmission using chaos," *Int. J. Bifur. Chaos*, vol. 5, pp. 1249-1254, 1995.

[11] A. S. Dmitriev, A. I. Panas, and S. O. Starkov, "Dynamic chaos as paradigm of communication," (in Russian) *Zarubezhnaja Radio Electronika, Uspekhi Sovremennoi Radioelektroniki*, vol. 10, pp. 4–26, 1997.

[12] T. Endo, "A review of chaos and nonlinear dynamics in phase-locked loops," *J. of Franklin Institute*, vol. 331B, pp. 859–902, 1994.

[13] A. V. Gaponov-Grekhov and M. I. Rabinovich, *Nonlinearities in Action*, Springer, Berlin, 1992.

[14] M. Hasler, "Engineering chaos for secure communication systems," *Phyl. Trans R. Soc. Lond. A*, vol. 353, pp. 1701–1709, 1995.

[15] S. Hayes, G. Grebogi, and E. Ott, "Communication with chaos," *Phys. Rev. Lett.*, vol. 70, pp. 3031-3034, 1993.

[16] M. P. Kennedy, "Communication with chaos: State of the art and engineering challenges," *Proc. NDES'96*, Seville, 1996, pp. 1–8.

[17] S. V. Kiyashko, A. S. Pikovsky, and M.I. Rabinovich, "Chaotic generator," *Radiotekhnika i Electronika*, vol. 25, pp. 336-343, 1980.

[18] L. Kocarev, K. S. Halle, K. Eckert, L. O. Chua, and U. Parlitz, "Experimental demonstration of secure communication via chaotic synchronization," *Int J. of Bifur. Chaos*, vol. 2, pp. 709-713, 1992.

[19] G. Kolumban and B. Vizvari, "Nonlinear dynamics and chaotic behaviour of the analog phase-locked loop," *Proc. NDES'95*, Dublin, 1995, pp. 99–102.

[20] M. Korzinova, V. Matrosov, and V. Shalfeev, "Communications using cascade coupled phase-locked loops chaos," *Int. J. of Bifur. Chaos*, vol. 9, 1999, in press.

[21] A. K. Kozlov and V. D. Shalfeev, Controlling chaotic oscillations in a generator with delayed phase-locked loop," (in Russian) *Izv. VUZov, Prikladnaya Nelineynaya Dinamika*, vol. 2, pp. 36–48, 1994.

[22] A. K. Kozlov and V. D. Shalfeev, "Chaos in controlled generators," *Proc. NDES'95*. Dublin, 1995, pp. 233–236.

[23] A. K. Kozlov and V.D. Shalfeev, "Selective suppression of deterministic chaotic signals," *Techn. Phys. Lett.*, vol. 19, pp. 769–770, 1993.

[24] A. J. Lichtenberg and M. A. Lieberman, *Regular and Stochastic Motion*, Springer, New York, 1983.

[25] W. Lindsey, *Synchronization Systems in Communication and Conrol*, Prentice–Hall Inc., Englewood Cliffs, New Jersey, 1972.

[26] R. Madan (Ed.), *Chua's Circuit: A Paradigm for Chaos*, World Scientific Pub. Co., Singapore, 1993.

[27] V. V. Matrosov, "Regular and chaotic oscillations of phase systems," *Techn. Phys. Lett.*, vol. 22, pp. 4–8, 1996.

[28] V. V. Matrosov, "Some particularities of dynamical behavior of two cascade coupled phase locked loops," (in Russian) *Izv. VUZov, Prikladnaya Nelineynaya Dinamika*, vol. 6, pp. 52–61, 1997.

[29] V. V. Matrosov, "Regular and chaotic oscillations of phase-locked loop with the second-order filter," *Proc. NDES'97*, Moscow, Russia, 1997, pp. 554–558.

[30] V. V. Matrosov and M. V. Korzinova, Modelling of nonlinear dynamic cascade phase-locked loop," *Radiophysics and Quantum Electronics*, vol. 36, pp. 555–559, 1993.

[31] V. V. Matrosov and M. V. Korzinova, "Collective dynamics of cascade coupled phase systems," (in Russian) *Izv. VUZov, Prikladnaya Nelineynaya Dinamika*, vol. 2, pp. 10–16, 1994.

[32] V. V. Matrosov and M. V. Shalfeeva, "Influence of coupling parameters on the nonlinear dynamics of two cascade-coupled phase-locked loops," *Radiophysics and Quantum Electronics*, vol. 38, pp. 180–184, 1995.

[33] V. V. Matrosov and M.V. Korzinova, "Synchronization and auto-oscillations in cascade coupled phase systems," *Proc. Nizhni Novgorod Univ. Nonlinear Dynamics*, 1996, pp. 77–92.

[34] A. L. Oppenheim, G. W. Wornell, S. H. Isabell, and K. M. Cuomo, "Signal processing in the context of chaotic systems," *Proc. IEEE ICASSP*, 1992, pp. VI117–120.

[35] M.I. Rabinovich and D. I. Trubetskov, *Introduction to the Theory of Oscillations and Waves*, Kluwer, Amsterdam, 1989.

[36] A. Sato and T. Endo, "Experiments of secure communications of phase-locked loops," *Proc. NDES'94*, 1994, pp. 117–122.

[37] V. Shakhgildyan and L. Belyustina (Eds.), *Phase Synchronization Systems*, Radio i Svyaz, Moscow, 1982 (in Russian).

[38] V. D. Shalfeev and V. V. Matrosov, "Pull–in and pull–out effects of synchronization of chaotic modulated oscillations," *Radiophysics and Quantum Electronics*, Vol. 41, No. 12, 1999, in press.

[39] V. Shalfeev, V. Matrosov, and M. Korzinova, Dynamical chaos in ensembles of coupled phase systems," (in Russian), *Zarubezhnaja Radio Electronika, Uspekhi Sovremennoi Radioelektroniki*, vol. 11, pp. 44–56, 1998.

[40] V. D. Shalfeev, V. V. Matrosov and M. V. Shalfeeva, "Chaos synchronization in coupled phase systems," *Proc. IEEE ISCAS'98*, New York, IEEE, 1998, pp. VI580–582.

[41] V. D. Shalfeev, G. V. Osipov, A. K. Kozlov, and A. R. Volkovskii, "Chaotic oscillations – generation, synchronization, controlling," (in Russian), *Zarubezhnaja Radio Electronika, Uspekhi Sovremennoi Radioelektroniki*, vol. 10, pp. 27–49, 1997.

[42] N. Smyth, C. Crowley, and M. P. Kennedy, "Improved receiver for CSK spread spectrum communications using analog phase locked loop chaos," *Proc. NDES'96*, Seville, 1996, pp. 27–32.

[43] A. R. Volkovskii and N. F. Rul'kov, "Synchronous chaotic response of a nonlinear oscillator system as a principle for detection of the information component of chaos," *Techn. Phys. Lett.*, vol. 19, pp. 71–75, 1993.

25

Chaos and Bifurcations in Feedback Control Systems

Joaquin Alvarez[1] and **Fernando Verduzco**[2]

[1] Scientific Research and Advanced Studies Center
of Ensenada (CICESE), 22860 Ensenada, BC, Mexico
jqalvar@cicese.mx

[2] Technological Institute of Sonora (ITSON)
85000 Cd. Obregon, Son, Mexico
verduzco@itson.mx

Abstract

> It is shown how simple control systems can exhibit complicated dynamics. Low dimensional systems with classical controllers are considered. Some kinds of bifurcations of equilibria, period-doubling, and chaotic behavior are analyzed by applying tools from dynamical systems theory and electrical engineering. Numerical and experimental results are included.

25.1 Introduction

Nonlinear behavior and feedback, two fundamental concepts for the control engineer, constitute essential properties of a system displaying chaotic dynamics. This behavior is sometimes denoted as complex, but the structure of dynamical processes displaying chaos may be very simple. In fact, for several years it has been known that simple feedback control systems can exhibit chaotic dynamics [3, 5].

At present, no general methods exist to predict chaotic behavior. In this way, two important routes to chaos are the cascade of period-doubling bifurcations and the generation of horseshoes produced by homoclinic bifurcations. Unfortunately, the analytical prediction of these mechanisms is not easy, and numerical experimentation is normally the only available way to analyze and predict chaos in physical systems.

An interesting feature that characterizes the generation of period-doubling bifurcations and horseshoes is the interaction between equilibrium points and limit cycles. Three techniques that consider these interactions to detect possible routes to chaos are the analytical methods of Melnikov/Smale and Shil'nikov, and the approximate method of harmonic balance. This chapter describes how these tools can be applied to analyze low dimensional plants fed back with classical controllers, showing how these simple systems can display bifurcations, closed orbits, and chaos. The analysis includes linear plants with controllers having Proportional (P), Integral (I), or Derivative (D) actions, and a static nonlinear block. Also, an important class of low dimensional robot manipulators, fed back with PD controllers, is analyzed. A more detailed treatment can be found in [1, 2, 15].

25.2 Prediction of Chaos

Given a dynamical system, $\dot{x} = f(x, t; \alpha)$, $x \in \mathbb{R}^n$, $\alpha \in \mathbb{R}$, a bifurcation occurs when a small variation in α produces a qualitative change in the system dynamics, for instance, a change in the number of equilibrium points or the appearance of periodic orbits. Elementary bifurcations are the saddle-node, transcritical, pitchfork, and Hopf [10]. Another important class of bifurcations are the so-called homoclinic. A heteroclinic orbit is a state trajectory that joins an equilibrium point to another equilibrium; a homoclinic orbit joins an equilibrium with itself. A bifurcation occurs when a small change in the parameter "breaks" the homoclinic orbit. This scenario may generate limit cycles or other more complex invariant sets.

A classical technique of analysis replaces the flow of a continuous system with a discrete system resulting from the intersection of the continuous flow with a transversal surface. This discrete system is called the Poincaré map, from which some properties of the continuous system may be inferred. In particular, if the map has a complex invariant set, then the continuous system also exhibits a complex behavior. The existence of these sets can be determined if the Poincaré map is equivalent to a discrete system for which the presence of a complex invariant set has been shown.

A typical system with an invariant set having a Cantor structure is the Smale horseshoe. This two-dimensional map stretches a set of initial conditions along one direction, compresses it along another direction, and folds the resultant elongated set. It is just this mechanism that can be induced when a homoclinic trajectory is "broken" by a small disturbance on the system (Figure 25.1).

Melnikov/Smale method. It is a technique to test the intersection of stable and unstable manifolds arising from small perturbations on homoclinic orbits. A transversal intersection may generate complex dynamics. The simplest case considers a Hamiltonian system given by

$$\dot{x} = X_H(x) + \varepsilon g(x,t), \ x \in \mathbb{R}^2, \ \varepsilon > 0, \qquad (25.2.1)$$

where g is periodic in t and X_H is a Hamiltonian vector field with Hamiltonian H. Suppose that, for $\varepsilon = 0$, system (25.2.1) has a homoclinic orbit $\gamma(t)$ to a saddle point p_0 (Figure 25.1). The Melnikov function of system (25.2.1) is defined as

$$\mathcal{M}(t_0) = \int_{-\infty}^{\infty} (X_H \wedge g)(\gamma(t - t_0))dt. \qquad (25.2.2)$$

Suppose also that (25.2.1) has a family of periodic orbits inside the region delimited by $\gamma(t)$, $t \in \mathbb{R}$. Let $p_\varepsilon^{t_0} = p_0 + \mathcal{O}(\varepsilon)$ be the unique saddle of the Poincaré map $P_\varepsilon^{t_0}$ associated to (25.2.1), and $W^{s,u}(p_\varepsilon^{t_0})$ be the stable and unstable manifolds of $p_\varepsilon^{t_0}$. If these manifolds intersect transversely for $\varepsilon \neq 0$, then (25.2.1) may display chaos. This is stated by the next theorem, where it is supposed that the conditions above hold [16].

THEOREM 25.1
If $\mathcal{M}(t_0)$ has simple zeros then, for ε small enough, there exists an integer $n \geq 1$ such that the system $y_{k+1} = (P_\varepsilon^{t_0})^n(y_k)$ has an invariant Cantor set on which it is equivalent (topologically conjugate) to a horseshoe map.

A small perturbation breaks the homoclinic orbit, giving rise to a homoclinic tangle. A rectangle near the equilibrium point is submitted, under the Poincaré map, to strong stretching and folding; these are fundamental ingredients in the generation of homoclinic chaos (Figure 25.1).

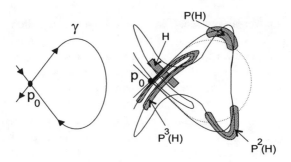

FIGURE 25.1
Homoclinic orbit and homoclinic tangle.

An approximate method. An alternative approach that establishes some conditions on equilibrium points and limit cycles for the possible existence of chaos has been proposed in [6, 7], where the authors used the harmonic balance method (HBM) and demonstrated the feedback canonical representation of some nonlinear systems.

The method to detect chaos considers a system in the Lur'e form, with a linear part represented by the transfer function $L(s)$ with input u and output y, and a static nonlinear part with input y and output $n(y) = -u$. If this system has an equilibrium point with a corresponding output $y(t) \equiv E$, then $E + L(0)n(E) = 0$ must be satisfied.

On the other hand, the existence of periodic orbits can be approximately predicted by the HBM [12]. Suppose that the system has a limit cycle of frequency ω, producing an output with Fourier expansion $y(t) = \sum_{k=-\infty}^{\infty} A_k \exp(jk\omega t)$. This signal drives the nonlinear block, whose output is $n(y(t)) = \sum_{k=-\infty}^{\infty} \tilde{N}_k \exp(jk\omega t)$. Using these expansions, the harmonic balance equations can be obtained,

$$A_k + L(jk\omega)\tilde{N}_k = 0, \ k \in Z. \tag{25.2.3}$$

These equations must be satisfied if the limit cycle indeed exists. If the higher harmonics are attenuated by the linear part, then the output can be given by $y(t) \cong y_0(t) = a \sin \omega t + b$. Also, if the Fourier expansion of the nonlinear block output can be approximated up to the first order, then (25.2.3) reduces to

$$b[1 + N_0(a,b)L(0)] = 0, \ \frac{1}{2j} + N_1(a,b)L(j\omega) = 0, \tag{25.2.4}$$

where $b = A_0$, $a = j2A_1$, $N_0 = \tilde{N}_0/b$, and $N_1 = \tilde{N}_1/a$. These equations give an approximate condition for a limit cycle to exist. Now, a linearization at $\omega/2$ of the HBM around this limit cycle yields the condition to have a double-period oscillation, given approximately by $y = y_0 + \varepsilon \sin(\omega t/2 - \phi)$,

$$1 + L(j\omega/2)N_{1/2}(a, b, \phi) = 0, \qquad (25.2.5)$$

where $N_{1/2} = \int_{-\pi}^{\pi} n'(y_0)\sin(\theta - \phi)e^{-j(\theta - \phi - \pi/2)}d\theta/\pi$, with $\theta = \omega t/2$.

The appearance of double-period orbits is not enough to guarantee a chaotic behavior, but usually indicates the onset of a sequence of similar events leading to a cascade of period-doubling (PD) bifurcations. Therefore, the conditions for possible existence of chaos via PD are the existence of a stable limit cycle and the onset of a related stable subharmonic term (Equations (25.2.4) and (25.2.5) and a stability test).

Another mechanism considered in [7] is the Shil'nikov scenario, where homoclinic trajectories in systems of dimension produce at least three horseshoes [10]. Consider the system $\dot{x} = f(x,\alpha)$, $x \in \mathbb{R}^3$, $\alpha \in \mathbb{R}$, with f smooth. Suppose that, for $\alpha = 0$, it has an orbit γ_0 homoclinic to the origin, with eigenvalues $\lambda > 0$ and $\sigma \pm j\omega$, $\sigma < 0$. The Shil'nikov theorem [10] establishes that, if $|\sigma| < \lambda$, then for $\alpha \neq 0$ it also has a homoclinic orbit γ near γ_0, and the Poincaré map of γ has a countable set of horseshoes.

The main difficulty in applying Shil'nikov-type results is to prove the existence of homoclinic trajectories. In fact, these trajectories are structurally unstable, therefore, very difficult to detect. Nevertheless, the existence of homoclinic orbits is often related to the collision of an equilibrium with a limit cycle [10]. This means that a homoclinic orbit may bifurcate to a limit cycle with the equilibrium persisting. Inversely, a homoclinic orbit may be produced when a limit cycle is close to a saddle equilibrium, so a small disturbance may produce a collision of these sets. In terms of the projection onto the output space, this condition may be given by $a = \theta |E - b|$, with $\theta \cong 1$. Therefore, the conditions to predict this scenario are the existence of a saddle equilibrium, the existence of a stable limit cycle, via (25.2.4), and the superposition of both sets, measured via θ [7].

25.3 Linear Plants with Classical Controllers

A large number of dynamical processes operating around an equilibrium point can be modeled by a standard linear second-order plant characterized by a natural frequency ω_n and a damping coefficient ζ. After a time scaling by ω_n, this plant can be described by

$$G(s) = \frac{k}{s^2 + 2\zeta s + 1}, \qquad (25.3.6)$$

which is proposed to be controlled with the compensator

$$C(s) = \frac{n_1 s + n_0}{s + d_0} \qquad (25.3.7)$$

and a static nonlinear block $u^* = \gamma(m)$ (Figure 25.2).

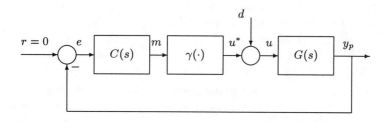

FIGURE 25.2
A classical controller with a nonlinear block.

In what follows, some control schemes will be analyzed. The existence, number, and type of equilibrium points, the existence of closed orbits, and the conditions for the existence of chaos will be established.

P-controller with saturation. If $n_0 = n_1 d_0$, then $C = n_1 = k_p$. A saturation $\gamma(m) = (|m+1| - |m-1|)/2$ leads to the closed-loop system

$$\begin{pmatrix} \dot{x}_1 \\ \dot{x}_2 \end{pmatrix} = \begin{pmatrix} 0 & 1 \\ -1 & -2\zeta \end{pmatrix} \begin{pmatrix} x_1 \\ x_2 \end{pmatrix} - K \begin{pmatrix} 0 \\ \gamma(x_1) - d \end{pmatrix}, \qquad (25.3.8)$$

where $x_1 = k_p y_p$, $x_2 = k_p \dot{y}_p$, and $K = k_p k$. Suppose a zero disturbance, and denote $\theta = (K, \zeta)$. The set of equilibrium points, which have the form $\bar{x} = (\bar{x}_1, 0)$, is shown in Figure 25.3.

From the characteristic polynomial around the equilibrium $(\bar{x}, \bar{\theta})$, the zones of nonhyperbolic equilibria can be obtained (Figure 25.3): (a) $|\bar{x}_1| > 1$, $\zeta = 0$, $\bar{K} < -1$ (I, $\lambda_{1,2} = \pm j$); (b) $|\bar{x}_1| < 1$, $\zeta > 0$, $\bar{K} = -1$ (V, $\lambda_1 = 0, \lambda_2 < 0$); (c) $|\bar{x}_1| = 0$, $\zeta = 0$, $\bar{K} > -1$ (VII, $\lambda_{1,2} = \pm j\omega$); (d) $|\bar{x}_1| < 1$, $\zeta < 0$, $\bar{K} = -1$ (X, $\lambda_1 = 0, \lambda_2 > 0$); and (e) $|\bar{x}_1| < 1$, $\zeta = 0$, $\bar{K} = -1$ (VIII, $\lambda_{1,2} = 0$). All these cases correspond to bifurcations of codimension one, except the last one, which is codimension two [10]. Cases (a) and (c) correspond to a trivial change of stability. For cases (b) and (d), Bendixson's criterion [13] shows that the system has no closed orbits in $\mathbb{R}^2$. The system has three equilibrium points for $K < -1$, one for $K > -1$, and a continuum of equilibria for $K = -1$; then it undergoes a degenerate pitchfork bifurcation.

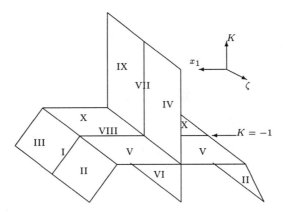

FIGURE 25.3
Equilibria surface of the P-saturation control system: II, IV (stable focus or nodes); III, IX (unstable focus or nodes); VI (saddle); I, V, VII, VIII, and X (nonhyperbolic equilibria).

When $\zeta = 0$, system (25.3.8) is Hamiltonian, with $H = x_2^2/2 + V(x_1)$, where V is the potential

$$V(x) = \begin{cases} \frac{1}{2}x^2 - K(x + \frac{1}{2}), & x < -1; \\ \frac{1}{2}(1+K)x^2, & |x| \leq 1; \\ \frac{1}{2}x^2 + K(x - \frac{1}{2}), & x > 1. \end{cases} \quad (25.3.9)$$

From the properties of V it can be concluded that, if $K > -1$, the unique equilibrium point $\bar{x} = (0,0)$ is a center. When $K = -1$ the trajectories are closed orbits if $x_2(0) \neq 0$, and a point if $x_2(0) = 0$ and $|x_1(0)| < 1$. Finally, when $K < -1$ the system has a saddle equilibrium at the origin, self-connected with two symmetric homoclinic orbits. Any other trajectory different from the three equilibria and the two homoclinic orbits is periodic. The homoclinic orbit x^h corresponding to $x_1 > 0$ is given by

$$x^h(t) = \begin{cases} (e^{k_1(t-t_1)}, k_1 e^{k_1(t-t_1)}), & t \leq t_1; \\ (-K + k_2 \cos t, -k_2 \sin t), & t_1 < t < t_2; \\ (e^{-k_1(t-t_2)}, -k_1 e^{-k_1(t-t_2)}), & t \geq t_2, \end{cases} \quad (25.3.10)$$

where $k_1 = \sqrt{-1-K}$, $k_2 = \sqrt{K(1+K)}$, $t_1 = \arcsin\sqrt{1/K+1} - \pi/2$, and $t_2 = -t_1$. The other orbit ($x_1 < 0$) is given by a similar expression. Figure 25.4 summarizes the different dynamical behaviors of this system for all values of parameters (ζ, K), which are directly related to the dissipation and the proportional gain.

The Melnikov/Smale method gives the conditions for an invariant strange set in this system. For small dissipation and perturbation terms $\zeta = \varepsilon\varsigma$,

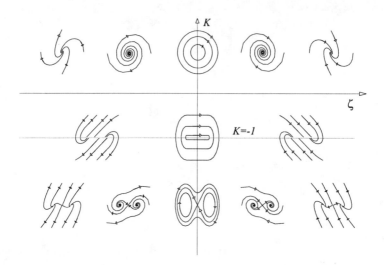

FIGURE 25.4
Phase portrait of the P-saturation control system.

$d(t) = \varepsilon \sin \omega t / K$, the Melnikov integral (25.2.2) is given by

$$\mathcal{M}(\varsigma, K, \omega, t_0) = -2k\left[\mathcal{M}_0(\varsigma, K) + \mathcal{M}_1(\omega, K) \cos \omega t_0\right], \quad (25.3.11)$$

$$\mathcal{M}_0 = -\varsigma(k_1^2 - 1 - Kk_1 t_1), \quad \mathcal{M}_1 = \frac{(2\omega^2 + k_1^2 - 1)(\omega \cos \omega t_1 - k_1 \sin \omega t_1)}{(\omega^2 - 1)(k_1^2 + \omega^2)}. \quad (25.3.12)$$

Therefore, $\mathcal{M}$ will have simple zeros, and the system will be chaotic, if

$$\left| \frac{\varsigma(k_1^2 - 1 - Kk_1 t_1)(1 - \omega^2)(k_1^2 + \omega^2)}{(\omega \cos \omega t_1 - k_1 \sin \omega t_1)(k_1^2 - 1 + 2\omega^2)} \right| < 1, \quad (25.3.13)$$

as shown in Figure 25.5. An electronic realization of this control system displays chaotic trajectories like those shown in Figure 25.6.

I-controller with saturation. When $n_0 = k_i$, $n_1 = k_p$, $d_0 = 0$, and the nonlinear block is a saturation, the control system has a unique equilibrium point at the origin, which is stable if $\zeta > 0$ and $0 < kk_i < 2\zeta\omega_n(1 + kk_i)$; otherwise, it is unstable. An interesting case appears for a pure integral controller and a saturation with slope one at the origin and that saturates at $m_l < 0$ if $m < m_l$ and $m_u > 0$ if $m > m_u$, with $|m_l| \neq |m_u|$.

This system is well adapted to be analyzed with the HBM. A direct application of this method yields $N_0 = 0$ (see Equation (25.2.4)). For the

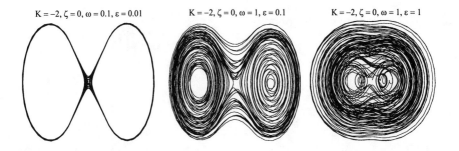

FIGURE 25.5
Chaotic trajectories of the P-saturation control system.

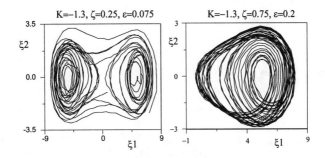

FIGURE 25.6
Chaotic trajectories of an electronic realization of the P-saturation control system.

particular values $m_l = -0.5$, $m_u = 5$, it is possible to obtain

$$N_1(a) \approx \frac{a\cos\alpha\sin\alpha + 2(b-m)\cos\alpha + a(\pi/2 - \alpha)}{\pi a}, \quad (25.3.14)$$

where $b \cong -0.8733a + 1.2294$ and $\alpha = \arcsin((m_l - b)/a)$. From (25.2.4) it can be shown that there exists a limit cycle with $\omega \cong 1$ and amplitude $|L(j)| \cong |K/2\zeta\omega_n|$, where $K = k_i k$. In fact, for some $a > 0$, there exists a stable limit cycle if $N_1(a) \approx 2\zeta/K$. In the same way, $\cos^2\alpha - [\alpha + \pi(\zeta/K - 1)/2]^2 - 9\pi^2/64K^2 = 0$ is a condition for a subharmonic of frequency $\omega/2$ to exist.

This system has a unique equilibrium, stable limit cycles, and double-period orbits; therefore, period-doubling bifurcations may be present, as shown in Figure 25.7. This Feigenbaum diagram is obtained by varying one parameter (K in this case) and sampling the maxima of the steady-state oscillation. A splitting of a branch indicates the appearance of a subharmonic that duplicates the period of the oscillation. Figure 25.8 shows a typical

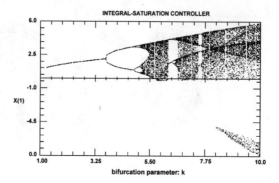

FIGURE 25.7
Bifurcation diagram of the I-saturation control system.

chaotic trajectory, while Figure 25.9 shows some attractors obtained from an electronic realization of this control system.

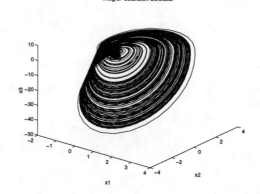

FIGURE 25.8
Chaos in the I-saturation control system. $\zeta = 0.3$, $K = 8$.

I-controller with a cubic polynomial. The last example of this section is given by an I-controller with a nonlinear block $\gamma = m(m^2 - m_0^2)$. This system has the equilibria $p_0 = 0$, $p_{1,2} = (0, 0, \pm m_0)$ for any point $(\zeta, k, k_i/\omega_n) \in \mathbb{R}^3$. The smooth nature of the polynomial and the application of the HBM permit one to show the existence of limit cycles and double-period oscillations; therefore, the period-doubling scenario may be present. However, the Shil'nikov scenario may also be present; in fact, it can be shown that the interaction equilibrium point-limit cycle is possible if

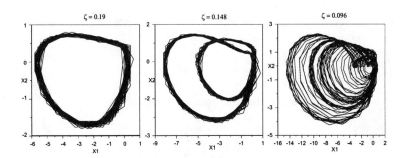

FIGURE 25.9
Experimental trajectories of the I-saturation control system. $K = 8$.

$b \neq 0$ (resp. $b = 0$) and $|\zeta/k_i| = m_0^2/3 - \varepsilon$ (resp. $\zeta/k_i = -m_0^2/6 - \varepsilon$), with ε a small positive number. This interaction produces strange attractors like those shown in Figure 25.10.

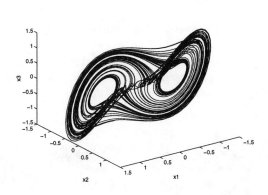

FIGURE 25.10
Chaos in the I-control system with a cubic polynomial. $m_0 = 1$, $\zeta = 0.5$, $kk_i = 1.4$.

25.4 PD-Controlled Robot Manipulators

Control of mechanical manipulators is a field of intense activity in robotics; to enhance their dynamical performance, many algorithms can be found in the literature (see, e.g., [14]). Among these algorithms, the Proportional-Derivative (PD) controller is widely used. For these systems, a complex

25.4.1 1-DOF System

Consider first a PD-controlled pendulum described by

$$I_a \frac{d^2 q}{dt^2} + e \frac{dq}{dt} + mgl_c \sin q = \tau, \tag{25.4.15}$$

$$\tau = k_p(q_d - q) + k_v(k_d \dot{q}_d - \dot{q}), \tag{25.4.16}$$

where q is the joint position, τ the driving torque, $I_a = I + ml_c^2$, I the moment of inertia, m the total mass, l_c the distance from the joint axis to the center of mass, e the frictional force coefficient, g the gravity acceleration, k_p and k_v the controller gains, q_d the desired position, and k_d permits to have velocity ($k_d = 0$) or velocity error feedback ($k_d = 1$). If $M = mgl_c/I_a$, $\mu_1 = k_p/MI_a$, $\mu_2 = (k_v + e)/I_a \sqrt{M}$, $\delta = k_v/I_a \sqrt{M}$, $x = q$, $y = \dot{q}/\sqrt{M}$, and $s = \sqrt{M}t$, then the next system is obtained,

$$\begin{aligned} \dot{x} &= y \\ \dot{y} &= -\sin x - \mu_1 x - \mu_2 y + \mu_1 q_d + \delta k_d \dot{q}_d. \end{aligned} \tag{25.4.17}$$

Bifurcations. If q_d is constant, the equilibrium points satisfy $y = 0$ and $\mu_1(q_d - x) = \sin x$. These points form a surface in the (μ_2, x, μ_1)-space, whose qualitative form is shown in Figure 25.11.

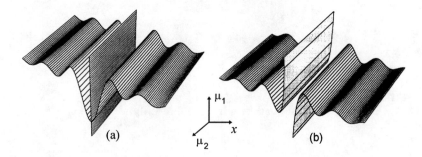

FIGURE 25.11
Qualitative form of the equilibrium surfaces of the PD-controlled pendulum, $q_d = k\pi$. (a) $k \in Z$; (b) $k \notin Z$.

The Jacobian matrix for these points has a characteristic polynomial $p(\lambda) = \lambda^2 + \mu_2 \lambda + \cos \bar{x} + \bar{\mu}_1$, which has roots with zero real part in the next cases: (a) $\cos \bar{x} + \bar{\mu}_1 = 0$, $\mu_2 \neq 0$ ($\lambda_1 = 0, \lambda_2 \neq 0$); (b) $\cos \bar{x} + \bar{\mu}_1 > 0$, $\mu_2 = 0$ ($\lambda_{1,2} = \pm i\omega$); (c) $\cos \bar{x} + \bar{\mu}_1 = 0$, $\mu_2 = 0$ ($\lambda_{1,2} = 0$). For case (a), a direct application of Sotomayor's theorem [13] permits one to establish the next result.

THEOREM 25.2
Let $(\bar{x}, \bar{\mu}_1)$ be a point satisfying $\cos\bar{x} + \bar{\mu}_1 = 0$, $\mu_2 \neq 0$, and $k \in Z$. Therefore, if $\bar{x} \neq k\pi$ (resp., $\bar{x} = k\pi$), then the PD-controlled pendulum (25.4.17) undergoes a saddle-node (resp., pitchfork) bifurcation.

When $\mu_2 = 0$ (cases (b) and (c)), system (25.4.17) is conservative, with potential $V = -\cos x + \mu_1 x^2/2 - \mu_1 q_d x + 1$. If $(\bar{x}, \bar{\mu}_1)$ satisfies $\cos\bar{x} + \bar{\mu}_1 > 0$, then $V' = 0$ and $V'' > 0$. Therefore, $\bar{x}$ is a minimum of V, and $(\bar{x}, 0)$ is a center. Finally, if $(\bar{x}, \bar{\mu}_1)$ satisfies $\cos\bar{x} + \bar{\mu}_1 = 0$, then $V' = 0 = V''$, so $\bar{x}$ is a degenerate critical point of V, and the next result follows (see Figure 25.12).

THEOREM 25.3
Suppose that $\mu_2 = 0$. Then, every point $\bar{p} = (\bar{x}, \bar{\mu}_1)$ satisfying $\cos\bar{x} + \bar{\mu}_1 > 0$ is a center of the PD-controlled pendulum (25.4.17). Moreover, when $\cos\bar{x} + \bar{\mu}_1 = 0$ then, for $\bar{x} = k\pi$, $k \in \mathbb{R}$, if k is not an integer, then the system undergoes a saddle-center bifurcation; if k is an even integer, then it has a degenerate saddle point; and if k is odd, then it has a degenerate center.

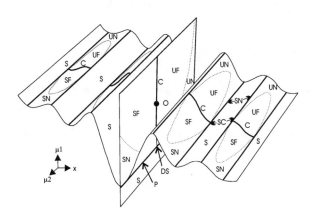

FIGURE 25.12
Bifurcation diagram of the PD-controlled pendulum for $q_d = 0$. — bifurcation curves, - - transition curves. O (origin), SN (stable node), SF (stable focus), UN (unstable node), UF (unstable focus), C (center), S (saddle), S-C (saddle-center), S-N (saddle-node), P (pitchfork), DS (degenerate saddle).

Homoclinic chaos. Now suppose that the desired position is periodic, $q_d = a\sin\omega t + b$, and that μ_1, μ_2, and δ can be given as $\mu_1 = \varepsilon\tilde{\mu}_1$, $\mu_2 = \varepsilon\tilde{\mu}_2$, $\delta = \varepsilon\tilde{\delta}$, with ε small. Then system (25.4.17) transforms to

$$\begin{aligned}\dot{x} &= y, \\ \dot{y} &= -\sin x + \varepsilon h(x,y,t),\end{aligned} \qquad (25.4.18)$$

where $h = -\tilde{\mu}_1 x - \tilde{\mu}_2 y + \tilde{\mu}_1 b + \tilde{N}\sin(\omega t + \phi)$, $\tilde{N} = a\sqrt{\tilde{\mu}_1^2 + \tilde{\delta}^2\omega^2 k_d^2}$, and $\phi = \arctan(\delta\omega k_d/\mu_1)$. When $\varepsilon = 0$, system (25.4.18) has two hyperbolic equilibria, $p_0^\pm = (\pm\pi, 0)$, connected by $\gamma_0^\pm(t) = 2(\pm\arctan(\sinh t), \pm\operatorname{sech} t)$. Furthermore, if $\Gamma_0 = \{\gamma_0^\pm(t)\,|\,t \in \mathbb{R}\} \cup \{p_0^\pm\}$, then the interior of Γ_0 contains a family of periodic orbits. The Melnikov function (25.2.2) is given by $\mathcal{M}^\pm(t_0) = \mathcal{A}\sin(\omega t_0 + \phi) + \mathcal{B}$, where $\mathcal{A} = \pm 2\pi\tilde{N}\operatorname{sech}(\omega\pi/2)$ and $\mathcal{B} = -8\tilde{\mu}_2 \pm 2\pi b\tilde{\mu}_1$. $\mathcal{M}^\pm$ will have simple zeros if

$$|\mathcal{A}| > |\mathcal{B}| \iff \left|\pm\pi N\operatorname{sech}\left(\frac{\omega\pi}{2}\right)\right| > |-4\mu_2 \pm \pi b\mu_1|, \qquad (25.4.19)$$

which can be solved in terms of the controller gains,

$$\begin{aligned}0 \;<\; & \pi^2 C_1^2(a^2\operatorname{sech}^2(\omega\pi/2) - b^2)k_p^2 + C_2^2(\pi^2 a^2\operatorname{sech}^2(\omega\pi/2)k_d^2\omega^2 - 16)k_v^2 \\ & \pm 8\pi C_1 C_2 bk_p k_v \pm 8\pi C_1 C_2 ebk_p - 32C_2^2 ek_v - 16C_2^2 e^2,\end{aligned} \qquad (25.4.20)$$

where $C_1 = 1/mgl_c$ and $C_2 = 1/\sqrt{mgl_c I_a}$. The analysis of this second degree polynomial in (k_p, k_v) yields the next result, which establishes that homoclinic chaos may always exist in a PD-controlled pendulum.

THEOREM 25.4
Consider the PD-controlled pendulum (25.4.17), driven by $q_d = a\sin\omega t + b$, and denote $\Delta = mgl_c/I_a$. Therefore, if $\Delta > \pi^2/4$ and $\omega \approx 0$, then for any pair (a,b), with $a \neq 0$, there exist gains (k_p, k_v) such that the closed-loop system has homoclinic chaos.

When $\Delta \leq \pi^2/4$, some regions exist in the (a,b)-plane (white zones in Figure 25.13) for which there are no values (k_p, k_v) that guarantee homoclinic chaos. Also, this figure shows (shadow) zones in both planes where chaos exists. Finally, some typical chaotic trajectories are displaying in Figure 25.14.

25.4.2 2-DOF System

For systems with two or more degrees of freedom, Holmes and Marsden [9] developed a technique to detect homoclinic chaos. This procedure can be applied to integrable Hamiltonians containing homoclinic and periodic orbits. When the perturbation is Hamiltonian, the technique is a straightforward combination of a reduction procedure that allows one to apply

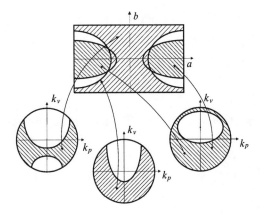

FIGURE 25.13
Parametric regions where chaos exists (PD-controlled, 1DOF robot).

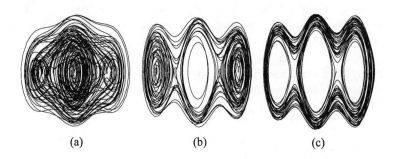

FIGURE 25.14
Chaotic trajectories $y(x)$ of the PD-controlled pendulum, for $\omega = 1$, $\theta = (k_d, k_p, k_v, a, b)$. **(a)** $\theta = (1, 0.2, 0.2, 0.1, 3)$; **(b)** $\theta = (1, 0.05, 0.004, 0.1, 5)$; **(c)** $\theta = (0, 0.05, -1, 1, 3)$.

Melnikov's method. For a non-Hamiltonian perturbation, it is necessary to establish an energy-transfer condition to ensure that at least one horseshoe survives near the energy balance point. This section describes the application of this technique to 2-DOF robot manipulators.

The motion equations of a two-link robot manipulator can be derived in the standard way from the Lagrangian formulation [14], using the generalized positions q and velocities $\dot{q}$. If $p = M\dot{q}$ is the generalized momentum, where M is the inertia matrix, then $H(q, p) = p^T M^{-1}(q) p / 2 + U(q)$ is the Hamiltonian of the undriven frictionless system, where U is the potential energy. Then the Hamiltonian equations of motion are

$$\begin{aligned}
\dot{q} &= \tfrac{\partial H}{\partial p}(q,p) = M^{-1}(q)p, \\
\dot{p} &= -\tfrac{\partial H}{\partial q}(q,p) - f(\dot{q}) + \tau \\
&= -\tfrac{1}{2}\tfrac{\partial}{\partial q}(p^T M^{-1}(q)p) - g(q) - f(M^{-1}(q)p) + \tau,
\end{aligned} \qquad (25.4.21)$$

where f and g are the viscous friction and gravity force terms, and τ is the torque. For the particular case of a 2-DOF robot manipulator moving in a vertical plane,

$$M(q) = \begin{pmatrix} \beta_1 & \beta_3 \cos(q_2 - q_1) \\ \beta_3 \cos(q_2 - q_1) & \beta_2 \end{pmatrix}, \qquad (25.4.22)$$

$$g(q) = \begin{pmatrix} \alpha_1 \sin q_1 \\ \alpha_2 \sin q_2 \end{pmatrix}, \quad f(\cdot) = E\dot{q} = \begin{pmatrix} e_1 & 0 \\ 0 & e_2 \end{pmatrix} M^{-1}(q)p, \qquad (25.4.23)$$

$$\begin{aligned}
\tau &= (\tau_1, \tau_2), \quad \alpha_1 = (m_1 l_{c_1} + m_2 l_1) g_0, \quad \alpha_2 = m_2 l_{c_2} g_0, \\
\beta_1 &= m_1 l_{c_1}^2 + m_2 l_1^2 + I_1, \quad \beta_2 = m_2 l_{c_2}^2 + I_2, \quad \beta_3 = m_2 l_1 l_{c_2},
\end{aligned} \qquad (25.4.24)$$

where the positions are taken with respect to a vertical line, m_i, l_i, l_{c_i}, and I_i are the mass, length, center of mass and inertia of link i, respectively, and g_0 is the acceleration of gravity.

Holmes/Marsden method. Consider a 2-DOF system composed by the integrable Hamiltonians (F, G) and the perturbation H_1,

$$H_\varepsilon(q, p, I, \theta) = F(q, p) + G(I) + \varepsilon H_1(q, p, I, \theta), \qquad (25.4.25)$$

where $(q,p) \in \mathbb{R}^2$ are generalized position and momentum, $(I, \theta) \in \mathbb{R}^2$ are action-angle coordinates, ε is a small parameter, and H_1 is 2π-periodic in θ. Suppose that $G(0) = 0$, $\omega(I) = G'(I) > 0$ (respectively, $\omega = G' < 0$) for $I > 0$ (resp. $I < 0$), and that the (q,p)-plane contains a homoclinic saddle $(\bar{q}, \bar{p})$, with homoclinic trajectory $\gamma^0(t) = (q^0(t), p^0(t))$. For ε small, $H_\varepsilon = h$ is invertible for I. These hypothesis permit one to reduce this 4th-dimensional system, on an energy level $H_\varepsilon = h$, to the 2nd-order system

$$\begin{aligned}
q' &= -\tfrac{\partial L_0}{\partial p}(q,p;h) - \varepsilon \tfrac{\partial L_1}{\partial p}(q, p, \theta; h) + \mathcal{O}(\varepsilon^2), \\
p' &= \tfrac{\partial L_0}{\partial q}(q,p;h) + \varepsilon \tfrac{\partial L_1}{\partial q}(q, p, \theta; h) + \mathcal{O}(\varepsilon^2),
\end{aligned} \qquad (25.4.26)$$

where $q' = \dot{q}/\dot{\theta}$, $p' = \dot{p}/\dot{\theta}$, $L_0 = G^{-1}(h - F)$, and $L_1 = -H_1/\omega$. It is possible to show that, for each h and θ, $\{L_0, L_1\} = \{F, H_1\}/\omega^2$, where $\{\cdot, \cdot\}$ denotes the Poisson bracket. Therefore, the Melnikov integral is given by

$$\mathcal{M}(\theta^0) = \int_{-\infty}^{\infty} \{F, H_1\}(t + \theta^0) dt, \qquad (25.4.27)$$

where $\{F, H_1\}$ is evaluated at $\gamma^0(t + \theta^0)$. If $\mathcal{M}$ has simple zeros, then for ε small enough, system (25.4.26) has a horseshoe on the surface $H_\varepsilon = h$.

Consider now a disturbance such that energy H is not conserved. For this case, the Hamiltonian is separated into an integrable H_0 with non-transversal homoclinic orbits, a small Hamiltonian εH_1 that may destroy the nominal geometry, giving rise to intersections, hence it may produce horseshoes, and a small, non-Hamiltonian perturbation that preserves some horseshoes. The energy function $H = H_\varepsilon = F + G + \varepsilon H_1$ is not conserved any longer; then $\dot{H} = \varepsilon h(q,p,I,\theta) + \mathcal{O}(\varepsilon^2)$, where $h = \gamma_1 f_1 \partial F/\partial q + \gamma_2 f_2 \partial F/\partial p + \omega(I)\delta_2 g_2$, the functions f_i and g_i being 2π-periodic in θ. Similar to before, it is possible to obtain a reduced three-dimensional system

$$\begin{aligned} q' &= -\frac{\partial L_0}{\partial p} - \varepsilon\left(\frac{\partial L_1}{\partial p} - \frac{\gamma_1 f_1}{\omega} - \frac{\partial L_0}{\partial p}\frac{\delta_1 g_1}{\omega}\right) + \mathcal{O}(\varepsilon^2), \\ p' &= \frac{\partial L_0}{\partial q} + \varepsilon\left(\frac{\partial L_1}{\partial q} + \frac{\gamma_2 f_2}{\omega} - \frac{\partial L_0}{\partial q}\frac{\delta_1 g_1}{\omega}\right) + \mathcal{O}(\varepsilon^2), \\ H' &= \varepsilon \bar{h}(q,p,H) + \mathcal{O}(\varepsilon^2), \end{aligned} \qquad (25.4.28)$$

where $\bar{h} = \delta_2 \bar{g}_2 - (\partial L_0/\partial q)\gamma_1 \bar{f}_1 - (\partial L_0/\partial p)\gamma_2 \bar{f}_2$, with the upper bar denoting the time average $\left[\int_0^{2\pi}(\cdot)d\theta\right]/2\pi$.

Let $\delta = (\gamma_1, \gamma_2, \delta_1, \delta_2)$ be the dissipation coefficient, and $P_{\varepsilon,\delta}^{\theta_0}$ the Poincaré map associated to system (25.4.28). Let $(\bar{q}, \bar{p})$ be the saddle equilibrium of the homoclinic orbit $(q^0(t), p^0(t))$ with energy h^0. Therefore, for each $H > h^0$, $(\bar{q}, \bar{p}, H)$ is a fixed point of $P_{0,0}^{\theta_0}$. Let $\gamma_{0,0}$ be such a curve of fixed points, and define

$$\Delta H = \varepsilon \int_{-\pi N}^{\pi N} \frac{\bar{h}}{\omega} d\theta = \varepsilon \int_{-\pi N/\omega}^{\pi N/\omega} \bar{h} dt, \qquad (25.4.29)$$

with $\bar{h}$ and ω evaluated on $\gamma^0(\theta)$ at an energy value H.

Condition (H) Assume that there exists a value $H_c > 1$ of H such that $\Delta H < 0$ if $H < H_c$, $\Delta H > 0$ if $H > H_c$, and $d(\Delta H(H_c))/dH \neq 0$.

It is possible to show that the Melnikov function of this system, at the level value H_c, is given by

$$\mathcal{M}_\delta(\theta_0) = \frac{1}{\omega^2}\left[M(\theta_0) + \int_{-\infty}^{\infty}\left(\frac{\partial F}{\partial q}\gamma_1 f_1 - \frac{\partial F}{\partial p}\gamma_2 f_2\right)(t - \theta_0)dt\right]. \qquad (25.4.30)$$

Then the next result can be established.

THEOREM 25.5
Suppose that it is possible to choose N, H_c, and δ such that condition (H) holds and the Melnikov integral (25.4.30) has simple zeros. Then, the iteration $(P_{\varepsilon,\delta}^{t_0})^N$ of the Poincaré map of system (25.4.28) has, for ε small enough, a Smale horseshoe that lies near the homoclinic orbit in the (q,p) variables, and close to the energy surface $H = H_c$.

Open-loop robot. Consider first the 2-DOF robot manipulator without friction, with $\tau = 0$ and $\varepsilon = l_{c_2}$. Then $H = F + G + \varepsilon H_1 + \mathcal{O}(\varepsilon^2)$, where $F = p_1^2/2\beta_1 + \alpha_1(1 - \cos q_1)$, $G = p_2^2/2\beta_2$, $H_1 = \bar{\alpha}_2(1 - \cos q_2) - \bar{\beta}_3 p_1 p_2 \cos(q_2 - q_1)/\beta_1\beta_2$, with $\alpha_2 = \varepsilon\bar{\alpha}_2$ and $\beta_3 = \varepsilon\bar{\beta}_3$. The system, called pendulum-oscillator, is integrable when $\varepsilon = 0$. The pendulum part has a heteroclinic cycle $\pm\gamma^0(t) = \pm(2\tan^{-1}(\sinh t), \sqrt{4\alpha_1\beta_1}\operatorname{sech} t)$. The Melnikov function is $\mathcal{M}(t_0) = \mathcal{A}(h)\sin\omega t_0$, where

$$\mathcal{A}(\omega(h)) = \frac{2\alpha_1\bar{\beta}_3\pi\omega^4}{\beta_1}\left[\frac{1}{\sinh(\omega\pi/2)} + \frac{1}{\cosh(\omega\pi/2)}\right], \omega = \sqrt{\frac{2(h - 2\alpha_1)}{\beta_2}}. \quad (25.4.31)$$

It is obvious that $\mathcal{M}$ has zeros, which can be proved to be simple; then the system can display chaos. Figure 25.15 shows some trajectories obtained from this system.

FIGURE 25.15
Typical trajectories of an open-loop 2DOF robot without dissipation.

The PD-controlled robot. Let $\tau = K_p(q_d - q) - K_v\dot{q}$ be the torque, with K_p and K_v diagonal matrices. Then the controlled system is given by

$$\begin{aligned}
\dot{q}_1 &= \frac{\partial F}{\partial p_1} + \varepsilon\frac{\partial H_1}{\partial p_1}, \quad \dot{q}_2 = \frac{\partial G}{\partial p_2} + \varepsilon\frac{\partial H_1}{\partial p_2}, \\
\dot{p}_1 &= -\frac{\partial F}{\partial q_1} - \varepsilon\frac{\partial H_1}{\partial q_1} + k_{p_1}(q_{d_1} - q_1) - (k_{v_1} + d_1)\dot{q}_1, \quad (25.4.32) \\
\dot{p}_2 &= -\varepsilon\frac{\partial H_1}{\partial q_2} + k_{p_2}(q_{d_2} - q_2) - (k_{v_2} + d_2)\dot{q}_2.
\end{aligned}$$

If $k_{p_1} = \varepsilon\bar{k}_{p_1}$, $k_{p_2} = \mathcal{O}(\varepsilon^2)$, and $k_{v_i} + d_i = \varepsilon(\bar{k}_{v_i} + \bar{d}_i) = \varepsilon\bar{\nu}_i$, then condition (**H**) is satisfied if

$$\omega = \omega_c = \frac{\pi\bar{k}_{p_1}q_{d_1}\sqrt{\alpha_1\beta_1} - 4\alpha_1\bar{\nu}_1}{\pi\beta_1\bar{\nu}_2 N} \neq 0, \quad (25.4.33)$$

for a large integer N. Then, $\mathcal{M}_\delta(\theta_0)$ will have simple zeros if

$$\mathcal{A}(\omega_c) > \left|2\pi\bar{k}_{p_1}q_{d_1}\sqrt{\alpha_1/\beta_1} - 8\alpha_1\bar{\nu}_1/\beta_1\right|, \quad (25.4.34)$$

where $\mathcal{A}$ is given by (25.4.31).

THEOREM 25.6
The PD-controlled, 2-DOF robot manipulator (25.4.32) will have a horseshoe if $k_{p_1} = \varepsilon^{\mu+1}\tilde{k}_{p_1}$, $k_{p_2} = \mathcal{O}(\varepsilon^2)$, $\nu_i = \varepsilon^{\mu+1}\tilde{\nu}_i$, $i = 1, 2$, $\tilde{\nu}_2 > 0$, and

$$0 < \frac{\pi \tilde{k}_{p_1} q_{d_1} \sqrt{\alpha_1 \bar{\beta}_1} - 4\alpha_1 \tilde{\nu}_1}{2\alpha_1 \bar{\beta}_3} < 1. \qquad (25.4.35)$$

Some trajectories obtained from this system are displayed by Figure 25.16.

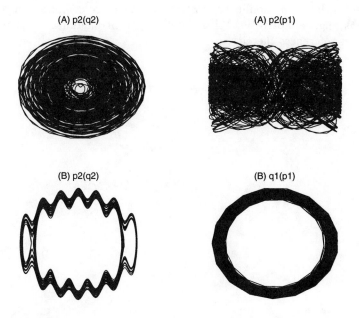

FIGURE 25.16
Projections of two typical trajectories (A and B) of a PD-controlled, 2DOF robot.

Remark. The same technique can be applied, with few modifications, to other similar algorithms, like the PD-controller with gravity compensation.

25.5 Conclusions

The control systems analyzed in this chapter have a very simple structure. Nevertheless, they are able to exhibit a diversity of complicated phenomena. The behavior displaying by these simple systems permits one to conjecture the presence of this same and even more complex, dynamical behavior in other classical control schemes including typical nonlinear blocks like dead zones, hysteresis, relays, and so on, or including time-delays in the plant. This complexity may be greater for discrete-time systems, which have not been considered in this short presentation. Obviously, this also applies to systems, linear or nonlinear, controlled with more complex control algorithms, e.g., adaptive [8, 11] or intelligent controllers (see [4] and the numerous references therein).

References

[1] J. Alvarez, E. Curiel, and F. Verduzco, "Complex dynamics in classical control systems," *Systems and Control Letters*, vol. 31, pp. 277-285, 1997.

[2] J. Alvarez and F. Verduzco, "Bifurcations and chaos in a PD-controlled pendulum," *Journal of Dynamic Systems, Measurement, and Control*, vol. 120, pp. 146-149, 1998.

[3] J. Bailleul, R. W. Brockett, and R. B. Washburn, "Chaotic motion in nonlinear feedback systems," *IEEE Trans. on Circ. Sys.*, vol. CAS-27, pp. 990-997, 1980.

[4] G. Chen and X. Dong, *From Chaos to Order. Methodologies, Perspectives, and Applications*, World Scientific Pub. Co., Singapore, 1998.

[5] P. A. Cook, "Simple feedback systems with chaotic behaviour," *Systems and Control Letters*, vol. 6, pp. 223-227, 1985.

[6] R. Genesio and A. Tesi, "Harmonic balance methods for the analysis of chaotic dynamics in nonlinear systems," *Automatica*, vol. 28, pp. 531-548, 1992.

[7] R. Genesio, A. Tesi, and F. Villoresi, "Models of complex dynamics in nonlinear systems," *Systems and Control Letters*, vol. 25, pp. 185-192, 1995.

[8] M. P. Golden and B. E. Ydstie, "Small amplitude chaos and ergodicity in adaptive control," *Automatica*, vol. 28, pp. 11-25, 1992.

[9] P. J. Holmes and J. E. Marsden, "Horseshoes in perturbations of Hamiltonian systems with two degrees of freedom," *Commun. Math. Phys.*, vol. 82, pp. 523-544, 1982.

[10] Y. A. Kuznetsov, *Elements of Applied Bifurcation Theory*, Springer-Verlag, New York, 1995.

[11] I. M. Y. Mareels and R. R. Bitmead, "Bifurcation effects in robust adaptive control," *IEEE Trans. on Circ. Sys.*, vol 35, pp. 835-841, 1988.

[12] A. I. Mees, *Dynamics of Feedback Systems*, Wiley, Chichester, 1981.

[13] L. Perko, *Differential Equations and Dynamical Systems*, Springer-Verlag, New York, 1991.

[14] M. W. Spong and M. Vidyasagar, *Robot Dynamics and Control*, Wiley, New York, 1989.

[15] F. Verduzco and J. Alvarez, "Bifurcation analysis of a 2-DOF robot manipulator driven by constant torques", to appear in *Int. J. of Bif. and Chaos*, 1999, in press.

[16] S. Wiggins, *Global Bifurcations and Chaos. Analytical Methods*, Springer-Verlag, New York, 1988.

26

Chaos and Bifurcations in Coupled Networks and Their Control

Tetsushi Ueta[1] and Guanrong Chen[2]

[1] Department of Information Science and Intelligent Systems
Tokushima University, Tokushima, 770-8506 Japan
tetsushi@is.tokushima-u.ac.jp

[2] Department of Electrical and Computer Engineering
University of Houston, Houston, Texas, 77204 USA
gchen@uh.edu

Abstract

This chapter investigates the complex dynamics and control of chaos in a system of strongly connected neural oscillators. More specifically, the existence and stability of some simple equilibria of a coupled-neuron system is first discussed. The system periodic solutions are then analyzed by using the Poincaré mapping method. Bifurcation diagrams and topological properties for these periodic solutions are obtained, where synchronization phenomena are classified. Finally, a simple feedback control method for stabilizing an in-phase synchronizing periodic solution embedded in the chaotic attractor of a higher-dimensional model of such coupled neural oscillators is presented.

26.1 Introduction

Oscillatory behavior is a basic phenomenon of the brain. When the brain is modeled by massive coupled neurons in a network structure, even in a much simpler version of it, the periodic orbits of the dynamical neural network can be used to describe cyclical processes in the nervous system. For example, they can be used to model reverberating memory in the cortex and thalamus [18], hallucinations in visual perception [5], circadian rhythm generation in the hypothalamus [2], and rhythmical movements in crustaceans [14]. Even the degenerate case, i.e., an equilibrium state, also corresponds to a possible stable or unstable activity pattern of the network. More sophisticated complex dynamics is chaos, which has been suggested to be a basis for behavioral variability of the brain activities [15, 11].

The dynamical system of a continuous-time neural network may be described by the following system of differential equations:

$$\dot{x}_i = -\alpha_i x_i + f\left(\rho_i + \sum_{j=1}^{n} c_{ij} x_j\right), \quad x_i \in \mathbf{R}, \quad i,j = 1, \ldots, n, \quad (26.1.1)$$

where all coefficients are constants, with $\alpha_i \geq 0$, and f is a sigmoid function defined by

$$f(x) = \frac{1}{1 + \exp(-\epsilon x)}, \quad \epsilon > 0. \quad (26.1.2)$$

In case that n is a big number, such a model has rich variety of oscillatory behaviors, various (in-phase, out-phase, or n-phase) synchronizations, tori, bifurcations, and chaos.

In [18], a model of neural oscillators, probably the simplest of its kind, is developed by considering a specific case of Equation (26.1.1), namely,

$$\begin{aligned} \dot{x} &= -\alpha x + f(ax - by + \rho_x) \\ \dot{y} &= -\beta y + f(cx - dy + \rho_y). \end{aligned} \quad (26.1.3)$$

This model describes the coupling of an excitatory and an inhibitory neuron via synapses, as illustrated by Figure 26.1 (a).

System (26.1.3) has some typical bifurcation phenomena, such as saddle-node and Hopf bifurcations from an equilibrium, generation and saddle connection of limit cycles, and other degenerate bifurcations [7, 6]. The bifurcation structure of Equation (26.1.3) in the ρ_x-ρ_y plane is somewhat similar to an averaged system of the van der Pol or the Duffing-Rayleigh oscillator. Therefore, coupled neurons can be thought of as a nonlinear oscillator. However, a network consisting of more than two such oscillators may not be assumed to be a mutually coupled system. This issue is different from the familiar coupled electric circuits, because the synaptic coupling

Introduction

here is one-directional and the coupling term is contained inside of the sigmoid function f.

In [7], weakly connected neural networks are studied via some analytical (averaging) methods under the assumption that the synaptic coefficients are small. In this chapter, we investigate coupled neural oscillators with strong synaptic connections. A motivation is that strong coupling between neurons in biophysical signaling mechanisms is also common, at least in many artificial neural networks (i.e., perceptions, recurrent neural networks, and cellular neural networks [16]).

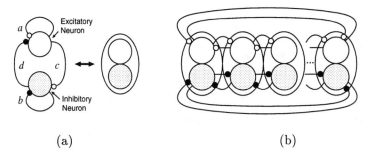

FIGURE 26.1
(a) Coupled neurons. (b) A ring configuration.

Observe that for the coupling configuration among all oscillators, the connections "excitatory → excitatory" and "inhibitory → inhibitory" are essential for in-phase and out-phase synchronizations [7]. Therefore, we consider this kind of coupling structure in our investigation of the change of synchronizing modes as the coupling coefficient is increased.

Figure 26.1 (b) shows a ring configuration of coupled neural oscillators. The corresponding system of equations are

$$\dot{x}_i = -\alpha x_i + f(ax_i - by_i + \rho_x + \delta_x(x_{i-1} + x_{i+1}))$$
$$\dot{y}_i = -\beta y_i + f(cx_i - dy_i + \rho_y - \delta_y(y_{i-1} + y_{i+1})) \quad (26.1.4)$$
$$i = 1, 2, \ldots, n \pmod{n},$$

where $\alpha > 0$, $\beta > 0$, and a, b, c, d, ρ_x, ρ_y, δ_x, and δ_y are constants.

In the following, we first study some simple equilibria existing in the system. Existence and stability of these equilibria are confirmed. Then, we investigate the network periodic solutions by using the Poincaré mapping method. Bifurcation diagrams and topological properties for these periodic solutions are obtained. Synchronization phenomena are also classified. Finally, we show a simple method for stabilizing an in-phase synchronizing periodic solution embedded in the chaotic attractor in a higher-dimensional model of coupled neural oscillators.

26.2 Some Simple Equilibria in the System

In this section, we analyze some simple equilibria in the ring configuration of the coupled oscillators, which is shown in Figure 26.1 (b). This kind of coupled oscillators have been studied as early as in [1].

THEOREM 26.1
Consider the coupled-neuron model (26.1.4). If

$$\alpha = \beta > 0, \quad \delta := \delta_x = \delta_y, \quad a + 2\delta = c,$$
$$d + 2\delta = b, \quad \rho := \rho_x = \rho_y > 0, \quad (26.2.5)$$

or if

$$\alpha = \beta > 0, \quad a = \delta_x = 0, \quad d = \delta_y = 0,$$
$$b = -c \neq 0, \quad \rho := \rho_x = \rho_y > 0, \quad (26.2.6)$$

then there is a unique equilibrium at $\theta^* := x_1^* = y_1^* = \cdots = x_n^* = y_n^*$. Moreover, if

$$|c - b| < 4\alpha/\epsilon \quad (26.2.7)$$

for the first case, or

$$|c| < 4\alpha/\epsilon \quad (26.2.8)$$

for the second case, then this equilibrium is a stable node.

PROOF We first observe that under conditions (26.2.5), we have the following equation to study for finding an equilibrium:

$$\dot{\theta} = -\alpha\theta + f(c\theta - b\theta + \rho) = -\alpha\theta + f((c-b)\theta + \rho).$$

We need to solve the algebraic equation

$$h(\theta) := -\alpha\theta + f((c-b)\theta + \rho) = 0. \quad (26.2.9)$$

Since $h(\theta)$ is continuous and $h(\theta) \to \infty$ as $\theta \to -\infty$ and $h(\theta) \to -\infty$ as $\theta \to \infty$, Equation (26.2.9) has at least one root, namely, the coupled-neuron model has at least one equilibrium.

Next, we notice that showing that θ^* is the unique root of Equation (26.2.9) is equivalent to showing that the function

$$h(\theta) = -\alpha\theta + \frac{1}{1 + e^{-\epsilon((c-b)\theta + \rho)}} \quad (26.2.10)$$

is strictly monotonic, so that $h(\theta) = 0$ has only one root.

To verify this, taking a derivative on the function (26.2.10) with respect to θ gives

$$h'(\theta) = -\alpha + f'\big((c-b)\theta + \rho\big) = -\alpha + \frac{\epsilon(c-b)e^{-\epsilon((c-b)\theta+\rho)}}{\left[1+e^{-\epsilon((c-b)\theta+\rho)}\right]^2}.$$

If $c \leq b$ then $h'(\theta) < 0$, so $h(\theta)$ is strictly decreasing, implying that $h(\theta) = 0$ has a single root. In this case, $f'(\theta) \leq 0$, as shown in Figure 26.2 (a), where $\frac{\rho}{b-c} > 0$. If $c > b$, then $f'(\theta) > 0$, $\frac{\rho}{b-c} < 0$, and $f''(\theta) < 0$ (for $\theta > 0$). This implies that $h(\theta) = 0$ has a single root, as shown in Figure 26.2 (b).

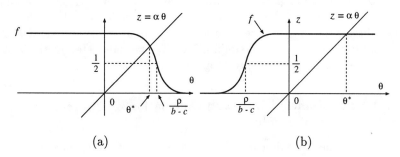

FIGURE 26.2
Existence of an equilibrium in the coupled-neuron model.

To determine the stability of θ^*, consider the system Jacobian

$$J\Big|_{x_i=y_i=\theta^*} = \begin{bmatrix} -\alpha & \partial f/\partial y_i \\ \partial f/\partial x_i & -\alpha \end{bmatrix}_{x_i=y_i=\theta^*},$$

which has eigenvalues

$$s_{1,2} = -\alpha \pm \frac{\epsilon(c-b)\,e^{-\epsilon((c-b)\theta+\rho)}}{\left[1+e^{-\epsilon((c-b)\theta+\rho)}\right]^2}.$$

Under condition (26.2.7), if $-(4\alpha/\epsilon) < c - b \leq 0$, then

$$s_1 \leq -\alpha < 0 \quad \text{and} \quad s_2 \leq -\alpha + \epsilon|c-b|/4 < 0;$$

if $0 < c - b < (4\alpha)/\epsilon$ then similarly we can verify that

$$s_2 \leq s_1 < 0.$$

Therefore, when $|c - b| \leq (4\alpha)/\epsilon$, this equilibrium θ^* is a stable node.

For the second case under conditions (26.2.6) and (26.2.8), the verification is similar.

26.3 Periodic Solutions and Chaos

In this section, we investigate the characteristics of the periodic solutions in the ring configuration of the coupled oscillators. In each individual oscillator, described by Equation (26.1.3), we fix the parameter values to be

$$\alpha = \beta = 1, \quad \epsilon = 1, \quad a = b = 10, \quad d = -2, \quad \rho_x = 0, \quad \rho_y = -6.$$

Based on the analysis about equilibria discussed above, we can calculate the Hopf bifurcation parameter and obtain $c = 6.62$. Then, we know that there are an unstable equilibrium and a stable limit cycle in the two-dimensional system. These repeller and attractor are created right after the Hopf bifurcation at a stable equilibrium. We are interested in the global behaviors of these two strongly coupled oscillators.

26.3.1 The Poincaré Map and Calculation of Bifurcation Parameter Values

Consider a general nonlinear autonomous system of the form

$$\frac{d\boldsymbol{x}}{dt} = \boldsymbol{f}(\boldsymbol{x}, \lambda), \quad \boldsymbol{x} \in \boldsymbol{R}^n, \quad \lambda \in \boldsymbol{R}, \tag{26.3.11}$$

where $\boldsymbol{f}$ is a C^∞ function, and λ is a real parameter. Suppose that system (26.3.11) has a solution φ, starting from an initial point, $\boldsymbol{x}_0$,

$$\boldsymbol{x}(t) = \varphi(t, \boldsymbol{x}_0), \quad \boldsymbol{x}(0) = \varphi(0, \boldsymbol{x}_0) = \boldsymbol{x}_0.$$

Also assume that there is a limit cycle of period $\tau > 0$ for an appropriate parameter value,

$$\varphi(\tau, \boldsymbol{x}_0) = \varphi(0, \boldsymbol{x}_0)$$

By transversely placing the Poincaré section Π described by a scalar function q to the limit cycle, we have

$$\Pi = \{\, \boldsymbol{x} \in \boldsymbol{R}^n \,|\, q(\boldsymbol{x}) = 0 \,\}. \tag{26.3.12}$$

Thus, we can define the Poincaré map T as follows:

$$\begin{aligned} T: \quad & \Pi \to \Pi \\ & \boldsymbol{x} \mapsto \varphi(\tau(\boldsymbol{x}), \boldsymbol{x}). \end{aligned} \tag{26.3.13}$$

To reduce the dimension of this map, we introduce the local coordinates

$$\boldsymbol{u} \in \Sigma \subset \boldsymbol{R}^{n-1},$$

Periodic Solutions and Chaos

with a projection and an embedding map

$$h: \Pi \to \Sigma, \quad h^{-1}: \Sigma \to \Pi.$$

Thereby we have the Poincaré map T_ℓ on the local coordinates, such that

$$\begin{aligned} T_\ell : \ & \Sigma \to \Sigma \\ & u \to h \circ T \circ h^{-1} = h(\varphi(\tau(h^{-1}(u)), h^{-1}(u))). \end{aligned} \quad (26.3.14)$$

In general, a point u_0 satisfying

$$u_0 = T_\ell(u_0) \quad (26.3.15)$$

is called a *fixed point of the map* (to distinguish it from an *equilibrium point of the system*) in this chapter. The stability of this fixed point is determined by the multipliers (roots) of the characteristic equation for the Jacobian of system (26.3.13) (see [17] for more details).

$$\chi(\mu) := \det(DT - \mu I_n) = 0, \quad (26.3.16)$$

where I_n is the $n \times n$ identity matrix.

To obtain bifurcation parameter values, we have to simultaneously solve Equations (26.3.12), (26.3.15), and (26.3.16), for u_0, λ, and τ.

26.3.2 A Simple Coupled-Neuron Model

In this section, we investigate a simple pair of coupled neurons.

$$\begin{aligned} \dot{x}_1 &= -x_1 + f(ax_1 - by_1 + \rho_x + \delta x_2) \\ \dot{y}_1 &= -y_1 + f(cx_1 - dy_1 + \rho_y - \delta y_2) \\ \dot{x}_2 &= -x_2 + f(ax_2 - by_2 + \rho_x + \delta x_1) \\ \dot{y}_2 &= -y_2 + f(cx_2 - dy_2 + \rho_y - \delta y_1) \end{aligned} \quad (26.3.17)$$

This model has a limit cycle and the Neimark-Sacker (NS) bifurcation occurs under a certain condition. After this bifurcation, the limit cycle changes its behavior to be quasi-periodic, phase-locking, and torus breakdown, as the connection coefficient δ is increased.

We investigate this system's dynamics on the c-δ plane. We choose the Poincare section $\Pi = \{x \mid x_1 - x_1^* = 0\}$, where x_1^* is the x-coordinate value of the equilibrium x^*. Limit cycle in this system is generated by the Hopf bifurcation of the equilibrium point x^*. Therefore, Π, which contains the point x^*, is adequate to be used as a transversal section.

Symmetry plays an important role in this system. For instance, we can easily see that the permutation

$$P: \mathbf{R}^4 \to \mathbf{R}^4; \ (x_1, y_1, x_2, y_2) \mapsto P(x_1, y_1, x_2, y_2) = (x_2, y_2, x_1, y_1)$$

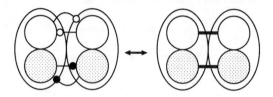

FIGURE 26.3
A couple of neuron oscillators and its abbreviation.

is invariant for Equation (26.3.17). A permutation P can be written as

$$P = \begin{pmatrix} 0 & I_2 \\ I_2 & 0 \end{pmatrix},$$

where I_2 is the 2×2 identity matrix. Let the system (26.3.17) be written in the form of (26.3.11). If the right-hand side of Equation (26.3.11) has the following relationship, then the equation is said to be P-symmetric:

$$f(Px) = Pf(x).$$

Due to this symmetry of the equation, all periodic solutions of the system have symmetrical properties. There typically exist three types of periodic solutions:

    in-phase synchronization (IN);

    almost out-phase synchronization (OUT);

    symmetric periodic solution (S).

For the IN solution, we have

$$\varphi(t, x_0) = P\varphi(t, x_0).$$

In this case, we always have $x_1(t) = x_2(t)$ and $y_1(t) = y_2(t)$ for all t. This means that both neural oscillators are completely synchronized without delay.

The OUT solution can be regarded as an almost out-phase synchronized solution for each oscillator, i.e., it has a half-period delay between two oscillators. Actually, this solution satisfies

$$\varphi(\tau/2, x_0) = Px_0,$$

where τ is the period.

For the solution S, there exist a pair of symmetric solutions, $\varphi_1(t)$ and $\varphi_2(t)$, satisfying

$$\varphi_1(t, x_0) = P\varphi_2(t, Px_0), \quad \varphi_2(t, x_0) = P\varphi_1(t, Px_0). \tag{26.3.18}$$

These solutions are generated by a pitchfork bifurcation of the IN solution.

Stabilities of periodic solutions are determined by the multipliers of the characteristic equation for the fixed point. We denote these stabilities by $_kD$ or $_kI$, with D and I indicating the orientation preserving and reversing maps on the unstable eigenspace of dimension k, respectively [8].

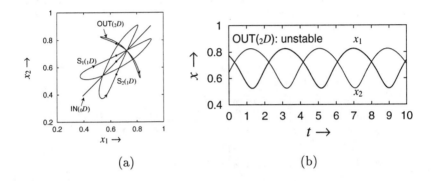

FIGURE 26.4
Coexistence of solutions. $c = 12.0$, $\delta = 2.747$. Black circles are fixed points of the Poincaré map.

Figure 26.4 (a) shows an example of the phase portraits (a projection of x_1-x_2) of the coexisting periodic solutions. In this figure, we can see the phase differences between two oscillators. For instance, IN($_0D$) is a stable in-phase synchronized solution since the state is always located on the diagonal $x_1 = x_2$. On the other hand, OUT($_2D$) indicates a two-dimensionally unstable and almost out-phase synchronized solution, as shown in Figure 26.4 (b). Moreover, S$_1$($_1D$) and S$_2$($_1D$) are one-dimensionally unstable saddles, and they are symmetrical about the diagonal $x_1 = x_2$ in this projection due to the relationship (26.3.18).

For system (26.3.17), bifurcation phenomena can be classified into five types as summarized in Table 26.1. In this table, $\emptyset$ indicates the disappearance of the fixed points, and $_0D^2$ indicates a two-periodic point for $_0D$.

Next, we show a bifurcation diagram in the c-δ parameter plane (see Figure 26.5). In Fig. 26.5 (b), each segment separated by the bifurcation curve has different topological properties of the corresponding periodic solutions.

TABLE 26.1
Typical local bifurcations for limit cycles (k is an integer).

Bifurcation	Symbol	Change of fixed points
Tangent	T	$_kD + {}_{k+1}D \iff \emptyset$
Period-doubling	Pd	$_kD \iff {}_{k+1}I + 2 \times {}_kD^2$
Neimark-Sacker	NS	$_kD \iff {}_{k+2}D + \text{torus}$
Pitchfork-I	Pf	$_kD \iff {}_{k+1}D + 2 \times {}_kD$
Pitchfork-II	Pf	$_kD \iff {}_{k-1}D + 2 \times {}_kD$

These segments are labeled by numbers (1)–(12) and their corresponding solutions are listed below:

(1) $\emptyset$
(2) IN($_0D$)
(3) IN($_0D$) + OUT($_2D$)
(4) OUT($_0D$)
(5) IN($_2D$) + OUT($_0D$)
(6) IN($_0D$) + OUT($_2D$) + 2 × S($_0D$) + 2 × S($_1D$)
(7) IN($_0D$) + OUT($_2D$) + 2 × S($_2D$) + 2 × S($_1D$)
(8) IN($_1D$) + OUT($_2D$) + 2 × S($_0D$)
(9) IN($_1D$) + OUT($_2D$) + 2 × S($_2D$)
(10) IN($_0D$) + OUT($_0D$)
(11) IN($_1D$) + OUT($_0D$) + 2 × S($_0D$)
(12) IN($_1D$) + OUT($_0D$) + 2 × S($_2D$)

In area (1), there exists only one stable equilibrium. After crossing two Hopf bifurcations (h_1 and h_2), the equilibrium changes to a completely (four-dimensionally) unstable one. The S-type of solutions are created by a pitchfork bifurcation (Pf) for an IN type of solution. In area (8), only two S-type solutions are observed, which are stable in the large.

Next, we focus on the Neimark-Sacker (NS) bifurcations. In Figure 26.5 (b), the half lines $\overline{AB}$, $\overline{AD}$, and $\overline{BC}$ are NS bifurcations of the IN, OUT, and S types of solutions, respectively. Thick arrows indicate that tori are generated by changing parameter values along the corresponding direction. In areas (7) and (9), many periodic solutions (without symmetrical property) and tori coexist, and they eventually collapse to chaos. There is no area in which IN, OUT, S solutions coexist as stable orbits; IN and OUT solutions are observed to be stable only in area (10), depending on initial conditions. From these diagrams and the above classification, the change between IN and OUT synchronizations is clarified.

Figure 26.6 shows the system Lyapunov exponents for the parameter range $2.25 < \delta < 2.42$, with $c = 11$. In the case of $\delta < 2.314$, there is an IN($_0D$) solution. At $\delta \approx 2.314$, an NS bifurcation occurs, then a

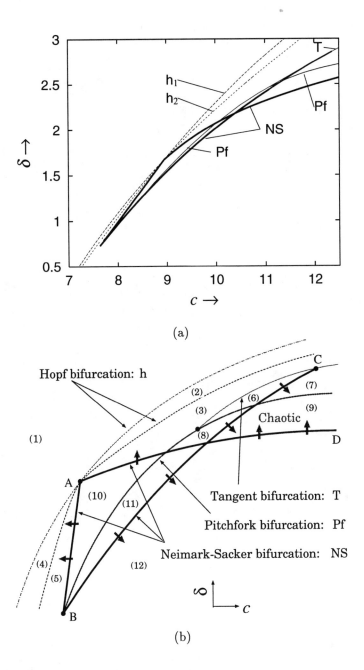

FIGURE 26.5
Bifurcation diagram (a) and its schematic diagram (b) $a = b = 10$, $d = -2$, $\rho_x = 0$, $\rho_y = -6$.

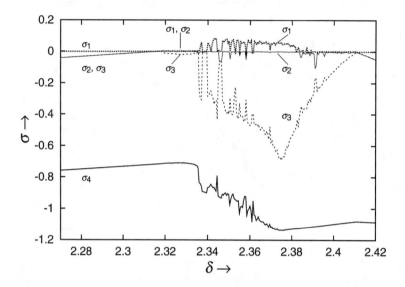

FIGURE 26.6
Lyapunov exponents. $c = 11$.

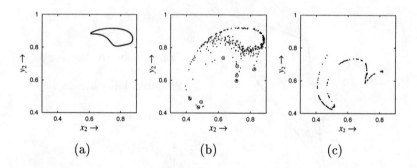

FIGURE 26.7
The Poincaré map in the x_2-y_2 plane. $c = 11$. (a) A torus. $\delta = 2.32$ (b) $\delta = 2.339$. The torus is broken, and the orbit behaves chaotically for a while. Finally, it is trapped into eight periodic points (marked by circles). (c) A chaotic attractor. $\delta = 2.34$.

quasi-periodic solution (torus) is generated. This can be confirmed by the fact that it has two zero Lyapunov exponents. For the parameter range $2.334 < \delta < 2.336$, a torus and many periodic attractors are alternatively observed (frequency locking). When $\delta \approx 2.335$, the system has a chaotic transient starting from any initial value within this parameter range. After the chaotic transient, the torus is broken and an eight-periodic attractor appears. By a slight increase in the value of δ, this eight-periodic attractor disappears via a tangent bifurcation; after that, periodic and chaotic responses are observed within a very narrow range of parameter perturbations. When $\delta \approx 2.336$, an NS bifurcation of a six-periodic attractor emerges. And yet, after this bifurcation, the generated torus is immediately absorbed into a coexisting chaotic attractor. Beyond the range of $\delta \approx 2.339$, chaos and many periodic windows are observed. We also found period-doubling bifurcations for some periodic solutions in these windows. From these observations, we conclude that a typical torus-breakdown scenario [9] exists in this system of coupled neurons.

26.3.3 Intermittency Responses

After a pitchfork bifurcation of $IN(_1D)$, in area (6) or (7) of Figure 26.5, we observe two saddle-type periodic solutions, $S(_1D)$s. Under certain initial conditions, we see a novel intermittency response: the orbit stays for a while near one of the two $S(_1D)$s (since an $S(_1D)$ has two stable manifolds and one unstable manifold whose corresponding multiplier is slightly larger than unity). Thus, after temporally staying at $S(_1D)$, the orbit escapes from $S(_1D)$. However, the orbit is immediately trapped into another $S(_1D)$ after a very short chaotic transient, and behaves periodically for a while again, as shown in Figure 26.8 (a). This switching between the two $S(_1D)$s is permanent, but each staying time at either $S(_1D)$ cannot be estimated, perhaps due to the chaotic nature of the motion (in fact, the maximum Lyapunov exponent of this attractor is 0.07).

This kind of intermittency is not related to tangent, period-doubling (flip) nor NS bifurcation. Instead, it is related to pitchfork bifurcation. This phenomenon does not conform to the types I–III intermittency classified in [12]. Figure 26.8 shows the manifolds of an $S(_1D)$. The Poincaré maps are slightly closer to the unstable manifold of the $S(_1D)$. Note that there also exists an $IN(_0D)$ (stable) solution at this parameter value, but it has another basin of attraction. Thus, this intermittency response can be everlastingly observed.

Similar phenomenon can be observed in the case of higher-dimensional ring configuration of coupled oscillators. Figure 26.9 shows some peculiar wave forms of six coupled oscillators. There are two oscillations of different amplitudes and the switching between these oscillators is chaotic. This phenomenon is also due to the intermittency response of $S(_1D)$s. In this

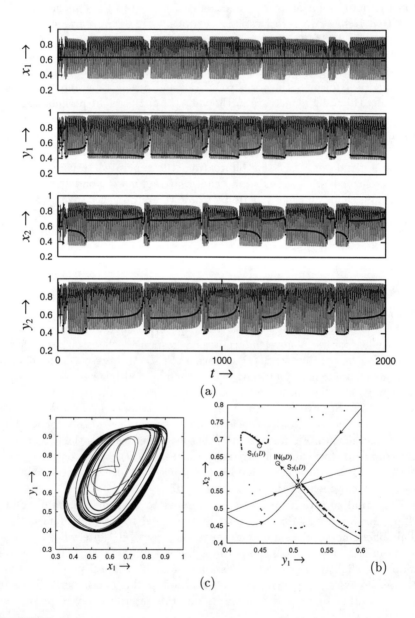

FIGURE 26.8
Intermittency response at $c = 12$ and $\delta = 2.668$. (a) Time response with an arbitrary initial condition. Small black circles are the Poincaré section points. (b) Phase portrait in the x_1-y_1 plane. (c) Stable and unstable manifolds of $S_2(_1D)$.

figure, we can see a temporal in-phase synchronization between two states of the system in a short period of time.

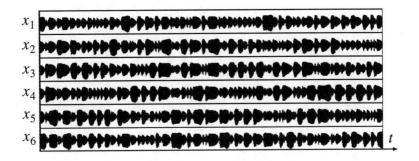

FIGURE 26.9
A snapshot of the wave form x_i in six neural oscillators. $c = 10$, $\delta = 1.09$, $1000 < t < 6000$. The average value of return time for a cycle is about 6.14 sec.

26.3.4 Coalesce of Tori

As another impressive phenomenon related to S solution, coalesce of tori (torus symmetry breaking bifurcation) is observed in Equation (26.3.17); see also Figure 26.10. This phenomenon is also found in another type of coupled neural oscillators [6].

Two tori in Figure 26.10 (a) are caused by crossing the NS bifurcation for $S(_0D)$ from area (11) to area (12) in Figure 26.5. In area (12) near the NS bifurcation curve $\overline{BC}$, there exist a unstable saddle $IN(_1D)$, a stable sink $OUT(_0D)$, and two $S(_2D)$s with two stable tori. Tori approach each other as the parameter δ decreases. In Figure 26.10 (c), they are coalesced with the saddle orbit, and finally a torus is generated and the saddle is separated from the torus (Figure 26.10 (d)). Note that the NS bifurcation for IN-type solution (curve $\overline{AB}$) does not affect this phenomenon.

In general nonlinear systems, these tori meet with torus-breakdown and are changed to chaotic solutions just before coalescing. To clarify this coalesce of plain tori, it is important to examine the manifolds of the saddle $IN(_1D)$. This remains a research problem for study.

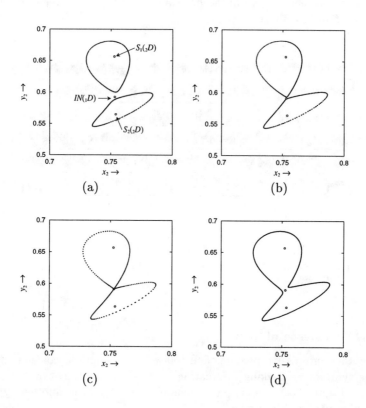

FIGURE 26.10
Coalesce of tori. All points are projections of the Poincaré mapping on the x_2-y_2 plane. (q): $\delta = 1.529$. (b): $\delta = 1.528$, (c): $\delta = 1.52774$, (d): $\delta = 1.527$, $a = b = 10$, $c = 9$, $d = -2$, $\rho_x = 0$, $\rho_y = -6$.

26.4 Controlling to Unstable Periodic Orbits

For some practical applications of artificial neural networks in information processing, such as solving optimization problems and searching for associative memory, chaos seems to be effective. In many other applications, however, chaotic system trajectory has to be stabilized, for example, to an unstable periodic orbit (UPO) embedded in the chaotic attractor [3, 4]. In particular, controlling UPOs is one of the many significant problems for neural networks [10].

In this section, we discuss how a UPO embedded in a chaotic attractor of the coupled-neuron model can be stabilized via feedback control. Since the mathematical model is known, we can calculate bifurcation parameter values or fixed/periodic points by using the Poincaré map and related analytic methods mentioned in Section 26.3.1, even though the target periodic solution is unstable.

In this attempt, the time-delayed feedback control [13] turns out to be suitable, since we have accurate information about the target UPO, which is used as the reference signal. Differing from [13], we feed the control input (stimuli) into the sigmoid function f based on the additive neural dynamics (see Equation (26.1.1) for a reason). For simplicity, the control is added only to the x-components of the system. Thus, the controlled system is

$$\dot{x}_i = -x_i + f(ax_i - by_i + \rho_x + \delta(x_{i-1} + x_{i+1}) + K(x_i^* - x_i))$$
$$\dot{y}_i = -y_i + f(cx_i - dy_i + \rho_x - \delta(y_{i-1} + y_{i+1})), \quad i = 1, 2, \cdots, n,$$

where x_i^* are the corresponding x-components of the UPO and $K > 0$ is a control gain to be determined (see Figure 26.12).

In the previous section, we have seen the chaotic response in the case of six-oscillators, with $c = 10$ and $\delta = 1.09$ (Figure 26.9). At these parameter values, there is a four-dimensionally unstable in-phase synchronized UPO (IN($_4D$)). The maximum multiplier of this UPO is 1.2058 (double roots), and it is embedded within the chaotic attractor.

Figure 26.11 shows the time series of the controlled system response, where the control gain $K = 2.0$ was determined by trial-and-error. This control gain can also be determined by local stability analysis which is usually too conservative. However, simulations have confirmed that by using conservative (sufficiently large) values of K, the control performance and its robustness can be remarkably improved.

Since we can accurately calculate the UPO by solving Equation (26.3.15), this control method might work as well for any UPO by using a sufficient large value of K. Yet for a long-periodic UPO with high precision, a huge memory is needed to store the UPO data. In the case of stabilizing an in-phase synchronized oscillation in the ring configuration of coupled oscillators, such memory is not needed in order to stabilize the UPO. That is, we can use the following equation as the reference oscillator:

$$\dot{x} = -x + f((a + 2\delta)x - by + \rho_x)$$
$$\dot{y} = -y + f(cx - (d + 2\delta)y + \rho_y).$$

This equation is part of Equation (26.1.4), with $x_{i-1} = x_i = x_{i+1}$ and $y_{i-1} = y_i = y_{i+1}$ therein. Using the x-component of the solution as the x^* of this equation, the same control performance has been verified. Figure 26.12 depicts this control scheme.

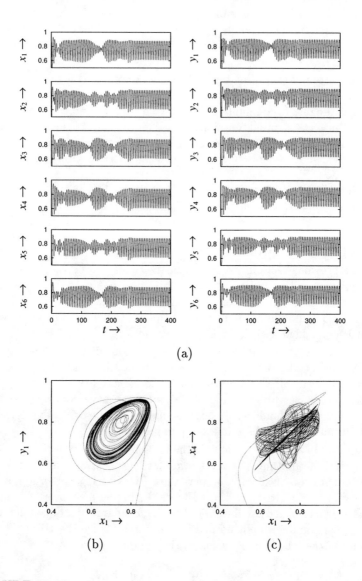

FIGURE 26.11
(a) Time response. The control is turned on at $t = 250$. (b) x_1-y_1 and (c) x_1-x_2 phase portraits. Under control, the system states converge to the target orbit (thick curve).

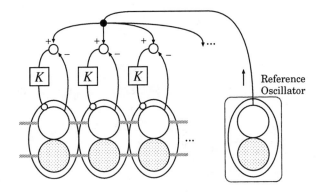

FIGURE 26.12
Coupled neurons with the controller for stabilization of in-phase synchronized orbits.

26.5 Concluding Remarks

In this chapter, we have investigated the rich dynamical behaviors and a simple but effective feedback chaos control method for a system of strongly connected neural oscillators. We have studied the existence and stability of some lower-dimensional system equilibria, and simulated some periodic orbits, bifurcation diagrams, chaotic motions, and synchronization phenomena. To this end, obtaining dynamic diagrams of higher-dimensional systems remains a challenge for future research.

Acknowledgment

The authors would like to thank Prof. H. Kawakami and Dr. T. Yoshinaga of Tokushima University, Dr. H. Nakajima of Kinki University, Dr. E. Izhikevich of Arizona State University, and Dr. Huaizhou Zhang of the University of Houston, for their valuable discussions and suggestions.

References

[1] A. A. Andronov and C. E. Chaikin, *Theory of Oscillations*, Princeton Univ. Press, Princeton, NJ, 1949.

[2] G. A. Carpenter and S. Grossberg, "A neural theory of circadian rhythms: split rhythms, after-effects, and motivational interactions," *J. of Theoretical Biology*, vol. 113, pp. 163–223, 1985.

[3] G. Chen, "Chaos, bifurcations, and their control," in J. Webster (Ed.), *Wiley Encyclopedia of Electrical and Electronics Engineering*, Wiley, New York, vol. 3, pp. 194-218, 1999.

[4] G. Chen and X. Dong, *From Chaos to Order: Methodologies, Perspectives and Applications*, World Scientific Pub. Co., Singapore, 1998.

[5] G. B. Ermentrout and J. D. Cowan, "Large scale spatially organized activity in neural nets," *SIAM J. of Applied Mathematics*, vol. 38, pp. 1–21, 1980.

[6] G. N. Borisyuk, R. M. Borisyuk, A. I. Khibnik, and D. Roose, "Dynamics and Bifurcations of Two Coupled Neural Oscillators with Different Connection Types," *Bulletin of Math. Biology,*, vol. 57, pp.809–840, 1995.

[7] F. C. Hoppensteadt and E. M. Izhikevich, *Weakly Connected Neural Networks*, Springer, AMS 126, 1997.

[8] H. Kawakami, "Bifurcation of periodic responses in forced dynamic nonlinear circuits," *IEEE Trans. on Circ. Sys.*, vol. 31, pp. 248–260, 1984.

[9] T. Matsumoto, M. Komuro, H.Kokubu, and R. Tokunaga, *Bifurcations*, Springer-Verlag, New York, 1993.

[10] S. Mizutani, T. Sano, T. Uchiyama, and N. Sonehara, "Controlling chaos in chaotic neural networks," *Electronics and Communication in Japan*, Part 3, vol. 81, pp. 73–81, 1998.

[11] G. J. Mpitsos, R. M. Burton, H. C. Creech, and S. O. Soinila, "Evidence for chaos in spike trains of neurons that generate rhythmic motor patterns," *Brain Research Bulletin*, vol. 21, pp. 529–538, 1988.

[12] Y. Pomeau, and P. Manneville, "Intermittent transition to turbulence in dissipative dynamical systems," *Commun. Math. Phys.*, vol. 74, pp. 189–197, 1980.

[13] K. Pyragas, "Continuous control of chaos by self-controlling feedback," *Phys. Lett. A*, vol. 170, pp. 421–428, 1992.

[14] A. Selverston, "A model system for the study of rhythmic behavior," in J. C. Fentress (Ed.), *Simpler Networks and Behavior*, Sinauwer, Sunderland, MA, 1976, pp. 83–98.

[15] C. Skarda and W. J. Freeman, "How brain makes chaos to make sense of the world," *Behavioral and Brain Sciences*, vol. 10, pp. 161–195, 1987.

[16] P. Thiran, *Dynamics and Self-Organization of Locally Coupled Neural Networks*, Presses Plytechniques et Universitaires Romandes, 1997.

[17] T. Ueta, M. Tsueike, H. Kawakami, T. Yoshinaga, and Y. Katsuta, "A computation of bifurcation parameter values for limit cycles," *IEICE Trans. Fundamentals*, vol. E80-A, pp. 1725–1728, 1997.

[18] H. R. Wilson and J. D. Cowan, "A mathematical theory of the functional dynamics of cortical and thalamic nervous tissue," *Kybernetik*, vol. 13, pp. 55–80, 1973.

27

Return Map Modulation in Nonautonomous Relaxation Oscillator

Toshimichi Saito and Hiroyuki Torikai

Department of Electronics and Electrical Engineering
Hosei University, Tokyo, 184-8584 Japan
tsaito@k.hosei.ac.jp, torikai@k.hosei.ac.jp

Abstract

As plural inputs are applicable to a nonlinear system and each input can cause useful function, effective fusion of the functions is desired. For this purpose, we study a circuit version of an integrate-and-fire model (IFM) to which three periodic inputs are applicable. The IFM is a relaxation oscillator and can be regarded as a simple neuron model: its state increases, reaches a threshold, is reset to a base, and repeats this manner. We then consider roles of the inputs using a one-dimensional return map. The first prime input is applied to the base and can control the basic shape of the return map: the IFM gives a variety of periodic and chaotic phenomena. The second and the third inputs have higher frequencies and smaller amplitudes than the first one, and are applied to the base and the threshold, respectively. The second one can modulate the return map to change the system stability: it changes from chaotic to periodic behavior or vise versa. The third input can modulate the return map to quantize the system state, and the IFM can give a variety of periodic phenomena. Dynamics of the quantized system can be simplified into an integer state map that is well suited for computer aided rigorous analysis.

27.1 Introduction

We consider a situation shown in Figure 27.1: a continuous-time nonlinear autonomous system can accept plural periodic inputs, $U_i(t)$, $i = 1, 2, ..., N$, and outputs Y. Here, an essential problem is to classify proper function caused by each input: heterogeneousness of each input. If we can find useful functions from the inputs and if we can fuse them effectively, we can obtain more useful functions. It contributes to the development of efficient control systems, communication systems, computational systems and so on. However, the inputs-outputs relationship is hard to clarify in nonlinear systems to which the superposition theorem cannot apply.

We consider such a problem for an integrate-and-fire model (IFM). The IFM is known as a simplified neuron model [24] and [8, 13, 20, 19, 3, 4]. The state of the IFM corresponds to a membrane potential and increases for a threshold level. When the state reaches the threshold, the IFM outputs a firing pulse, and the state is reset to a base level. Repeating this manner, the IFM generates a firing pulse-train. It is suggested that such pulse-train deeply relates to information processing functions in our brain [36]. In the study on the IFM, response characteristics form single periodic input is fundamental and some interesting results have been published [24] and [8, 13, 20]. They suggest that the IFMs can exhibit various periodic or chaotic pulse-trains. However, the systems with plural inputs have not been sufficiently studied so far.

In this paper, we provide a simple circuit model of the IFM to which three periodic inputs are applicable. Basically, the IFM is a nonautonomous relaxation oscillator. Its implementation example is shown in [34] and [32, 33] with some laboratory data. The first input is prime and applied to the base. The second and the third inputs have higher frequencies and smaller amplitudes than the first one, and are applied to the base and the threshold, respectively.

We investigate the roles of the inputs using a mapping procedure: we can derive one-dimensional return maps that describe the pulse-trains. At

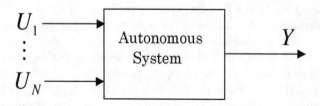

FIGURE 27.1
System with plural inputs.

present, one-dimensional return map is a simple and useful tool for analysis and synthesis of nonlinear dynamics: analysis of a higher-dimensional return map is very hard so far [30, 22, 21, 25].

Then we can see that the first input controls a basic system dynamics. We can obtain various shapes of the return map by changing the shape of the input: the IFM generates various periodic and chaotic pulse-trains. We refer to this return map as a basic return map.

As the second input is applied to the base, it modulates the basic return map: the macroscopic (respectively, the microscopic) shape of the return map is determined by the first one (respectively, the second one). Then the second input can change the stability or the instability of the pulse-train, e.g., it changes from chaotic to periodic behavior or vise versa. This fact is deeply relevant to a number of chaos control techniques [26, 27, 12, 17, 18, 15, 9, 35] and their applications [7, 28], because the techniques use weak perturbations in order to change stability or instability of the objective systems.

As both the first and the third inputs are applied, the third one modulates the basic return map to quantize its state. It may be a typical behavior for oscillators with weak periodic perturbations [2, 29]. In this case, the system exhibits only periodic pulse-train, however, it can be changed into various forms as the third input varies. Dynamics of the quantized system can be simplified into an integer state map which is well suited for computer aided rigorous analysis [32, 33, 11].

27.2 Unit Shape Function

In order to analyze the system of Figure 27.1, an autonomous system with plural inputs, we introduce a tool to describe the periodic inputs. As we discuss afterward, the shapes of the inputs determine the shape of the return map: the shapes of the inputs are important to consider the system behavior. Hence we describe the periodic inputs by unit shape function (USF), although there exist various methods to describe them, e.g., Fourier series. The USF is defined as a function from the unit interval $I = [0, 1)$ to reals $\boldsymbol{R}$ with some normalizations (see Figure 27.2):

$$S : I \to \boldsymbol{R} \qquad (27.2.1)$$

1) Normalization for amplitude:

$$\|S(t)\| \equiv \sup_{t \in I} S(t) - \inf_{t \in I} S(t) = 1. \qquad (27.2.2)$$

2) Normalization for DC component:

$$\int_0^1 S(t)dt = 0. \tag{27.2.3}$$

3) Normalization for period and phase:

$$C_1 = \int_0^1 S(t)e^{-j2\pi t}dt \neq 0, \quad \angle C_1 = -\frac{\pi}{2}, \tag{27.2.4}$$

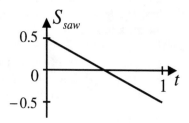

FIGURE 27.2
An example of the unit shape function.

Using the USF, a periodic signal is described by

$$U(t) = \sum_{n=-\infty}^{\infty} KS\left(\frac{1}{P}(t-\Phi)-n\right), \tag{27.2.5}$$

where $0 \leq \Phi < P$. The parameters K, P, and Φ control amplitude, period, and phase, respectively. In other words, the USF represents all the signals described by formula (27.2.5). The USF may be regarded as a simplified version of the Radial Basis function [23]. Note that we treat the DC term independently from the periodic signals. As N inputs are applicable to a system, they are described by

$$U_i(t) = \sum_{n_i=-\infty}^{\infty} K_i S_i\left(\frac{1}{P_i}(t-\Phi_i)-n_i\right), \quad i = 1 \sim N. \tag{27.2.6}$$

Without loss of generality, we assume $P_1 \geq P_i$. Since the inputs are applied to an autonomous system, we determine the time origin based on U_1. Applying the time shift $t' = t - \Phi_1$, we obtain

$$U_1(t') = \sum_{n_1=-\infty}^{\infty} K_i S_i\left(\frac{1}{P_1}t' - n_1\right). \tag{27.2.7}$$

For the other inputs U_i, $i \geq 2$, there must exist some nonnegative integer m_i such that

$$\Phi_i + (m_i - 1)P_i \leq \Phi_1 \leq \Phi_i + m_i P_i.$$

Let $\phi_i = \Phi_i + m_i P_i - \Phi_1$ and let $l_i = n_i - m_i$. Applying the time shift $t' = t - \Phi_1$ again, we obtain

$$U_i(t') = \sum_{l_i=-\infty}^{\infty} K_i S_i \left(\frac{1}{P_i}(t' - \phi_i) - l_i \right), \quad \text{for } i \geq 2. \qquad (27.2.8)$$

Hereafter, we use symbol t instead of t' and identify integers n_1 and l_i: the periodic inputs are described by

$$U_i(t) = \sum_{n=-\infty}^{\infty} K_i S_i \left(\frac{1}{P_i}(t - \phi_i) - n \right), \qquad (27.2.9)$$

where $\phi_1 = 0$, $0 \leq \phi_j < P_j$, and $P_1 \geq P_j$ ($j \geq 2$).

27.3 Integrate-and-Fire Model with Three Inputs

Figure 27.3 shows the circuit model of the IFM, where $B(t)$ is a base signal, $Th(t)$ is a threshold signal, and $B(t) < Th(t)$ is assumed. If the state v increases and reaches the threshold, the monostable multivibrator MM outputs a voltage pulse Y. The pulse closes the switch and the state jumps to the base. The jump and the output pulse correspond to the firing of a neuron. Repeating this manner, the IFM generates a pulse-train of Y. If both the base and the threshold are constant, $B(t) = B_0$ and $Th(T) = Th_0$, this IMF is an autonomous relaxation oscillator with sawtooth waveform. It corresponds to the autonomous system in Figure 27.1. We then consider

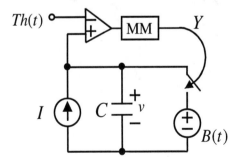

FIGURE 27.3
Integrate-and-fire model.

the case where three periodic inputs, U_1 to U_3 in (27.2.9), are applicable to the system:

$$B(t) = B_0 + U_1(t) + U_2(t), \quad Th(t) = Th_0 + U_3(t). \qquad (27.3.10)$$

As is clarified afterward, the first input U_1 can determine the basic system dynamics, the second input U_2 can control the system stability or instability, and the third input U_3 can quantize the system state, where U_2 and U_3 are assumed to have smaller amplitudes and higher frequencies than U_1. The case where more inputs are applied to the system will be discussed elsewhere.

The circuit dynamics are described by

$$\begin{cases} C\dfrac{dv}{dt} = I, & \text{for } v < Th(t), \\ v(t) = B(t), & \text{if } v(t^-) = Th(t^-), \end{cases} \qquad (27.3.11)$$

$$Y(t) = \begin{cases} V_{DD}, & \text{if } v(t^-) = Th(t^-), \\ V_{SS}, & \text{otherwise,} \end{cases}$$

where V_{DD} and V_{SS} are the high and low output levels of the MM, respectively. Note that we have omitted dynamics for $v > Th(t)$. Using the following dimensionless variables and parameters:

$$\tau = \frac{t}{P_1}, \quad x = \frac{C(v - B_0)}{IP_1}, \quad y = \frac{Y - V_{SS}}{V_{DD} - V_{SS}}, \quad a_0 = \frac{C(Th_0 - B_0)}{IP_1},$$

$$k_i = \frac{K_i Th_0}{IP_1}, \quad T_i = \frac{P_i}{P_1}, \quad \theta_i = \frac{\phi_i}{P_1},$$

Equation (27.3.11) is transformed into

$$\begin{cases} \dfrac{dx}{d\tau} = 1, & \text{for } x < a(\tau), \\ x(\tau) = b(\tau), & \text{if } x(\tau^-) = a(\tau^-), \end{cases} \qquad (27.3.12)$$

$$y(\tau) = \begin{cases} 1, & \text{if } x(\tau^-) = a(\tau^-), \\ 0, & \text{otherwise,} \end{cases}$$

$$b(\tau) = u_1(\tau) + u_2(\tau), \quad a(\tau) = a_0 + u_3(\tau),$$

$$u_1(\tau) = \sum_{n=-\infty}^{\infty} k_1 S_1(\tau - n),$$

$$u_j(\tau) = \sum_{n=-\infty}^{\infty} k_j S_j \left(\frac{1}{T_j}(\tau - \theta_j) - n \right),$$

where $a(\tau) > b(\tau)$, $0 \leq \theta_j < T_j$, and $j = 2, 3$. Note that this normalized system includes IFM-type neuron models in [8, 13, 20]. Hereafter, for simplicity, let $T_i = \frac{1}{M_i}$, where M_i is a positive integer.

27.4 Basic Return Map by the First Prime Input

The IFM described by (27.3.12) generates the pulse-train $y(\tau)$. The pulse-train dynamics can be analyzed using the pulse positions as the following. For a given first pulse position $\tau_1 > 0$, let τ_n denote the n-th pulse position. As discussed before, the system is a relaxation oscillator when the threshold and the base are constants, i.e., $a(\tau) = a_0$, $b(\tau) = 0$ ($u_1 = u_2 = u_3 = 0$). In this case, the state started from $(\tau_n, 0)$ is given by $x(\tau) = \tau - \tau_n$ for $\tau_n < \tau < \tau_{n+1}$, and the pulse position is governed by

$$\tau_{n+1} = f_0(\tau_n) \equiv \tau_n + a_0, \quad f_0 : \mathbf{R}^+ \to \mathbf{R}^+, \tag{27.4.13}$$

where $\mathbf{R}^+$ denotes nonnegative reals. It describes sawtooth-type periodic oscillation with period a_0.

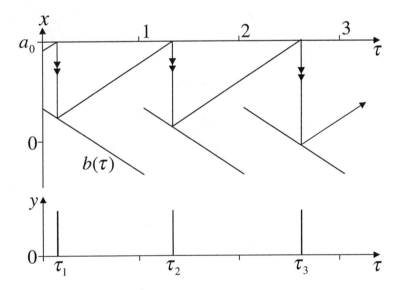

FIGURE 27.4
IFM with the first base input.

We then consider the role of the first inputs using mapping procedure. As shown in Figure 27.4, we consider the case where $b(\tau) = u_1(\tau) \neq constant$ and $u_2 = u_3 = 0$. In this case, the state started from $(\tau_n, b(\tau_n))$ is given by $x(\tau) = \tau - \tau_n + u_1(\tau_n)$ for $\tau_n < \tau < \tau_{n+1}$. Since the state reaches the threshold a_0 at $\tau = \tau_{n+1}$, the pulse-train is governed by

$$\tau_{n+1} = f_b(\tau_n) \equiv \tau_n + a_0 - u_1(\tau_n). \quad f_b : \mathbf{R}^+ \to \mathbf{R}^+, \tag{27.4.14}$$

$$u_1(\tau) = \sum_{m=0}^{\infty} k_1 S_1(\tau - m).$$

An example of this pulse position map is depicted in Figure 27.5. Note that f_b represents dynamics given by $h(\tau_{n+1}) = f_b h(\tau_n)$, where $h(\tau) = \tau + \theta_1$ is a shift operator for $0 \le \theta_1 < 1$. Equation (27.4.14) implies that the pulse-train characteristics are controllable by u_1. In fact, as we desire a pulse-train described by

$$\tau_{n+1} = f_d(\tau_n), \quad f_d(\tau_n + n) = f_d(\tau_n) + n, \quad \text{for } 0 \le \tau_n < 1,$$

it can be realized by the first input of the form

$$u_1(\tau) = f_d(\tau) - \tau_n - a_0,$$

where note that $\tau_{n+1} > \tau_n$ is a necessity. In order to reduce the dynamics

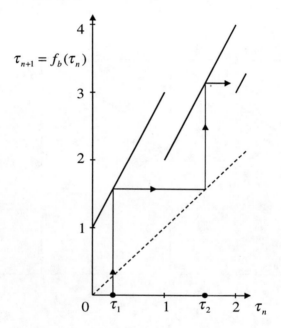

FIGURE 27.5
Basic pulse position map.

into a map from an interval into itself, we introduce a transformation

$$z_n = \tau_n \quad (\text{mod } 1). \tag{27.4.15}$$

Then we define

$$z_{n+1} = F_b(z_n) \equiv f_b(z_n) \quad (\text{mod } 1), \quad F_b : I \to I. \tag{27.4.16}$$

Using a USF, F_b is described by

$$F_b(z_n) = z_n + a_0 - k_1 S_1(z_n), \quad (\text{mod } 1). \tag{27.4.17}$$

For convenience, we refer to F_b and f_b as a basic return map and a basic pulse position map, respectively. We then define periodic points, chaos, and related objects.

DEFINITION 27.1 [21] *Let F be a mapping from I to I. A point $p \in I$ is said to be a fixed point or a periodic point with period 1 if $F(p) = p$. For $k \geq 2$, a point $p \in I$ is said to be a periodic point with period k if $F^k(p) = p$ and $F^l(p) \neq p$ for $1 \leq l < k$, where F^k denotes the n-fold composition of F.*

Let DF^k denote the derivative of F^k. A periodic point p with period k is said to be stable, critical, and unstable, if $|DF^k(p)| < 1$, $|DF^k(p)| = 1$ and $|DF^k(p)| > 1$, respectively.

A sequence of periodic points, $\{F(p), \cdots F^k(p)\}$, $k \geq 1$ is said to be a periodic sequence. A periodic sequence is said to be a stable periodic sequence (respectively, unstable periodic sequence) if it consists of stable (respectively, unstable) periodic point(s), and we abbreviate them by SPS (respectively, UPS).

A point $p_e \in I$ is said to be an eventually periodic point, if not p_e but $F^m(p_e)$ is a periodic point for some positive integer m.

F is said to be chaotic if there exists some positive integer n such that $|DF^n| > 1$ for almost everywhere in I. In this case, there exists some subinterval $J \subset I$ where F is ergodic and has a positive Lyapunov exponent [22].

Figure 27.6 shows an example of the chaotic return map with a periodic point and an eventually periodic point.

Example 27.2

We consider the case where the sawtooth-type USF is applied to the first base input:

$$u_1(\tau) = k_1 \sum_{m=0}^{\infty} S_{saw}(\tau - m), \quad u_2(\tau) = u_3(\tau) = 0. \tag{27.4.18}$$

$$S_{saw}(\tau) = -(\tau - 0.5), \quad \text{for } \tau \in I.$$

The S_{saw} is shown in Figure 27.2 and the dynamics is shown in Figure 27.4. Substituting Equation (27.4.18) into Equations (27.4.14) and (27.4.16), we

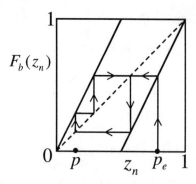

FIGURE 27.6
Basic return map. p is an unstable periodic point with period 3, and p_e is an eventually periodic point.

obtain the basic pulse position map and the basic return map.

$$\tau_{n+1} = f_b(\tau_n) \equiv \tau_n + a_0 - \sum_{m=0}^{\infty} k_1 S_{saw}(\tau_n - m), \quad \tau_n \in \mathbf{R}^+, \quad (27.4.19)$$
$$z_{n+1} = F_b(z_n) \equiv z_n + a_0 - k_1 S_s(z_n), \quad (\text{mod } 1), \quad z_n \in I.$$

Figure 27.7 shows two examples of the basic return maps. Noting

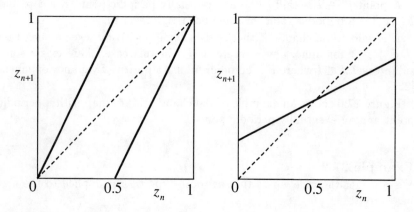

FIGURE 27.7
Basic return maps for sawtooth-type u_1. Left: $k_1 = 1$, $a_0 = 1.5$. Right: $k_1 = -0.5$, $a_0 = 1$.

$$DF_b(z) = k_1 + 1, \quad \text{for almost all } z \in I,$$

we can say

F_b exhibits chaos for $|1+k_1| > 1$ (Figure 27.7 left): the pulse-train is chaotic.

F_b exhibits SPS for $|1+k_1| < 1$ (Figure 27.7 right): the pulse-train is periodic.

Remark. For $0 < 1+k_1 < 1$, F_b is into and is identical with a return map from a neuron model discussed in [24]. In this case, F_b can exhibit various SPSs as parameters vary and can exhibit nonperiodic phenomenon in a measure-zero parameter subset. However, we omit the measure-zero case. Also, for $-1 < 1+k_1 \le 0$, F_b must have at least one fixed point because a segment $\tau_{n+1} = f_b(\tau_n)$, $0 \le \tau_n < 1$, must intersect line $\tau_{n+1} = \tau_n + m$ on the (τ_n, τ_{n+1}) plane for some positive integer m. The case $k_1 = 0$ corresponds to the autonomous oscillator.

In the following sections, we consider effects of the second and third inputs to the basic dynamics: how the return map F_b is modulated by u_2 or u_3.

27.5 Role of the Second Base Input

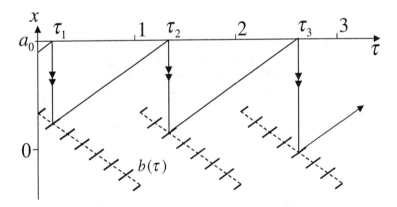

FIGURE 27.8
IFM with the first and the second base inputs.

As shown in Figure 27.8, we consider the case where $b(\tau) = u_1(\tau) + u_2(\tau)$ and $a(\tau) = a_0$, $(u_1 \ne 0, u_2 \ne 0, u_3 = 0)$, the pulse position is governed by

$$\tau_{n+1} = f_{b2}(\tau_n) \equiv \tau_n + a_0 - u_1(\tau_n) - u_2(\tau_n), \quad f_{b2}: \mathbf{R}^+ \to \mathbf{R}^+, \quad (27.5.20)$$

$$u_2(\tau) = \sum_{m=-1}^{\infty} k_2 S_2(M_2(\tau - \theta_2) - m),$$

where M_2 is a positive integer and $0 \leq \theta_2 < \frac{1}{M_2}$. Note that f_{b2} is given by replacing u_1 in Equation (27.4.14) with $u_1 + u_2$. Applying the mod 1 transformation (27.4.15), we obtain the return map

$$z_{n+1} = F_{b2}(z_n) \equiv f_{b2}(z_n) \pmod{1}, \qquad (27.5.21)$$

$$F_{b2}(z_n) = z_n - a_0 - k_1 S_1(z_n) - \sum_{m=-1}^{M_2} k_2 S_2(M_2(z_n - \theta_2) - m) \pmod{1}.$$

We refer to f_b and F_b as a second pulse position map and a second return map, respectively. Here, we can see that the basic return map F_b is modulated by the second base input u_2: u_2 can change microscopic shape of the basic return map F_b. It suggests that u_2 may control the stability. If u_1 and u_2 are sawtooth-type, we can give an almost complete answer as shown in the following.

Example 27.3

We apply a sawtooth-type second base input to the system in Example 1:

$$u_2(\tau) = \sum_{m=-1}^{\infty} k_2 S_{saw}(M_2(\tau - \theta_2) - m), \qquad (27.5.22)$$

where M_2 is a positive integer and $0 \leq \theta_2 < \frac{1}{M_2}$. u_1 is the same as Equation (27.4.18) and $u_3 = 0$. The system dynamics is shown in Figure 27.8. Substituting Equation (27.5.22) into Equations (27.5.20) and (27.5.21), we obtain the second pulse position map and the second return map.

$$\tau_{n+1} = f_{b2}(\tau_n) \equiv f_b(\tau_n) - \sum_{m=-1}^{\infty} k_2 S_{saw}(M_2(\tau - \theta_2) - m), \quad \tau_n \in \mathbf{R}^+,$$
$$z_{n+1} = F_{b2}(z_n) \equiv f_{b2}(z_n) \pmod{1}, \quad z_n \in I,$$
$$(27.5.23)$$

Figure 27.9 shows two examples where chaotic and periodic return maps in Figure 27.7 are changed into periodic and chaotic, respectively. That is, this second input can control stability of the basic return map. In fact, we have

$$DF_{b2}(z) = k_1 + k_2 M_2 + 1, \text{ for almost all } z \in I,$$

and we can say

F_{b2} exhibits SPS for $|1 + k_1 + k_2 M_2| < 1$, provided F_{b2} has a periodic point.

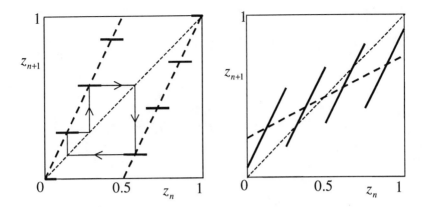

FIGURE 27.9
Second return maps modulated by sawtooth-type u_2. Left: $k_1 = 1$, $a_0 = 1.5$, $k_2 M_2 = -2$, $M_2 = 7$, $M_2 \theta_2 = 0.5$. Right: $k_1 = -0.5$, $a_0 = 1$, $k_2 M_2 = 1.5$, $M_2 = 4$, $\theta_2 = 0$.

TABLE 27.1
Phenomena from two base inputs.

IFM with u_1: condition and phenomenon		IFM with u_1 and u_2: condition and phenomenon	
$\|1 + k_1\| > 1$	Chaos	$\|1 + k_1 + k_2 M_2\| > 1$	Chaos
$\|1 + k_1\| > 1$	Chaos	$\|1 + k_1 + k_2 M_2\| < 1$	SPS
$\|1 + k_1\| < 1$	SPS	$\|1 + k_1 + k_2 M_2\| > 1$	Chaos
$\|1 + k_1\| < 1$	SPS	$\|1 + k_1 + k_2 M_2\| < 1$	SPS

F_{b2} exhibits chaos for $|1 + k_1 + k_2 M_2| > 1$.

That is, as F_b exhibits either chaos or SPS, u_2 can change the phenomenon into either chaos or SPS. This manner is summarized in Table 27.1.

If $1 + k_1 + k_2 M_2 = 0$, the modulated map F_{b2} is flat almost everywhere and the state is discretized. Although this discretization is possible only for the sawtooth type u_1, the threshold input u_3 can realize the discretization for any shape of u_1 as discussed in the next section.

27.6 Role of the Threshold Input

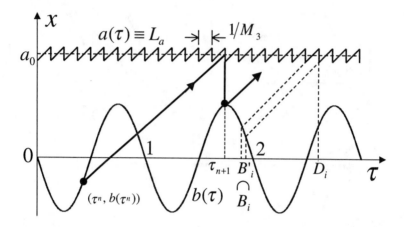

FIGURE 27.10
IFM with the first and the third inputs.

As shown in Figure 27.10, we consider the case where $b(\tau) = u_1(\tau)$ and $a(\tau) = a_0 + u_3(\tau)$, ($u_1 \neq 0$, $u_2 = 0$, $u_3 \neq 0$), the pulse position is governed by the following, provided $u_3(\tau)$ is continuous:

$$\tau_{n+1} - u_3(\tau_{n+1}) = \tau_n + a_0 - u_1(\tau_n), \quad \tau_n \in \mathbf{R}^+, \quad (27.6.24)$$

$$u_3(\tau) = \sum_{m=-1}^{\infty} k_3 S_3(M_3(\tau - \theta_3) - m),$$

where M_3 is a positive integer and $0 \leq \theta_3 < \frac{1}{M_3}$. Note that Equation (27.6.24) is given by replacing τ_{n+1} in Equation (27.4.14) with $\tau_{n+1} - u_3(\tau_{n+1})$. However, if u_3 has discontinuity points (defined afterward), Equation (27.6.24) is void. Then the discontinuity points cause an important quantization function discussed afterward. In order to make the return map valid for the discontinuity, we define some objects.

Let $DS = \{D_1, D_2, \cdots\}$, $0 \leq D_i < D_{i+1}$, be a set of points such that $\tau \in DS$ gives "falling" discontinuity: $a(D_i^-) - a(D_i) > 0$. We omit "rising" discontinuity, $a(D_i^-) - a(D_i) < 0$, because it cannot cause the firing. We refer to D_i and $(D_i, a(D_i))$ as a discontinuity base and a discontinuity point, respectively. We then describe the threshold $a(\tau)$ as a figure in the

(τ, x) plane:

$$a(\tau) \equiv L_a, \quad L_a = L_c \cup L_d, \quad L_d = \bigcup_i L_{di},$$
$$L_c = \{(\tau, x) | \tau \neq D_i, x = a(\tau)\},$$
$$L_{di} = \{(\tau, x) | \tau = D_i, a(D_i) \leq x \leq a(D_i^-)\},$$
(27.6.25)

where i is a positive integer. L_{di} characterizes each discontinuity base. We then define a function,

$$\alpha(\tau, x) = \begin{cases} 0, & \text{for } (\tau, x) \in L_a, \\ 1, & \text{otherwise.} \end{cases}$$
(27.6.26)

On the (τ, x) plane, we consider a trajectory started form $(\tau_n, b(\tau_n))$. The trajectory is given by $x = \tau - \tau_n + u_1(\tau_n)$ and it must intersect L_a. Let (τ_{n+1}, X) be the intersection, where $X \equiv \tau_{n+1} - \tau_n + u_1(\tau_n)$. Also, let B_i be a subset in the τ axis such that the trajectory intersects L_{di} only if it starts from $(\tau_n, b(\tau_n))$, $\tau_n \in B_i$. Let $B_d = \cup_i B_i$. Using the above, the pulse position map, $\tau_{n+1} = f_a(\tau_n)$, is described by

$$\alpha(f_a(\tau_n), f_a(\tau_n) - \tau_n + u_1(\tau_n)) = 0:$$

$$\begin{cases} f_a(\tau_n) - u_3(f_a(\tau_n)) = \tau_n + a_0 - u_1(\tau_n), & \text{for } \tau_n \notin B_d, \\ f_a(\tau_n) = D_i & \text{for } \tau_n \in B_i. \end{cases}$$
(27.6.27)

The corresponding return map is given by

$$z_{n+1} = F_a(z_n) \equiv f_a(z_n) \pmod{1} \quad z_n \in I.$$

We refer to $f_a : \mathbf{R}^+ \to \mathbf{R}^+$, and $F_a : I \to I$ as a third pulse position map and a third return map, respectively. It goes without saying that Equation (27.6.27) is consistent with Equation (27.6.24) if $u_3(\tau)$ is continuous. However, our interests are in the case where all the trajectories cannot hit L_c. In fact, if the slope of $u_3(\tau)$ is larger than 1 for $\tau \neq D_i$, the trajectory cannot intersect L_c and it must hit L_d. In this case, the pulse position is restricted in the discontinuity base and the system state is discretized: u_3 can cause quantization function. We demonstrate this function in the next example.

Example 27.4

We consider the quantization function caused by a sawtooth-type threshold input u_3.

$$u_1(\tau) = \sum_{m=0}^{\infty} k_1 S_1(\tau - m), \qquad (27.6.28)$$
$$u_3(\tau) = \sum_{m=-1}^{\infty} k_3 S_{saw}(M_3(\tau - \theta_3) - m),$$

where u_1 is any periodic input, $k_3 < 0$, M_3 is a positive integer and $0 \leq \theta_3 < \frac{1}{M_3}$. Note that $k_3 < 0$ prohibits "rising" discontinuity. Recalling Equation (27.6.25), the threshold $a(\tau)$ can be described by the following L_a:

$$L_a = L_c \cup L_d, \quad L_d = \bigcup_i L_{di},$$
$$L_c = \{(\tau, x) | \tau \neq D_i, x = a_0 + u_3(\tau)\}, \quad D_i = \theta_3 + \frac{i}{M_3}, \qquad (27.6.29)$$
$$L_{di} = \{(\tau, x) | \tau = D_i, |x - a_0| \leq \frac{|k_3|}{2}\},$$

where i is a positive integer. For simplicity, we focus on the case $|k_3|M_3 > 1$: all the trajectories cannot hit L_c but L_d. In this case, the intersection is restricted in L_d and the system state is discretized. In order to describe the dynamics, we define a line and its division by

$$L_b = \bigcup_i L_{bi}, \quad L_{bi} = \{(\tau, x) | x = a_0 - \frac{|k_3|}{2}, D_i - \frac{1}{M_3} < \tau \leq D_i\}. \qquad (27.6.30)$$

Assuming $a_0 - \frac{|k_3|}{2} > b(\tau)$, trajectory started from L_{bi} must hit L_{di}. Hence we can define a map

$$Q : L_b \to L_d, \quad \tau_{n+1} = Q(\tau'_n) = D_i, \text{ for } \tau'_n \in L_{bi}.$$

Also, trajectory started from $(\tau_n, b(\tau_n))$ must hit L_b. Letting $(a_0 - \frac{|k_3|}{2}, \tau'_n)$ be the hit point, we obtain

$$\tau'_n = f_b(\tau_n) - \frac{|k_3|}{2}.$$

That is, the quantization is described by a third pulse position map or a third return map as the following:

$$\tau_{n+1} = f_a(\tau_n) \equiv Q(f_b(\tau_n) - \frac{|k_3|}{2}), \quad \tau_n \in \mathbf{R}^+, \qquad (27.6.31)$$
$$z_{n+1} = F_a(z_n) \equiv f_a(z_n) \pmod 1, \quad z_n \in I.$$

We refer to a pair of D_i and its preimage as a discontinuity point of the mappings (27.6.31). The number of discontinuity points increases as M_3 increases. Since the slopes of F_a are flat almost everywhere, F_a can exhibit only SPS, provided F_a is differentiable on the SPS. In Figure 27.11, we can see that the basic return is quantized by the threshold input u_3. Note that

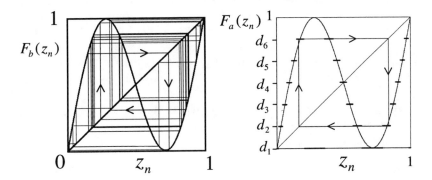

FIGURE 27.11
Quantization by the threshold input u_3. Left: Basic return map for $a_0 = 1$, $S_1(\tau) = \frac{1}{2}\sin 2\pi\tau$ for $\tau \in I$ and $k_1 = -1.46$. Right: Quantized map for $\theta_3 = 0$, $M_3 = 7$ and $|k_3|M_3 = 1$.

the pulse positions from f_a are to be restricted in the set of the discontinuity bases $DS = \{D_1, D_2, \cdots\}$. Also, letting $DI = DS \cap f_a(I) \pmod{1}$, the state of the return map F_a is restricted in DI. Letting m denote the number of the discontinuity bases in DI and relabeling all the discontinuity bases by $DI = \{d_1, \cdots, d_m\}$, $d_i < d_{i+1}$, the dynamics of the discretized state is described by the following integer state map:

$$F_I : Z_I \to Z_I, \quad F_I(i) = j, \quad \text{for } F_a(d_i) = d_j, \tag{27.6.32}$$

where $Z_I = \{1, \cdots, m\}$. As M_3 increases, this map may mimic the original dynamics or may cause different dynamics from the original one.

Note that this simplification is applicable to various quantized maps and that any element in Z_I must be either a periodic point or an eventually periodic point. This integer map (27.6.32) is well suited for computer-aided rigorous analysis. In order to clarify the number and the period of the SPSs, we have a simple and efficient algorithm [32, 33]:

(Analysis algorithm for the integer map)

Step 1: Set the counter $NP = 0$ and select an element $l_1 \in Z_I$. The selected element l_1 is either periodic point or eventually periodic point. Then go to Step 3.

Step 2: If all the elements in Z_I are "don't care," go to Step 4. Otherwise, select an element $l_1 \in Z_I$, which is not "don't care" and go to Step 3.

Step 3: Iterate F_I from l_1 and get a sequence $(l_1, F_I(l_1), ..., F_I^k(l_1))$ until it satisfies one of the following two cases. (At the beginning of the algorithm, Case 1 is satisfied.)

Case 1: The sequence is a periodic sequence or it consists of both periodic sequences and eventually periodic points, i.e., $F_I^r(l_1) = F_I^k(l_1)$, $0 \leq r \leq k$. In this case, renew the counter as $NP = NP+1$ and store the period of the sequence in PE_{NP}. Also, declare all the points in the sequence as "don't care."

Case 2: The sequence consists of eventually periodic points, i.e., $F_I^k(l_1)$ is "don't care." In this case, declare all points in the sequence as "don't care."

After one of these operations, go to Step 2. Note that one of them must be satisfied for any $l_1 \in Z_I$.

Step 4: The total number of the periodic sequences is NP and their periods are PE_i, $i = 1, 2, ..., NP$. Then terminate the algorithm.

Example 27.5

We consider the pulse-train dynamics for the sawtooth-type base and threshold:

$$u_1(\tau) = \sum_{m=0}^{\infty} S_{saw}(\tau - m), \quad a_0 = 1.5,$$
$$u_3(\tau) = -\frac{1}{M_3} \sum_{m=-1}^{\infty} S_{saw}(M_3 \tau - m), \tag{27.6.33}$$

and $u_2 = 0$, where M_3 is a positive integer. We then have obtained the quantized return maps as shown in Figure 27.12. Using the algorithm, the total number NP of the SPSs for M_3 can be clarified as shown in Figure 27.12.

27.7 Conclusions

We have considered a simple circuit model of the IFM to which three periodic inputs are applicable. Using one-dimensional return maps, we have clarified roles of the inputs. The first input controls shape of the basic return map and gives various periodic or chaotic pulse-trains. The second input modulates the basic return map and can change the stability or the instability of the pulse-train, e.g., it changes from chaotic to periodic behavior or vise versa. The third input modulates the basic return map to quantize the system state and the IFM exhibits various periodic pulse-trains. Dynamics of the quantized system can be simplified into an integer state map which is well suited for computer-aided rigorous analysis.

Based on the above results, we enumerate some of the future problems in the following:

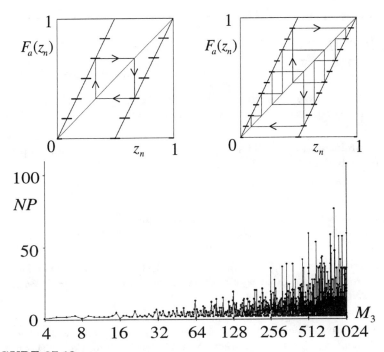

FIGURE 27.12
Quantized return maps and total numbers of the SPSs. Left upper: $M_3 = 6$. Right upper: $M_3 = 11$. Lower: Total numbers NP of the SPSs for M_3.

1. Analysis of bifurcation phenomena from the circuit model.
2. Systematic analysis of the quantized return map itself.
3. Effective fusion of useful function by each input and its application to information processing systems [6, 1].
4. Synthesis and analysis of a network of IFMs, and its application to digital multiplex communication systems [34] and [14, 10].
5. Analysis of other relaxation oscillators, e.g., human sleep-wake cycles model [5] and chaotic relaxation oscillators with hysteresis [31].

References

[1] Y. V. Andreyev, A. S. Dmitriev, and S. O. Starkov, "Information processing in 1–D systems with chaos," *IEEE Trans. on Circ. Sys.*, I, vol. 44, pp. 21-28, 1997.

[2] Y. Braiman and I. Goldhirsch, "Taming chaotic dynamics with weak periodic perturbations," *Phys. Rev. Let.*, vol. 66, pp. 2545-2548, 1991.

[3] E. Catsigeras and R. Budelli, "Limit cycles of a bineural network model," *Physica D*, vol. 56, pp. 235-252, 1992.

[4] S. Coombes and S. H. Doole, "Neuronal populations with reciprocal inhibition and rebound currents: Effects of synaptic and threshold noise," *Phys. Rev. E*, vol. 54, pp. 4054-4065, 1996.

[5] S. Daan, D. G. M. Beersma, and A. A. Borbely, "Timing of human sleep: recovery process gated by a circadian pacemaker," *American Physiological Society*, R161-R182, 1984.

[6] A. C. Davies, "Digital counters and pseudorandom number generators from a perspective of dynamics," in A. C. Davies and W. Schwarz (Eds.), *Nonlinear Dynamics of Electronic Systems*, World Scientific Pub. Co., 1994, pp. 373-382.

[7] W. L. Ditto, S. N. Rauseo, and M. L. Spano, "Experimental control of chaos," *Phys. Rev. Lett.*, vol. 65, pp. 3211-3214, 1990.

[8] L. Glass and M. C. Mackey, "A simple model for phase locking of biological oscillators," *J. Math. Biol.*, vol. 7, pp. 339-352, 1979.

[9] P. Gora and A. Boyarsky, "A new approach to controlling chaotic systems," *Physica D*, vol. 111, pp. 1-15, 1998.

[10] M. Hasler, M. D.-Restituto, and A. R.-Vazquez, "Markov maps for communications with chaos," *Proc. of NDES*, 1996, pp. 161-166.

[11] C. S. Hsu and R. S. Guttalu, "An unravelling algorithm for global analysis of dynamical systems: an application of cell-to-cell mappings," *Trans. ASME*, vol. 47, pp. 940-948, 1980.

[12] E. R. Hunt, "Stabilizing high-period orbits in a chaotic system: The diode resonator," *Phys. Rev. Lett.*, vol. 67, pp. 1953-1955, 1991.

[13] J. P. Keener, F. C. Hoppensteadt, and J. Rinzel, "Integrate-and-fire models of nerve membrane response to oscillatory input," *SIAM J. Appl. Math.*, vol. 41, pp. 503-517, 1981.

[14] M. P. Kennedy, "Communication with chaos: State of the art and engineering challenges," *Proc. of NDES*, 1996, pp. 1-8.

[15] T. Kousaka, T. Ueta, and H. Kawakami, "Destabilizing control of stable orbits," *Proc. of NOLTA*, Las Vegas, 1995, pp. 997-1000.

References

[16] H. Li and J. Chern, "Goal-oriented scheme for taming chaos with a weak periodic perturbation: Experiment in a diode resonator," *Phys. Rev. E*, vol. 54, pp. 2118-2120, 1996.

[17] M. A. Matias and J. Guemez, "Stabilization of chaos by proportional pulses in the system variables," *Phys. Rev. Lett.*, vol. 72, pp. 1455-1458, 1994.

[18] K. Mitsubori and T. Saito, "Control of piecewise linear chaos by occasional proportional feedback," in J. Awrejcewics (Ed.), *Nonlinear Dynamics: New Theoretical and Applied Results*, Akademie Verlag, 1995, pp. 361-375.

[19] R. P. Pascual and J. Lomniz-Adler, "Coupled relaxation oscillators and circle maps," *Physica D*, vol. 30, pp. 61-82, 1988.

[20] R. Perez and L. Glass, "Bistability, period doubling bifurcations and chaos in a periodically forced oscillator," *Phys. Lett. A*, vol. 90, pp. 441-443, 1982.

[21] A. Lasota and M. C. Mackey, *Chaos, Fractals, and Noise*, 2nd Ed., Springer-Verlag, 1994.

[22] T. Y. Li and J. A. Yorke, "Ergodic transformations from an interval into itself," *Trans. Amer. Math. Soc.*, vol. 235, pp. 183-192, 1978.

[23] A. I. Mees, M. F. Jackson and L. O. Chua, "Device modeling by radial basis functions," *IEEE Trans. on Circ. Sys.*, I, vol. 39, pp. 19-27, 1992.

[24] J. Nagumo and S. Sato, "On response characteristics of a mathematical neuron model," *Kybernetik*, vol. 10, pp. 155-164, 1972.

[25] E. Ott, *Chaos in Dynamical Systems*, Cambridge Univ. Press, 1993.

[26] E. Ott, C. Grebogi and J. A. Yorks, "Controlling chaos," *Phys. Rev. Lett.*, vol. 64, pp. 1196-1199, 1990.

[27] K. Pyragas and A. Tamasevicius, "Experimental control of chaos by delayed selfcontrolling feedback," *Phys. Lett. A*, vol. 180, pp. 99-102, 1993.

[28] R. Roy, T. W. Murohy, T. D. Masier, Z. Gills, and E. R. Hunt, "Dynamic control of a chaotic laser: Experimental stabilization of a globally coupled system," *Phys. Rev. Lett.*, vol. 67, pp. 1953-1955, 1991.

[29] N. F. Rul'kov and A. R. Volkovskii, "Threshold synchronization of chaotic relaxation oscillations," *Phys. Lett. A*, vol. 179, pp. 332-336, 1993.

[30] A. N. Sharkovsky and L. O. Chua, "Chaos in some 1-D discontinuous maps that appear in the analysis of electrical circuits," *IEEE Trans. on Circ. Sys.*, I, vol. 40, pp. 722-731, 1993.

[31] H. Torikai and T. Saito, "Synchronization of chaos and its itinerancy from a network by occasional linear connection," *IEEE Trans. on Circ. Sys.*, I, vol. 45, pp. 464-472, 1998.

[32] H. Torikai and T. Saito, "Computer aided analysis of chaotic oscillators with periodic perturbation," *Proc. of NDES*, Budapest, 1998, pp. 75-78.

[33] H. Torikai and T. Saito, "Return map quantization from an integrate-and-fire model with two periodic inputs," *IEICE Trans. Fundamentals*, E82-A, 1999, in press.

[34] H. Torikai, T. Saito, and W. Schwarz, "Synchronization via multiplex pulse-train," *IEEE Trans. on Circ. Sys.*, I, 1999, in press.

[35] T. Tsubone and T. Saito, "Stabilizing and destabilizing control for a piecewise linear circuit," *IEEE Trans. on Circ. Sys.*, I, vol. 45, pp. 172-177, 1998.

[36] M. Watanabe and K. Aihara, "What functional connectivity can do: software driven neural networks," *Proc. of ICONIP*, pp. 1370-1373, 1998.

28

Controlling Chaos in Discrete-Time Computational Ecosystems

Toshimitsu Ushio, Takehiro Imamori, and Tatsushi Yamasaki

Department of Systems and Human Science
Osaka University, Toyonaka, Osaka, 560-8531 Japan
ushio@sys.es.osaka-u.ac.jp

Abstract

First, we introduce a discrete-time version of Hogg-Huberman's model which is a model of computational ecosystems with a reward mechanism. Second, we examine its dynamic properties related to strategies. Finally, we propose a packet routing method in computer networks based on the Hogg-Huberman model.

28.1 Introduction

Recently, many approaches to resource allocation problems in multi-agent systems have been proposed. One approach called market based control is inspired by economical behavior and introduces concepts such as prices and auctions. In this approach, each agent submits a price to use or supply a resource, and negotiations among agents are achieved by an auction [5]. Market based control is applicable in various fields. Kurose and Simha proposed a decentralized algorithm for optimal file allocation in distributed computer systems and gives conditions for the convergence of the algorithm [13]. Huberman and Clearwater proposed market based distributed controllers for saving energy in buildings, and it is shown by experiment that this approach is more effective than a conventional control scheme such as PID control [5, 8].

The other approach called computational ecology is inspired by a biological or social organization and uses a macro model of behaviors of agents. Each agent makes a decision under incomplete knowledge and delayed information, and tries to improve his benefit or payoff accruing from the environment. Huberman and Hogg introduced a simple model of such a multi-agent system described by a 1st-order differential equation with delay [9]. It is shown that this model exhibits various nonlinear phenomena such as bifurcations and chaos [11, 12].

It is well-known that infinitely many periodic orbits are embedded in chaotic attractors [6]. Ott et al. proposed a method, called the OGY method, for stabilizing target periodic orbits embedded in chaotic attractors [16]. By the OGY method, stabilization is achieved by a very small input signal, and applied to many chaotic systems [3, 4]. While the OGY method is a very general method applicable for chaotic systems whose linearized systems are stabilizable, control method for specified systems have also been proposed. For example, Ushio et al. deal with stabilization of chaotic switched arrival systems [19, 20].

Hogg and Huberman introduce a reward mechanism for stabilization of a fixed point of the computational ecosystems [7]. Such a model of a computational ecosystem with a reward mechanism will be called the Hogg-Huberman model. Billard proposed a model of resources with queue based on the Hogg-Huberman model, and discussed a reward mechanism with stochastic learning automata [2].

On the other hand, how to design multi-agent systems in engineering fields is an important problem [14, 22]. Ushio and Motonaka applied a "modified" Hogg-Huberman model to a resource allocation problem in a manufacturing system [18] We apply the computational ecosystem model in communication networks. Routing is an important problem for large network management [1]. Each packet has many routes to its destination.

So traffic controllers in each node achieve high performance of networks by selecting an appropriate route for each packet or by determining a selection rate for each outgoing link. But traffic is changed dynamically and traffic controllers determine rates of selections for outgoing links depending on current traffic. Ideally, such determination requires global information on traffic. But it is impossible to get complete global information on-line. Thus, traffic controllers are forced to make a decision on the selection rates with incomplete information on traffic. So the routing problem can be formulated using the Hogg-Huberman model.

We introduce a discrete-time version of the Hogg-Huberman model where reward mechanism is based on net bias. The existence of fixed points and their properties are discussed. By computer simulation, we will investigate effects of net bias on stability and transient behaviors. Then, we apply the model to a routing problem in a packet network.

28.2 The Discrete Time Hogg-Huberman Model

A discrete-time computational ecosystem with two resources is described by [9]

$$f(k+1) = f(k) + \alpha \left(\rho(k) - f(k) \right) \tag{28.2.1}$$

$$\rho(k) = \frac{1}{2}\left(1 + \mathrm{erf}\left(\frac{G_1(f(k-\tau)) - G_2(f(k-\tau))}{\sqrt{2}\sigma}\right)\right) \tag{28.2.2}$$

where f is the fraction of agents using resource 1, α is the rate at which agents reevaluate their decision, ρ is the probability that an agent will prefer resource 1 over resource 2, $G_i (i = 1, 2)$ is a payoff of resource i, τ is a delayed time, and σ represents imperfection of information, and $\mathrm{erf}(\cdot)$ is the error function defined by

$$\mathrm{erf}(x) = \frac{2}{\sqrt{\pi}} \int_0^x e^{-t^2} dt$$

In this model, if delay time of information is large or imperfection of information is small, then its behavior becomes chaotic. Hogg and Huberman introduced a reward mechanism for stabilizing the chaotic behavior [7]. A system with a reward mechanism is called the Hogg-Huberman model [7] and described by

$$f_{rs}(k+1) = f_{rs}(k) + \alpha \left(f_s^{str}(k)\rho_{rs}(k) - f_{rs}(k) \right)$$
$$+ \gamma \left(f_r^{res}(k)\eta_s(k) - f_{rs}(k) \right) \tag{28.2.3}$$

$$f_r^{res} = \sum_s f_{rs}, \quad f_s^{str} = \sum_r f_{rs} \qquad (28.2.4)$$

$$\eta_s = \frac{\sum_r f_{rs} G_r}{\sum_r f_r^{res} G_r} \qquad (28.2.5)$$

where f_{rs} is a fraction of agents with strategy s using resource r, f_r^{res} is a fraction of all agents using resource r, f_s^{str} is a fraction of all agents with strategy s, ρ_{rs} is the possibility that an agent with strategy s will prefer resource r, γ is the rate at which an agent will change its strategy, and η_s represents the possibility that an agent will prefer strategy s.

In Equation (28.2.3), the second term represents the dynamics of a model of computational ecosystems and the third term represents a reward mechanism.

By summing Equation (28.2.3), $f_r^{res}(k)$ and $f_s^{str}(k)$ are governed by

$$f_r^{res}(k+1) = f_r^{res}(k) + \alpha\left(\sum_s f_s^{str}(k)\rho_{rs}(k) - f_r^{res}(k)\right)$$

$$f_s^{str}(k+1) = f_s^{str}(k) + \gamma\Big(\eta_s(k) - f_s^{str}(k)\Big) \qquad (28.2.6)$$

There are many ways of giving strategies. We consider a reward mechanism based on net bias [7]. Then, $\rho_{1s}(k)$ and $\rho_{2s}(k)$ are given by the following equations:

$$\rho_{1s}(k) = \frac{1}{2}\left(1 + \mathrm{erf}\left(\frac{G_1(f_1^{res}(k-\tau)) + s - G_2(f_1^{res}(k-\tau))}{\sqrt{2}\sigma}\right)\right) \qquad (28.2.7)$$

$$\rho_{2s}(k) = 1 - \rho_{1s}(k) \qquad (28.2.8)$$

where each strategy s corresponds to net bias given by a real number and represents preference of resources. In other words, the agent which has a strategy of positive net bias s tends to prefer resource 1 over resource 2 while the agent which has a strategy of negative net bias s tends to prefer resource 2 over resource 1.

We show an example of the time response of $f_1^{res}(k)$ in Figure 28.1 where payoffs and parameters are set to be as follows: $G_1(f) = 4 + 7f - 5.333f^2$, $G_2(f) = 4 + 3f$, $\alpha = 0.85$, $\sigma = 0.25$, $\tau = 1$, $\gamma = 1$, the number of strategies is 21 with $s \in \{-1.0, -0.9, -0.8, \cdots 0.8, 0.9, 1.0\}$, and all initial values of f_{rs} are set to be equal.

Figure 28.1 shows that the system becomes stable by a reward mechanism. Moreover, it is shown that $G_1(f_1^{res}) = G_2(f_1^{res})$ in the steady state. Thus, this steady state provides an optimal fraction in the sense that all payoffs from resources are the same if all agents select resources according to the fraction, whereas the system behavior given by Equation (28.2.1) does not converge to it.

Analysis of Fixed Points

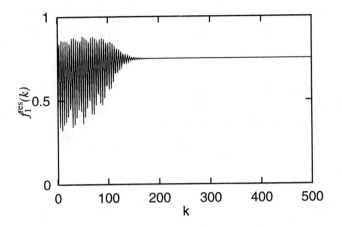

FIGURE 28.1
Stabilization by a reward mechanism.

In the following, we investigate existence and stability of a fixed point at which $G_1(f_1^{res}) = G_2(f_1^{res})$ in the Hogg-Huberman model with net bias.

28.3 Analysis of Fixed Points

In this section, we investigate a fixed point in the discrete-time Hogg-Huberman model with net bias [17]. Let the number of strategies be n. In this paper, f_1^{res*}, f_2^{res*}, f_s^{str*}, ρ_{1s}^*, ρ_{2s}^*, and η_s^* represent values of f_1^{res}, f_2^{res}, f_s^{str}, ρ_{1s}, ρ_{2s}, and η_s at a fixed point, respectively. Then, Equations (28.2.6) and (28.2.6) imply the following equations:

$$f_r^{res*} = \sum_s f_s^{str*} \rho_{rs}^* \quad (r = 1, 2) \tag{28.3.9}$$

$$\eta_s^* = f_s^{str*} \tag{28.3.10}$$

By Equations (28.2.3) and (28.3.10), we have

$$(\alpha \rho_{rs}^* + \gamma f_r^{res*}) f_s^{str*} = (\alpha + \gamma) f_{rs}^* \quad (r = 1, 2) \tag{28.3.11}$$

By the definition of η_s and Equation (28.3.10), we have

$$\eta_s^* = \frac{f_{1s}^* G_1^* + f_{2s}^* G_2^*}{f_1^{res*} G_1^* + f_2^{res*} G_2^*} = f_s^{str*} \tag{28.3.12}$$

where $G_i^* = G_i(f_i^{res*}), (i=1,2)$. Therefore, we have the following equation:

$$(f_s^{str*} f_r^{res*} - f_{rs}^*)(G_1^* - G_2^*) = 0 \quad (r=1,2) \tag{28.3.13}$$

and we consider the following two cases.

<u>Case 1. $G_1^* - G_2^* \neq 0$</u>

In this case, we have $f_s^{str*} f_r^{res*} - f_{rs}^* = 0$, and the following equation is obtained from Equations (28.2.3) and (28.3.10).

$$\rho_{rs}^* = f_r^{res*} \quad \forall s \quad (r=1,2) \tag{28.3.14}$$

Thus, a fixed point exists only when $n=1$.

<u>Case 2. $G_1^* - G_2^* = 0$</u>

By Equation (28.2.7), we have

$$\rho_{1s}^* = \frac{1}{2}\left(1 + \text{erf}\left(\frac{s}{\sqrt{2}\sigma}\right)\right) \tag{28.3.15}$$

Moreover, from Equation (28.3.9),

$$f_1^{res*} = \frac{1}{2}\sum_s f_s^{str*}\left(1 + \text{erf}\left(\frac{s}{\sqrt{2}\sigma}\right)\right) \tag{28.3.16}$$

Recall that $\sum_s f_s^{str*} = 1$. Thus, Equations (28.3.11) and (28.3.16) imply that f_{rs}^* lies on an $(n-2)$-dimensional manifold. It is also shown that a fixed point at which $G_1^* = G_2^*$ exists if the number of strategies is equal to or more than two and it is unisolated if the number of strategies is equal to or more than three.

28.4 Net Bias and Transient Behavior

How the system behavior changes according to net bias is an important problem. In this section, we investigate the relationship between net bias and the transient behavior by computer simulation.

28.4.1 Unisolation of Fixed Points

We consider the case $n=3$, and let $s \in \{-1.0, 0, 1.0\}$. We use the same parameter values as those in Section 28.2 except s. The relationship between initial conditions and convergent points of f_1^{res*} and f_s^{str*} is shown in Table 28.1.

Table 28.1 shows that every behavior converges to a different fixed point, but the value of f_1^{res} at every convergent point is same ($f_1^{res} = 0.750047$).

TABLE 28.1
The relationship between initial conditions and convergent points

	initial values	convergent points
$f^{str}_{s=-1}$	0.542581	0.234669
$f^{str}_{s=0}$	0.022203	0.030536
$f^{str}_{s=1}$	0.435216	0.734795
f^{res}_1	0	0.750047

	initial values	convergent points
$f^{str}_{s=-1}$	0.121855	0.161314
$f^{str}_{s=0}$	0.778492	0.177247
$f^{str}_{s=1}$	0.099652	0.661439
f^{res}_1	0	0.750047

	initial values	convergent points
$f^{str}_{s=-1}$	0.333333	0.161391
$f^{str}_{s=0}$	0.333333	0.177093
$f^{str}_{s=1}$	0.333333	0.661516
f^{res}_1	0	0.750047

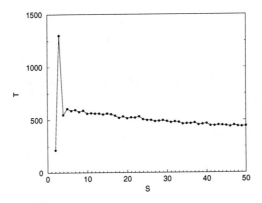

FIGURE 28.2
Number of strategies and average time of convergence.

28.4.2 Number of Strategies and Average Time of Convergence

We investigate the relationship between number of strategies n and the average time of convergence T. Now, we set net bias uniformly with equal interval lengths in the interval $[-1, 1]$, that is, $s \in \{-1,\ -1+\frac{2}{n-1},\ \cdots,\ 1\}$. In this setting, chaotic behaviors are stabilized and converge to the optimal points if n is greater than or equal to two. We take all initial conditions randomly, and the average time of convergence to a fixed point is shown in Figure 28.2, where the horizontal axis represents the number of strategies and the vertical axis represents the average time of convergence. It is an interesting result that the average time of convergence is the largest when $n = 3$, and the smallest when $n = 2$. When n is greater than three, it decreases slowly as n increases.

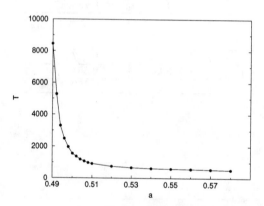

FIGURE 28.3
Width of strategies and average time of convergence.

28.4.3 Width of Strategies and Average Time of Convergence

We investigate the system behavior when we set net bias $s \in \{-a,\ 0,\ a\}$, where a is a positive real number. In the following, we will call a the width of strategies. In simulation, we use the same parameter values as those in Section 28.2 except s.

Shown in Figure 28.3 is the relationship between the width of strategies and average time of convergence, where the horizontal axis represents the width of strategies and the vertical axis represents the average time of convergence. The average time of convergence increases exponentially as a is close to 0.5. Moreover, it is shown that the system behavior does not converge to any fixed point and becomes chaotic when a is smaller than 0.48.

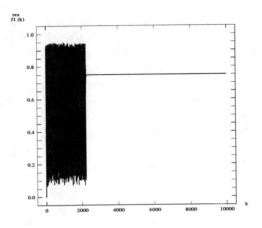

FIGURE 28.4
Time response ($s \in \{-0.5, 0, 0.5\}$).

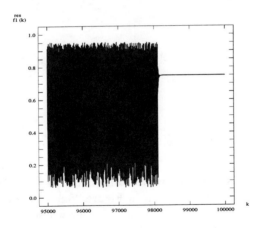

FIGURE 28.5
Time response ($s \in \{-0.48, 0, 0.48\}$).

Shown in Figures 28.4 and 28.5 are system behaviors when $a = 0.5$ and $a = 0.48$, respectively. These figures show that, in both cases, before behaviors converge to a fixed point, transient behaviors oscillate quasi-randomly with large magnitude, and suddenly they converge to a fixed points. Moreover, f_1^{res} converges to the same value ($f_1^{res} = 0.750047$) where $G_1 = G_2$. Such a transient behavior persists for a very long time. For example, as shown in Figure 28.5, during the first 98000 steps, the system behaves quasi-randomly. Shown in Figure 28.6 is the relationship between $f_1^{res}(k+1)$ and $f_1^{res}(k)$ from $k = 50000$ to $k = 55000$ for the behavior. There is an empty region surrounded by a transient state set. The fixed point $f_1^{res} = 0.750047$ is in the empty region. Shown in Figure 28.7 is a system behavior when the initial conditions are set in the empty region. It converges to the fixed point rapidly. Thus, it takes much time for the system behavior to get into the empty region, and once it enters the empty region, it converges to the fixed point exponentially. Such a phenomenon is called a chaotic transient [10, 15].

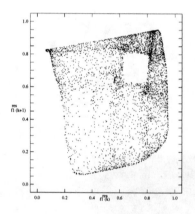

FIGURE 28.6
Chaotic transient ($s \in \{-0.48, 0, 0.48\}$).

In the same way, we set the initial conditions near the fixed point at $a = 0.47$, and its behavior is shown in Figure 28.8. It does not converge to a fixed point, but becomes a chaotic behavior whose attractor is quite similar to that in Figure 28.6. Moreover, as the width a becomes smaller, a chaotic attractor is split into several islands. For example, shown in Figure 28.9 is a chaotic attractor when $a = 0.3$, which consists of four islands.

From these observations, it is concluded that, as the width of strategies is smaller, a chaotic transient appears, and chaos appears when a fixed point becomes unstable. Thus, in order to make the system converge to a fixed point at which payoff functions take the same value, it is necessary

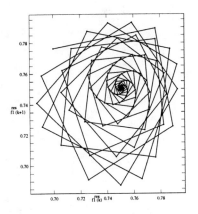

FIGURE 28.7
Convergence process ($s \in \{-0.48, 0, 0.48\}$).

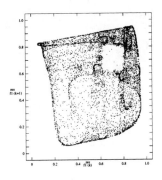

FIGURE 28.8
System behavior ($s \in \{-0.47, 0, 0.47\}$).

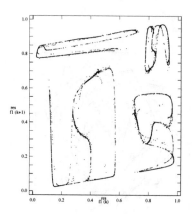

FIGURE 28.9
System behavior ($s \in \{-0.3, 0, 0.3\}$).

to set the width of strategies to a sufficiently large value.

28.5 Application to a Routing Problem

In this section, we apply the Hogg-Huberman model to a packet routing problem [21].

28.5.1 Network Model

We consider a packet network, which sends data by dividing several blocks with specified length called "packet" [1]. This network consists of nodes, links, and buffers. Let N be a number of nodes. Figure 28.10 illustrates a model of link selection. We assume the following conditions:

1. Each node knows only its state at real-time, and has no information on the other nodes.
2. All packets have the same length.
3. In each node, every arriving packet is assigned to an outgoing link, waits in the corresponding buffer for transmission, and is transmitted to the next node by FIFO control.

When a packet is generated in a node, both destination and allowable delay time are given. Every packet will be discarded when time that has

Application to a Routing Problem

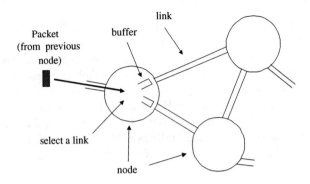

FIGURE 28.10
Model of a link selection.

elapsed during its delivery is larger than its allowable delay time.

There may be several routes to the destination so that, in each node on transmission, there are several links possible for the assignment. Packets are classified into (at most) N groups of agents depending on their destinations. Each group determines the assignment without knowing those of other groups, and there are no negotiations between groups. In each group, packets are assigned an outgoing link according to the selection rate which is determined by a controller or a decision maker for each group. This chapter will propose a method for the determination of the selection rate based on the Hogg-Huberman model.

Since it is assumed that routing is done in each node independently, for simple notation, we do not indicate a node number at each system parameter, and we use the following parameters:

l_r: label of outgoing links($r = 1, \ldots, L$), where L is the total number of outgoing links.

C_r: transmission time per one packet at link l_r.

B_r: number of packets waiting at link l_r.

B_{MAXr}: buffer capacity of link l_r.

V_{ri}: valuation index of packet group $P_i(i = 1, \ldots, N$) for link l_r, where P_i is a group of packets whose destination node is i.

Moreover, each packet has the following two parameters: allowable delayed time W and elapsed time t. It is assumed that any packet satisfying $W < t$ is discarded.

28.5.2 Packet Routing Method

In this section, we propose a packet routing algorithm. Each packet is classified into a packet group according to its destination. Each outgoing link is regarded as a resource, and each packet selects a link. This selection problem is formulated as a conflict resolution problem in resource allocation among packet groups. In each node, rates of link selections in each packet group is determined based on the Hogg and Huberman model. Each packet selects a link according to the selection rate, and waits in the corresponding buffer for transmission. Determinations of the rates are done using payoffs from links so that the total amount of payoffs is improved.

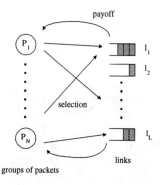

FIGURE 28.11
Computational ecology model with conflict among groups of agents.

Hogg and Huberman deal with one agent group. But in our systems, each packet group has different payoffs for links. Thus, there is conflict over the groups as shown in Figure 28.11, and we modify the Hogg-Huberman model as follows:

$$\lambda_{irs}(k+1) = \lambda_{irs}(k) + \alpha(\lambda_{is}^{str}(k)\rho_{irs} - \lambda_{irs}(k))$$
$$+ \gamma(\lambda_{ir}^{res}(k)\eta_{is} - \lambda_{irs}(k)) \qquad (28.5.17)$$

$$\lambda_{ir}^{res} = \sum_s \lambda_{irs}, \quad \lambda_{is}^{str} = \sum_r \lambda_{irs}$$

$$\eta_{is} = \frac{\sum_r \lambda_{irs} G_{ir}}{\sum_r \lambda_{ir}^{res} G_{ir}}$$

where

λ_{irs}: current link selection rate at which packets in P_i select link l_r with strategy s.

G_{ir}: payoff function for link l_r in P_i.

ρ_{irs}: desired link selection rate at which packets in P_i select link l_r with strategy s.

η_{is}: desired strategy selection rate at which packets in P_i take strategy s.

γ: degree at which packets in P_i change their strategies.

λ_{ir}^{res}: total rate at which agents in P_i select link l_r.

λ_{is}^{str}: total rate at which agents in P_i take strategy s.

Note that the desired link selection rate ρ_{irs} and strategy selection rate η_{is} which are determined by using only local information are regarded as possibilities of agents' preference with respect to resources and strategies in Equation (28.2.3), respectively. A fundamental idea of Hogg-Huberman approach is that a link selection rate is determined dynamically and independently based on local information so that payoffs from links are equal. But an optimal selection rate is not obtained on-line. So each agent subgroup with strategy s in group P_i estimates the optimal selection rate for link l_r, which is denoted by ρ_{irs}, and the rate will be controlled to converge the estimated values. Moreover, ρ_{irs} is given by the following equation:

$$\rho_{irs} = \frac{\text{erf}\left(\frac{G_{ir}+s-\frac{\sum_{j=1}^{L} G_{ij}}{L}}{\sqrt{2}\sigma}\right)+1}{\sum_{k=1}^{L}\left\{\text{erf}\left(\frac{G_{ik}+s-\frac{\sum_{j=1}^{L} G_{ij}}{L}}{\sqrt{2}\sigma}\right)+1\right\}} \quad (28.5.18)$$

Finally, we will determine payoff functions. Intuitively, payoff functions should reflect both waiting time in buffers and accessibility to destination. In other words, their values become larger if waiting time in buffers are smaller and/or numbers of nodes on route to the destination are fewer. So we use the following payoff functions:

$$G_{ir} = \frac{a \times V_{ri}}{C_r \times (\bar{B}_r + 1)}$$

$$\bar{B}_r = \frac{\sum_{t=0}^{T-1} B_r(k-t)}{T}$$

where a is a constant parameter, $\bar{B}_r$ denotes an average of packet numbers over the last T time stored in the buffer corresponding to link l_r, and V_{ri} is the number of the shortest selectable routes after packets is transmitted via link l_r. So the denominator and numerator of G_{ir} represent the valuations about a waiting time and selectability of routes, respectively. Thus, payoff functions reflect network topology, and, by Equation (28.5.17), load balance among links is achieved.

28.5.3 Example

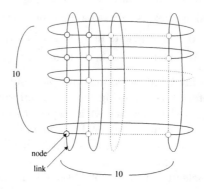

FIGURE 28.12
Simulation model.

In this section, we compare the proposed method with other methods by simulation. We use a network which consists of 10 × 10 torus nodes as shown in Figure 28.12. We will compare performance of the proposed method with that of the following two well-known methods.

1. **A static routing method**: We search for the shortest path for packet's transmission based on the transmission rate at each link in advance, and send packets along the path.

2. **An adaptive routing method**: We search the shortest path based on both current states of the network reported from other nodes and the transmission rate of each link at every fixed time interval [1]. The routing is done according to a performance index Len_r of link l_r in each node defined by

$$\text{Len}_r = C_r + \frac{1}{b}\left(C_r \times \frac{\sum_{t=0}^{T-1} B_r(k-t)}{T}\right) \quad (28.5.19)$$

For simplicity, we ignore influences of overhead caused by transmission of global information on the network, and we use the Dijkstra's algorithm to search the shortest path in both algorithms.

We set the transmission time C_r of link in the range from 40 to 60 at random. In the Hogg-Huberman model, each packet group can select only links which are on one of the shortest path routes. We use five strategies with net bias $s = -0.004, -0.002, 0, 0.002, 0.004$ for each packet group. Let the initial selection rates of links be equal within selectable links. Let $B_{MAXr} = 200$ for each buffer, and it is assumed that packets are assigned to the second selected link when the first selected buffer overflows. Each parameter is set as follows:

$\alpha = 0.70$, $\gamma = 1.0$, $\sigma = 0.1$, $a = 1.0$, $T = 100$, $W = 10000$, and $b = 15$

For each method, simulation runs for 500,000 unit-times, and, for the Hogg-Huberman and the adaptive method, the selection rate λ_{ir}^{res} is updated at every 2000 unit-times.

Shown in Figure 28.13 is the relationship between the packet loss rate and the packet generation rate, and shown in Figure 28.14 is the relationship between average of transmission delay (average time which packets have taken until their arrival to its destination) and the packet generation rate. These two figures show that the packet loss rate and the average of transmission delay are improved by the proposed method as compared with the other two methods.

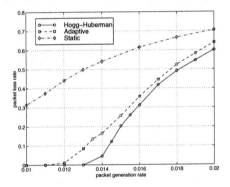

FIGURE 28.13
Packet loss rate.

28.6 Conclusions

We consider the discrete-time model of a computational ecosystem with a reward mechanism. It is shown that its fixed point is optimal with respect to payoffs if a reward mechanism based on net bias is used. By computer

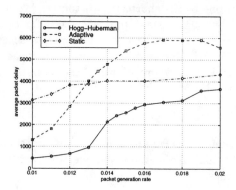

FIGURE 28.14
Average of packet delay.

simulation, stability and transient behaviors are examined. We also deal with its application to a packet routing problem. This model is a very useful tool for designing a decision maker with incomplete information.

References

[1] D. Bertsekas and R. Gallager, *Data Networks*, Prentice-Hall, New Jersey, 1987.

[2] E. A. Billard, "Controlling instability in distributed queuing systems," *Proc. of IEEE Int. Symp. on Intelligent Control*, Columbus, Ohio, 1994, pp. 111–117.

[3] G. Chen and X. Dong, "From chaos to order: perspectives and methodologies in controlling chaotic nonlinear dynamical systems," *Int. J. Bifur. Chaos*, vol. 3, pp. 1363–1409, 1993.

[4] G. Chen and X. Dong, *From Chaos to Order: Methodologies, Perspectives and Applications*, World Scientific Pub. Co., Singapore, 1998.

[5] S. H. Clearwater (Ed.), *Market-Based Control*, World Scientific Pub. Co., Singapore, 1996.

[6] J. Guckenheimer and P. Holmes, *Nonlinear Oscillations, Dynamical Systems, and Bifurcations of Vector Fields*, Springer, New York, 1983.

[7] T. Hogg and B. A. Huberman, "Controlling chaos in distributed systems," *IEEE Trans. on Sys. Man, Cybern.*, vol. 21, pp. 1325–1332, 1991.

[8] B. Huberman and S. H. Clearwater, "A multi-agent system for controlling building environments," *Proc. of ICMAS'95*, San Francisco, 1995, pp. 171–176.

[9] B. A. Huberman and T. Hogg, "The behavior of computational ecologies," in B. A. Huberman (Ed.), *The Ecology of Computation*, Elsevier Science Pub., 1988, pp. 77–115.

[10] J. L. Kaplan and J. A. Yorke, "Preturblence: a regime observed in a fluid flow model of Lorenz," *Commun. Math. Phys.*, vol. 67, pp. 93-108, 1979.

[11] J. O. Kephart, T. Hogg, and B. A. Huberman, "Dynamics of computational ecosystem," *Phys. Rev. A*, vol. 40, pp. 404–421, 1989.

[12] J. O. Kephart, T. Hogg, and B. A. Huberman, "Collective behavior of predictive agents," *Physica D*, vol. 42, pp. 48–65, 1990.

[13] J. F. Kurose and R. Simha, "A microeconomic approach to optimal resource allocation in distributed computer systems," *IEEE Trans. on Computers*, vol. 38, pp. 705–717, 1989.

[14] K. Kuwabara, T. Ishida, Y. Nishibe, and T. Suda, "An equilibratory market-based approach for distributed resource allocation and its applications to communication network control, " in S. H. Clearwater (Ed.), *Market-Based Control: A Paradigm For Distributed Resource Allocation*, World Scientific Pub.Co., 1996, pp. 53–73.

[15] E. Ott, *Chaos in Dynamical Systems*, Cambridge University Press, 1993.

[16] E. Ott, C. Grebogi, and J. A. Yorke, "Controlling chaos," *Phys. Rev. Lett.*, vol. 64, pp. 1196–1199, 1990.

[17] T. Ushio and T. Imamori, "Hogg-Huberman strategy with net bias in chaotic discrete-time computational ecosystems," *Pro. of IEEE Contr. Decis. Conf.*, San Diego, CA, 1997, pp. 389–394.

[18] T. Ushio and N. Motonaka, "Controlling chaos in a Hogg-Huberman model of a manufacturing system," *IEICE Trans. Fundamentals*, vol. E81-A, pp. 1507–1511, 1998.

[19] T. Ushio, H. Ueda, and K. Hirai, "Controlling chaos in a switched arrival system," *Sys. and Contr. Lett.*, vol. 26, pp. 335–339, 1995.

[20] T. Ushio, H. Ueda, and K. Hirai, "Stabilization of periodic orbits in switched arrival system with N buffers," *Proc. of IEEE Contr. Decis. Conf.*, Kobe, Japan, 1996, pp. 1213–1214.

[21] T. Ushio and T. Yamasaki, "A packet routing method based on a Hogg-Huberman strategy," *Proc. of IFAC LSS'98*, Patras, Greece, 1998, pp. 786–791.

[22] H. Yamaki, M. P. Wellman, and T. Ishida, "A market-based approach to allocating QoS for multimedia applications," *Proc. of ICMAS'96*, Nara, Japan, 1996, pp. 385–392.

Index

A
absorber 83
anticontrol of bifurcation 302
anticontrol of chaos 48
arrhythmia 65, 327
autonomous circuit 62
axial flow compressor 348

B
bifurcation 137, 164, 196, 205, 299, 347, 369, 391
bifurcation control 299, 312, 347, 369, 391
breathing pattern 36

C
CDMA 275
CO_2 laser 184
cardiac model 325
chaos control 45, 71, 89, 107, 179, 233, 255, 275, 625
chaos in coupled networks 581
chaos in control system 559
chaos synchronization 107, 167, 275, 417, 439, 501, 529
chaotic attractor 48, 99, 269, 282, 290, 399
chaotic control algorithm 92
chaotic LQR control 102
chaotic signal 457
chaotic vibration, 131
chemical reactor 226
Chua's circuit 48, 275, 285, 397, 407, 505
coupled chaotic circuits 501
coupled map lattices 119
coupled networks 581
coupled ODEs 115
cubic polynomial system 219
curvature coefficient 207

D
DC-DC converter 32, 398, 408
DDE 242
diffusion control 110

E
ecosystem 625
electronics 155, 235
embedding 3, 25
energy 133
entropy 124

F
feedback control 109, 121
fibrillation/defibrillation 65
fractal 166
frequency domain method 179

G
graph coloring 524

H
harmonic balance 208
halo-chaos 117
heart 65, 327
Hogg-Huberman model 627

I
IFM 603
identification 23
implementation 45, 292
impulsive control 275
information 13, 553
initial condition 155
input-output system 1
intermittency 593
invariant 374
inverted pendulum 95

J
jet engine 347

L
Lur'e system 281, 425, 442, 445, 448
limit cycle 205, 214
Lyapunov exponent 124, 165, 519

M
Moore-Greitzer model 353, 384
mechanical system 71
modeling 2, 23

Index

return map modulation 603

N
NARMAX 23
NARX 28
noise 247, 457
nonlinear feedback control 109
normal form 214, 369, 374

O
OGY 45, 52, 73, 90, 196, 325
OPF 45, 59
observer 417, 439
odd number multiplier limitation 255
operating condition 78

P
PLL 156, 531
PDE 108, 131
PWS 391
packet routing problem 636
phase system 529
pinning 112
prediction of chaos 560

Q
quadratic map 328
quadratic normal form 378

R
Rayleigh-Bénard convection 299, 313
reconstruction 2, 24, 429
representation 23
robot manipulator 569
Rössler system 268, 290
rotating stall control 359

S
secure communication 292, 471, 477
sensitivity 155
sliding mode 260, 397
spatiotemporal chaos 107
stability 137, 187, 277, 349

T
TDF, TDFC 255, 325
time delay 57, 111, 233, 255, 325
time series 1, 23, 109
tracking 112

U
UFP 329
UPO 242, 256, 596
utilizing chaos 89, 477

W
wave equation 131